ESO ASTROPHYSICS SYMPOSIA
European Southern Observatory

Series Editor: Bruno Leibundgut

A. Richichi F. Delplancke F. Paresce
A. Chelli (Eds.)

The Power of Optical/IR Interferometry: Recent Scientific Results and 2nd Generation Instrumentation

Proceedings of the ESO Workshop held in Garching, Germany, 4-8 April 2005

 Springer

Volume Editors

Andrea Richichi
Francoise Delplancke
European Southern Observatory
Karl-Schwarzschild-Str. 2
85748 Garching
Germany

Alain Chelli
Observatoire de Grenoble
Lab. d'Astrophysique
Grenoble CX 38041
France

Francesco Paresce
IASF Bologna
Area Ricerca di Bologna
Via Piero Gobetti 101
Bologna 40129
Italy

Series Editor

Bruno Leibundgut
European Southern Observatory
Karl-Schwarzschild-Str. 2
85748 Garching
Germany

ISBN 978-3-642-09360-9   e-ISBN 978-3-540-74256-2

Springer is a part of Springer Science+Business Media
springer.com
© Springer-Verlag Berlin Heidelberg 2008
Softcover reprint of the hardcover 1st edition 2008

Cover design: WMXDesign, Heidelberg

Printed on acid-free paper   55/3180/Integra   5 4 3 2 1 0

Preface

In the ancient times the Romans had an expression, *Per Aspera ad Astra*. It is not clear whether they already experimented with interferometry at those times or whether they enjoyed clairvoyant abilities. Certainly, this expression captures the essence of modern optical and near-infrared long-baseline interferometry, where technical challenges and engineering problems are the necessary ingredients of each single day of work. It is, then, only natural that so many of the workshops and meetings dedicated to interferometry until now have incorporated large fractions of technical and engineering papers.

Back in 2003, when we started the preparations for what was to become the workshop published in this book, it was clear that the situation was changing. Alas, the technical and engineering problems were not finished, but by then interferometry had reached a status of scientific maturity that warranted a workshop fully dedicated to results, and many of them absolutely ground-breaking. If a whole generation of interferometers such as the GI2T, Mark III, PTI, SUSI, IOTA, NPOI and others had attracted attention and shown the way forward with convincing results, new powerful facilities such as the VLTI, the Keck-I and CHARA were finally making heads turn, paper after paper and press release after press release.

These new interferometers along with their sophisticated instrumentation have provided new levels of accuracy, spectral resolution and access to various spectral bands from the optical to the thermal infrared. Investigations are now enabled on a wealth of astrophysical sources with unprecedented levels of angular resolution and sensitivity, producing a considerable body of new, exciting scientific results. The ESO VLTI stands among the most powerful of such facilities. In celebrating the completion of the first phase of the VLTI development, we saw an opportunity to invite the community to come together to review and discuss not just interferometers, but science with interferometers and its impact on astronomy as a whole. This workshop was thus a celebration not just of the VLTI, of which Garching is the European home, but of the scientific success of interferometry. The workshop was also intended to showcase ideas and concepts for the future of interferometry, and in particular for the second generation of VLTI instrumentation. This second part of the workshop was organized in collaboration with the European Interferometry Initiative (EII).

With about 170 registered participants from 14 countries in addition to ESO and with a programme that was bulging at the seams with excellent presentations, it is perhaps not too indulgent of us to consider the workshop, which took place in Garching on 4–8 April, 2005, a success. For this, we have to thank also the excellent organizational support at ESO, and at the Max-Planck-Institut for Extraterrestrial Physics that kindly provided the large room required for this oversized audience.

Looking back to that week of some years ago, we feel a deep satisfaction with the outcome of the workshop, with the overwhelming participation, with the richness of results presented and above all with the fertile exchange of ideas and discussions which took place. The only regret is that, due to a number of technical problems, the publication of so many important contributions has been delayed for such a long time. We can only hope for your indulgence, while looking forward to the next occasion to celebrate the scientific achievements of astronomical interferometry.

Garching
June 2007

Alain Chelli
Françoise Delplancke
Francesco Paresce
Andrea Richichi

Contents

Part II Science: Stars — circumstellar matter, IR objects

Part III Science: Stars — binaries and multiples

**Part IV Science: Stars — Galactic centre, AGNs, astrometry,
exo-planets and future targets**

Part VII Instrumentation: Concepts for future interferometric
instrumentation

Part VIII Posters — Science

Understanding Cross Talk on the NPOI Multibeam Combiner

Part IX Posters — Instrumentation

A Model Experiment for APreS-MIDI

From the VLBI to the VLTI: An APreS-MIDI Image Reconstruction Study

A Numerical Simulator for VITRUV

Multiway Beam Combiners for VITRUV

A Laboratory Interferometer for VITRUV

Image Reconstruction with VITRUV

VITRUV Precursors: IONIC2T/IONIC3T

First Results on Integrated Optics Developments for Mid-Infrared Interferometry

APreS-MIDI, a 4 Beam Recombiner

List of Contributors

ABSIL, Olivier
Institut d'Astrophysique et de
Géophysique - Liège University
(Belgium)
absil@astro.ulg.ac.be

AKESON, Rachel
Michelson Science Center -
Caltech (USA)
rla@ipac.caltech.edu

ALBRECHT, Simon
Leiden Observatory
(The Netherlands)
albrecht@strw.leidenuniv.nl

ANTONIUCCI, Simone
Università di Tor Vergata - INAF
Osservatorio Astronomico di Roma
(Italy)
antoniucci@mporzio.astro.it

AUBERT, Coralie
Laboratoire d'Astrophysique de
Grenoble (France)
caubert@enserg.fr

AUFDENBERG, Jason
NOAO - Tucson (USA)
jasona@noao.edu

BAKKER, Eric
Leiden Observatory
(The Netherlands)
bakker@strw.leidenuniv.nl

BARON, Fabien
University of Cambridge - Cavendish
Laboratory (UK)
baron@mrao.cam.ac.uk

BARRY, Richard
NASA Goddard Space
Flight Center - Exo-planets &
Stellar Astrophysics Lab. (USA)
rkbarry@iri1.gsfc.nasa.gov

BECKERS, Jacques
University of Chicago - Astronomy
and Astrophysics Department (USA)
jbeckers@rcn.com

BENISTY, Myriam
Laboratoire d'Astrophysique de
Grenoble (France)
Myriam.Benisty@obs.ujf-
grenoble.fr

BERGER, Jean-Philippe
Laboratoire d'Astrophysique de
Grenoble (France)
jean-philippe.berger@obs.ujf-
grenoble.fr

BODEN, Andy
California Institute of Technology -
Michelson Science Center (USA)
bode@ipac.caltech.edu

BONNEAU, Daniel
Observatoire de la Côte d'Azur -
Gemini (France)
daniel.bonneau@obs-azur.fr

BORKOWSKI, Virginie
Collège de France - LISE (France)
borkowski@obs-hp.fr

BUSCHER, David
University of Cambridge -
Cavendish Laboratory (UK)
dfb@mrao.cam.ac.uk

CARUSO, Fabio
European Southern Observatory -
Paranal Observatory (Chile)
fcaruso@eso.org

CASALI, Mark
European Southern Observatory -
Instrumentation Division (Germany)
mcasali@eso.org

CASSAING, Frédéric
ONERA - DOTA (France)
frederic.cassaing@onera.fr

CHELLI, Alain
Laboratoire d'Astrophysique de
Grenoble and Jean-Marie Mariotti
Center (France)
Alain.Chelli@obs.ujf-
grenoble.fr

CHESNEAU, Olivier
Observatoire de la Côte d'Azur -
GEMINI (France)
Olivier.Chesneau@obs-azur.fr

CORREIA, Serge
Astrophysikalisches Institut
Postdam (Germany)
scorreia@aip.de

COUDÉ DU FORESTO,
Vincent
Observatoire de Paris-Meudon
LESIA (France)
vincent.foresto@obspm.fr

COYNE, Julien
University of Cambridge - Cavendish
Laboratory (UK)
jc466@mrao.cam.ac.uk

CRUZALÈBES, Pierre
Observatoire de la Côte d'Azur -
GEMINI (France)
pierre.cruzalebes@obs-azur.fr

DANCHI, William
NASA Goddard Space
Flight Center - Infrared
Astrophysics (USA)
wcd@iri1.gsfc.nasa.gov

DAVIS, John
University of Sydney - School of
Physics (Australia)
j.davis@physics.usyd.edu.au

DE JONG, Jeroen
European Southern Observatory -
VLTI (Germany)
jdejong@eso.org

DE VRIES, Cor
SRON - HEA (The Netherlands)
C.P.de.Vries@sron.nl

DELBÓ, Marco
INAF - Osservatorio
Astronomico di
Torino (Italy)
delbo@to.astro.it

DELPLANCKE, Françoise
European Southern Observatory -
VLTI (Germany)
fdelplan@eso.org

DEMORY, Brice-Olivier
Observatoire de Genève & EPFL
(Switzerland)
brice-olivier.demory@epfl.ch

DEN HARTOG, Roland
European Space Agency - ESTEC
(The Netherlands)
rdhartog@rssd.esa.int

DEN HERDER, Jan-Willem
SRON - HEA (The Netherlands)
J.den.Herder@sron.nl

DI BENEDETTO, G. Paolo
INAF - Astrofisica Spaziale e Fisica
Cosmica, Milano (Italy)
pdibene@mi.iasf.cnr.it

DI FOLCO, Emmanuel
European Southern Observatory -
VLTI (Germany)
Emmanuel.Difolco@obspm.fr

**DOMICIANO DE SOUZA JR,
Armando**
MPI für Radioastronomie
Bonn (Germany)
adomicia@mpifr-bonn.mpg.de

DRIEBE, Thomas
MPI für Radioastronomie - IR
Interferometry, Bonn (Germany)
driebe@mpifr-bonn.mpg.de

DUGUÉ, Michel
Observatoire de la Côte d'Azur -
GEMINI (France)
dugue@obs-nice.fr

DUTREY, Anne
Université de Bordeaux (France)
Anne.Dutrey@obs.u-bordeaux1.fr

ECKART, Andreas
Universität zu Köln - Physikalisches
Institut (Germany)
eckart@ph1.uni-koeln.de

EIROA, Carlos
Universidad Autonoma de Madrid -
Facultad de Ciencias (Spain)
carlos.eiroa@uam.es

EISENHAUER, Frank
MPI für Extraterrestrische Physik -
Infrared/Submm Astronomy
Garching (Germany)
eisenhau@mpe.mpg.de

ERGENZINGER, Klaus
EADS-Astrium Friedrichshafen
(Germany)
klaus.ergenzinger@astrium.
eads.net

FEDELE, Davide
European Southern Observatory -
Garching (Germany)
dfedele@eso.org

FELDT, Markus
Max-Planck Institut
für Astronomie - Heidelberg
(Germany)
mfeldt@mpia.de

FLAMENT, Sébastien
Observatoire de la Côte d'Azur -
GEMINI (France)
flament@obs-nice.fr

FRIDLUND, Malcolm
European Space Agency/ESTEC
(The Netherlands)
malcolm.fridlund@esa.int

GAI, Mario
INAF - Osservatorio
Astronomico di
Torino (Italy)
gai@to.astro.it

GARCIA, Paulo J. V.
Porto University, Centro de
Astrofisica (Portugal)
pgarcia@astro.up.pt

GENZEL, Reinhard
Max-Planck-Institut für
Extraterrestrische Physik
Garching (Germany)
genzel@mpe.mpg.de

GIL, Carla
European Southern Observatory -
Santiago (Chile)
cgil@eso.org

GLINDEMANN, Andreas
European Southern Observatory -
VLTI (Germany)
aglindem@eso.org

GOLDMAN, Bertrand
Max-Planck Institut für
Astronomie - Heidelberg (Germany)
goldman@mpia-hd.mpg.de

GONDOIN, Philippe
ESA - Science Payload & Advanced
Concept Office (The Netherlands)
pgondoin@rssd.esa.int

GRASER, Uwe
Max-Planck Institut für
Astronomie - Heidelberg (Germany)
graser@mpia.de

GROENEWEGEN, Martin
University of Leuven (Belgium)
groen@ster.kuleuven.ac.be

GULL, Theodore
Goddard Space
Flight Center - Lab.
for Astronomy
and Solar Physics (USA)
theodore.r.gull@nasa.gov

HANIFF, Christopher
University of Cambridge - Physics
Dept. (UK)
cah@mrao.cam.ac.uk

HARVEY, Paul
University of Texas, Astronomy
Department (USA)
pmh@astro.as.utexas.edu

HEMPEL, Marc
AIU Jena (Germany)
marc@astro.uni-jena.de

HENNING, Thomas
Max-Planck Institut für
Astronomie - Heidelberg (Germany)
henning@mpia.de

HERNANDEZ, Oscar
Laboratoire d'Astrophysique de
Grenoble (France)
Oscar.Hernandez@obs.ujf-
grenoble.fr

HERWATS, Emilie
Laboratoire d'Astrophysique de
Grenoble (France)
Emilie.Herwats@obs.ujf-
grenoble.fr

HRON, Joseph
Institute of Astronomy -
University of Vienna (Austria)
hron@astro.univie.ac.at

HUMMEL, Christian
European Southern
Observatory - VLTI (Chile)
chummel@eso.org

ISELLA, Andrea
Osservatorio Astrofisico
di Arcetri (Italy)
isella@arcetri.astro.it

JAFFE, Walter
Leiden Observatory
(The Netherlands)
jaffe@strw.leidenuniv.nl

JANKOV, Slobodan
Université de Nice Sophia
Antipolis - Lab.
Astrophysique (France)
Slobodan.Jankov@unice.fr

JOCOU, Laurent
Laboratoire d'Astrophysique de
Grenoble (France)
laurent.jocou@obs.ujf-
grenoble.fr

JOERGENS, Viki
Leiden Observatory
(The Netherlands)
viki@strw.leidenuniv.nl

KAUFER, Andreas
European Southern Observatory -
Paranal Observatory (Chile)
akaufer@eso.org

KELLERER, Aglaë
European Southern Observatory -
Garching (Germany)
akellere@eso.org

KERN, Pierre
Laboratoire d'Astrophysique de
Grenoble (France)
Pierre.Kern@obs.ujf-
grenoble.fr

KERVELLA, Pierre
Observatoire de Paris Meudon -
LESIA (France)
pierre.kervella@obspm.fr

KIM, Sam
Canada-France-Hawaii-Telescope
(USA)
kim@cfht.hawaii.edu

KOEHLER, Bertrand
European Southern Observatory -
VLTI (Germany)
bkoehler@eso.org

KÖHLER, Rainer
Leiden Observatory
(The Netherlands)
koehler@strw.leidenuniv.nl

KOTANI, Takayuki
Observatoire de Paris Meudon -
LESIA (France)
takayuki.kotani@obspm.fr

KOUBSKY, Pavel
Astronomical Institute of Ondrejov
(Czech Republic)
koubsky@sunstel.asu.cas.cz

KRAUS, Stefan
Max-Planck Institut für Radioastronomie - Bonn (Germany)
skraus@mpifr-bonn.mpg.de

KRAWCZYK, Rodolphe
Alcatel Space - Systems & Advanced
Projects Directorate (France)
rodolphe.krawczyk@space.
alcatel.fr

LABADIE, Lucas
Laboratoire d'Astrophysique de
Grenoble (France)
lucas.labadie@obs.ujf-
grenoble.fr

LACHAUME, Régis
Max-Planck Institut für
Radioastronomie - IR Interferometry
Bonn (Germany)
lachaume@mpifr-bonn.mpg.de

LACOUR, Sylvestre
Observatoire de Paris Meudon -
LESIA (France)
sylvestre.lacour@obspm.fr

LAGARDE, Stéphane
Observatoire de la Côte d'Azur -
GEMINI (France)
lagarde@obs-nice.fr

LARDIÈRE, Olivier
Collège de France - LISE (France)
lardiere@obs-hp.fr

LAUNHARDT, Ralf
Max-Planck Institut
für Astronomie - Heidelberg
(Germany)
rl@mpia.de

LE COARER, Etienne
Laboratoire d'Astrophysique de
Grenoble (France)
lecoarer@obs.ujf-grenoble.fr

LEBOUQUIN, Jean-Baptiste
Laboratoire d'Astrophysique de
Grenoble (France)
jean-baptiste.lebouquin@obs.
ujf-grenoble.fr

LEIBUNDGUT, Bruno
European Southern Observatory -
Garching (Germany)
bleibund@eso.org

LEINERT, Christoph
Max-Planck Institut
für Astronomie - Heidelberg
(Germany)
leinert@mpia-hd.mpg.de

LÉNA, Pierre
Université Denis Diderot, LESIA
Paris (France)
pierre.lena@obspm.fr

LI CAUSI, Gianluca
INAF - Osservatorio
Astronomico di
Roma (Italy)
licausi@mporzio.astro.it

LIGORI, Sebastiano
INAF - Osservatorio
Astronomico di
Torino (Italy)
ligori@to.astro.it

LOPEZ, Bruno
Observatoire de la Côte d'Azur -
GEMINI (France)
lopez@obs-nice.fr

LORENZEN, Dirk H.
Deutschlandfunk - Forschung aktuell
(Germany)
DLorenzen@compuserve.com

LUDWIG, Hans-Guenter
Lund Observatory - Lund University
(Sweden)
hgl@astro.lu.se

MALBET, Fabien
Laboratoire d'Astrophysique de
Grenoble (France)
Fabien.Malbet@obs.ujf-
grenoble.fr

MARCONI, Alessandro
INAF - Osservatorio
Astrofisico di
Arcetri (Italy)
marconi@arcetri.astro.it

MASSONE, Giuseppe
INAF - Osservatorio
Astronomico di
Torino (Italy)
massone@to.astro.it

MATHIAS, Philippe
Observatoire de la Côte d'Azur -
GEMINI (France)
mathias@obs-nice.fr

MEILLAND, Anthony
Observatoire de la Côte d'Azur -
GEMINI (France)
anthony.meilland@obs-azur.fr

MEISENHEIMER, Klaus
Max-Planck Institut
für Astronomie - Heidelberg
(Germany)
meise@mpia.de

MEISNER, Jeffrey
Leiden Observatory
(The Netherlands)
meisner@strw.leidenuniv.nl

MENUT, Jean-Luc
Observatoire de la Côte d'Azur -
GEMINI (France)
menut@obs-nice.fr

MÉRAND, Antoine
Observatoire de Paris-Meudon
LESIA (France)
Antoine.Merand@obspm.fr

MILLOUR, Florentin
LUAN/LAOG (France)
Florentin.Millour@obs.ujf-
grenoble.fr

MONNET, Guy
European Southern Observatory -
Garching (Germany)
gmonnet@eso.org

MONTMERLE, Thierry
Laboratoire d'Astrophysique de
Grenoble (France)
montmerle@obs.ujf-grenoble.fr

MOORWOOD, Alan
European Southern Observatory -
Instrumentation Division (Germany)
amoor@eso.org

MOSONI, Laszlo
Max-Planck Institut
für Astronomie - Heidelberg
(Germany)
mosoni@szombat.konkoly.hu

MOURARD, Denis
Observatoire de la Côte d'Azur -
GEMINI (France)
Denis.Mourard@obs-azur.fr

MUGRAUER, Markus
AIU Jena (Germany)
markus@astro.uni-jena.de

NARDETTO, Nicolas
Observatoire de la Côte d'Azur -
GEMINI (France)
Nicolas.Nardetto@obs-azur.fr

NEUHÄUSER, Ralph
AIU Jena (Germany)
rne@astro.uni-jena.de

OHNAKA, Keiichi
Max-Planck Institut für
Radioastronomie - Infrared
Interferometry, Bonn (Germany)
kohnaka@mpifr-bonn.mpg.de

PARESCE, Francesco
IASF, Bologna (Italy)
paresce@iasfbo.inaf.it

PASTORINI, Guia
Universitá degli Studi di Firenze -
Astronomia e scienza dello spazio
(Italy)
guia@arcetri.astro.it

PATRU, Fabien
Observatoire de la Côte d'Azur -
GEMINI (France)
fabien.patru@obs-azur.fr

PAUMARD, Thibaut
Max-Planck-Institut für
Extraterrestrische Physik
Garching (Germany)
paumard@mpe.mpg.de

PERCHERON, Isabelle
European Southern Observatory -
DMD (Germany)
ipercher@eso.org

PERRIER, Christian
Laboratoire d'Astrophysique de
Grenoble (France)
christian.perrier@obs.ujf-
grenoble.fr

PERRIN, Guy
Observatoire de Paris-Meudon
LESIA (France)
guy.perrin@obspm.fr

PETERSON, Deane
Stony Brook University - Physics
and Astronomy (USA)
dpeterson@astro.sunysb.edu

PETR GOTZENS, Monika
European Southern Observatory -
Garching (Germany)
mpetr@eso.org

PETROV, Romain
University of Nice - Sophia Antipolis,
LUAN (France)
petrov@unice.fr

PONCELET, Anne
Observatoire de Paris-Meudon
LUTH/LESIA (France)
anne.poncelet@obspm.fr

POTT, Jörg-Uwe
European Southern Observatory -
Garching (Germany)
jpott@eso.org

PREIBISCH, Thomas
Max-Planck Institut für
Radioastronomie - Infrared
Interferometry, Bonn (Germany)
preib@mpifr-bonn.mpg.de

PREIS, Olivier
Laboratoire d'Astrophysique de
Grenoble (France)
Olivier.Preis@obs.ujf-
grenoble.fr

QUANZ, Sascha Patrick
Max-Planck Institut
für Astronomie - Heidelberg
(Germany)
quanz@mpia.de

QUELOZ, Didier
Observatoire de Genève
(Switzerland)
Didier.Queloz@obs.unige.ch

QUIRRENBACH, Andreas
Leiden Observatory
(The Netherlands)
quirrenb@strw.leidenuniv.nl

RABIEN, Sebastian
Max-Planck-Institut
für Extraterrestrische
Physik - IR Group
Garching (Germany)
srabien@mpe.mpg.de

RAJAGOPAL, Jayadev
NASA Goddard Space
Flight Center - Infrared
Astrophys./Univ.
of Maryland (USA)
jayadev@iri1.gsfc.nasa.gov

RATZKA, Thorsten
Max-Planck Institut
für Astronomie - Heidelberg
(Germany)
ratzka@mpia.de

RIAUD, Pierre
Institut d'Astrophysique et de
Géophysique, Liège University
(Belgium)
riaud@astro.ulg.ac.be

RIBAK, Erez N.
Technion - Israel Institute of
Technology (Israel)
eribak@physics.technion.ac.il

RICHARDSON, Jeremy
NASA Goddard Space Flight Center
- Infrared Astrophysics (USA)
lee.richardson@colorado.edu

RICHICHI, Andrea
European Southern Observatory -
VLTI (Germany)
arichich@eso.org

RIVINIUS, Thomas
Landessternwarte, Heidelberg
(Germany)/ESO Paranal (Chile)
T.Rivinius@lsw.uni-
heidelberg.de

ROCCATAGLIATA, Veronica
European Southern Observatory -
VLTI, Garching (Germany)
vroccata@eso.org

SACUTO, Stéphane
Observatoire de la Côte d'Azur -
GEMINI (France)
stephane.sacuto@obs-azur.fr

SCHAEFER, Gail
Space Telescope Science Institute
(USA)
gschaefer@stsci.edu

SCHIEDER, Rudolf
Universität zu Köln - Physikalisches
Institut (Germany)
schieder@ph1.uni-koeln.de

SCHMITT, Henrique
Naval Research Observatory -
Remote Sensing Division (USA)
hschmitt@ccs.nrl.navy.mil

SCHNEIDER, Jean
Observatoire de Paris-Meudon
LUTH (France)
jean.schneider@obspm.fr

SCHÖLLER, Markus
European Southern Observatory -
VLTI (Chile)
mschoell@eso.org

SCHUHLER, Nicolas
European Southern Observatory -
VLTI (Germany)
nschuhle@eso.org

SÉGRANSAN, Damien
Observatoire de Genève
(Switzerland)
Damien.Segransan@obs.unige.ch

SETIAWAN, Johny
Max-Planck-Institut
für Astronomie - Heidelberg
(Germany)
setiawan@mpia.de

SIMON, Michal
State University of New York - Dept.
of Physics and Astronomy (USA)
michal.simon@sunysb.edu

SNIJDERS, Bart
TNO, Optics, Delft
(The Netherlands)
bart.snijders@tno.nl

STECKLUM, Bringfried
Thüringer Landessternwarte,
Tautenburg - Star Formation
(Germany)
stecklum@tls-tautenburg.de

SURDEJ, Jean
Institut d'Astrophysique et de
Géophysique, Liège University
(Belgium)
surdej@astro.ulg.ac.be

SWAIN, Mark
Laboratoire d'Astrophysique de
Grenoble (France)
mark.swain@obs.ujf-grenoble.fr

TISSERAND, Patrick
Max-Planck Institut
für Astronomie - Heidelberg
(Germany)
tisseran@hep.saclay.cea.fr

TUBBS, Robert
Leiden Observatory
(The Netherlands)
tubbs@strw.leidenuniv.nl

TÜNNERMANN, Andreas
Fraunhofer Inst. of Applied Optics
& Precision Engineering - Jena
(Germany)
tuennermann@iof.fhg.de

TUTHILL, Peter
Sydney University - School of
Physics (Australia)
p.tuthill@physics.usyd.edu.au

VALAT, Bruno
European Southern Observatory -
VLTI (Germany)
bvalat@eso.org

VAN DER AVOORT, Casper
SRON & TU Delft
(The Netherlands)
c.vanderavoort@tnw.tudelft.nl

VANNIER, Martin
European Southern Observatory -
Paranal Observatory (Chile)
mvannier@eso.org

VENEMA, Lars
ASTRON, Research & Development
(The Netherlands)
swillens@astron.nl

WALLACE, Debra
NASA Goddard Space Flight Center
- Infrared Astrophysics (USA)
wallace@exo1.gsfc.nasa.gov

WALLANDER, Anders
European Southern Observatory -
Garching (Germany)
awalland@eso.org

WATERS, Rens
Sterrenkundig Instituut "Anton
Pannekoek", Univ. Amsterdam
(The Netherlands)
rensw@science.uva.nl

WEIGELT, Gerd
Max-Planck-Institut
für Radioastronomie - Bonn
(Germany)
weigelt@mpifr-bonn.mpg.de

WINKLER, Walter
MPG - Albert-Einstein Institute,
Hannover
walter.winkler@rzg.mpg.de

WITTKOWSKI, Markus
European Southern Observatory -
Garching (Germany)
mwittkow@eso.org

WOITKE, Peter
Leiden Observatory
(The Netherlands)
woitke@strw.leidenuniv.nl

WOLF, Sebastian
Max-Planck-Institut
für Astronomie - Heidelberg
(Germany)
swolf@mpia.de

The Power of Optical/IR Interferometry

Francesco Paresce

European Southern Observatory, Karl-Schwarzschild-Strasse 2, D-85748 Garching, Germany and
Istituto Nazionale di Astrofisica, Via del Parco Mellini 84, 00136 Rome, Italy
fparesce@inaf.it

Summary. Much has been accomplished in the last few years since the first fringes of the large interferometers like Keck and VLTI. Many technical challenges have been met and tamed and a surprising amount of good science produced even with instrumentation that was far from its optimum configuration. I briefly review some of these accomplishments and describe what can and, hopefully, will be accomplished in the next decade by interferometry in the optical and IR provided the astronomical community continues to support this endeavour in the manner it deserves.

1 Introduction

An important event occurred in June 1996 in Garching where the interferometry community gathered to discuss the state and prospects of the discipline that at the time was still mired on both sides of the Atlantic in a rather parlous state. In Europe in particular, the situation was far from rosy as ESO had just a few years earlier suspended practically all work on interferometry for the VLT because of financial problems. The history of this period is properly reviewed by Beckers and Lena in these proceedings and I do not intend to repeat it here but it is worthwhile to remember that Riccardo Giacconi had just decided at that time to restart the VLTI program as best as could be done with the limited resources available. In hindsight, we all owe Riccardo a big debt of gratitude for his courage and foresight in this area.

Therefore, the June 1996 meeting was intended in large part to rally the troops and convince the skeptics among the astronomical community of which there were quite a few of the great potential of interferometry. The meeting was a definite success on both fronts and the quantity and quality of the presentations was exceptional. The proceedings were elegantly published in ESO's astrophysics symposia series soon after [1]. In these proceedings, one can obtain a faithful picture of where we stood then and, especially, of where we wanted to be in the next few years.

Inevitably, it took a little bit (but not really all that much) longer than prophesied then for most areas of investigation except that of extragalactic science which proceeded much more rapidly than anyone expected. In the following, I will try to measure that progress in the intervening 9 years with a few well chosen examples of the results so far and to try to forecast where we should be in the next 9 years. I apologize up front for concentrating on VLTI and Keck Interferometer (KI) results and ignoring the excellent recent results from facilities such as PTI and SUSI etc that have been well reviewed elsewhere.

2 A few examples of current results

Approximately 35 peer reviewed papers have appeared in the astronomical literature from the VLTI since its first fringes in 2001. A half dozen have also appeared from the Keck Interferometer. In general, these papers presented work in areas clearly predicted and expected by the speakers at the June 1996 meeting but with results that were, in many cases, quite unexpected. The VLTI results in particular have been described in some detail in a series of ESO Messenger articles [2],[3]. I pick a few here for purposes of illustration.

Some of the most exciting recent results have come in the area of stellar rotation. For example, only direct measures of Be star photospheres by interferometry can overcome the challenge of proving whether these objects rotate close to a few percent of their critical velocity or not. This proof will have a profound impact on the dynamical models for Be star disc formation due to rapid rotation combined with mechanisms like pulsation, radiation pressure of photospheric hot spots or expelled plasma by magnetic flares.

The southern star Achernar (α Eridani) is the brightest Be star in the sky and, therefore, a perfect target for the VLTI. It also represents a convenient object to test the validity of the concepts briefly described above. Analysis of the data by Domiciano de Souza et al. [4] reveals an extremely oblate shape from the distribution of equivalent UD diameter values on an ellipse. The results of this fit are: major axis $2a = 2.53 \pm 0.06$ mas, minor axis $2b = 1.62 \pm 0.01$ mas and minor-axis orientation $\alpha_0 = 39° \pm 1°$. It was shown that, in the particular case of Achernar, its observed asymmetry reflects its true photospheric distortion with a negligible CSE contribution.

Using this data, it was found that the commonly adopted Roche approximation (uniform rotation and centrally condensed mass) fails to explain Achernar's extreme oblateness. This result opens new perspectives in basic problems in stellar physics such as rotationally enhanced mass loss of Early-type stars. In addition to its intimate relation with magnetism and pulsation, rapid rotation thus provides a key to understanding the Be phenomenon, which is one of the outstanding non-resolved problems in theoretical astrophysics.

Another extremely interesting case is that of Eta Car described by Van Boekel et al., 2003 [5]. The most important conclusion from the VLTI data is that the central object is not spherically symmetric. In fact, its major axis is aligned with that of the large-scale structure of the homunculus. This alignment on all scales means that the 1840 outburst looks like a scaled-up version of the present-day wind, and that this wind is stronger along the poles than in the equatorial plane.

This can be understood in the framework of radiation-driven winds from rapidly rotating stars: centrifugal forces favor mass-loss in the equatorial plane, but the radiation pressure in these massive stars is stronger in the polar regions because of the von Zeipel effect (the stronger gravity near the poles leads to a higher temperature). For Eta Car, the von Zeipel effect is more important than the centrifugal levitation leading to a polar wind. The VLTI observations, thus, favor a model that interprets the morphology of Eta Car on all scales with a radiation-driven wind from a rapidly rotating star.

Turning to proto-planetary or stellar accretion discs, it has been known for some time that most of the dust in discs around newborn stars is made up of silicates. In the natal cloud this dust is amorphous, i.e. the atoms and molecules that make up a dust grain are put together in a chaotic way, and the grains are fluffy and very small, typically about 10 V4 mm in size. However, near the young star where the temperature and density are highest, the dust particles in the circum-stellar disc tend to stick together so that the grains become larger. Moreover, the dust is heated by stellar radiation and this causes the molecules in the grains to re-arrange themselves in geometric (crystalline) patterns.

Accordingly, the expectation is that the dust in the disc regions that are closest to the star is soon transformed from Tpristine T (small and amorphous) to Tprocessed T (larger and crystalline) grains. Model calculations show that crystalline grains should be abundant in the inner part of the disc at the time of formation of the Earth. Spectral observations of silicate grains in the mid IR around $10\,\mu$m should tell whether they are Tpristine T or Tprocessed T. Earlier observations of discs around young stars have shown a mixture of pristine and processed material to be present, but it was impossible to tell where the different grains resided in the disc.

Thanks to a hundred-fold increase in angular resolution with the VLTI and the highly sensitive MIDI instrument, analysis of detailed infrared spectra of the various regions of the proto-planetary discs around the three young Herbig Ae stars, only a few million years old by Van Boekel et al., 2004 [6] now show that the dust close to the star is much more processed than the dust in the outer disc regions. In one star, the dust is processed in the entire disc. In the central region of this disc, it is extremely processed, consistent with completely crystalline dust. In the other two stars, the dust in the inner disc is fairly processed whereas the dust in the outer disc is nearly pristine.

An important conclusion from the VLTI observations is, therefore, that the building blocks for Earth-like planets are present in circum-stellar discs from the very start. This is of great importance as it indicates that planets of the terrestrial type like the Earth are most probably quite common in planetary systems, also outside the solar system.

Further steps towards our better understanding of the stellar mass loss process can be taken with interferometric measurements of post-AGB stars. One of the basic unknowns in the study of late-type stars is the mechanism by which spherically symmetric AGB stars evolve to form axi-symmetric planetary nebulae (PNe). Antoniucci, Paresce and Wittkowski [7] presented the first detection of the envelope surrounding the post-AGB binary source HR 4049 by K-band VINCI observations. The physical size of the envelope in the near-infrared K-band is about 15 AU. These measurements provide information on the geometry of the emitting region and cover a range of position angles of about 60 deg. They show that there is only a slight variation of the size with position angle covered within this range. These observations are, thus, consistent with a spherical envelope at this distance from the stellar source, while an asymmetric envelope cannot be completely ruled out due to the limitation in azimuth range, spatial frequency, and wavelength range. Further investigations using AMBER can reveal the geometry of this near-infrared component in more detail, and MIDI observations can add information on cooler dust at larger distances from the stellar surface.

Finally, I should mention the truly ground breaking first observations of extragalactic objects with interferometry specifically of the AGNs NGC 4151 by the KI [8] and of NGC 1068 [9], [10] by the VLTI. In the case of NGC 4151, the KI found a marginally compact source of <0.1 pc diameter while in the case of NGC 1068, the VLTI resolved a complex structure consisting of a hot partially resolved component of approximately 0.7 pc in size and a warm resolved component of a few pcs in size. In summary, it is obvious that already the first two years of VLTI observations of AGN have provided completely new views into the cores of these still mysterious objects. New facilities like the phase closure instrument AMBER or the external fringe tracking system FINITO will soon extend the VLTI capabilities to a level that interferometric observations of galactic nuclei will become both routine and the most important factor for our understanding of AGN physics

3 Near future possibilities

The near term future is rich with scientific possibilities as, I hope, the truly exceptional potential of interferometry with large telescopes become reality. These possibilities are described in considerable detail in a number of ESO documents [11], [12], [13]. Moreover, we not only expect the VLTI to be soon equipped with PRIMA but also for the LBT, the KI and CHARA facilities to come on line in their full glory in the next few years adding to the panoply of observing capabilities at milliarcsecond (mas) resolution. All these instruments will provide unprecedented capabilities for major discoveries. For example, images of the surface of stars will provide fresh insight into convection cells, spots, flares, and other phenomena driven by convection and magnetic fields. These have been studied in detail on the Sun, but they also offer a rich field for interferometric observations, because their character varies substantially across the HR diagram and with stellar age. Direct interferometric imaging, and combinations of interferometric and Doppler tomographic techniques, can thus substantially broaden our knowledge of stellar activity, convection, and magnetism, and provide new stimuli for the theoretical modeling of these phenomena.

Astrometric searches for extra-solar planets will, of course, be one of the key activities in the next decade with large interferometers like VLTI and KI. The planets we are observing now are certainly not anything like those in our own solar system. This is mainly due to the strong observational bias still present in the peculiarities of the RV technique that favours detection of large planets in orbits very near the parent star that give rise to the highest velocities.

It is evident how the two techniques are complementary to each other. The RV technique is particularly sensitive to close-in massive planets around stars at any distance but only those whose spectra is particularly rich in absorption lines that can be used to determine the radial velocity to the exquisite precision required. In contrast, the astrometry technique favours the intermediate mass planets orbiting at larger separations. Both techniques still have a wide region of overlap, however, that can be conveniently exploited to get to a unique determination of their mass.

Since the closest star forming regions are located between ∼50 (TW Hyd) and ∼140 pc (Taurus-Auriga) from the Sun, interferometers like the VLTI would be in the enviable position of easily finding Jupiter-sized planets around stars of any spectral type and, most significantly from the physical point of view, of any age up to the MS.

This capability would, in fact, allow them to explore, for the first time, both the precise time of formation and the subsequent evolution of planets in all PMS stages leading into and well up along the MS. Since astrometry is not limited to absorption line rich cool stars as is the RV technique, it can also quickly attack and resolve the burning issue related to the strongly implied possibility that the IR excess MS stars like Beta Pic, Fomalhaut and Vega have or have not already formed planets in their observed debris disks.

The VLTI, by exploiting its dual mode of operation with phase referencing, can, at the same time, probe in considerable detail not only the planet itself through its effect on the motion of the parent star but also the evolution and structure of the proto-planetary/proto-stellar disk from which it eventually must emerge and interact. This affords another unprecedented opportunity to finally understand and pin down the presently uncertain connection between the two phenomena. At the typical distance of the MS early type star IR excess disks, for example, the VLTI IR spatial resolution of a few mas ensures a clear view of these disks down to the level of a few tenths of AU resolution at a few AU radii where most of the interaction between a possible planet and the edge of the small particle disk is expected to occur.

Interferometry with the VLTI can also be expected during the coming years to contribute decisively to our knowledge of the structure and evolution of the circum-stellar environment of young low- and intermediate-mass stars, which is an important part of the route to planet formation. Studies of the structure of debris disks around young main sequence stars will illuminate the last phases of planet formation when the accretion disks are being cleared out by the combined action of planetary resonances, solar radiation pressure, and collisions. Such images will also be used to deduce the presence and main characteristics of possible planets that would be impossible to determine any other way.

The VLTI will be able to spatially resolve the displacement of the photo center of microlensing events, which can be observed in time. This will allow us to break the parameter degeneracy which is inherent in photometric observations of microlensing events and permit a unique determination of the distance of the lensing source with important implications in our understanding of the mass and distribution of the lenses.

Astrometric observations of the galactic center cluster with PRIMA could reduce the error bars of current data by two orders of magnitude (from a few mas to a few tens of μas). In addition to improvements of the orbital parameters, this would enable searches for small deviations from Keplerian motion. The mass and distance of the central black hole, general relativistic precession of the orbital pericenter and the distribution of matter within about 100 AU from the central black hole could be obtained from such data.

Finally, we will be able, for the first time, to address several questions which are fundamental to our understanding of AGNs. Do the putative tori exist? That is, does the dust indeed settle into a symmetrical, geometrically thick structure? What is the size of the hottest parts of the dust distribution radiating at infrared wavelength and how does it depend on AGN luminosity and size of the stellar core? Is the apparent size of the hot dust distribution in the cores of AGN compatible with the assumed physical torus model? Can one find direct observational evidence for the torus orientation postulated by various unified schemes – namely that tori are seen face-on in Seyfert I galaxies and edge-on in Seyfert IIs? Does the torus axis align with other indicators for the source axis like outflow phenomena or polarization? Is the dust distributed smoothly or in clumps? What is the maximum dust temperature and where is it located?

4 Far future possibilities

There are a number of really exciting possibilities lying just outside our current planned vision for the next decade or so in optical IR interferometry that I would like to briefly mention in closing. The first is that of interferometry in Antarctica where, for example, in principle, one can get an improvement over Paranal of a factor of over 1000 in the backgound normalized hypervolume

[14]. Several proposals to exploit this exceptional capabilitiy are described in these proceedings.

Another area is the exploitation of the space environment that is particularly favorable to interferometry. SIM, of course, is a well known program due to launch hopefully in the next decade but also Darwin whose specific purpose is to detect extrasolar terrestrial-like planets is another. In the latter, a flotilla of free flying satellites would be used to provide extremely sensitive nulling capabilities for this purpose. This idea is also described in greater detail in these proceedings. Finally, of course, I should mention the on going efforts to plan for much larger ground based arrays than the existing ones that would reach kilometer to tens of kilometer baselines needed to probe ever more deeply the far reaches of the universe.

In conclusion, I hope and expect that when a similar symposium to these first two will take place a decade from now, all the scientific possibilities mentioned above and, hopefully, many more we have not been imaginative enough to think up now will be discussed in detail with exciting and unexpected results galore.

A word of caution, however, is warranted in leaving you here. It is always sobering for me to think that at ESO itself the VLTI has had and in some ways still has a lower priority than the VLT and the next program ALMA. The VLTI has survived many an attempt to kill or limit it by producing a series of stunning new results even in its infancy when the instrumentation available had not yet reached anywhere near its peak performance. This is certainly very encouraging but I would like to point out that we are not out of the woods yet especially in these difficult times. Unless we can keep up the pressure on our funding agencies to support us properly, these fantastic opportunities may dissolve into thin air especially when the ELTs will come to the fore. It is only our ability to provide ever more sophisticated and scientifically compelling results (especially images) that we can stave off this unfortunate consequence.

Acknowldegments I owe an enormous debt of gratitude to all those that have participated with me in this great adventure of the VLTI in the last ten years both on the technical and scientific sides. I cannot mention them all here as it would take up more pages than I am allowed but everything I said in this talk was made possible only by their great commitment and perseverance.

References

1. Science with the VLT Interferometer, F. Paresce, Ed., Springer-Verlag, 1997
2. Wittkowski, M., et al. Recent scientific results from the VLTI, The ESO Messenger, **119**, 36, 2005
3. Richichi, A., Paresce, F., Harvesting results with the VLTI, The ESO Messenger, **114**, 26, 2003.
4. Domiciano de Souza, A., et al., The spinning-top Be star Achernar from VLTI-VINCI, Astron. Astrophys., **407**, L47, 2003

5. van Boekel, R., et al., Direct measurement of the size and shape of the present-day stellar wind of eta Carinae, Astron. Astrophys., bf 410, L37, 2003

6. van Boekel, R., et al., The building blocks of planets within the 'terrestrial' region of protoplanetary disks, Nature, **432**, 479, 2004

7. Antoniucci, S., Paresce, F., Wittkowski, M., VINCI-VLTI measurements of HR 4049: The physical size of the circumbinary envelope, Astron. Astrophys., **429**, L1, 2005

8. Swain, M., et al., Interferometer Observations of Subparsec-Scale Infrared Emission in the Nucleus of NGC 4151, Astrophys J., **596**, L163, 2003

9. Jaffe, W., et al., The central dusty torus in the active nucleus of NGC 1068, Nature, **429**, 47, 2004

10. Wittkowski, M., et al., VLTI/VINCI observations of the nucleus of NGC 1068 using the adaptive optics system MACAO, Astron. Astrophys., **418**, L39, 2004

11. Scientific objectives of the VLTI, F. Paresce, March 15, 2001,

12. Long term strategy for VLTI up to 2020, Glindemann, A., Paresce, F., Quirrenbach, A., Richichi, A., 17 June 2003 VLT-PLA-15000-3085

13. Reference missions for PRIMA, Perrin, G., et al., 10 August, 2004, ESO/STC-362.

14. Swain, M., The Potential for Exoplanet Science with Infrared Interferometry at Dome C, EAS Publications Series, **14**, 147, 2005

The Early Days of the *Very Large Telescope* Interferometer

Pierre Léna

Université Paris VII & Observatoire de Paris
pierre.lena@obspm.fr
Observatoire de Paris 92195 Meudon, France

Summary. This recollection begins in 1978: ESO was building its future, considering the need and technical possibility of a very large telescope. In 1974, the success of optical interferometry had opened the way to a daring concept for the VLT, soon conceived as an array of four 8-m telescopes. In 1987, the VLT project was approved and began: it included a coherent mode, where the four telescopes, plus two smaller and movable ones, could be combined interferometrically at near-infrared wavelengths. This became possible because adaptive optics, developed for this purpose in France and ESO, succeeded to phase individual pupils, hence providing the full interferometric sensitivity of large pupils. After recalling briefly the early steps which led to the highly successfull operation of a VLTI in 2001, some suggestions may be drawn for the future of optical interferometry in Europe.

The author of this short VLTI story would like to share some of his personal recollections and references. By no means could he pretend to give here an historical account, which would require a thorough search of documentation, interviews of the early actors and confrontation of viewpoints. The author simply hopes to help the readers, especially the younger ones, to find in a reading of the past, even biased, some good reasons to think and build the future. He also apologizes to the many people, often friends, he does not quote and who have been so decisive in the VLTI success.

1 1974-1987: a dream takes shape

In a night of August 1974, on the beautiful hill of Mont-Gros where stands the Observatoire de Nice, with its dome conceived by Gustave Eiffel and Charles Garnier, Antoine Labeyrie, age 31, observed the star Vega with a small but remarkably conceived optical interferometer of his construction (Fig. 1, *left*, [1]). With a baseline of 12 meters, he demonstrated that the concept brillantly imagined a century before by Hippolyte Fizeau could be revived. Indeed, Edouard Stephan, at the Observatoire de Marseille in 1874, could obtain an upper limit (0.158") of a stellar diameter by applying the idea, Albert Michelson had measured the diameter of Jupiter's Galilean satellites in 1881, then the first stellar diameters in 1920 at the Hale telescope on Mt. Wilson (California), but the Labeyrie success re-opened the field in an entirely new way: since independent telescopes were used, their distance B, which sets the angular resolution λ/B, was no longer a critical parameter and the combination of fully independent telescopes of any size became conceivable.

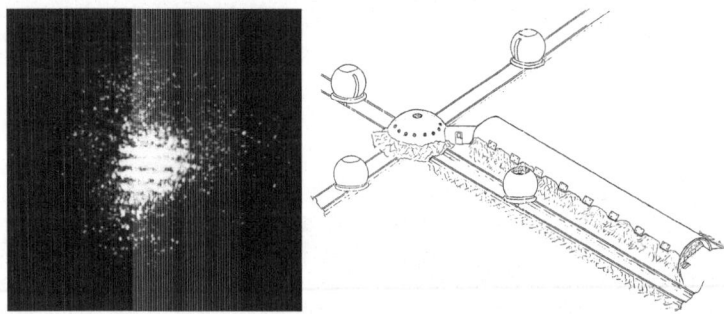

Fig. 1. *Left:* The first interferometric observation of Labeyrie [1]. Interference fringes are obtained on Vega with a 12-meter baseline, 25 cm telescopes and a bandwidth of 50 nm. *Right:* Antoine Labeyrie's presentation at the 1977 Geneva Conference, with this caption: *"Coherent array in the construction stages at CERGA. Starting with two spherical telescopes, the array is expected to grow progressively into a full-size system involving many telescopes."* [2].

Three years later, Lodewijk Woltjer, director of the *European Southern Observatory*, still building the 3.6-m telescope at La Silla (Chile) but looking way ahead, called in Geneva a Conference on *Optical telescopes of the future* [2]. Which concept should ESO select for the next generation? The young Labeyrie immediatly proposed to envisage an array of telescopes, and showed the system (Fig. 1, *right*) he was beginning to build at the new site of Plateau de Calern, near Grasse (France). Because of their promise of superior angular resolution, already well established at radio-wavelengths, arrays were warmly discussed with contrasted views. Harry van der Laan, with the great experience of radio-interferometry obtained at Dwingeloo (Netherlands), said: *"I must skip, for lack of time but not for lack of admiration, any direct comment on (...) the advances of optical interferometry (...) I doubt if the fringe benefits of ground-based coherent arrays (...) outweigh the additional constraints and costs..."*. Robert Hanbury-Brown, considered as the most knowledgeable person of the time in optical intensity interferometry, commented that an array *"...might be cheaper to build. However, we have serious doubts about its accuracy in the presence of atmospheric scintillation and also about its performance at long baselines"*.

Regarding the future telescope site, the seeing question was indeed omnipresent: Harold Babcock raised it with great force, but amazingly he was not the one to mention adaptive optics, although he had been the first to propose it twenty years earlier [3]. This technique, which later will prove so critical for the VLTI, was discussed by G. Bourdet with this prophetic comment:

"*...another possible and very efficient technique for improving the quality of atmospheric turbulence limited image forming instruments, is the real time phase control system*". In the Conference conclusions, which promoted the study of a large aperture telescope for the ESO European countries, L. Woltjer carefully stated: "*...I have concentrated here exclusively on the benefits for optical studies of a large collecting area (...) Concerning spatial resolution, it seems clear that there is no point in trying to compete with the Space telescope* [at this time in construction] *in the range 0.1 to 1 arcsec. For smaller sizes, ground-based arrays might offer interesting possibilities*".

The seeds were in the ground, and a new ESO Conference was planned for 1981 on *The scientific importance of high angular resolution at optical and infrared wavelengths*. The minutes of the ESO Council of May 1980 read: "*To a question by Prof. Blaauw about the significance of the (...) conference in the context of the VLT, the Director general replied that ESO was organizing a conference (...) which should have an immediate impact on the question as to whether the VLT - if it were built as an array - should be built as a coherent or as an incoherent array*". An incoherent array would simply add up, in a common image or spectrum, the light intensities coming from several telescopes, while a coherent one would preserve the phases and provide the angular resolution.

The 1981 Conference was a success [4] and the VLT Study group, chaired by Jean-Pierre Swings with a term to cover all aspects of the project, included "*...an ambition of a VLT (...) which would keep one large avenue for the future, i.e. the capability for spatial interferometry*". Discussions began to explore all kinds of possibilites (see below) and at the Cargèse Study week on VLT, held by ESO in 1983, the Council President Paul Ledoux quietly commented: "*...it had been stated by some persons in Cargèse that it would be desirable to put all these telescopes on rails in order to move them about for interferometry (...) a slightly risky undertaking*". In June of this year, the Council approved the creation and staffing of a VLT Project group, and one year later an Interferometry working group (IWG) was installed, with the term "*...to define realistic scientific objectives for interferometry with large telescopes (in the optical and especially in the infrared) and to assess the implications of interferometry on the specifications (and thereby the cost !) of a VLT*" [5].

From science and finance considerations, the target diameter for the VLT was set to 16 m, but as monolithic mirrors were preferred, industry fixed a feasibility limit of 8 m to their diameter and the VLT became an array of four telescopes in order to be a "16-m equivalent", as expressed by its first logo. The IWG worked in this frame and produced its final report *Interferometric imaging with the VLT* [6] in time for the ESO Venice Conference [7] in 1986. The *Proposal for the construction of the 16-m VLT*, published for the ESO Council in March 1987 [8], fully included the coherent, interferometric combination of the four 8-m telescopes. During this whole period, an excellent cooperation was achieved between Europe and the United States, in order to explore in detail all aspects of optical interferometry and a possible complementarity between each side of the Atlantic [9], [10], [11].

Finally, on December 6, 1987, the ESO Council unanimously decided to implement the VLT Program, which fully included optical interferometry. The remarkable point was that VLTI was not to be considered as a separate or optional development of the VLT, but as an integral part of a single global instrument, having different modes of observation. This early choice would later prove its virtue, as the VLTI specifications would remain present all along the development of the project.

2 The VLTI conception: hot issues (1981-1987)

The six years which lead to the approval of the VLT interferometric mode, as it was called, were indeed full of heavy work and discussions. Let us give here a glimpse into these questions, which included the grounds for the scientific program of an interferometer, its sensitivity, its accuracy, its imaging capability and, last but not least, the diversity of technical challenges associated with them.

2.1 Which science for an interferometer?

After Labeyrie's successful demonstration on Vega (1974), it took more than a decade for the young optical interferometry to produce results of scientific relevance, obtained with small apertures (less than 50 cm). At Cerga, the French group published in 1986 a beautiful result on the Hα envelope of the star γ Cass, measuring a gaussian shape with an extension of 3.6 mas (Fig. 2, [12]). With the SUSI interferometer at Narrabri (Australia), Davis & Tango determined a Sirius diameter of 5.63 ± 0.08 mas with outstanding accuracy [13].

In the infrared, speckle interferometry, developed by Léna and coworkers in the late 1970s, continued to explore atmospheric properties and to demonstrate the value of high angular resolution, with Mel Dyck discovering an IR companion to T Tau [14], or Chris Leinert measuring the orbit of the M dwarf GL866 [15]. In 1980, Roger Angel and Nick Woolf had also pointed out the advantages offered by infrared for interferometry, thanks to the larger value of the Fried parameter r_o and of the atmospheric coherence time τ_o, a point also stressed at the 1981 Conference by Léna ("*The ease of aperture synthesis in the infrared, already explored (...) should be soon confirmed and developped*" [4]). In 1984, at the KPNO 4-m telescope, Jean-Marie Mariotti and Stephen Ridgway obtained speckled fringes and published in 1988 double-Fourier spatio-spectral interferometry on the late-type star IRC+10216 [16].

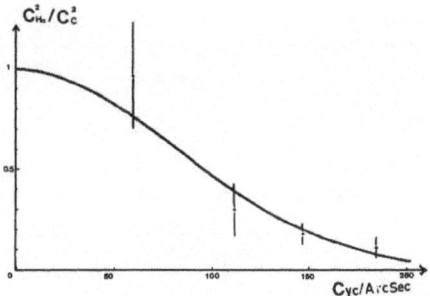

Fig. 2. Visibility values obtained with the I2T interferometer at Plateau de Calern on the star γ Cass at Hα. The gaussian fit gives, for the envelope size, an angular value of 3.9 mas at $1/e$ of maximum [12].

2.2 Sensitivity issues: speckled images or adaptive optics?

No interferometry science program could be proposed without a clear understanding of its sensitivity. As early as 1984, it was obvious that the presence of speckles in the images, even at near IR wavelengths, was detrimental to the high sensitivity one could hope from large apertures. In a paper published in 1986, Mel Dyck (considered as an expert in high angular resolution at IR wavelengths) and Ed Kibblewhite wrote: *"...Arrays consisting of telescopes with individual apertures larger than the atmospheric correlation length are probably not workable and certainly not cost effective* [17].The early working papers on VLTI do not mention adaptive optics (AO), and this elegant solution first appears in the ESO context in 1984 (Fig. 3), while at a Workshop called by the Canada-France-Hawaii telescope in 1986, AO was discussed as a way to suppress the speckles and improve the single telescope resolution. At a Conference held in Tucson in 1986, Jacques Beckers and François Roddier presented the adaptive optics program they had just started at NOAO [18].

The same year, with the specific aim to make VLTI possible, France began an astronomical AO program, soon joined by ESO and Fritz Merkle. This program led to the highly successful *ComeOn* instrument on the 3.6-m telescope at La Silla (Fig. 4, [19]). Indeed, the 1987 VLT Proposal refers to the *ComeOn* instrument and states: *"In the interferometric mode, the full gain of the 8-m single apertures of the VLT is (...) only obtained if adaptive optics is applied for a real-time partial or full phase compensation of the degradations due to atmospheric turbulence"*. The knowledge of Earth atmosphere properties for imaging had been constantly progressing since 1974 and began to provide firm grounds to evaluate the sensitivity, after the publication of the fundamental paper of François Roddier in 1981 [20]. Danchi's work in 1974 with the MacMath Solar telescope 5.5m baseline in an heterodyne mode at 10.6 μm, or Mariotti & Di Benedetto's work with the I2T interferometer at Cerga in 1984 provided useful data for the atmospheric piston

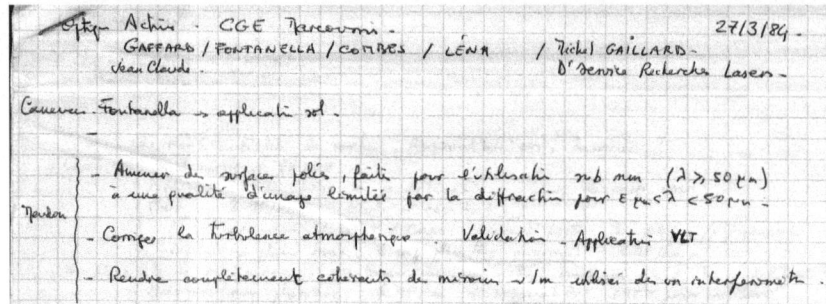

Fig. 3. An extract of P.Léna's notebook, dated 27/3/1984 and showing the mention of AO for VLT and interferometry, discussed at an ONERA meeting long before the *ComeOn* instrument was decided. Participants (top line) were M. Combes, J.-C. Fontanella, M. Gaillard, J.-P. Gaffard and P. Léna.

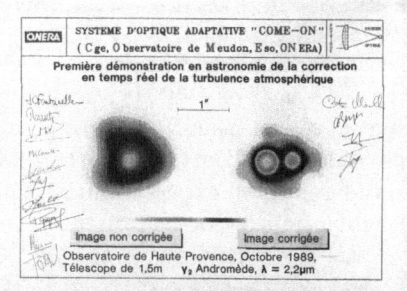

Fig. 4. The first astronomical adaptive optics observation: the double star γ_2 And is resolved at Observatoire de Haute-Provence in October 1989, with the *ComeOn* instrument. Signatures of J.-C. Fontanella, P. Gigan, P. Léna, F. Merkle, G. Rousset and others are present.

caracteristic time τ_o or the turbulence outer scale L_o, both important quantities for interferometry.

On the basis of all this, it became possible in 1987 to give a prospective sensitivity of the VLT interferometric mode, and to derive a potential scientific program, which focused on stellar surfaces and diameters, stellar environments, star formation and protoplanetary discs...and the galactic center. Fig. 5 was included in the proposal and can reasonably be compared to today's actual VLTI performances. These values were later (1989) refined by Beckers in the final VLTI Implementation Plan [21]. The feasibility of highly wanted extragalactic observations (AGNs) remained debated, although their value had been vigorously pointed out at the 1981 Conference by Martin Rees (in [4]). It is remarkable that the ESO Council, with a great sense of vision, dared to accept at face value the solid scientific and technical arguments given and, despite the quite small amount of actual interferometry scientific

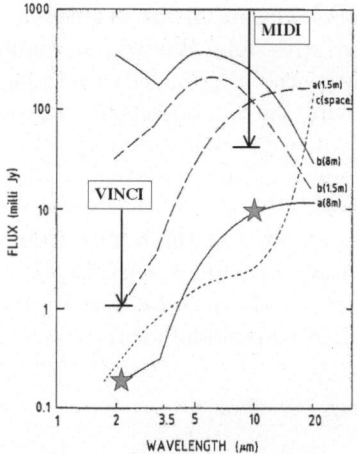

Fig. 5. The expected sensitivity of VLTI, as it was estimated in 1987 in the VLT Proposal ([8], page 97). The curves give respectively: a/ performances for VLT telescopes (8m & 1.5 m) with perfect adaptive optics ; b/ performances without adaptive optics, again for 1.5 m and 8 m ; c/ performance of an interferometer in space, with 1.5 m telescopes. The most critical assumption was the maximum integration time which atmospheric piston fluctuations could allow. The adopted value was 1s, too large a value for Paranal (but not unlikely now in Antarctica !). The arrows allow to compare these predictions with the actual performances obtained since 2001 on the UTs, with the instruments VINCI (K- band at 2.2 μm, limit $m_K = 12$) and MIDI (N-band at 10 μm, limit $N = 4$). Since the exposure time of current observations is much shorter (7 ms), the arrows point to the level one would reach with a 1 s exposure time. This level is about a factor 5 above the 1987 prediction ($\star$), and this is due to current residual transmission losses, insufficient Strehl ratio and other factors.

results yet obtained in 1987, agreed to include interferometry into the VLT program.

2.3 Imaging with an interferometer?

The great debate during all these years was about imaging: what would be the scientific value of VLTI if it could not provide images with a decent number of pixels, not to speak of the questionable impact of visibility curves on the general public in press releases? Would atmospheric piston phase shifts definitely prevent Fourier phases determination and good imaging? Radio-astronomy was indeed challenging, with arrays like the VLA, but also brought many ideas and confidence in the ultimate feasibility of VLTI. Indeed, it was radio techniques which had inspired, in 1978, Gert Weigelt to explore speckle holography [22], and publish his triple-correlation method for the restoration of speckled images [23]. This method opened the way to image reconstruction in the visible. Actual phase closure in the visible would actually take more than a decade to succeed, with John Baldwin at Cambridge (UK), [24].

To properly cover the spatial frequencies (*u-v plane*), many configurations were discussed. Even moving the 8-m telescopes was considered [25], although not too seriously. Finally, two movable auxiliary telescopes of ca.1.5 m were adopted, as this *sub-array* would be operative all the time and provide, although with reduced sensitivity, the much wanted *u-v* coverage. Finally, a compromise design was presented at the Venice Conference, where the four 8-m telescopes were still aligned (Fig.6, *left*). Later, after many discussions (Fig.6, *right*) it will belong to Jacques Beckers to make this unsatisfactory scheme evolve into the final trapezoidal array.

Fig. 6. *Left*: Artist drawing of the VLT linear array, with two movable interferometric auxiliary telescopes on the summit (and slope !) of Paranal, as described in the final VLT Proposal before the 1987 decision of the ESO Council [8].*Right*: Warm discussion on VLTI at Observatoire de Haute-Provence in 1989. From left to right: Daniel Enard (back), Fritz Merkle, Massimo Tarenghi, Antoine Labeyrie, Michel Faucherre (Picture by the author).

3 Parallel developments of interferometers (1974-1987)

The Labeyrie's success in 1974 led a number of groups to conceive or build optical interferometers of modest size (Fig. 7): these efforts explored many technical and operational aspects of interferometry and their results greatly helped the genesis of VLTI. In Europe emerged the *Interféromètre à 2 Télescopes* (I2T, 1974) and its successor the GI2T (1986) at Plateau de Calern (France), the Erlangen (West Germany, at the time) and COAST (Cambridge, UK) interferometers (end 1980s), in United States the Mark III at Mt. Wilson (1986) and the original Fizeau configuration of the *MultiMirror Telescope* MMT (1983), in Australia the SUSI interferometer (1985). Since 1974, Charles Townes and his collaborators were developing heterodyne 10.6

µm interferometry, first at the McMath telescope (Arizona, 1974), then later on Mt. Wilson.

Parallel to the VLTI, larger interferometers were also discussed, as in the *National New Technology Telescope* (NNTT) project and its four 8-m telescopes with Beckers in the USA (1986), or the emergence of an interferometric mode for the Keck Telescopes, with Michael Shao (ca. 1987).

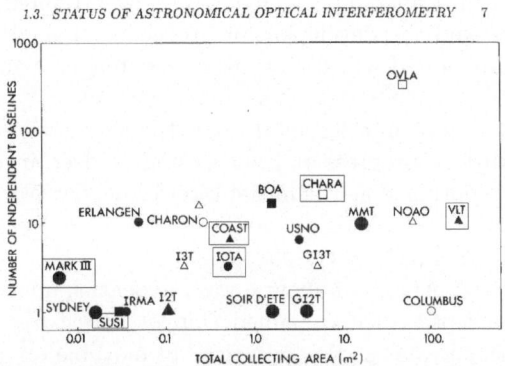

Fig. 7. A 1989 summary of the optical interferometers in existence or being developed: circles are for baselines < 100 m, triangles for 100-300 m, squares for > 300 m ; large filled symbols for instruments in operation, small ones under construction, open in planning (From the VLTI Implementation plan, p.7, 1989). Frames have been added to designate instruments in operation in 2005

A critical issue for the scientific use of interferometry was the accuracy at which a complex visibility could be measured, despite the atmospheric phase perturbations. At this time, the best values for the visibility amplitude uncertainties were in the range 1 to 5 %, quite unsufficient to properly constrain models when a very limited number of baselines is available. In 1988, Coudé du Foresto began to introduce single mode fibers to filter out the spatial modes of the wavefronts [26], a progress which immediatly improved the visibility amplitude accuracies by nearly an order of magnitude, giving an increased confidence in the potential VLTI performances.

In this busy context, the small group of enthusiastic engineers (Daniel Enard, Fritz Merkle, Ray Wilson) working at ESO since 1983 to define the VLTI feasibility had to consider a long list of challenging technical problems, while most of them would be solved in detail only after the 1987 approval of the VLT (see J. Beckers's contribution in this Volume): sub-micron stability of telescopes and Coudé train, movable auxiliaries, transport of light in vacuum or in the air, long delay lines with nanometric accuracy, phase shifts due to micro-seismicity, light losses due to numerous reflections, polarization, pupil transport and reconfiguration, etc. Nevertheless, these preliminary technical studies all converged in 1987 toward a positive assessment of the VLTI feasibility.

4 From 1987 to present, a success story

As Jacques Beckers, in this Volume, details the VLTI story from 1988 to 1993, when under his leadership the final design was adopted, adapted to the selected VLT site and the developement proceeded, I shall only recall here some events of this period which have also contributed to shape the present VLTI.

Given the novelty of interferometry and the scarcity of competences, it was out of question to progress at ESO without a deep involvement of the community. This continued as the initial Interferometry Working Group was pursued under the successive responsabilities of Jacques Beckers, then of the regretted Jean-Marie Mariotti († 1998).

Even before 1986, Antoine Labeyrie was advocating an extremely elegant interferometer concept, which he named *Optical Very Large Array* (OVLA), made of 27 telescopes, 1 to 2 meter in size, all movable on an ellipse (intersection of the ground surface with a paraboloid aimed at the source), hence without delay lines, able in principle to provide an excellent *u-v* coverage. This met with other interests in Germany (Genzel, Weigelt) and in France (Guilloteau, Foy, Léna), and in 1988 an *European Wide Band Multiaperture Interferometer*, called VISIR, was advocated to precede and prepare the VLTI.

Finally, after ESO help to a feasibility study at IRAM for movable 1.5-m telescopes, the whole matter diverged from OVLA and converged in 1989-1980 on the joint support of CNRS (Insu) in France and Max-Planck Gesellschaft in Germany, to enhance the VLTI with an extra auxiliary telescope (AT) to be added to the two approved ones, merging most of European efforts (except UK, which was not a member of ESO at this time) in a single interferometric project, namely the VLTI. Signed at the end of 1992, the Memorandum of agreement opened the so-called *VLTI Enhancement policy*, which today has led, on top of the ESO normal budget, to many additional contributions from ESO Member states, like an external fringe tracker (Italy), a fourth AT (Belgium, Switzerland, Italy), a dedicated astrometric configuration (Germany, Netherlands, Switzerland), star separators for faint objects observations (France, Germany), differential delay lines (Switzerland, Germany, Netherlands) in order to get a flexible, full time operating imaging interferometer with several movable telescopes, aside the main array of 8.2-m units. This is indeed to the benefit of the whole ESO community, and not exclusively to the one of these contributors.

Due to a strong financial pressure in France and Germany, a deep VLTI crisis occured in november 1993, which can be summarized by quoting the ESO Council minutes: "*Following the presentation and discussion of different alternatives for cost reduction, Council adopts further modifications to the VLT programme plan. This includes the postponement of the implementation of VLTI, VISA [the ATs sub-array], Coudé train and associated adaptive optics for all telescopes (...) Furthermore, the executive will endeavour to reintroduce full Coudé and interferometric capabilities at the earliest possible date".*

This blow could have been deadly to VLTI. The morale of the small European interferometric community in Europe went down, Jacques Beckers left ESO and Pierre Léna, then French representative at Council, gave his resignation to the French minister. It took all the patience and cleverness of Oskar von der Luehe and Jean-Marie Mariotti, supported by the ESO Visiting Committee, for this community to obtain in 1996 from Ricardo Giacconi, then Director general, and from Council a recovery plan which first resumed adaptive optics and led to the success of the NAOS instrument in Dec 2001, then resumed interferometry and led to the fringes first obtained with siderostats on March 2001, then with the two 8.2-m telescopes on Oct 2001, then with three UTs in May 2004, finally with the two first ATS on March 2005. The wealth of scientific VLTI results to date can indeed be appreciated in the present Volume.

5 A few lessons for the future

Today, in 2005, it is clear that many new ideas are emerging, in order to develop a VLTI which still has a long way to go. In this respect, one should be careful to remember that the VLTI is a unique facility of a new kind, which has never existed before: in the opinion of the author, its stepwise development, the way the community interacts with it should not be judged exactly with the same criteria as the ones used for well established instruments. Yet, VLTI was approved as an integral part of the VLT and should not be treated in the future as a somewhat different or independent project.

What can the VLTI story teach us about the future plans of Europe? Indeed the *Overwhelmingly Large Telescope* (OWL)[1] is now present in the landscape and there may never exist an OWL-Interferometer, as the OWL concept seems to focus on a single dish telescope. Yet, it is already obvious that resolutions beyond 1 mas will be required in the future [27] and will call for an imaging interferometer having both the sensititvity of large (4 m? or more?) pupils and baselines longer than the VLTI (1 to 10 km?). A future European interferometer will certainly capitalize on the lessons still being learned from VLTI ; it might also benefit from the industrial concepts developed for OWL (e.g. mirror production) ; the question of its site will probably be decoupled from OWL's requirements, and this leads to the current consensus to explore in depth the astronomical potential of the Antarctica plateau

---

[1] OWL may not happen exactly as its present design, but somehow Europe will implement its own ELT in the future.

(Dome C). This future interferometer will use novel techniques, which are currently demonstrated at different scales, recalling us of the early VLTI parallel developments: integrated optics, or single-mode fiber optics which is just demonstrating its capability for kilometric baselines with the OHANA success in June 2005 [28].

Indeed, as in the early VLTI days, there are voices today to express that interferometry on the ground shall never reach the same performances as in space, a point which is certainly well made. Yet, what was also true twenty-five years ago, and still remains, are the facts that a/ an interferometer in space is extremely expensive ; b/ ground-based interferometry is far from having yet reached the limits in sensitivity, imaging capability and wavelength coverage that ground-based observations will ultimately encounter. If these views in exclusive favor of space had prevailed at the time, no VLTI would have been built.

A final point should be made: during these 25 years, ESO has been able, with great care and cost, to internally establish a group of interferometric expertise, which is unique in the world. It would seem wise that any future European interferometric project be developed with this expertise, therefore within ESO - which does not mean without possible external partners.

References

1. Labeyrie, A., Interference fringes obtained on VEGA with two optical telescopes, Ap.J. **61**, 33 (1995). For the history of Labeyrie's work at Nice Observatory, see Colloquium for the 30 years of interferometry at Nice, Oct. 2004, Observatoire de Nice (2005).
2. Optical telescopes of the future, Pacini, F., Richter, W. & Wilson, R.N. Eds. Proceedings of an ESO Conference, Geneva, 12-15 Dec 1977. The quotations from H. Babcock, G. Bourdet, H. van der Laan, A. Labeyrie, L. Woltjer are from their respective contributions to this Conference.
3. Babcock, H. The possibility of compensating astronomical seeing, Publ.Astr.Soc.Pac. **65**, 229 (1953).
4. Scientific importance of high angular resolution at infrared and optical wavelengths, Ulrich, M.H. & Kjaer, K. Eds., Proceedings of an ESO Conference, Garching, 24-27 March 1981.
5. Letter from Lodewijk Woltjer to Pierre Léna, author's archives, 1984. The composition of this first Interferometry Working Group was: O. Citterio, D. Downes, A. Labeyrie, P. Léna, J. Noordam, F. Roddier, J.P. Swings, G. Weigelt (who joined later), with ESO members D. Enard, F. Merkle, M. Tarenghi (joining later), R. Wilson.
6. Interferometric imaging with the Very Large Telescope, Final Report of the ESO/VLT Working group on interferometry (VLT Report No.49), ESO (1986).
7. Second Workshop on ESO's Very Large Telescope, D'Odorico, S. & Swings, J.-P. Eds., Proceedings No. 24, ESO (1986).
8. Proposal for the construction of the 16-m Very Large Telescope, ESO, March 1987.

9. Interferometric imaging in astronomy, Goad, J.W. Ed., Proceedings of the Joint Workshop on High-resolution imaging from the ground using interferometric techniques, Oracle (Arizona), ESO & NOAO, Jan 12-15, 1987.

10. Cambridge Workshop on imaging interferometry, Cambridge (Mass), Oct 28-30, 1987, Batelle Ed. (1987).

11. High-resolution imaging by interferometry, 2 vol., Merkle, F. Ed., Joint NOAO - ESO Conference held at Garching, March 15-18, 1998.

12. Thom, C., Granes, P., Vakili, F. Optical interferometric measurements of Gamma Cassiopeiae's envelope in the H-alpha line, Astron.Astroph. **165**, L13 (1986).

13. Davis, J., Tango, W.J., New determination of the angular diameter of Sirius, Nature, **323**, 234 (1986).

14. Dyck, H.M., Simon, T., Zuckerman, B. Discovery of an IR companion to T Tau, Ap.J. **225**, L103 (1982).

15. Leinert, C., Haas, M., Jahreiss, M., Gliese 866 - A double M dwarf, Astron.Astroph. **164**, L29 (1986).

16. Mariotti, J.M., Ridgway, S., Double Fourier spatio-spectral interferometry at KPNO 4-m telescope, **195**, 350-363 (1988).

17. Dyck, H.M., Kibblewhite, E.J., Giant infrared telescopes for astronomy: a scientific rationale, Publ. Astr.Soc.Pac. **98**, 260 (1986).

18. Beckers, J.M., Roddier, F. et al, National Optical Astronomy Observatories (NOAO) infrared adaptive optics program. I - General description, Advanced technology optical telescopes III, SPIE, Tucson, March 3-6, 1986.

19. Rousset, G. et al. First diffraction-limited astronomical images with adaptive optics, Astron.Astrophys., **230**, L29 (1990).

20. Roddier, F., The effects of atmospheric turbulence in optical astronomy, Progr.Opt., **XIX**, 281-276 (1981).

21. The VLT Interferometer implementation plan, Report by the ESO/VLT Interferometry Panel (VLT Report No. 59b), ESO (1989).

22. Weigelt, G.P., Speckle holography measurements of the stars Zeta Cancri and ADS 3358, Appl.Opt. **17**, 266 (1978).

23. Lohmann, A.W., Weigelt, G., Wirnitzer, B., Speckle masking in astronomy - triple correlation theory and applications, Appl.Opt. **22**, 402 (1983).

24. Baldwin, J.E. et al, The first images from an optical aperture synthesis array: mapping of Capella with COAST at two epoch, Astron.Astroph. **306**, L13 (1996).

25. Citterio, O., Considerations on the movability of the telescopes for the VLT interferometry configuration, Working paper within the Interferometry working group (P. Léna's archives), June 1985.

26. Coudé du Foresto, V., Ridgway, S., Mariotti, J.M., FLUOR: a stellar interferometer using single-mode infrared fibers,*in* High resolution imaging by interferometry II, Beckers, J. & Merkle, F. Eds. ESO (1991).

27. "Science Case for A Next Generation Optical/Infrared Interferometric Facility", 37th Liège International Astrophysical Colloquium, Surdej, J. Ed., 23-26 Août 2004.

28. Perrin, G. et al., Science **311**, 194 (2006).

1988–1993: The Final Definition of the Very Large Telescope Interferometer, its Site and Configuration

Jacques M. Beckers

Department of Astronomy and Astrophysics University of Chicago Chicago, IL, 60637, USA jbeckers@astro.washington.edu

Summary. This paper focuses on the period during which the VLTI evolved from its conceptual phase to the way it is largely implemented now. That phase did not include the VLTI instrumentation, which was defined later and which did not make use of the homothetic beamcombining system that was part of the 1988 – 1993 design.

1 Introduction

As already described before by Pierre Léna [1], the "Proposal for the Construction of the 16-M Very Large Telescope" was accepted in December, 1987. The approved project included in addition to the construction of 4 individual 8-meter (eventually 8.2-meter) telescopes, funding for the implementation of a combined incoherent focus (hence the "16-M Very Large Telescope" in the title) as well as a combined coherent, interferometric focus. The 16-M aspect of the VLT has not been implemented. It was abandoned early on since the light losses involved in combining the 4 light beams were judged detrimental and since the vast improvement in read-out noise of CCD and near-IR detectors made the need to raise the photo-electron levels well above the earlier high read-out electron levels less needed. The interferometric combination was implemented and constitutes what is now called the VLT Interferometer (VLTI).

I was most fortunate to be chosen in early 1988 to head the VLT Interferometer program shortly after ESO had been given the go-ahead for the construction of the VLT. At that time the individual 8-m telescope were reasonably well defined, using the ESO New Technology Telescope on La Silla as its prototype. The definition of the VLTI was still in the conceptual stage with substantial disagreement in the small European interferometry community as to the configuration and siting requirements of the VLTI. Conflicting opinions involved, for example, the desirability of a linear, redundant array of 8-m telescopes or a 2-D, non-redundant array. Some even proposed making the 8-m telescopes movable, whereas others preferred leaving them stationary and adding small relocatable auxiliary telescopes (ATs) to expand the (u,v) plane coverage. At that time even the site of the VLT had not yet been chosen; potential sites being considered were Cerro Paranal, Cerro Vizcachas (near La Silla) and Cerro La Montura (near Paranal). The preference of the interferometry community was for Cerro Vizcachas since it seemed better suited for the linear VLTI; the rest of the VLT community preferred Cerro Paranal because of its better astronomical quality. Since the start of the VLT construction, including site development, was an urgent task after the proposal approval, there was high pressure to make major decisions with regard to the VLTI. The items to be addressed included: (i) the final definition of the VLTI requirements and especially its configuration, (ii) related to that the choice of the VLT(I) site, (iii) a definition of the VLTI in sufficient detail to assure that VLT site development and 8-m telescope design and procurement took the interferometric mode needs fully into account, and (iv) the specifications of the 8-m telescopes that took their interferometric requirements into account.

A small VLTI team including, in addition to myself, 1 – 3 scientists and engineers was formed within ESO that worked closely together with the VLT Interferometry Panel and with the VLT project staff to resolve these issues in a timely matter. In another paper [2] I described in more detail than is possible here the resulting VLT(I) definition both of the site lay-out and the optical configuration. I refer to that paper for a more complete description of the VLTI definition in the 1988 – 1993 interval. In the present paper I will focus on some of the details of those studies. Those include: (i) the original choice of a homothetic beamcombiner with an 8 arcsec FOV, (ii) my plans for the "blind" interferometric observations of very faint objects (e.g. high-z galaxies) and (iii) my view for the future use and development of the VLTI.

2 Why a Homothetic Beam Combiner with an 8 arcsecond FOV?

The 1993 design of the VLTI included a so-called homothetic beamcombiner which made the VLTI entrance and exit pupil configurations to be homothetic to one another. As shown at the MMT interferometer [3] such beamcombining results in a large interferometric FOV, like in a Fizeau interferometer, even though one tends to think of an interferometer with separate feeds as a Michelson interferometer. In the post-1993 revision of the VLTI [4] the homothetic beamcombiner was eliminated and replaced by the then standard way of doing interferometry using beam combination with overlapping exit pupils of the individual telescopes (the extreme form of Michelson interferometry). Doing that reduces the FOV of the interferometer to zero (or better: to the size of the Airy disk of the individual telescope apertures). Recently there appears to be a renewed interest to include in the future VLTI plans an extended interferometric FOV capability [5, 6]. Both homothetic mapping and mosaicing by scanning the Airy disk sized FOV across the desired total FOV are being considered, with mosaicing apparently being preferred [5] by the ESO-VLTI staff.

My preference remains with the homothetic beamcombiner. Although often thought of as a beamcombiner for image plane interferometry, it was more than that. It is a means of delivering to the interferometric focus a FOV over which the beams are combined in phase (with precise metrology) or coherently (within the coherence length with moderate accuracy metrology). Any point (of Airy Disk size) within the FOV can subsequently be studied with pupil plane interferometry, including nulling. But in addition to giving a large co-phased/coherent FOV the proposed beamcombiner could be used as small FOV device in which the exit pupil configuration can be configured in any configuration like e.g. in a linear redundant configuration useful for spectroscopic interferometry or as a "densified pupil" as proposed by A. Labeyrie for his "hypertelescope" concept [7].

The optical and interferometric FOV for the homothetic beamcombiner was chosen to be 8 arcsec in diameter for the 8-m telescope array (or VIMA = VLT Main Array in contrast to VISA = VLT Sub-Array consisting of the Auxiliary Telescopes). With that FOV a large enough area was available around the Galactic Center to include the point source IRS7 for phase tracking and adaptive optics (AO) purposes. In addition it allows, for example, (i) coverage of a large area of the βPic envelope which uses βPic itself as a phase and AO reference and (ii) imaging of the red supergiant Antares which has a blue, unresolved companion 2.9 arcsec away which can be used as reference [8]. Antares' photosphere has a diameter of 0.041 arcsec and its chromosphere/corona is possibly significantly larger. For the 8.2-m telescopes that is larger than the Airy disk at visible and near-IR wavelengths.

A larger than 8 arcsec FOV would have required a substantial increase in the diameter of the delay line cat's eye reflectors. However, in the post-1993 de-scoping of the VLTI even these diameters were decreased so that the optical FOV of the VIMA was reduced to 2 arcsec (and the interferometric FOV to the Airy disk). In the original 1993 design both FOVs for the VISA were already taken as 2 arcsec.

3 What is the VLTI Blind Operation Concept?

For most of the sky there will be no bright enough (in the K-Band brighter than $m_K = 13$) unresolved object available to co-phase the VLTI telescopes. That is even the case after the commissioning of PRIMA which can relay a bright enough phase reference object as far as 60 arcsec (= the radius of the VLT 8-m unvignetted coudé foci) away into the 2 arcsec diameter VLTI optical FOV. At high galactic latitudes that results in sky coverage of about 1%. All current interferometers rely on the availability of a nearby bright, unresolved object for their operation. Since optical interferometric imaging is still in its childhood phase, and since there is plenty of interesting astrophysics to be done even with such very limited sky coverage, this limitation is currently not considered to be a major issue. Nonetheless in the 1988 –

1993 design of the VLTI I examined ways for eventually obtaining full sky VLTI coverage [9]. I named the resulting observing mode *"blind operation"*.

One way to achieve blind operation would be by co-phasing the VLTI on a bright, unresolved object first, followed by an offset pointing to the area of interest. It would require a very precise and accurate open-loop control of the VLTI opto-mechanical systems over the duration of the observation. Although such an option can not be excluded, especially after the installation of appropriate metrology systems, I did not see it as the preferred way to achieve blind operation. Instead I proposed a system similar to that being implemented in PRIMA, but one that would relay a reference object from as far as $\geq$600 arcsec away to the VLTI FOV thus giving close to 100% sky coverage at high galactic latitudes. Such a relay can only be at the VLT 8-m Nasmyth foci where the optical FOV has a radius of 900 arcsec. The proposed relay mechanism consisted out of a hinged pair of identical quartz rhombs whose exit and entrance normals respectively rotate around a common axis thus making their combined entrance and exit normals adjustable in distance. All four 8-m telescopes would have identical rhomb assemblies (RA) build into their image de-rotator/guidance systems thus introducing identical OPD and polarization changes in all 4 light paths. The displaced exit beam from the RA would be fed into the 60 arcsec radius coudé optical FOV where another optical assembly would adjust for the OPD differences between the reference and science object beams while also moving both beams into the 8 arcsec interferometric FOV of the VIMA. The latter device is quite similar to the PRIMA system being built for the current VLTI. The blind operation device was not part of the 1993 VLTI implementation plan. However, the 1993 design of the Nasmyth image de-rotators/guidance systems by Francis Franza took the eventual inclusion of the RAs into account.

At these large angular separations the science and reference objects experience different OPD differences between the telescopes. For a 130 meter base line, 900 arcsec separation and zenith viewing these are: (i) a slowly changing common geometrical OPD due to the different sky positions of $\leq$57 cm depending on their sky position, (ii) fast changing common atmospheric OPD fluctuations which depend strongly on the outer of turbulence L0. For L_0 = (∞, 25 and 10 meters these are 59, 34 and 5 microns RMS, and (iii) fast changing differential OPD changes between science and reference object amounting to $\sim$9 microns. The first two OPD variations are removed by the fringe tracking on the reference object; the third prevents the device to keep the 4 science beams co-phased. In the blind operation the science object will therefore not exhibit constantly phased fringes. Provided that the spectral resolution R $(= \lambda/\Delta\lambda)$ of the observations is large enough to make the coherence length R*λ > OPD variations, one is, however, assured that the science object has fringes. For short enough exposure times ($\sim$0.1 sec) those fringes are present even in a photon starved object image. One can recover both fringe visibility and closure phases by integrating the triple correlations or bi-spectra of the images over time as was done in speckle interferometry in the past. My estimate of achievable sensitivity was magnitude 20 for the R through K bands for point sources, 10 minutes of total integration times, 0.7 arcsec seeing and spectral bandwidth equal to the coherence lengths (R = 10 and 4 respectively at I- and K-Bands).

Blind operation as described above might be implemented at the present VLTI with its 2 arcsec optical FOV in a similar way by combining such a Nasmyth focus device with the PRIMA star combiner (STS) and an extended stroke differential delay line (DDL).

4 My View for the Future Use and Development of the VLTI.

My personal preferences for the future scientific use of the current VLTI including PRIMA are very much like the science projects mentioned in section 2 as the justification for the wide interferometric FOV of the 1993 VLTI:

- The study of the physics of galactic centers by observing the center of our galaxy in a detail that is unattainable elsewhere. The center of the Milky Way plays in that sense the same role as the Sun for the study of stars.
- The imaging and spectroscopy of Antares using the blue companion as a phase reference [8]. In the NIR about 1000 pixels could be resolved on its photospheric disk.
- The study of the envelope of βPic using βPic as the phase reference itself.

In all three cases PRIMA can be used to give a phase reference. A "large" FOV could be obtained by a mini-homothetic beamcombiner or mosaicing.

For future development of the VLTI I would propose to add 4 more ATs (as was already foreseen in the 1993 design [2]). Together with the 8 delay lines the VLTI tunnel is able to accommodate, that would provide a powerful interferometric imaging device. Like Pierre Léna1 I would use the 4 extra telescopes to give larger baselines beyond the 200 meter baseline available in the present VLTI configuration and use the OHANA concept [10] to couple the remote ATs to the present interferometric beam combination setup. The exact location of the additional AT stations has to be chosen to optimize the "super-VLTI" imaging capability. They, of course, can be located at different heights from the current VLT(I) Paranal platform.

Acknowledgements I acknowledge the positive interactions with the ESO VLTI staff and attendants to this workshop.

References

1. P. Léna, 2007, "The Early Days of the Very Large Telescope Interferometer", these proceedings.

2. J. M. Beckers, 2004, "Interferometric Imaging in Astronomy: A personal Retrospective", Reviews in Modern Astronomy (Ed. R.E. Schielicke) 17, 239.

3. J.M. Beckers, 1986, "Field of View Considerations for Telescope Arrays", SPIE Proceedings 628, 255.

4. O. von der Lühe, et al., 1997, "A New Plan for the VLTI", The ESO Messenger 87, 8.

5. A. Glindemann, et al., 2003, Astrophysics and Space Science 286, 35.

6. C. van der Avoort, 2004, "Experimental Performance of Homothetic Mapping for Wide Field Interferometric Imaging", SPIE Proceedings 5491, 1587.

7. A. Labeyrie, 2003, "Hypertelescope Imaging: from Exo-Planets to Neutron Stars", SPIE Proceedings 4852, 236.

8. H.-G. Ludwig and J. Beckers, 2007, "Towards Interferometric Imaging of Red Supergiants", these proceedings.

9. J.M. Beckers, 1991, "Blind Operation of Optical Interferometers: Options and Predicted Performance", Experimental Astronomy 2, 57.

10. J.M. Mariotti, et al., 1996, "Interferometric Connection of Large Ground-Based Telescopes", A&A Suppl. 116, 381.

Part I

Science: Stars — stellar diameters, limb darkening, flattening, surface structures

Stellar Diameters: Breaking the Barriers

A. Richichi

European Southern Observatory, Karl-Schwarzschildstr. 2, 85748 Garching b.M., Germany, arichich@eso.org

Summary. The first ideas on the use of interferometry and lunar occultations to measure angular diameters were proposed about one century ago, the first observations were attempted about seventy-five years ago, and the mass production of angular diameters came to reality about 25 years ago. Since then, this kind of measurements has been considered a well-established, even slightly boring, astronomical knowledge. But exactly, how many angular diameters do we know? How well do we know them? For which stars, and with which implications? Will we break the milliarcsecond barrier soon? Last but not least, are angular diameters really interesting? These are some of the questions that I address in my presentation.

1 Introduction

Since the beginning of modern astronomy, and in fact already when humans first looked up to the sky, our imagination has been wondering at the nature of those flickering points of light which are the stars. One of the keys to understanding their enigmatic and fascinating appearance is the knowledge of their angular size. Are they really infinitely small points of light? Or do they have an extension which, with sufficient magnification, can be revealed to be finite? What does this tell us about their nature, their distance? Though the answer might seem obvious today, in the past numerous minds were occupied by this problem, and experiments of various complexity were devised over the course of centuries. All attempts were, needless to say, unsuccessful.

It is now almost exactly one century after the first modern ideas were formulated on how to possibly measure stellar angular diameters, and about 80 years since Michelson and Pease [2] succeeded in putting one of these ideas into practice. Such measurements have now become an accepted routine. It is perhaps worthwhile, in a workshop dedicated to the success of interferometry, to ask ourselves what the actual statistics are on available stellar diameters, and ponder whether we should continue to invest efforts and resources in a field of research which to most astronomers has the same appeal of determining what could be the menu at the next fast food restaurant.

2 Three fundamental questions

I would like to pose, and attempt to answer, three brief but rather crucial questions concerning the investigation of stellar angular diameters, namely: Why? How? And provocatingly: Enough?

It is generally widely accepted that stellar angular diameters are some kind of fundamental pillar of modern astronomy, but this is often considered in the same league as the length of one astronomical unit, or the duration of a sinodal month. It is perhaps useful to recollect that angular diameters represent the only means to determine the effective temperature T_{eff} of stars. This in turn provides the most critical check for models of stellar atmospheres. In the presence of a distance determination, angular diameters can be converted of course to linear sizes. We will not enter here the somewhat thorny subject of the definition of the angular diameter of a gaseous stellar sphere, which is discussed for example by Scholz [5]. Things get more interesting when one considers that many stars have angular diameter which are in fact changing considerably in time. From irregular to semiregular late-type giants, to precise pulsational clocks as in Cepheid stars, the stage is open to follow diameter changes as a function of wavelength, of optical depth, of variability phase, of changes in spectral type. The more than occasional presence of companions and of complex circumstellar environments provide additional spice, and those with a sense of astromasochism might want to remember the effects of rotation, dark and hot spots, surface structure. In short, angular diameters are a rich realm which can still keep astronomers busy.

How to measure angular diameters? Alpha Centauri A is a good equivalent of our own Sun, and closer than any other star. Yet its angular diameter is less than 10 milliarcseconds (mas). Giant stars have linear diameters which are 2 orders of magnitude bigger, but they also are relatively rare and thus at larger distances. Indeed, a systematic study of angular diameters is faced with measurements at the ≈ 1 mas level. This is beyond the reach of all standard techniques. Even equipped with adaptive optics, the diffraction limit of the largest telescopes is not sufficient to resolve any star, except for a few isolated cases of very large, very nearby, and very atypical objects. The answer is then to use ad-hoc techniques. These include methods sometimes quite diverse from each other, such as stellar eclipses, lunar occultations (LO), and interferometry. In the recent years long-baseline interferometry (LBI) has established itself as the approach of choice, involving delicate and challenging technical requirements but offering at the same time the widest possibilities of investigation. While modern LBI has been used at a number of facilities for some decades already, it is only in the last few years that it has reached sufficient maturity and user-friendliness to be offered on a regular basis. In this respect, ESO with the VLTI is paving the way for observations open to the whole community and not only to the black-belts of interferometry.

Do we have enough of angular diameters yet? The reader might suspect that I ask this provocative question only because I have a firm "yes" as an answer in my pocket: I do not. The CHARM2 catalogue (Richichi et al. [3]) lists 6718 angular diameters determinations of which 1975 are direct measurements, including 1270 by LBI, 502 by LO, and a few from other mixed methods. At least one angular diameter measurement exists for 668 stars. These numbers are sufficiently imposing to make anyone wonder what additional information are we to gain by measuring some more hundreds of diameters. In fact, the situation is less satisfactory than it would seem, as I try to illustrate below:

1. We have extensive measurements for giant stars of various types, but very few (of order 2% of the total) for cool main sequence stars. Yet these are the most common constituent of stellar populations.
2. We have virtually no direct measurements for pre-main sequence stars, yet these are the objects for which our theoretical predictions are the most uncertain.
3. In order to test theoretical models, accuracies of less than 5% are often quoted as the minimum requirement on angular diameters. Much less than half of the available measurements comply with this.
4. The majority of late-type giant stars, which are the most common target for angular diameter investigations, are subject to diameter variability. Yet the available measurements are often limited to one snapshot in time, and often no sufficient correlated information exists on the variability status of the object.

These are some of the reasons which in fact convince me that it is not only desirable, but crucially important to continue investigations of angular diameter. This effort should be directed in a coordinated approach towards the investigation of selected types of stars, in close synergy with photometric and spectroscopic investigations at various wavelengths.

3 Statistics of available angular diameters

It is useful to take a closer look at our present knowledge on angular diameters, and perhaps the best way to do this is by means of graphs and histograms. In this section I present a number of them, which I have produced using CHARM2.

Figure 1 shows how many stars have more than one measurement, by either LO or LBI. The integral of this histogram is 668, and it can be appreciated that more than half of the stars have only one measurement available. About 10% of the stars have been measured ten or more times, however one should be careful that in many cases this does not imply totally independent measurements, but rather simultaneous measurements at different wavelengths.

The distribution of the brightness of stars with available measurements is represented in Figs. 2 to 4 by means of V and K magnitudes and $12\,\mu$m flux,

34   A. Richichi

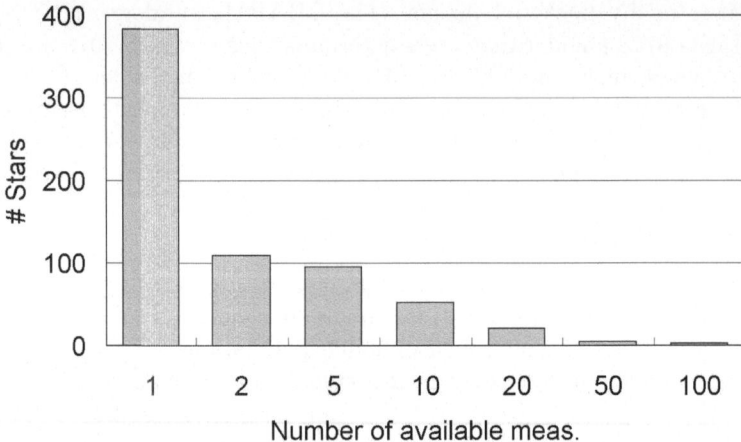

Fig. 1. Statistics of repeated angular diameter measurements available for a same star.

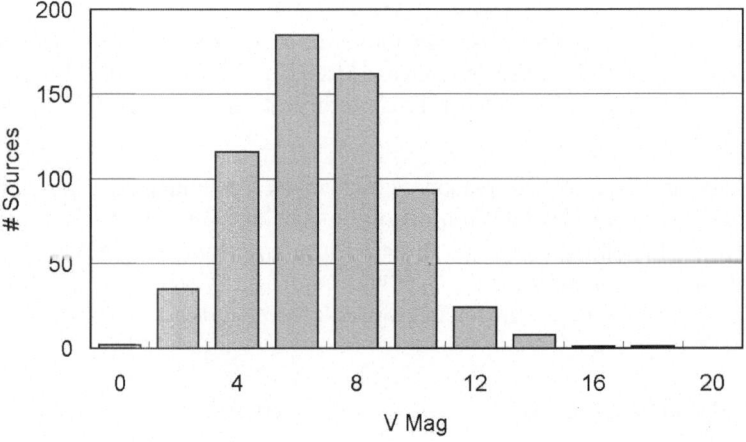

Fig. 2. Distribution of V-band magnitudes for stars with available angular diameter measurements.

respectively. It is apparent that the target stars are all quite bright, with a peak around $V \approx 6$ and $K \approx 2$ magnitudes. This is to be expected, since the angular diameter is obviously smaller for fainter stars. The histograms also imply that the target stars are typically red, another consequence of the limit in angular resolution: for a fixed angular diameter, a redder star is angularly larger than a bluer one.

The distribution of spectral types, shown in Fig. 5, confirms that the targets consist mostly of K and M stars. We can add that these are typically giants, though no comprehensive statistics is available on the luminosity class.

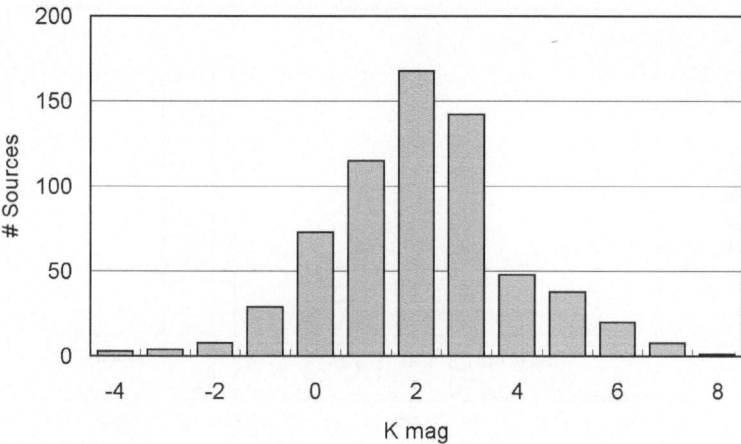

Fig. 3. Distribution of K-band magnitudes for stars with available angular diameter measurements.

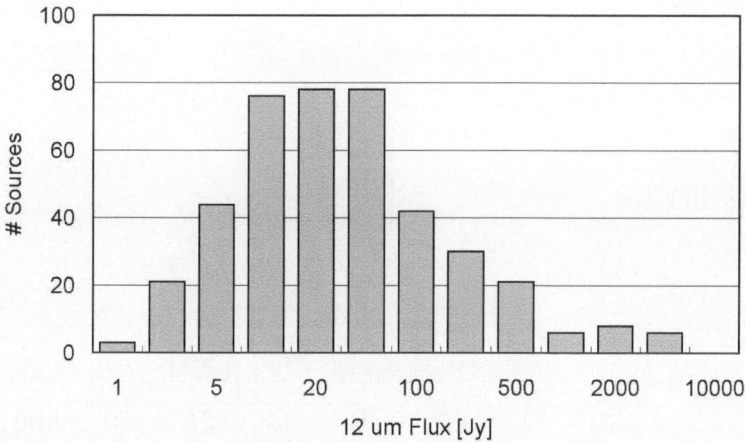

Fig. 4. Distribution of 12μm fluxes for stars with available angular diameter measurements.

Of particular interest is of course the distribution of measured angular diameters, shown in Fig. 6. The shape of this distribution can be interpreted as the result of two trends. On one side we have the steep increase of sources with decreasing angular diameter limit: this is shown by the increase in the statistics moving from right to left in Fig. 6. On the other side, we have the technical difficulty of measuring smaller and smaller angular diameters, which cuts into the distribution on the left of Fig. 6. As a result, we have a distinct peak around 5 mas. Stars significantly larger than this are quite rare (and have basically been mostly measured already), while below 1-2 mas the

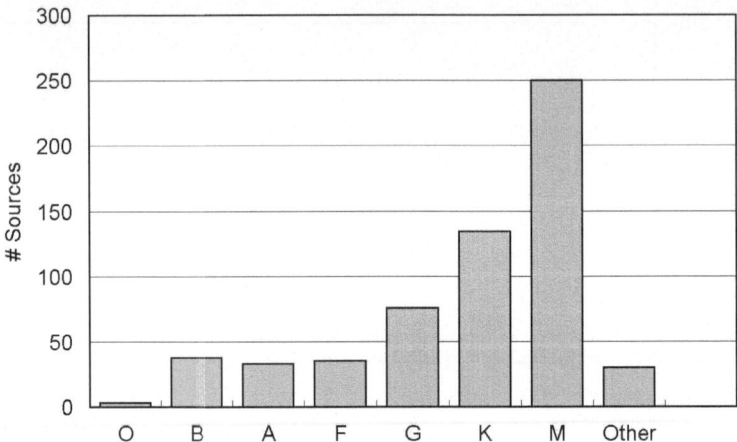

Fig. 5. Distribution of spectral types for stars with available angular diameter measurements.

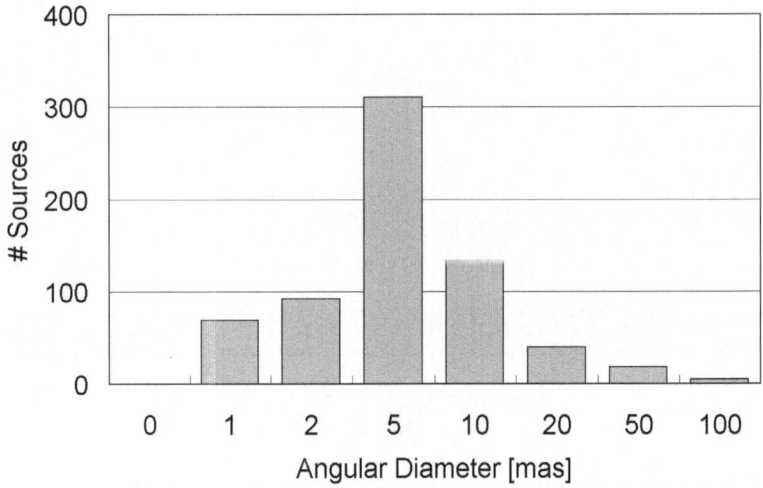

Fig. 6. Statistics of measured angular diameter.

measurements are limited by our current technical abilities. I will return on this point in Sect. 4.

Of course, an angular diameter does not tell us much if we do not have an error associated with it. In this respect, it is interesting to inspect Figs. 7-8, which illustrate the distribution of relative errors by number of measurements and as a function of the angular diameter. It can be seen that only about half of the measurements reach a level of accuracy that can be considered interesting, i.e. at or below 5%. There is a trend of decreasing accuracy with

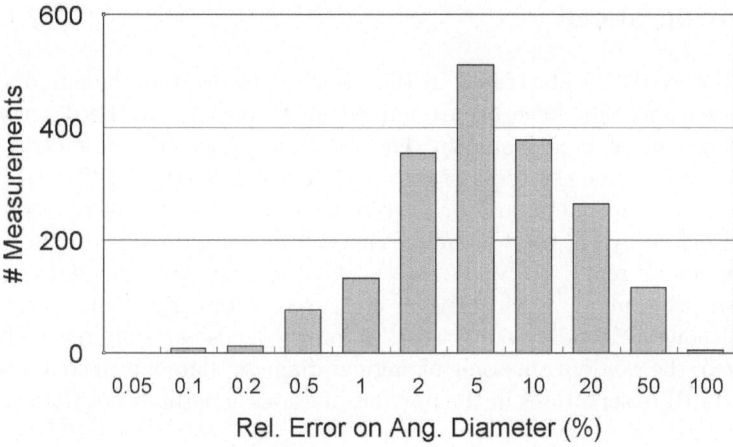

Fig. 7. Statistics of relative errors on the measured angular diameters.

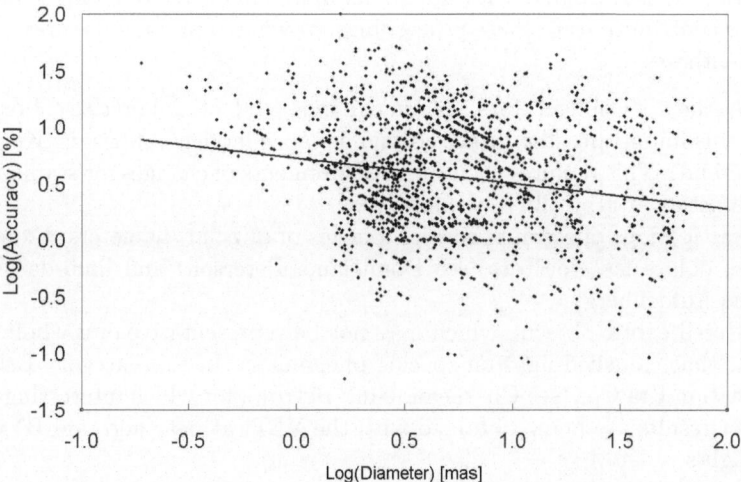

Fig. 8. Distribution of relative errors as a function of the measured angular diameters.

smaller diameters, which is to be expected if one recalls the technical difficulties of pushing the measurements to their limit. However, the trend is not very strong and a very large scatter of accuracies exists for all angular diameter values. This indicates the intrinsic scatter of accuracy in the LO and LBI methods, due to the fact that atmospheric conditions, and for LBI also the particular interferometer used, can significantly alter the quality of results for the same star.

4 Moving ahead

From the statistics illustrated in the previous pages, the angular diameter determinations that have been obtained up to present can be described as numerous but of mixed quality. The use of modern interferometric facilities can easily increase these statistics. For example, the VLTI operates in the southern emisphere, and has access to a good fraction of the sky never probed before by an interferometer with similar performance. Already in its commissioning phase the VLTI has collected angular diameter data on hundreds of stars, many of which never measured before (Richichi & Percheron [4]). Although extending our statistical knowledge is certainly not to be discouraged, the random amassing of angular diameter data which characterized LO and LBI observations in the previous decades is perhaps not the best way forward.

There are whole classes of stars, and of high angular resolution measurements, which are extremely interesting and that are coming into the reach of modern facilities. I name below a few examples, all of which are illustrated in various contributions to these proceedings for which I provide the first author in parentheses:

- the study of Cepheid stars, yielding an independent and precise calibration of this important step in the cosmic distance ladder (see Davis, Kervella, Mérand). This requires accurate measurements of changes for stars having diameters of typically less than 2 mas.
- investigations of second-order properties of angular diameters. Examples are: oblate fast rotators (see Domiciano, Peterson) and limb-darkening (see Aufdenberg).
- selected exotic objects, which may not be representative of a whole class but that can shed light on specific phenomena. In this category I should mention Eta Car (see Chesneau, Gull, Petrov), for which interesting first-ever results are being obtained with the VLTI at near and mid-IR wavelenghts.
- classes of stars for which statistics of angular diameters are poor or entirely lacking. Examples are: investigations of the low-mass end of the main sequence (Sègransan), which are proving of fundamental importance to assess the effective temperature of stars closer and closer to the hydrogen-burning limit and that are in strong demand by theoreticians; and investigations on pre-main sequence stars (Simon), where a direct diameter determination is the key to place such stars correctly in the H-R diagram, providing a key contraint to evolutionary tracks and stellar models.

While I have apparently neglected from the above shopping list the traditional fodder of high angular resolution studies, namely late-type giants and cool variables, these stars can indeed greatly benefit too from the renaissance of interferometric observations brought about by powerful and service-oriented facilities. The key factor for an advancement in this field will come not so much from an inordinate increase of diameter determinations, but rather from an improvement in their quality. In particular, the monitoring of stellar characteristics along the variability cycle and at various wavelengths appears to be of crucial importance.

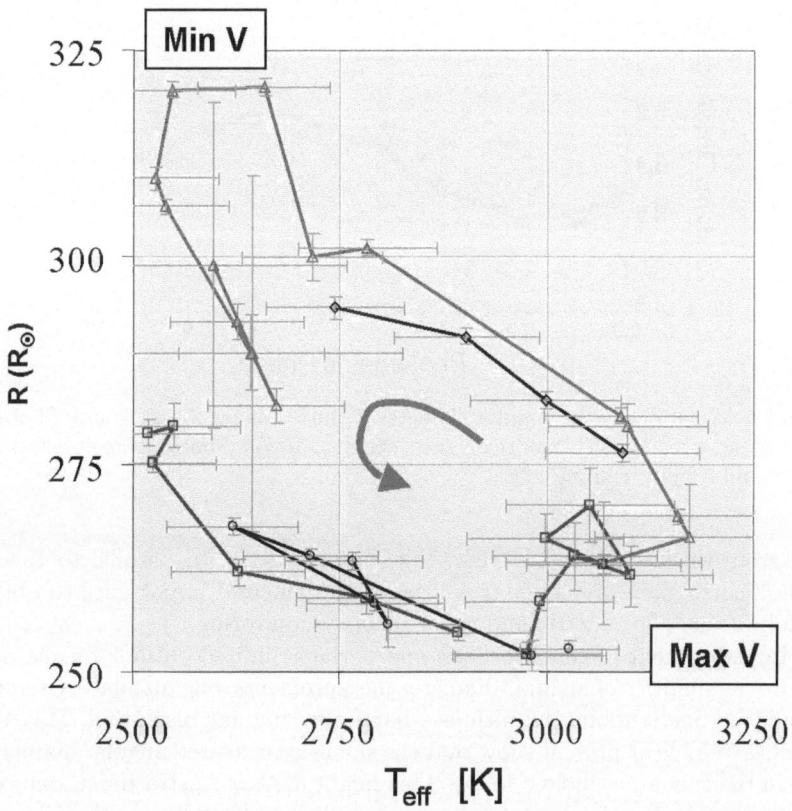

Fig. 9. Pulsational characteristics of the Mira star R Boo as revealed by PTI observations (adapted from G. Van Belle, priv. comm.). The lines connect data points observed in four campaigns from 1999 through 2002.

As illustrative examples, Figure 9 shows the preliminary detection of a clear cycle of radius and temperature variations in R Boo, while Figure 10 shows a clear phase shift with wavelength in the diameter changes of S Lac along its photometric variability curve. Both stars are Miras, and have been observed with the PTI interferometer (courtesy G. Van Belle). Such observations will provide the key for a thorough theoretical understanding of the global picture in late-type and AGB stars, namely that of dynamically active

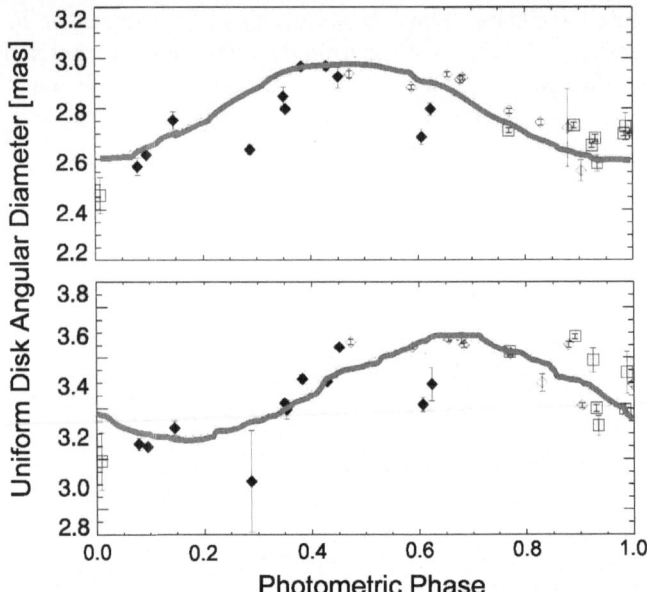

Fig. 10. Variations of the angular diameter of the Mira star S Lac, from PTI observations (adapted from G. Van Belle, priv. comm.). Top and bottom are observations at 2.0 and 2.4 μm, respectively.

pulsators with continuously changing characteristics. In parallel to the interferometric measurements, it will be of fundamental importance to obtain simultaneous photometric and spectroscopic monitoring.

In the context of this synthetic and perhaps limited outlook on the near future possibilities of angular diameter measurements, one mandatory remark should be made about the ultimate barrier in angular resolution. Measurements available at present show that the smallest measured angular diameters are in the range just below 1 mas. This might appear frustrating, if one considers that in fact such performance was already available (although on a limited number of stars) with a facility such as the intensity interferometer almost half a century ago (see Hanbury Brown [1]).

In fact, both main techniques considered here, namely LO and LBI, are susceptible to considerable improvements in their limiting angular resolution. In the case of LO, the main limitation is caused by the difficulty of measuring small amplitude changes in the diffraction fringes, against the signal fluctuation induced by atmospheric turbulence and in particular by scintillation. For some bright stars at least, it should be possible to use narrow filter bandpasses (thus increasing the intrisinc fringe amplitude and especially the number of fringes), and to observe simultaneously at various wavelengths. Since scintillation is chromatic, while hopefully the angular diameter does not change too much with wavelength, it would be possible to reduce significantly a major

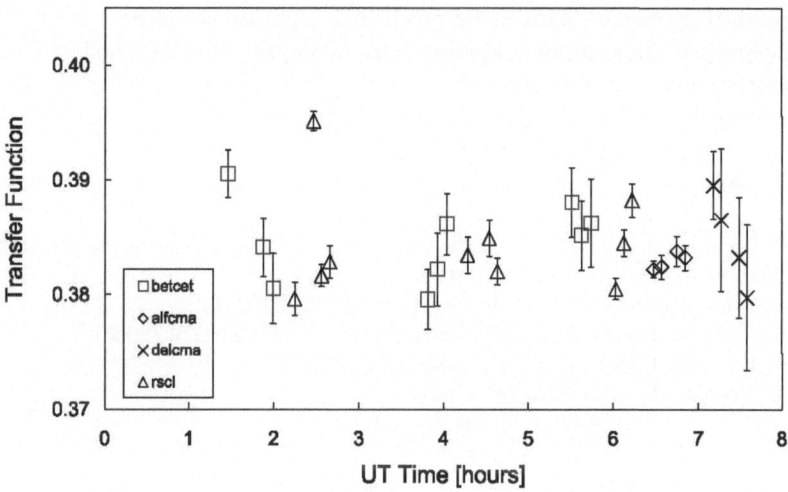

Fig. 11. VLTI transfer function observed at the VLTI on the night of 14 November 2001. Adapted from Richichi & Percheron [4].

noise contribution. Simulations, that I do not report here, show that limiting resolutions between 0.01 and 0.1 mas could be achieved with ≈2 m-class telescopes.

Concerning LBI, one might argue that pushing the limiting resolution is only a matter of increasing the baselines, and perhaps decreasing the wavelength of observation. This approach would certainly lead to an increased potential resolution, but it is not clear that this potential would be immediately translated in smaller angular diameter determinations. Indeed, after atmospheric turbulence the main limitation to the resolution and accuracy of an interferometer is set by our knowledge of suitable calibrators.

Figure 11 is adapted from a recent study of over 10,000 calibrator observations performed with the VLTI during its initial commissioning (Richichi & Percheron [4]). It shows an example of a good night of observations with the VLTI, extracted randomly from a database of hundreds available. Under good conditions, the VLTI shows a stable transfer function with a standard deviation of just 0.5% over a whole night. Note that R Scl is a carbon star and not a calibrator. In principle, this kind of performance would allow to discriminate between a visibility of, say, 0.90 and unity. This would correspond in principle to measuring angular diameters of just 0.25 mas.

If one could realize an interferometer with kilometric baseline operating under similar conditions of instrumental and atmospheric stability, it would be conceivable to push angular diameters significantly below the 0.1 mas limit. This requires however that calibrators are available with a matching accuracy. This is at present not the case, although efforts are undergoing in many interferometric communities to improve the situation, either by analyzing and consolidating existing data or by producing accurate theoretical predictions of the angular diameter of field stars. This can aptly be called the last barrier of interferometry.

References

1. R. Hanbury Brown: *The intensity interferometer. Its applications to astronomy*, (London: Taylor & Francis, 1974)
2. A.A. Michelson A.A., F.G. Pease: ApJ. **51**, 257 (1920)
3. A. Richichi, I. Percheron, M. Khristoforova: A&A **431**, 773 (2005)
4. A. Richichi, I. Percheron: A&A **434**, 1201 (2005)
5. M. Scholz: MNRAS **321**, 347 (2001)

Imaging the Effects of Rotation in Altair and Vega

D. M. Peterson[1], C. A. Hummel[2], T. A. Pauls[3], J. T. Armstrong[3], J. A. Benson[4], C. G. Gilbreath[3], R. B. Hindsley[3], D. J. Hutter[4], K. J. Johnston[5], and D. Mozurkewich[6]

[1] Department of Physics and Astronomy, Stony Brook University, Stony Brook, NY 11794-3800 DPeterson@astro.sunysb.edu

[2] European Southern Observatory (ESO), Casilla 19001, Santiago 19, Chile CHummel@eso.org

[3] Naval Research Laboratory, Code 7215, 4555 Overlook Ave. SW, Washington, DC 20375 Pauls@nrl.navy.mil, Tom.Armstrong@nrl.navy.mil, Charmaine.Gilbreath@nrl.navy.mil, Hindsley@nrl.navy.mil

[4] U.S. Naval Observatory, Flagstaff Station, 10391W. Naval Observatory Rd., Flagstaff, AZ 86001-8521 JBenson@nofs.navy.mil, DJH@nofs.navy.mil

[5] U.S. Naval Observatory, 3450 Massachusetts Ave. NW, Washington, DC, 20392-5420 KJJ@astro.usno.navy.mil

[6] Seabrook Engineering, 9310 Dubarry Rd., Seabrook, MD 20706 Dave@Mozurkewich.com

Summary. After a brief review of rotation among upper main sequence stars and von Zeipel's [1] theory for the interiors, we describe our interferometric measurements of two bright A stars, Altair and Vega. The Navy Prototype Optical Interferometer [2] (jointly operated by the US Naval Observatory, the Naval Research Laboratory and Lowell Observatory) which works at visible wavelengths has implemented baselines of sufficient length to initiate true imaging of the disks of the brightest A stars. We report here measurements of Altair, the third brightest A star in the sky. "Closure phase" techniques show that Altair deviates dramatically from a normal limb-darkened disk, indicating a strongly asymmetric intensity distribution. A Roche model provides a good fit to the data, indicating that Altair is rotating at about 90% of its breakup (angular) velocity. We find that a gravity darkening law exponent appropriate for a radiative star is required by the observations and we describe the potential of this object for testing the assumption of solid body rotation throughout its envelope. We will also describe recent measurements of Vega which confirm the proposed interpretation of spectral line measurements indicating that this star is also rapidly rotating, but seen nearly pole on.

1 Introduction: Rotation Among Early-Type Stars

It has been known almost since the invention of spectroscopy, that the chemically normal, early-type stars, stars earlier than F5, rotate rapidly [3]. When allowance is made for the growing effects of gravity darkening [4] at the later spectral types, the onset of high rotation is quite sudden. This sudden break has been assumed to be associated with the rapid onset of the convective envelope around $1.3\,M_\odot$ and the ability of the later spectral types to shed angular momentum through their solar winds [5].

The effects of rapid rotation on the structure of stars has not escaped the attention of theorists. The problem was solved in elegant form over 80 years ago by von Zeipel [1], under the simplifying assumptions of a centrally concentrated mass (no quadrupole contributions to the gravity), rigid rotation and a fully radiative envelope (appropriate for hotter stars), who showed these stars adopted Roche spheroid figures. Subsequent investigations, summarized for example in [6], showed that so long as solid body rotation held, the deviations from a point source gravity law were indeed very small, even at the surface of stars rotating at critical velocity.

Two important predictions came out of this work. First, so long as rigid rotation holds there is a maximum rotational velocity a star may attain given by balancing centrifigual and gravitational accelerations at the equator, $\Omega_c^2 = GM/R^3$. The second was that rotation would induce both flattening and a temperature variation over the surface. Defining an effective gravity as just the net of gravity and centrifugal acceleration as a function of latitude, von Zeipel [1] showed that $T_{\rm eff}^4(\theta) \propto g_{\rm eff}(\theta)$ where θ is the colatitude, an effect known as gravity darkening.

Over the decades this last result has been tested successfully in close binary systems which can be shown to induce equivalent effects [6]. However, these effects had never been demonstrated directly in an isolated star until the report from the Palomar Testbed Interferometer (PTI [7]) of detection of significant oblateness in Altair [8]. The measurement was quickly followed by a measurement of oblateness on Achernar by the VLTI [9]. These were striking results, with the PTI measurments completely in agreement with the star's measured projected rotational velocity and the von Zeipel theory. As important as these first measurements were, they went only part of the way to a full test of the von Zeipel's theory, particularly gravity darkening.

We describe below the first measurements in imaging mode by the NPOI [2] that directly detects gravity darkening on Altair and preliminary results that appear to do the same on Vega. To understand these results we next describe the measurement of "closure" phases.

2 Interferometry and Closure Phases

A number of groups are now operating multi-element Michelson interferometers at visual and near infrared wavelengths [10]. In a Michelson configuration the interference between beams is obtained pairwise [11]. In this case the measurable is the complex visibility, $\tilde{V} = |V|\exp(i\theta)$, which is the Fourier Transform of the brightness distribution along the baseline (the "strip-brightness distribution). In general the phase, θ, is not a useful measurement at optical wavelengths because the atmosphere introduces differential phase shifts between the individual mirrors, the shifts being large, random and rapidly varying. The situation is fully summarized in the review by Monnier [10].

Measures of the amplitudes of the complex visibilities, refered to simply as "visibilities", are the main staple of interferometry but are limited in important ways. Even when an object is well sampled, spatially, one only obtains the autocorrelation of the intensity distribution. Specifically, the deduced intensities are forced to be centro-symmetric, asymmetries are encoded only in the phases.

It was recognized many years ago by the radio interferometry community [12] that the lost phase information could be partially retained if, in times short compared to the atmospheric time scales, one were able to sum the phases from three separate baselines whose individual telescopes formed a triangle. In this circumstance the induced phase errors for each telescope entered the sum twice but with opposite signs and drop out of the result. This "closure" phase looses some information and is not so open to simple interpertation. But it can be easily calculated if one is fitting simple models and immediately removes quadrant ambiguities, etc.

Closure phases and phases in general have some structure that one needs to be aware of. The phasor representation of the complex visibility, above, defines the amplitude as always being positive. Even in the simplest symmetric case of a point source, the complex visibility, the sinc function, changes signs as it goes through the various zeros. This is represented in the phasor by having the phases go through 180 deg shifts at the appropriate points. In turn closure phases exhibit the same behavior, but now the shifts accomodate the zeros on all three baselines.

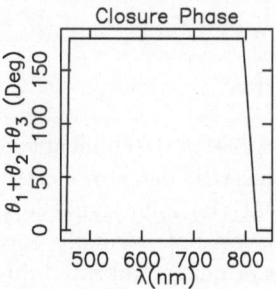

Fig. 1. A simulated closure phase for a well resolved stellar disk. In this case the disk is limb-darkened but no other structure is present. The centro-symmetric nature of the intensity distribution leads to two zero crossings on the longest baselines, giving two 180 deg shifts in the closure phase, but no other structure

A more realistic example of this behavior is shown in Fig. 1, which graphs the closure phases in a simulation of a $T_{eff} = 7800\,$K star with an angular diameter of 3.37 mas as would be viewed by the AW-W7-AE triangle of the

NPOI at the zenith. The fourier plane sampling is a function of the baseline lengths divided by the wavelength. The range of wavelengths effectively scans through two zeros on the longest baseline, triggering the transitions in the (closure) phase. It is this simple but abrupt behavior that characterizes centro-symmetric objects. Significant deviations from this behavior, usually seen near the transitions, indicate asymmetries in the intensity distribution.

3 Altair

Altair was observed in "imaging", i.e. closed triangle, mode for the first time at NPOI in late May and early June 2001. The light from three stations, AW, AE, and W7[2], were brought together to form three baselines, each recorded at a separate spectrometer. Thirty two channels were recorded simultaneously at each spectrograph, running from 443–852 nm, nominally. However, the wavelength region around the HeNe (metrology) laser wavelength is avoided, the channel just shortward of that was not operating and the 4 shortest wavelengths on the spectrometer recording the AW–W7 baseline were not functioning, which affects the triple products.

On examination of these early data it was recognized immediately [13] that the intensity across the disk was distinctly asymmetric. In the meantime Altair had been scheduled for additional observations with potentially more complete uv plane coverage using 4-station and even 6-station combinations. These turned out to have some unanticipated problems. In what follows we have focused instead on the earliest, 2001, data set and of those nights, on the one night with the most data and most extended hour angle range (i.e. uv coverage), $25-May-2001$.

3.1 Detector nonlinearities

Even this limited data set presented significant challenges. Subsequent investigation indicated that the APD detector electronics had significant deadtimes that became serious at the high signal levels encountered with this first magnitude star (flux levels not normally encountered). The nonlinear responses are in the process of being calibrated, but are not yet available.

Fortunately, because of the manner in which the signal is sampled at NPOI and because in this case each baseline was recorded on separate spectrometers, we have been able to show that the phases, and hence the closure phase, are not affected to first order [14]. By adding a free, multiplicative scaling factor (independent of wavelength) to the visibilities from each baseline we have also removed most of the effects from the amplitudes - see [14] for details.

3.2 Coherent averaging

On the positive side, these data have been subjected to a reduction algorithm which substantially reduces the noise in phase measurement in the low photon per sample limit [15]. The effect is that phase measurements of reasonable S/N are recovered in the bluest channels where earlier algorithms limited the useful range to longer than about 540 nm.

3.3 Model fitting

We show in Figs. 2 and 3 the triple amplitudes, $V_1 V_2 V_3$, and phases, $\phi_1 + \phi_2 + \phi_3$ for the observations taken on the night indicated. Also shown by a solid line is the model fit to these data. We have limited the amplitude information in the fits to just the triple amplitudes, even though the individual visibilities contain even more signal, because of the remaining chromatic biases that are surely left in the amplitude data. We do require some amplitude data, the zero crossings mainly fix the overall angular scale of the object. By limiting ourselves to just the triple amplitudes we hope to minimize any biases in the deduced parameters.

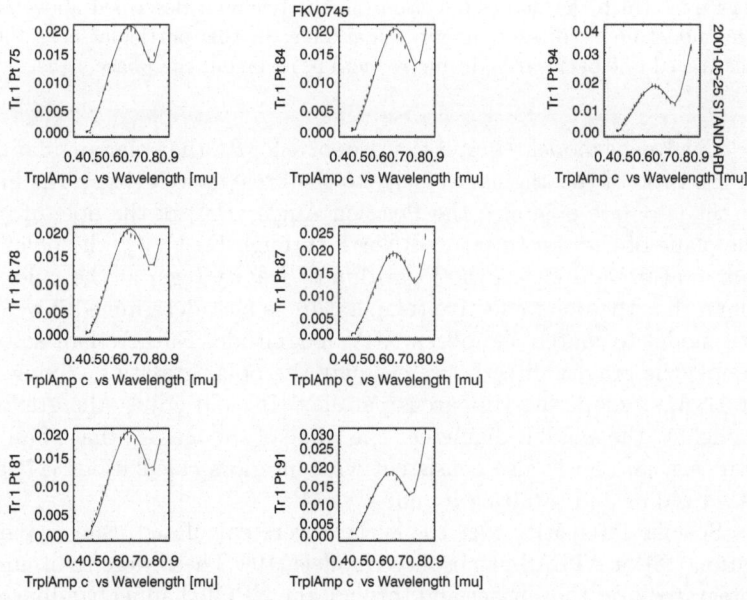

Fig. 2. Triple amplitudes, observed and calculated (solid line) for Altair. The extended wavelength coverage provided by the coherent integration algorithm allow us to cover nearly the entire first ring of the visibility

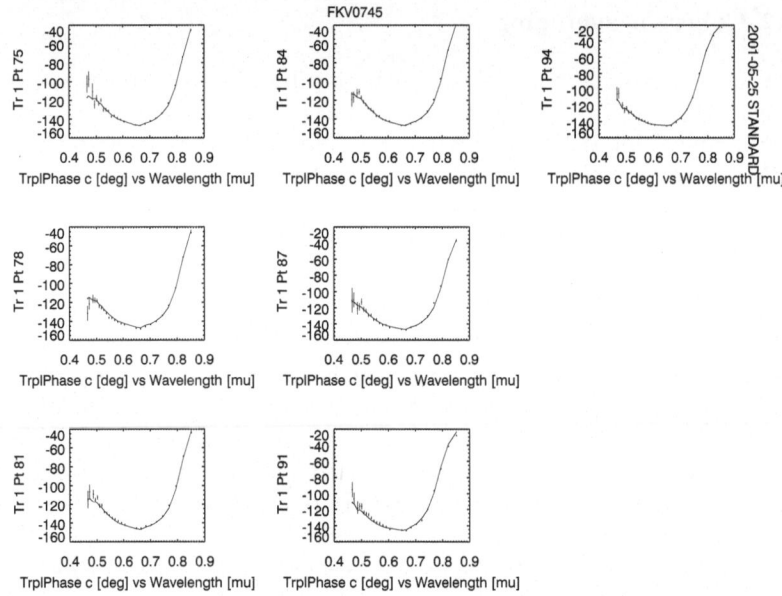

Fig. 3. Closure phase observations for Altair and the model fit (solid line). The degree to which the phases deviate from the behavior described above is striking. The model fit is correspondingly remarkable. In this particular case the triangle was traced out backwards during reduction, reversing the phase signs.

The Roche model requires that we solve for θ_p, the polar angular diameter, ω, the ratio of the angular velocity to the breakup velocity, i, the inclination or tilt ($i = 0$ is pole on), the Position Angle (PA) of the pole projected on the plane of the sky (measured North through East), T_p, the polar effective temperature and $\log g_p$, the logarithm of gravity (cgs) at the pole. We have found that the polar effective temperature is best determined if we constrain the model to match the observed V magnitude. Rather than solve for (or adopt) the gravity directly, we evaluate the polar gravity by fixing the mass at 1.8 $M_\odot$, adopt the Hipparcus parallax [16] and adjust the gravity as the model fits the angular diameter. The value of the mass is that of an A7 VI-V star, e.g. [8] and is also consistent with interiors calculations [17] given the fitted radius and resulting luminosity [14].

Specific intensities over the surface were calculated using a parameterization [18] of ATLAS atmosphere models [19]. These were then numerically integrated over the surface and through the NPOI channel bandpasses to create the complex visibilities to be fitted to the observations. Standard fitting algorithms were used [14].

3.4 Altair model

Figure 4 shows a gray-scale rendering of Altair as seen at 500 nm. Model parameters and some derived quantities are summarized in Table 1. The reductions are at a somewhat preliminary state so we are not able to give robust error estimates and the parameters appear to be strongly correlated by these data. However, as an estimate the rotation parameter, ω is probably no greater than 0.92 and no less than 0.85, that being arguably the single most critical parameter in the fits.

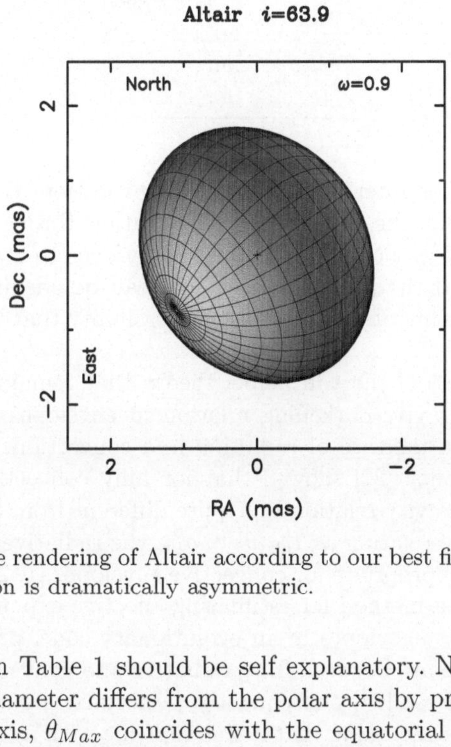

Fig. 4. A gray-scale rendering of Altair according to our best fit Roche model. The intensity distribution is dramatically asymmetric.

Most entries in Table 1 should be self explanatory. Note that θ_{min} the apparent minor diameter differs from the polar axis by projection while the apparent major axis, θ_{Max} coincides with the equatorial angular diameter. The reduced χ^2 is substantially higher than we would like to see for a good fit. Most of this stems from calibration noise, which we estimate from the variation of the calibrator scans. Had we added this noise in the χ^2 would have dropped to around 1.5. That is still uncomfortably large - we suspect the remainder comes from chromatic effects which are not removed by the scaling factors.

3.5 Altair as a laboratory

These results are interesting on several fronts. First we note the recent announcement [20] that Altair is a low amplitude δ Scuti pulsator. This has important potential. One of the two main uncertainties in applying von Zeipel's

Table 1. Altair model parameters

Parameter	Units	Value
$\omega = \Omega/\Omega_c$	–	0.90
i	deg	63.9
PA	deg	124
θ_p	mas	2.96
θ_{min}	mas	3.06
θ_{Max}	mas	3.60
T_p	K	8740
T_{eq}	K	6890
V_{eq}	$\mathrm{km\,s^{-1}}$	273
$V\sin i$	$\mathrm{km\,s^{-1}}$	245
χ^2/ν	–	3.9

theory is whether the interior undergoes rigid rotation. Helioseismology has determined [21] that the Sun is in rigid rotation throughout its radiative interior, but this does not automatically carry over to stars with outer radiative envelopes. If the pulsation spectrum can be adequately modelled in such a rapidly rotating object there is the possibility that this issue could be resolved.

The other aspect of the von Zeipel theory that remains uncertain is the adequacy of the gravity darkening relation in envelopes where the energy is transported by a mixture of radiation and convection, as is the case for the late A stars. Lucy [22] showed that for fully convective envelopes that the temperature–gravity relation was quite different from the radiative case. Parameterizing this relation as $T_{\mathrm{eff}}(\theta) \propto g_{\mathrm{eff}}^{\beta}$, the radiative case has $\beta = 0.25$ while $\beta = 0.08$ is appropriate for convective envelopes [1, 22]. Claret [23] has recently suggested a method for estimating effective exponents by analyzing the structure of the enevlopes in an evolutionary code, providing values for β in cases intermediate between fully radiative and convective envelopes.

Reiners and Royer [24] subsequently performed a rotation analysis on Altair's spectrum, concluding that it was rotating with a total linear velocity of $V_{eq} \leq 245$ at the one sigma level. From Claret's work [23] they adopted a gravity darkening exponent of $\beta = 0.09$. We have redone our analysis adopting the lower value and find that it does not provide a very good fit: the projected velocity became $V\sin i = 295\,\mathrm{km\,s^{-1}}$ (observed values are in the range 200–240 kms), the reduced χ^2 increased to 7.6 and the model predicted a much redder color ($B - V = 0.26$ whereas using the radiative exponent produced 0.21, the observed value). The radiative value is clearly the better choice.

Finaly, we note that Altair (along with α Cep, also A7 VI-V) are the two hottest main sequence objects displaying chromospheric emission lines of C II and Lyα [25, 26]. This is taken to indicate that convection of sufficient violence to generate the required acoustic fluxes and magnetic fields is present in late A stars. Since α Cep is measured to have a projected rotation velocity almost idential to Altair and since our model shows Altair to have an extensive region of 7000 K gas, these conclusions probably need to be reexamined.

4 Vega

Vega is *the* standard. It is the standard for spectral type A0 V [27], it is the abundance standard for this spectral region owing to very sharp lines, e.g. [28] and it is the absolute photometric standard for the visible and near IR [29]. Even so Vega has always stood out as somewhat unusual. For example, most sharp lined B and A stars are chemically peculiar, yet Vega appears to be (nearly) normal [28], but the list of its pecularities is much too long to reproduce here.

Following up on one of those pecularities – that weak lines appeared to have a boxy shape while strong lines appeared more rounded as would be expected from simple rotational broadening – led Gulliver, Hill and Adelman [30] to the striking conclusion that Vega was a rapid rotator seen nearly pole-on. This would be a fairly remarkable conclusion and could require reconsideration of some of Vega's roles as a standard.

4.1 Observations

On the same dates that NPOI took the observations described earlier it was also observing Vega, nominally as a "check" star – to make sure the measurements reported a circular object to be indeed, circular. Vega being twice as bright, the data were even more affected by problems described for Altair, §3.1, and were set aside. After realizing that phase measurements could survive the detector limitations we examined those data more closely, confirming that the phases did appear to be intact.

Figure 5 shows the closure phases for two of the scans along with a preliminary model fit (solid line). The phase transition associated with the zero on the long baseline is clearly much softer than would be the case for a purely symmetric intensity profile: measurements in the transition deviate many sigmas from 0° or 180°, the two permissible values in the symmetric case.

4.2 Vega Model

Fitting a model to the Vega observations is more difficult than for Altair in several ways. First the additional flux exacerbates the detector problems described in §3.1; we ultimately have not used any of the amplitude data from these runs. Secondly, the asymmetry is very mild and apparent only

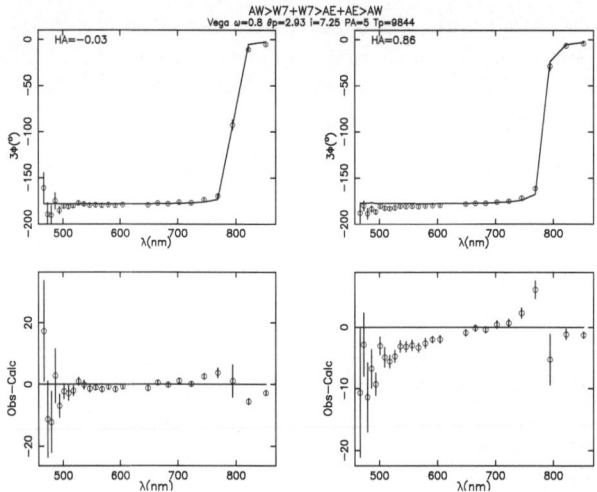

Fig. 5. Representative closure phase observations of Vega and a model fit. The phases show clear signs of asymmetry, although much more subtle than for Altair.

because of the high S/N in the phase data, consistent with an object seen almost perfectly pole-on. This creates a near degeneracy among the fitted parameters, notably among ω, i and PA. Finally, since Vega is nearly at the same declination as NPOI's latitude and the long baseline lies mostly East-West, there is almost no change in effective position angle through an evening's observations. Without long baselines at other position angles or amplitude data from these observations there is little to be concluded other than that we can add considerable weight to the rapid rotator hypothesis.

To make some headway, we added the original PTI observations of Vega [31] to the reductions. This stabilized the fits somewhat and we present the results here as being very preliminary. There are still strong correlations among the parameters, but the PTI data establish the rotation to be near $\omega = 0.80$ rather broadly. In this case the inclination comes out near $i = 7.5°$ and the position angle near $PA = 5°$. The angular diameters (mas) are then: $\theta_{min} = 3.33$, $\theta_{Msx} = 3.34$ and $\theta_p = 2.93$, with the apparent oblateness less than 0.3%. This model predicts $V \sin i = 27 \, \mathrm{km \, s^{-1}}$, a bit fast compared to $21 \, \mathrm{km \, s^{-1}}$ observed [30].

We show a gray scale rendering of the disk as seen at 500 nm in Fig. 6. The asymmetry is hardly noticable. However, compared to a non-rotating star the intensity rolls off much more steeply at the limbs. This is due to the local effective temperature changing from 9750 K at the pole to 8400 K at the equator – a situation one would clearly want to account for in an abundance standard.

These results are clearly preliminary, we hesitant to even guess at realistic errors for the above parameters. But between these two A stars we clearly

Vega *i*=7.25

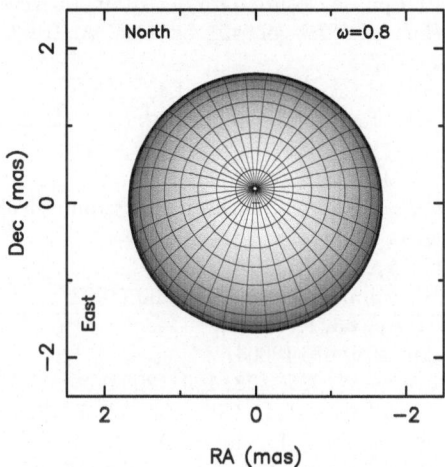

Fig. 6. A gray scale representation of Vega's disk at 500 nm.

see the potential of high angular resolution *imaging* to contribute to our understanding of rotation on the upper main sequence.

Acknowledgment D.M.P. would like to acknowledge the assistance of Robert Zagarello and Jinmi Yoon.

References

1. H. von Zeipel: Monthly Notices R. A. S. **84**, 684 (1924)
2. J. T. Armstrong, et al.: Ap. J. **496**, 550 (1998)
3. N. Walker: Observatory **85**, 245 (1965)
4. J. Hardorp, M. Scholz: A. & Ap. **13**, 353 (1971)
5. R. P. Kraft: Ap. J. **150**, 551 (1967)
6. J.-L. Tassoul: *Theory of Rotating Stars*, (Princeton University Press, Princeton New Jersey 1978)
7. M. M. Colavita, et al.: Ap. J. **510**, 505 (1999)
8. G. T. van Belle, G. T. Ciardi, R. R. Thompson, R. L. Akeson, E. A. Lada: Ap. J. **559**, 1155 (2001)
9. A. Domicino de Souza, P. Kervella, S. Jankov, E. Janot-Pacheco, L. Abe: A. & Ap. **407**, L47 (2003)
10. J. D. Monnier: Reports on Prog. in Phys. **66**, 789 (2003)
11. M. Born, E. Wolf: *Principles of Optics* 5th edn (Oxford: Pergamon Press 1965)
12. A. R. Thompson, J. M. Moran, G. W. Swenson: Jr. *Interferometry and Synthesis in Radio Astronomy* (Wiley, New York 1986. Reprint by Krieger, Malabar, Florida 1994)
13. N. Ohishi, T. E. Nordgren, D. J. Hutter: Ap. J. **612**, 463 (2004)
14. D. M. Peterson et al.: Ap. J. 636, 1087 (2006)

15. C. A. Hummel, D. Mozurkewicn, J. A. Benson, D. M. Peterson: Coherent integration of NPOI phase closure data on Altair. In *New Frontiers in Stellar Interferometry: Proc. of SPIE*, vol 5491, ed by W. A. Traub (SPIE, Bellingham, Washington 2004), 707
16. M.A.C. Perryman et al.: A. & Ap. **323**, L49 (1997)
17. G. Schaller, G. Schaerer, G. Meynet, A. Maeder: A. & Ap. Suppl. **96**, 269 (1992)
18. W. Van Hamme: A. J. **106**, 2096 (1993)
19. R. L. Kurucz: *CD-Rom No. 13* (Smithsonian Astrophysical Observatory, Cambridge Massachussets 1993)
20. D. L. Buzasi et al.: Ap. J. **619**, 1072 (2005)
21. E. A. Spiegel, J. P. Zahn: A. & Ap. **265**, 106 (1992)
22. L. B. Lucy: Zeit. f. Ap. **65**, 89 (1967)
23. A. Claret: A. & Ap. **406**, 623 (2003)
24. A. Reiners, F. Royer: A. & Ap. **428**, 199, (2004)
25. T. Simon, W. B. Landsman, R. L. Gilliland: Ap. J. **428**, 319 (1994)
26. F. Walter, L. D. Matthews, J. L. Linsky: Ap. J. **447**, 353 (1995)
27. W. W. Morgan, P. C. Keenan: Ann. Rev. A. Ap. **11**, 29 (1973)
28. D. Gigas: A. & Ap. **165**, 170 (1986)
29. R. C. Bohlin, R. L. Gilliland: A. J. **127**, 3508 (2004)
30. A. F. Gulliver, G. Hill, S. J. Adelman: Ap. J. **429**, L81 (1994)
31. D. R. Cardi, G. T. van Belle, R. L. Akeson, R. Thompson, E. A. Lada, S. B. Howell: Ap. J. **559**, 1147 (2001)

Rapid Rotation across the HR Diagram with VLTI: Achernar and Altair

A. Domiciano de Souza

Max-Planck-Institut für Radioastronomie, Auf dem Hügel 69, 53121 Bonn, Germany adomicia@mpifr-bonn.mpg.de

Summary. Rapid rotation effects can nowadays be directly measured by stellar interferometry. We present our recent high-quality interferometric data recorded with VLTI/VINCI on two rather different rapidly rotating stars: Achernar (B3Vpe star) and Altair (A7IV-V star). By using a first-order analysis (equivalent uniform disc model) we estimate the oblateness of these stars. For Altair, we also perform a comparison with observations from other modern interferometers (PTI and NPOI), finding a good agreement between the data from these 3 interferometers.

1 Introduction

Rotation is nowadays accepted as one of the fundamental parameters governing the structure and evolution of stars. In particular, rapid rotation leads to two main physical modifications: (1) the star becomes oblate because of a strong centrifugal force and (2) it exhibits a latitudinal-dependent effective temperature distribution, i.e. a gravity darkening effect (after the seminal work of von Zeipel 1924). Present optical/IR long-baseline interferometers allow us to directly measure these two important physical effects (Domiciano de Souza et al. 2002).

2 Observations and results

2.1 Achernar (α Eri, HR 472, HD 10144)

Dedicated observations of Achernar have been carried out during the ESO period 70, from 11 September to 12 November 2002, with quasi-uniform time coverage, on the VLTI equipped with the VINCI[1] instrument (e.g. Kervella et al. 2003). For these observations VINCI recombined the light from two telescopes (40 cm siderostats) in the astronomical K band, which is centered at $\lambda \simeq 2.2$ μm and covers $\simeq 0.4\mu$m. Two interferometric baselines were used, 66 m and 140 m in ground length as shown by Fig. 1 (top). After the data processing of raw interferograms (see Kervella et al. 2003 for details) we obtained 60 measures of calibrated squared visibilities V^2 of Achernar at different baseline projections on the sky B_{proj} (Fig. 1 bottom).

---

[1] V(LT) IN(terferometer) C(ommissioning) I(nstrument)

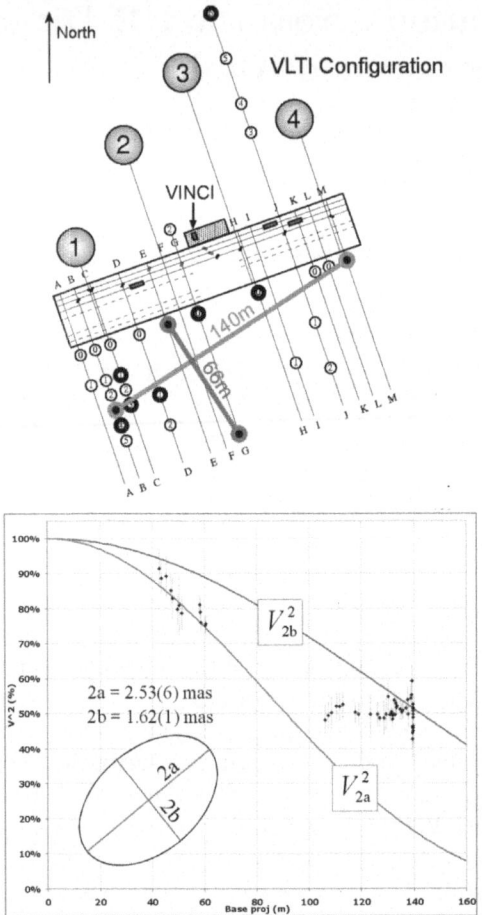

Fig. 1. *Top:* VLTI/VINCI ground baselines used for Achernar observations with 40 cm siderostats. *Bottom:* Squared visibilities V^2 measured on Achernar as a function of the baseline projections on the sky B_{proj}. Thanks to these high precision V^2 measures (mean errors of 3.5%) and to an efficient baseline synthesis effect due to Earth-rotation it is clear that these observations cannot be reproduced by any circularly symmetric model. As a convenient first approximation we derived from each V^2 an equivalent uniform disc (UD) angular diameter $\oslash_{\mathrm{UD}}$ from the relation $V^2 = |2\mathrm{J}_1(z)/z|^2$. Here, $z = \pi \oslash_{\mathrm{UD}} (\alpha) B_{\mathrm{proj}} (\alpha) \lambda^{-1}$, J_1 is the Bessel function of the first kind and of first order, and α is the azimuth angle of B_{proj} at different observing times due to Earth-rotation. By fitting an ellipse to the $\oslash_{\mathrm{UD}}(\alpha)$ points we obtain a major axis $2a = 2.53 \pm 0.06$ milliarcsec (mas), a minor axis $2b = 1.62 \pm 0.01$ mas, and a minor-axis orientation $\alpha_0 = 39° \pm 1°$. The best fit ellipse and the V^2 curves corresponding to $2a$ and $2b$ are also shown in the right panel. In this first-order analysis the stellar oblateness is given by the ratio $2a/2b = 1.56 \pm 0.05$.

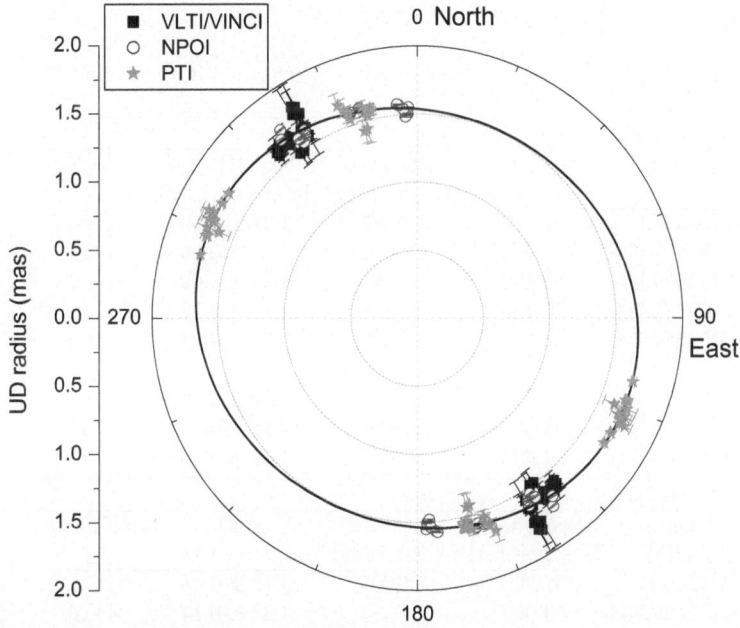

Fig. 2. Equivalent UD angular size of Altair as a function of azimuth α (in degrees), derived from the V^2 for VLTI/VINCI, PTI, and NPOI. There is a good agreement between all these distinct $\oslash_{\mathrm{UD}}$. By fitting an ellipse to the $\oslash_{\mathrm{UD}}(\alpha)$ points we obtain a major axis $2a = 3.39 \pm 0.02$ milliarcsec (mas), a minor axis $2b = 3.02 \pm 0.03$ mas, and a minor-axis orientation $\alpha_0 = 114° \pm 3°$. In this first-order analysis the stellar oblateness is given by the ratio $2a/2b = 1.12 \pm 0.02$. We note that for this analysis the error bars from NPOI where multiplied by an estimated fixed factor $(= 4)$ to account for the fact that NPOI V^2 probably underestimate long-term errors (Ohishi, private communication; see also Hummel et al. 2004).

2.2 Altair (α Aql, HR 7557, HD 187642)

The visibility measurements were all recorded on the E0-G1 baseline of the VLTI/VINCI (ground length of 66 m). We combined the stellar light using a classical fiber-based triple coupler (MONA) for the K band observations, and an integrated optics beam combiner (IONIC, Lebouquin et al. 2004) in the H band. Standard K and H ($1.5 < \lambda < 1.8\,\mu\mathrm{m}$) band filters were used for these observations. Observations are summarized in Table 1.

Altair have also been observed with PTI (V^2 in the K band; van Belle et al. 2001) and NPOI (V^2, triple amplitudes, and closure phases in the visible; Ohishi, Nordgren & Hutter 2004). In order to test the agreement between

Artair's observations from VLTI/VINCI, PTI, and NPOI, we show in Fig. 2 the equivalent UD angular diameters as a function of azimuth, derived from the V^2 measured by these three interferometers.

Table 1. VLTI/VINCI observations of Altair performed in the H and K bands.

H band				
Date (JD)	Projected Baseline (m)	Position Angle^a (deg)	UD diameter (mas)	Calibrator
2452477.655	62.110	139.77	3.22 ± 0.09	α Ind
2452477.659	61.702	139.97	3.18 ± 0.10	α Ind
2452479.561	65.950	140.66	3.23 ± 0.09	α Ind
2452479.706	55.760	144.48	3.26 ± 0.12	α Ind
2452482.726	52.131	148.77	3.28 ± 0.24	α Ind
2452483.645	61.490	140.08	3.20 ± 0.10	α Ind
2452484.699	54.905	145.37	2.99 ± 0.18	α Ind
2452485.594	64.851	139.08	3.15 ± 0.13	α Ind
2452485.598	64.627	139.07	3.21 ± 0.13	α Ind
K band				
Date (JD)	Projected Baseline (m)	Position Angle (deg)	UD diameter (mas)	Calibrator
2452469.722	57.285	143.05	3.18 ± 0.13	24 Cap
2452469.755	53.065	147.53	3.26 ± 0.14	24 Cap
2452469.763	51.957	149.02	3.26 ± 0.15	24 Cap
2452531.587	52.790	147.89	3.52 ± 0.44	χ Phe
2452531.592	52.204	148.67	3.64 ± 0.47	χ Phe
2452531.596	51.624	149.49	3.49 ± 0.48	χ Phe
2452536.511	60.454	140.68	3.28 ± 0.15	70 Aql
2452536.543	56.759	143.52	3.32 ± 0.18	70 Aql
2452536.547	56.226	144.03	2.98 ± 0.20	70 Aql
2452536.578	52.212	148.66	3.14 ± 0.33	70 Aql
2452536.582	51.738	149.33	3.60 ± 0.27	70 Aql

^{a}: $0°$ is North and $90°$ is East.

3 Conclusions and perspectives

Long baseline interferometry is a powerful tool for studying rapidly rotating stars, providing important clues to the many unanswered questions concerning their structure and evolution.

As shown here, we can estimate the stellar oblateness by using a first-order analysis (equivalent uniform disc model). However, only a physically consistent modelling allows us to fully understand the impact of rapid rotation across the HR diagram from interferometric observations. A model dedicated to stellar interferometry (Domiciano de Souza et al. 2002) was used to interpret the observations of Achernar and Altair. A complete description of our analysis is given by Domiciano de Souza et al. (2003) for the case of Achernar and by Domiciano de Souza et al. (2005) for the case of Altair (see also our paper dedicated to Altair in these proceedings).

References

1. Domiciano de Souza, A., Vakili, F., Jankov, S., Janot-Pacheco, E. & Abe, L. 2002, A&A, 393, 345
2. Domiciano de Souza, A., Kervella, P., Jankov, S., Abe, L., Vakili, F., di Folco, E., & Paresce, F. 2003, A&A, 407, L47
3. Domiciano de Souza, A., Kervella, P., Jankov, S., Vakili, F., Ohishi, N., Nordgren, T. E. & Abe, L. 2005, A&A, 442, 567
4. Hummel, C. A., Mozurkewich, D., Benson, J. A., & Peterson, D. M. 2004, SPIE, 5491, 707
5. Kervella, P., Gitton, Ph., Ségransan, D., et al. 2003, SPIE, 4838, 858
6. Lebouquin, J. B., Rousselet-Perraut, K., Kern, P., Malbet, F., Haguenauer, P., Kervella, P., Schanen, I., Berger, J. P., Delboulbé, A., Arezki, B., Schöller, M. 2004, A&A, 424, 719
7. Ohishi, N., Nordgren, T. E. & Hutter, D. J. 2004, ApJ, 612, 463
8. van Belle, G. T., Ciardi, D. R., Thompson, R. R., Akeson, R. L. & Lada, E. A. 2001, ApJ, 559, 1155

Multi-Wavelength Interferometry of Evolved Stars Using VLTI and VLBA

M. Wittkowski[1], D. A. Boboltz[2], T. Driebe[3], and K. Ohnaka[3]

[1] European Southern Observatory, Garching, Germany mwittkow@eso.org
[2] U.S. Naval Observatory, Washington, DC, USA dboboltz@usno.navy.mil
[3] Max-Planck-Institut für Radioastronomie, Bonn, Germany
   driebe@mpifr-bonn.mpg.de,kohnaka@mpifr-bonn.mpg.de

Summary. We report on our project of coordinated VLTI/VLBA observations of the atmospheres and circumstellar environments of evolved stars. We illustrate in general the potential of interferometric measurements to study stellar atmospheres and envelopes, and demonstrate in particular the advantages of a coordinated multi-wavelength approach including near/mid-infrared as well as radio interferometry. We have so far made use of VLTI observations of the near- and mid-infrared stellar sizes and of concurrent VLBA observations of the SiO maser emission. To date, this project includes studies of the Mira stars S Ori and RR Aql as well as of the supergiant AH Sco. These sources all show strong silicate emission features in their mid-infrared spectra. In addition, they each have relatively strong SiO maser emission. The results from our first epochs of S Ori measurements have recently been published [5] and the main results are reviewed here. The S Ori maser ring is found to lie at a mean distance of about 2 stellar radii, a result that is virtually free of the usual uncertainty inherent in comparing observations of variable stars widely separated in time and stellar phase. We discuss the status of our more recent S Ori, RR Aql, and AH Sco observations, and present an outlook on the continuation of our project.

1 Introduction

The evolution of cool luminous stars, including Mira variables, is accompanied by significant mass-loss to the circumstellar environment (CSE) with mass-loss rates of up to $10^{-7} - 10^{-4}\,M_\odot$/year (e.g. [18]). The detailed structure of the CSE, the detailed physical nature of the mass-loss process from evolved stars, and especially its connection with the pulsation mechanism in the case of Mira variable stars, are not well understood. Furthermore, one of the basic unknowns in the study of late-type stars is the mechanism by which usually spherically symmetric stars on the asymptotic giant branch (AGB) evolve to form axisymmetric or bipolar planetary nebulae (PNe). Possible origins of asymmetric structures include, among others, binarity, capture of substellar companions, stellar rotation, or magnetic fields. While it is generally believed that the observed pronounced asymmetries of the envelopes of PNe form when the star evolves from the tip of the AGB branch toward the blue part of the HR diagram, there is evidence for some asymmetric structures already around AGB stars and supergiants (e.g., [27, 28, 29, 21, 4]).

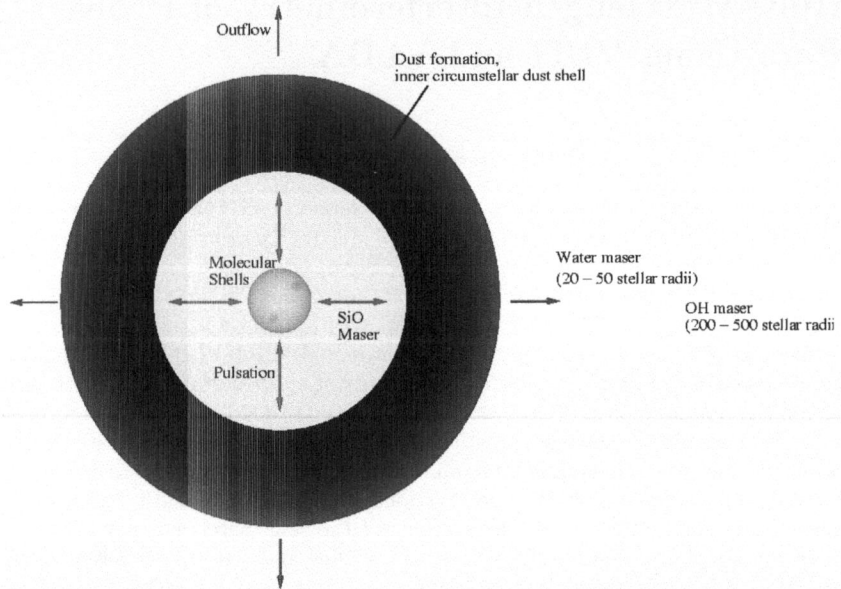

Fig. 1. Sketch of a Mira variable star and its circumstellar envelope (CSE). A multi-wavelength study (MIDI/AMBER/VLBA) is well suited to probe the different regions shown here. Near-infrared observations (using AMBER) are most sensitive to the center-to-limb intensity variation across the stellar disk itself, to surface features, and to the molecular shell outward the photospheric disk. Mid-infrared observations (using MIDI) probe the molecuar shell, as well as the dust formation zone. Maser observations using the VLBA probe the spatial distribution and kinematics of maser radiation of different molecules (SiO, OH, and H_2O) at the distances indicated in the figure. Owing to the stellar variability, only contemporaneous observations are meaningful.

Coordinated multi-wavelength studies (near-infrared, mid-infrared, radio, millimeter) of the stellar surface (photosphere) *and* the CSE at different distances from the stellar photosphere and obtained at corresponding cycle/phase values of the stellar variability curve are best suited to improve our general understanding of the atmospheric structure, the CSE, the mass-loss process, and ultimately of the evolution of symmetric AGB stars toward axisymmetric or bipolar planetary nebulae. Fig. 1 shows a schematic view of a Mira variable star, indicating the different regions that can be probed by different techniques/wavelength ranges (VLTI/AMBER, VLTI/MIDI, VLBA/maser, ALMA). Near-infrared observations (using AMBER) are most sensitive to the center-to-limb intensity variation across the stellar disk itself, to surface features, and to the molecular shell outward the photospheric disk. Mid-infrared observations (using MIDI) probe the molecuar shell, as well as the dust formation zone. Maser observations using the

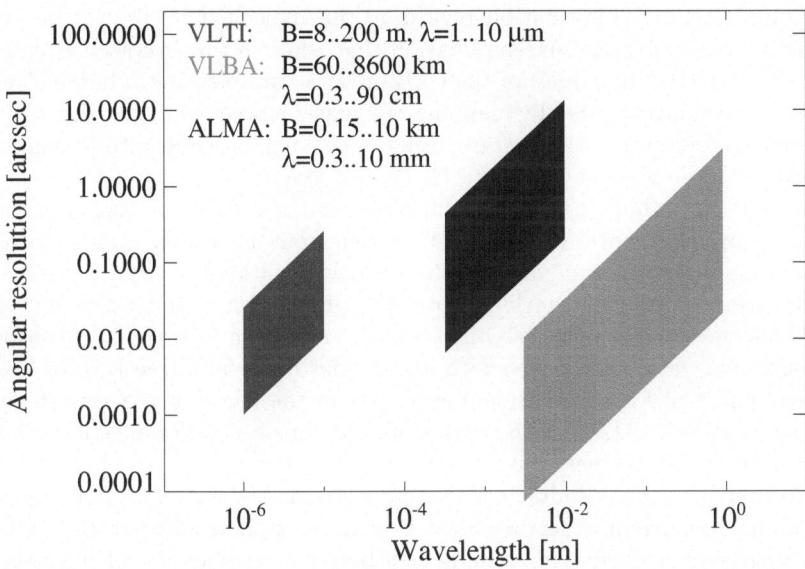

Fig. 2. Comparison of resolution and wavelength ranges of the infrared, millimeter, and radio interferometric facilities VLTI, ALMA, and VLBA.

VLBA probe the spatial distribution and kinematics of maser radiation of different molecules (SiO, OH, and H_2O) at the distances indicated in Fig. 1. Fig. 2 shows a comparison of the VLTI, VLBA, and ALMA interferometric facilities in terms of wavelength ranges and angular resolution. VLTI, VLBA, and ALMA allow us to observe the same evolved stars in terms of sensitivity, and reach a comparable angular resolution at their respective wavelength ranges.

The conditions near the stellar surface can best be studied by means of optical/near-infrared long-baseline interferometry. This technique has provided information regarding the stellar photospheric diameter, asymmetries/surface inhomogeneities, effective temperature, and center-to-limb intensity variations including the effects of close molecular shells, for a number of non-Mira and Mira giants (see, e.g., [12, 30, 26, 15, 32, 33, 5, 9]).

The structure and physical parameters of the molecular shells located between the photosphere and the dust formation zone, as well as of the dust shell itself can be probed by mid-infrared interferometry (e.g. [7]). This has also recently been demonstrated by using the spectro-interferometric capabilities of the VLTI/MIDI instrument to study the Mira star RR Sco [22]. The model obtained in this work includes a warm molecular (SiO and H_2O) layer as well as a dust shell of corundum and silicate, and can well reproduce the obtained MIDI visibility values.

Complementary information regarding the molecular shells can be obtained by observing the maser radiation that some of these molecules emit. The structure and dynamics of the CSE of Mira variables and other evolved stars has been investigated by mapping SiO maser emission at typically about 2 stellar radii toward these stars using very long baseline interferometry (VLBI) at radio wavelengths (e.g., [3, 19, 5]).

Results regarding the relationships between the different regions mentioned above and shown in Fig. 1 suffer often from uncertainties inherent in comparing observations of variable stars widely separated in time and stellar phase (see the discussion in [5]). Both, the photospheric stellar size as well as the mean diameter of the SiO maser shell are known to vary as a function of the stellar variability phase with amplitudes of 20-50% (see [17] for theoretical and [26] for observational estimates of the variability of the stellar diameter; as well as [16] for theoretical and [8] for observational estimates of the variability of the mean SiO maser ring diameter).

To overcome these limitations, we have established a program of coordinated and concurrent observations at near-infrared, mid-infrared, and radio wavelengths of evolved stars, aiming at a better understanding of the structure of the CSE, of the mass-loss process, and of the triggering and formation of asymmetric structures.

In the following, we describe recent results obtained with optical/infrared interferometry on the stellar atmospheric structure of regular non-Mira (Section 2) and Mira (Section 3) giants. Our joint VLTI/VLBA observations of the Mira star S Ori are discussed in Section 4. Finally, we give an outlook (Section 5) on further measurements and future ideas. The latter includes desirable 2nd generation instrumentation based on the requirements of this particular project alone.

2 The atmospheric structure of non-Mira giants

Fundamental parameters, most importantly radii and effective temperatures, of regular cool giant stars have frequently been obtained with interferometric and other high angular resolution techniques, thanks to the favorable brightness and size of these stars. Further parameters of the stellar structure, as the strength of the limb-darkening effect, can be studied when more than one resolution element across the stellar disk is employed. Through the direct measurement of the center-to-limb intensity variation (CLV) across stellar disks and their close environments, interferometry probes the vertical temperature profile, as well as horizontal inhomogeneities. However, the required direct measurements of stellar intensity profiles are among the most challenging programs in modern optical interferometry. Since more than one resolution element across the stellar disk is needed to determine surface structure parameters beyond diameters, the long baselines needed to obtain this resolution also produce very low visibility amplitudes corresponding to vanishing

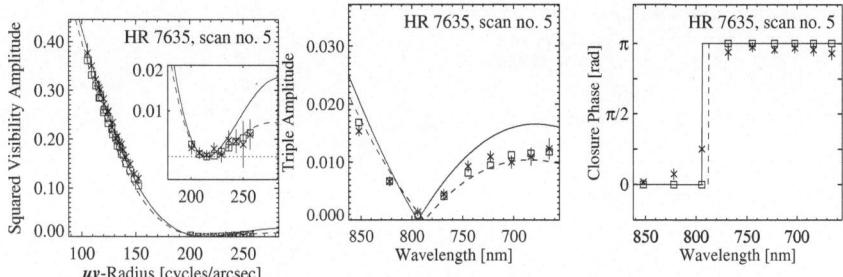

Fig. 3. NPOI limb-darkening observations (squared visibility amplitude, triple amplitude, closure phase) of the M0 giant γ Sge, together with a comparison to the best fitting ATLAS 9 model atmosphere prediction (squares). For comparison, the solid line denotes a uniform disk model, and the dashed line a fully-darkened disk model. ATLAS 9 models with variations of T_{eff} and $\log g$ result in significantly different model predictions. From [30].

fringe contrasts. Such direct limb-darkening studies have been accomplished for a relatively small number of stars using different interferometric facilities (including, for instance, [11, 23, 10, 30, 32]).

Recent optical multi-wavelength measurements of the cool giants γ Sge and BY Boo [30] succeeded not only in directly detecting the limb-darkening effect, but also in constraining ATLAS 9 ([20]) model atmosphere parameters. Fig. 3 shows one dataset including squared visibility amplitudes, triple amplitudes, and closure phases of the M0 giant γ Sge obtained with NPOI, together with a comparison to the best fitting ATLAS 9 model atmosphere prediction. ATLAS 9 models with variations of T_{eff} and $\log g$ result in significantly different model predictions. By this direct comparison of the NPOI data to the ATLAS 9 models alone, the effective temperature of γ Sge is constrained to 4160 ± 100 K. The limb-darkening observations are less sensitive to variations of the surface gravity, and $\log g$ is constrained to 0.9 ± 1.0 [30]. These constraints are well consistent with independent estimates, such as calibrations of the spectral type. Furthermore, it was shown that these interferometric and spectroscopic measurements of γ Sge both compare well with predictions by the same spherical PHOENIX [13] model atmosphere [2].

The first limb-darkening observation that was obtained with the VLTI succeeded in the early commissioning phase of the VLTI [32]. Using the VINCI instrument, K-band visibilities of the M4 giant ψ Phe were measured in the first and second lobe of the visibility function. These observations were found to be consistent with predictions by PHOENIX and ATLAS model atmospheres, the parameters for which were constrained by comparison to available spectrophotometry and theoretical stellar evolutionary tracks (see Fig. 4). Such limb-darkening observations also result in very precise and accurate radius estimates because of the precise description of the CLV. Future use of the spectro-interferometric capabilities of AMBER and MIDI will enable

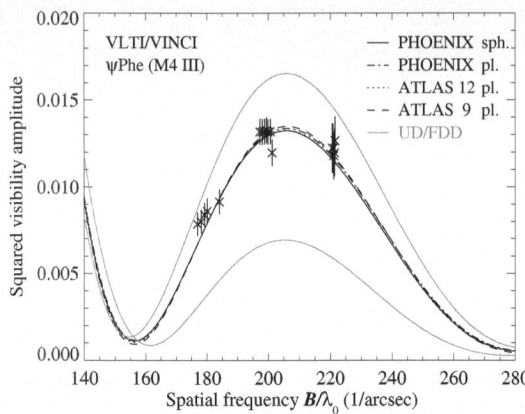

Fig. 4. VLTI limb-darkening observations of the M4 giant ψ Phe [32].

us to study the wavelength-dependence of the limb-darkening effect, which results in stronger tests and constraints of the model atmospheres than these broad-band observations (cf. the wavelength-dependent optical studies with NPOI as described above).

Another strong test of model atmospheres is the direct comparison of spectro-photometry, high-resolution spectra, and limb-darkening observations to predictions by the same model atmosphere. Such studies are presented in these proceedings (V. Roccatagliata et al.). Available spectrophotometry, high-resolution UVES ultraviolet/optical spectra, as well as near-infrared VLTI/VINCI K-band limb-darkening measurements are compared to predictions by PHOENIX model atmospheres, and good agreement is found (see Figs. 1 and 2 in Roccatagliata et al., these proceedings).

3 The atmospheric structure of Mira giants

For cool pulsating Mira stars, the CLVs are expected to be more complex than for non-pulsating M giants due to the effects of molecular layers close to the continuum-forming layers. Based on self-excited hydrodynamic model atmospheres of Mira stars ([14, 25, 17], Scholz & Wood, private communication), broad-band CLVs may indeed appear as Gaussian-shaped or multicomponent functions, and to exhibit temporal variations as a function of stellar phase and cycle, in accordance with observations (see introduction). These complex shapes of the CLV make it difficult to define an appropriate stellar radius. Different radius definitions, such as the Rosseland mean radius, the continuum radius, or the radius at which the filter-averaged intensity drops by 50%, may result in different values for the same CLV. For complex CLVs at certain variability phases these definitions can result in differences

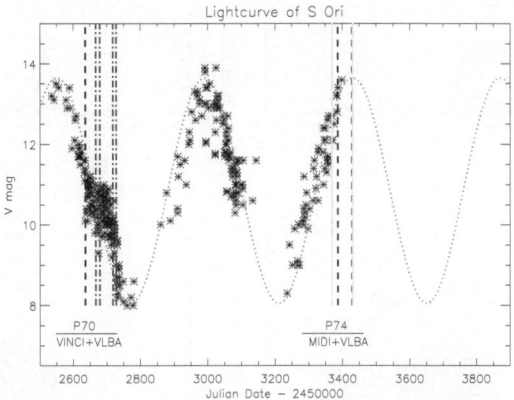

Fig. 5. Lightcurve of S Ori together with the epochs of our joint VLTI/VLBA measurements obtained so far. Note that the y-axis is given with increasing V magnitude, i.e. the stellar maximum is at the bottom and stellar minimum at the top. The study of S Ori was started in ESO period P70 (Dec. 2002/Jan. 2003) including near-infrared K-band VINCI and VLBA/SiO maser observations [5]. In December 2004/January 2005, we obtained concurrent mid-infrared VLTI/MIDI and VLBA/SiO maser observations.

of up to about 20% (on these topics, see also [24]). However, interferometric measurements covering a sufficiently wide range of spatial frequencies can directly be compared to CLV predictions by model atmospheres without the need of a particular radius definition. At pre-maximum stellar phases, when the temperature is highest, the broad-band CLVs are less contaminated by molecular layers, and different radius definitions agree relatively well (Scholz & Wood, private communication).

K-band VINCI observations of the prototype Mira stars o Cet and R Leo have been presented by [33] and [9], respectively. These measurements are also desribed in more detail elsewhere in these proceedings (Driebe et al., Fedele et al.). These measurements at post-maximum stellar phases indicate indeed K-band CLVs which are clearly different from a uniform disk profile already in the first lobe of the visibility function. The measured visibility values were found to be consistent with predictions by the self-excited dynamic Mira model atmospheres described above that include molecular shells close to continuum-forming layers.

4 Joint VLTI/VLBA observations of the Mira star S Ori

We started our project of joint VLTI/VLBA observations of Mira stars in December 2002/January 2003 with coordinated near-infrared K-band VLTI/VINCI observations of the stellar diameter of the Mira variable S Ori

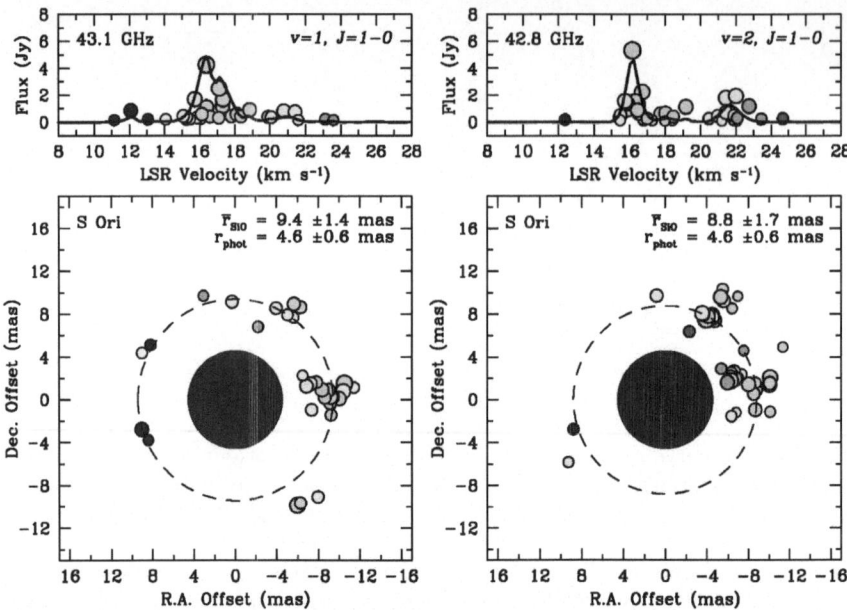

Fig. 6. First-ever coordinated observations between ESO's VLTI and NRAO's VLBA facilities: SiO maser emissions toward the Mira variable S Ori measured with the VLBA, together with the near-infrared diameter measured quasi simultaneously with the VLTI (red stellar disk). The left panels shows the 43.1 GHz maser transition, and the right panel the 42.8 GHz transition. From [5].

and quasi-simultaneous VLBA observations of the 43.1 GHz and 42.8 GHz SiO maser emissions toward this star [5]. We obtained in December 2004/January 2005 further concurrent observations including mid-infrared VLTI/MIDI observations to probe the molecular layers and the dust shell of S Ori, and new epochs of VLBA observations of the 43.1 GHz and 42.8 GHz SiO maser rings.

The December 2002/January 2003 observations represent the first-ever coordinated observations between the VLTI and VLBA facilities, and the results from these observations were recently published [5]. Analysis of the SiO maser data recorded at a visual variability phase 0.73 show the average distance of the masers from the center of the distribution to be 9.4 mas for the $v = 1, J = 1 - 0$ (43.1 GHz) masers and 8.8 mas for the $v = 2, J = 1 - 0$ (42.8 GHz) masers. The velocity structure of the SiO masers appears to be random with no significant indication of global expansion/infall or rotation. The determined near-infrared, K-band, uniform disk (UD) diameters decreased from ~ 10.5 mas at phase 0.80 to ~ 10.2 mas at phase 0.95. For the epoch of our VLBA measurements, an extrapolated UD diameter of $\Theta_{\mathrm{UD}}^K = 10.8 \pm 0.3$ mas was obtained, corresponding to a linear radius of $R_{\mathrm{UD}}^K = 2.3 \pm 0.5$ AU or $R_{\mathrm{UD}}^K = 490 \pm 115\ R_\odot$. The model predicted difference between the continuum and K-band UD diameters is relatively low in the pre-maximum region of the visual variability curve as in the case of our observations (see above). At this phase of 0.73, the continuum diameter may be smaller than the K-band UD diameter by about 15% [17]. With this assumption, the continuum photospheric diameter for the epoch of our VLBA observation would be Θ_{Phot}(VLBA epoch, phase $= 0.73$) ≈ 9.2 mas. Our coordinated VLBA/VLTI measurements show that the masers lie relatively close to the stellar photosphere at a distance of ~ 2 photospheric radii, consistent with model estimates [16] and observations of other Mira stars [6]. This result is virtually free of the usual uncertainty inherent in comparing observations of variable stars widely separated in time and stellar phase.

The new 2004/2005 VLTI and VLBA data are currently being reduced and analyzed.

5 Outlook

We are concentrating on a few stars in order to understand the CSE for a few sources in depth. In addition to the S Ori data described above, we have to date VLTI/MIDI observations of the supergiant AH Sco (Jul. 2005, Aug. 2005), and of the Mira star RR Aql (Jul. 2005, Aug. 2005), as well as concurrent VLBA observations for each of these targets/epochs. These data are currently being analyzed. There may be hints toward an inherent difference in the structure between Mira variables and supergiants, in particular regarding the relative distances of photosphere, SiO maser ring, and inner dust shell boundary (cf. [7, 5]).

Further studies will aim at including more detailed near-infrared studies of the stellar atmospheric structure (close to the photosphere) employing VLTI/AMBER, concurrent with VLTI/MIDI and VLBA observations as discussed above. Making use of the spectro-interferometric capabilities of AMBER, and also of the closure-phase information, these studies can in principle also reveil horizontal surface inhomogeneities (see, e.g. [31]).

A further step toward our better understanding of the stellar mass-loss process are interferometric measurements of post-AGB stars. The first detection of the envelope which surrounds the post-AGB binary source HR 4049, by K-band VINCI observations, was recently reported by [1]. A physical size of the envelope in the near-infrared K-band of about 15 AU (Gaussian FWHM) was derived. These measurements provide information on the geometry of the emitting region and cover a range of position angles of about 60 deg. They show that there is only a slight variation of the size with position angle covered within this range. These observations are, thus, consistent with a spherical envelope at this distance from the stellar source, while an asymmetric envelope cannot be completely ruled out due to the limitation in azimuth range, spatial frequency, and wavelength range. Further investigations using the near-infrared instrument AMBER can reveal the geometry of this near-infrared component in more detail, and MIDI observations can add information on cooler dust at larger distances from the stellar surface.

In the more distant future, when second generation instruments at the VLTI become available, a very valuable addition and continuation of this project would be an improved imaging capability of the VLTI, both at near-infrared as well as at mid-infrared wavelengths. This would enable us to detect and correlate asymmetric structures at the stellar surface and dust shell in a much more precise way, and hence to better understand the transition from spherically symmetric AGB stars to axisymmetric or bipolar planetary nebulae. Improved imaging capabilities can be reached by an increased number of simultaneously combined beams. Furthermore, an improved spatial resolution (by using longer baselines) would be desirable to better match the high angular resolution of the VLBA.

References

1. S. Antoniucci, F. Paresce, & M. Wittkowski: A&A **429**, L1 (2005)
2. J. P. Aufdenberg, & P. H. Hauschildt: Proc. SPIE **4838**, 193 (2003)
3. D. A. Boboltz, P. J. Diamond, & A. J. Kemball: ApJ **487**, L147 (1997)
4. D. A. Boboltz, & P. J. Diamond: AAS **197**, 4507 (2000)
5. D. A. Boboltz, & M. Wittkowski: ApJ **618**, 953 (2005)
6. W. D. Cotton, B. Mennesson, P. J. Diamond, et al.: A&A **414**, 275 (2004)
7. W. C. Danchi, M. Bester, C. G. Degiacomi, et al.: AJ **107**, 1469 (1994)
8. P. J. Diamond, & A. J. Kemball: ApJ **599**, 1372 (2003)
9. D. Fedele, M. Wittkowski, F. Paresce, et al.: A&A **431**, 1019 (2005)
10. A. R. Hajian, J. T. Armstrong, C. A. Hummel et al.: ApJ **496**, 484 (1998)
11. R. Hanbury Brown, et al.: MNRAS **167**, 475 (1973)
12. C. A. Haniff, M. Scholz, & P. G. Tuthill: MNRAS **276**, 640 (1995)
13. P. H. Hauschildt, F. Allard, J. Ferguson, et al.: ApJ **525**, 871 (1999)
14. K.-H. Hofmann, M. Scholz, & P. R. Wood: A&A **339**, 846 (1998)
15. K.-H. Hofmann, U. Beckmann, T. Blöcker, et al.: New Astronomy **7**, 9 (2002)
16. E. M. L. Humphreys, M. D. Gray, J. A. Yates, et al.: A&A **386**, 256 (2002)
17. M. J. Ireland, M. Scholz, & P. R. Wood: MNRAS **352**, 318 (2004)
18. M. Jura, & S. G. Kleinmann: ApJS **73**, 769 (1990)
19. A. J. Kemball, & P. J. Diamond: ApJ **481**, L111 (1997)
20. R. Kurucz: Kuurcz CD-ROM No. 17., Cambridge, Mass. (1993)
21. J. D. Monnier, P. G. Tuthill, B. Lopez, et al.: ApJ **512**, 351 (1999)
22. K. Ohnaka, J. Bergeat, T. Driebe, et al.: A&A **429**, 1057 (2005)
23. A. Quirrenbach, D. Mozurkewich, D. F. Buscher, et al.: A&A **312**, 160 (1996)
24. M. Scholz: Proc. SPIE **4838**, 163 (2003)
25. A. Tej, A. Lancon, M. Scholz, & P. R. Wood: A&A **412**, 481 (2003)
26. R. R. Thompson, M. J. Creech-Eakman, & R. L. Akeson: ApJ **570**, 373 (2002)
27. G. Weigelt, Y. Balega, K.-H. Hofmann, & M. Scholz: A&A **316**, L21 (1996)
28. G. Weigelt, Y. Balega, T. Blöcker, et al.: A&A **333**, L51 (1998)
29. M. Wittkowski, N. Langer, G. Weigelt: A&A **340**, L39 (1998)
30. M. Wittkowski, C. A. Hummel, K. J. Johnston, et al.: A&A **377**, 981 (2001)
31. M. Wittkowski, M. Schöller, S. Hubrig, et al.: AN **323**, 241 (2002)
32. M. Wittkowski, J. P. Aufdenberg, & P. Kervella: A&A **413**, 711 (2004)
33. H. C. Woodruff, M. Eberhardt, T. Driebe, et al.: A&A **421**, 703 (2004)

Limb Darkening: Getting Warmer

J. P. Aufdenberg[1], H.-G. Ludwig[2], P. Kervella[3], A. Mérand[3],
S. T. Ridgway[1,3], V. Coudé du Foresto[3], T. A. ten Brummelaar[6],
D. H. Berger[6], J. Sturmann[6], and N. H. Turner[6]

[1] National Optical Astronomy Observatory, 950 N. Cherry Ave Tucson, AZ, USA
`jasona@noao.edu, sridgway@noao.edu`
[2] Lund Observatory, Lund University, Box 43, 22100 Lund, Sweden
`hgl@astro.lu.se`
[3] LESIA, UMR 8109, Observatorie de Paris-Meudon, 5 place Jules Janssen, 92195
Meudon Cedex, France `pierre.kervella@obspm.fr,`
`antoine.merand@obspm.fr, vincent.foresto@obspm.fr`
[4] The CHARA Array, Mt. Wilson Observatory, Mt. Wilson, CA 91023, USA
`theo@chara-array.org, berger@chara-array.org, judit@chara-array.org,`
`nils@chara-array.org`

Summary. We present interferometric observations and model atmosphere analyses of three stars: the F-type subgiant Procyon, the A-type supergiant Deneb, and the B-type supergiant Rigel. We use VLTI/VINCI and Mark III observations of Procyon to test recent multiwavelength limb-darkening predictions from 3-D hydrodynamic atmosphere simulations with no free parameters for convection. We also investigate the effects of different 1-D atmospheric convection treatments on limb-darkening predictions. We show that the 3-D model predictions are confirmed and we find that 1-D models fail to reproduce Procyon's UV spectral energy distribution, a result consistent with models of granulation for Procyon's surface. We use observations employing the longest baselines of the CHARA Array together with the FLUOR beam combiner to determine precise angular diameters for the two early-type supergiants and test limb-darkening predictions from expanding atmosphere models of these stars' stellar winds. For Deneb, we derive angular diameters consistent with previous measurements, but which vary with position angle at the $\simeq 3\%$ level. Observations of the 2nd lobe of Deneb's visibility curve are more consistent with expanding atmosphere predictions than hydrostatic atmosphere predictions. For Rigel, we derive from the CHARA/FLUOR observations a limb-darkened angular diameter consistent with a recent VLTI/IONIC measurement and 8% larger than reported from the Intensity Interferometer.

1 Procyon: Limb Darkening and Convection

Procyon (α CMi = HR 2943 = HD 61421) is an F5 IV-V star at a distance of 3.5 parsecs, and has arguably the most precise angular diameter measurements in both the visible [1] *and* near-IR [2] of any dwarf star other than the Sun. Recent multi-band microlensing observations [3] are also providing precise limb-darkening measurements of distant solar-type stars. Procyon, however, is special not only for its proximity, but its membership in a visual binary with a white dwarf [4]. The orbital solution coupled with the measured angular diameter of the primary (formally Procyon A) and the measured bolometric flux provide well-constrained values for the mass, radius, and effective temperature. With these fundamental parameters at hand, sophisticated 3-D hydrodynamical atmosphere models [5] have been constructed that make limb-darkening predictions significantly different from standard 1-D model atmospheres. This is a case where "observations meet theory". The model predictions have been published, then later compared with observations, not fine-tuned to match the observations. The 3-D models have no free parameters for convection. The predictions are a result of a computational solution to a radiation-hydrodynamics problem. Thus, limb-darkening measurements provide a test of state-of-the-art models of atmospheric convection. For context, we turn first to our Sun, where a connection between limb darkening and convection was first investigated nearly 100 years ago.

1.1 Solar Limb Darkening

As we approach 100 years since Karl Schwarzschild's groundbreaking 1906 paper [6] on the theory of solar limb darkening, it is remarkable to see how far the field has come and to acknowledge there is still room for substantial improvement. Particularly striking is the recent revision in solar oxygen abundance [7] by nearly a factor of two with the application of 3-D hydrodynamical atmosphere models. Now, as optical interferometry enters the mainstream of astrophysics, we are beginning to investigate with high precision the limb darkening of stars other than the Sun. Such investigations allow us to test spatially resolved model *intensities* and to complement traditional tests of model *fluxes* from spectroscopy, the backbone of astrophysical analysis [8]. Such tests are giving us confidence in the predictive power of model atmospheres and at the same time pointing the way toward further improvements in these models.

Schwarzschild's theoretical work followed upon more than three decades of photometric observations which established quantitatively the center-to-limb intensity variation of the solar photosphere and its wavelength dependence [9, 10]. In 1877, for example, H.C. Vogel made visual spectrophotometric observations that clearly showed limb darkening to be stronger in the blue relative to the red. Schwarzschild assumed a wavelength-independent opacity for the solar atmosphere; the recognition that hydrogen was the principle constituent of the solar atmosphere did not come until 1925 [11]. He compared models to the mean bolometric center-to-limb profile established by Müller and showed that a model temperature structure based on radiative equilibrium, rather than adiabatic equilibrium, was consistent with these observations. In other words, he concluded that in the visible layers of the Sun's atmosphere the transport of energy is dominated by radiation, not by convection.

Subsequent work (for a brief review see [12]) also concluded that the effects of convection on limb darkening were subtle or insignificant. This conclusion held until less than a decade ago. In 1997, standard 1-D stellar atmosphere models of the Sun [13] were shown to be inconsistent with increasingly discriminating observations [14]. New 1-D models [13] with convective overshooting, the depth of convective penetration into layers of the atmosphere stable against convection, provide a better match to observations. More recently, 3-D models for solar limb darkening [15] have led to the same conclusion: 1-D model atmospheres with standard mixing-length convection are too limb darkened, particularly in the blue ($\lambda = 500$ nm), relative to observations.

1.2 Predictions and Observations for Procyon

A similar relationship between 3-D and standard 1-D limb-darkening predictions has been established for Procyon [5]: at 1 μm, the normalized center-to-limb predictions are quite comparable; while at 450 nm, the 1-D intensity profile is up to 20% fainter than the 3-D intensity profile at intermediate limb angles. The predicted monochromatic limb-darkening correction, a model-dependent scale factor between the angular size derived from the visibility data assuming a uniformly bright stellar disk and the true wavelength-independent angular size corresponding to a physical radius, is either 1.081 (1-D model) or 1.064 (3-D model) at 450 nm, a difference of 1.6%. This can be tested interferometrically.

The broad wavelength coverage needed for the interferometric test requires data from more than one interferometer. We use 500 nm and 800 nm data from the Mark III interferometer [16] and 2.2 μm data from VLTI/VINCI. The Mark III interferometric data presented here comprise visibility measurements at 800 nm [1] and unpublished visibility measurements at 500 nm from D. Mozurkewich (2004, private communication). The VLTI/VINCI observations were obtained by combining coherently the light coming from the two VLTI test siderostats (0.35 m aperture) on the B3-D3 and E0-G1 baselines (24 m and 66 m in ground length) using the VINCI beam combiner [17] equipped with a standard K band filter ($\lambda = 2.0-2.4\,\mu$m). While earlier VLTI/VINCI observations [2] were limited to a maximum baseline of 24 m, the 22 new squared visibility measurements cover the 42-64 m range, significantly lower in the first lobe of the Procyon visibility curve. Figure 1 shows all these data in comparison with a best fitting model.

The construction of model atmospheres for Procyon and the subsequent calculation of synthetic radiation fields for comparison with interferometry and photometry is done three ways: (1) Stand-alone 1-D PHOENIX [18] structures, radiation fields and spectra; (2) CO^{5}BOLD [19] 3-D structures temporally and spatially averaged to 1-D, then read by PHOENIX for computation of the corresponding radiation fields; (3) Stand-alone 1-D ATLAS 12 structures, radiation fields and spectra (R. Kurucz 2004, private communication). Table 1 lists the most important model parameters, see [12] for more details.

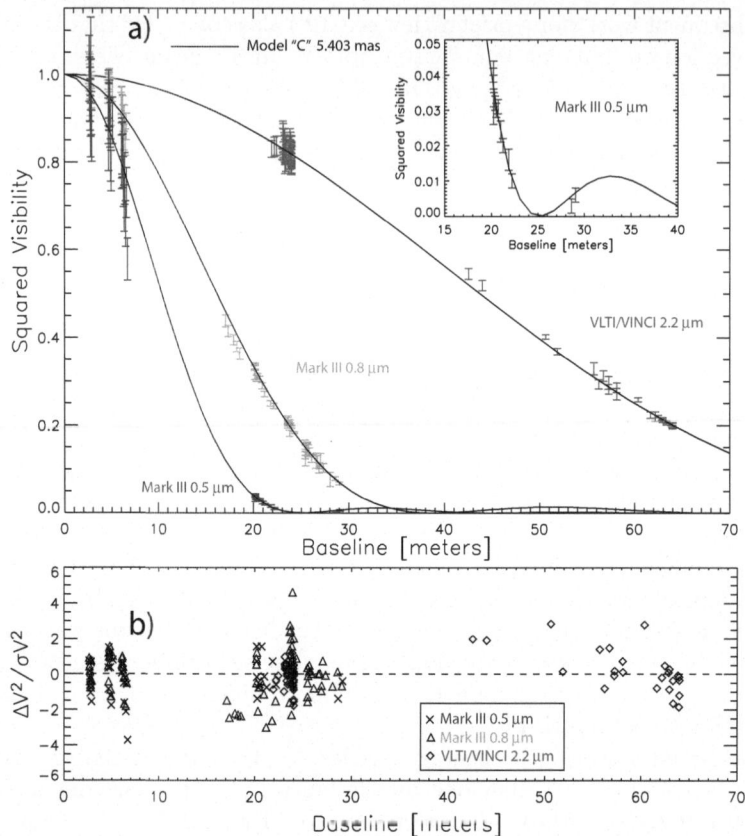

Fig. 1. (a) The squared visibility data from Mark III and VLTI/VINCI as a function of baseline compared to the synthetic visibilities from the 3-D CO^5BOLD + PHOENIX model "C2" with an angular diameter of 5.403 mas. (b) The deviations of each data set from the model.

The computation of the synthetic visibilities from the model radiation fields simulates the bandwidth-smeared squared visibility. At a projected baseline B and mean wavenumber λ_0^{-1}, the synthetic squared visibility,

$$V(B, \lambda_0)^2 = \frac{\int_0^\infty V(B, \lambda)^2 \, \lambda^2 \, d\lambda}{\int_0^\infty S(\lambda)^2 \, F_\lambda^2 \, \lambda^2 \, d\lambda}, \qquad (1)$$

is computed at each wavelength from a Hankel transform,

$$V(B, \lambda) = \int_0^1 S(\lambda) I(\mu, \lambda) J_0 \left[\pi \theta_{\mathrm{LD}} (B/\lambda)(1 - \mu^2)^{1/2} \right] \mu \, d\mu, \qquad (2)$$

where $I(\mu, \lambda)$ is the model radiation field (in photons cm^{-2} s^{-1} sr^{-1}), $S(\lambda)$ is the instrument sensitivity curve, μ is the cosine of the angle between the

Table 1. Stellar Atmosphere Models

Star	Model	Code(s)	$T_{\text{eff}}(K)$	$\log(g)$	R $(R_\odot)$	Additional Parameters
Procyon	A	PHOENIX	6530	3.95	2.1	α =1.25 (mixing length)
Procyon	B	PHOENIX	6530	3.95	2.1	α =0.5
Procyon	C1	CO^{5}BOLD/PHOENIX	6500	4.00	2.1	global mean structure
Procyon	C2	CO^{5}BOLD/PHOENIX	6500	4.00	2.1	12 weighted structures
Procyon	D	ATLAS 12	6530	3.95	2.1	no overshooting
Procyon	E	ATLAS 12	6530	3.95	2.1	100% overshooting
Procyon	F	ATLAS 12	6530	3.95	2.1	50% overshooting
Deneb	G	PHOENIX	8600	1.30	173	hydrostatic
Deneb	H	PHOENIX	9000	1.30	173	10^{-7} M$_\odot$/yr; 225 km/s
Rigel	J	PHOENIX	12100	2.00	66	hydrostatic
Rigel	K	PHOENIX	12500	2.00	66	10^{-7} M$_\odot$/yr; 250 km/s
All	U	Uniform Disk	$V^2 = \left\| 2J_1(\pi\theta_{\text{UD}}B/\overline{\lambda})/(\pi\theta_{\text{UD}}B/\overline{\lambda}) \right\|^2$			

line of sight and the surface normal, and θ_{LD} is the limb-darkened angular size. The mean wavenumber is computed from

$$\lambda_0^{-1} = \frac{\int_0^\infty \lambda^{-1} \, S(\lambda) \, F_\lambda \, d\lambda}{\int_0^\infty S(\lambda) \, F_\lambda \, d\lambda}, \tag{3}$$

where

$$F_\lambda = 2\pi \int_0^1 I(\mu, \lambda) \, \mu \, d\mu \tag{4}$$

is the flux. For each model radiation field, non-linear least squares fits are performed to the visibility data at 500 nm, 800 nm, and 2.2 μm yielding three values for θ_{LD}. The fit results are shown in Figure 2.

We find, as predicted [5], that standard 1-D models are too limb darkened in the blue (500 nm) relative to the near-IR (K-band). Furthermore, 3-D models yield the same angular diameter at these wavelengths indicating a degree of relative limb darkening consistent with observations. This result provides some confidence in 3-D model predictions which until now had been tested predominantly by high-resolution spectroscopy of the Sun [7] and Procyon [5]. Although Figure 2 shows that model "F", a 1-D model with 50% overshooting, provides the most consistent fit to the interferometric observations, we emphasize that the 3-D models have no free parameters for convection. In addition, 1-D models with overshooting are recognized not to reproduce Procyon's Strömgren photometric indices [13, 20] and all 1-D models fail to produce Procyon's photospheric continuum below 160 nm, consistent with models of granulation for Procyon's surface [12].

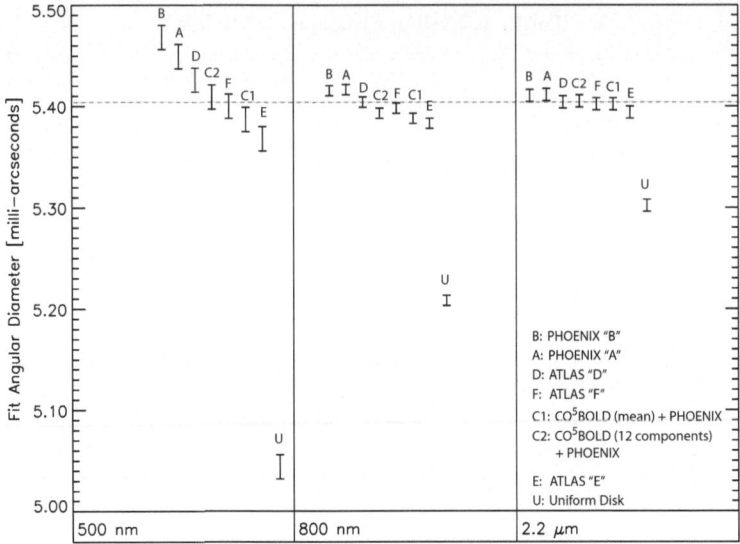

Fig. 2. A comparison of best fit angular diameters at 500 nm, 800 nm (Mark III) and 2.2 μm (VLTI/VINCI) for seven atmosphere models of Procyon and a uniform disk model (see Table 1). The atmosphere models all have essentially the same effective temperature and surface gravity, but differ in their treatment of convection. 3-D model "C2" and 1-D model "F" yield the same diameter at 500 nm and 2.2 μm within the errors. Two important points: (1) Derived K-band limb-darkened angular diameters are less model-dependent than diameters at shorter wavelengths and (2) observations in the visible are critical for testing model atmospheres.

2 Deneb: Mass Loss and Asymmetry

The brightest, nearest, and best-studied A-type (A2 Ia) supergiant is Deneb (α Cygni = HR 7924 = HD 197345) [21]. A-type supergiants are the brightest stars at visual wavelengths (up to $M_V \simeq -9$) and are therefore among the brightest single stars visible in galaxies. In addition, these supernovae Type-II progenitors show potential as independent distance indicators via the Wind Momentum-Luminosity Relationship [22]. For this role, the mass-loss rates for these stars' radiation driven winds must be very well constrained. The standard mass-loss rate diagnostics are spectroscopic (P-Cygni profiles) and spectrophotometric (infrared-millimeter-radio excess). Deneb's wind has been theoretically predicted [21] to enhance limb darkening relative to hydrostatic models, models without mass loss. If true, the amplitude of the 2nd lobe of Deneb's visibility curve should be less than hydrostatic models predict.

To test this prediction, we obtained 25 precise measurements [23] of Deneb's 1st and 2nd lobes with four baselines (E2-W2 152 m, S2-W2 178 m, E2-W1 251 m, and E1-W1 313 m) of the CHARA Array [24] using the FLUOR beam combiner [25] during 2004. The 2nd lobe visibility

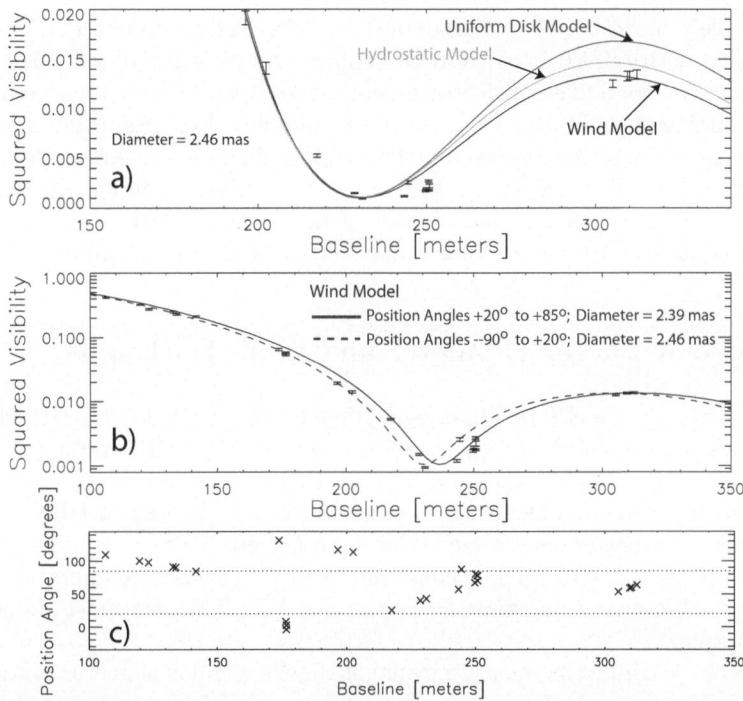

Fig. 3. CHARA/FLOUR squared visibility data for Deneb (A2 Ia) in the 1st and 2nd lobes as a function of projected baseline. (a) A close-up of the data near the 1st null and in the 2nd lobe in comparison to Deneb models from Table 1. (b) The full CHARA/FLUOR data set (note the logarithmic y-axis). Model "H" fit to subsets of data with different position angle ranges (as indicated) yield angular diameters that differ by $\simeq 3\%$. (c) The corresponding position angles as a function of projected baseline.

measurements clearly show limb darkening in excess of that expected from a hydrostatic model atmosphere (see Figure 3). We conclude that Deneb's stellar wind has been detected interferometrically. While these results are quite encouraging and give us some confidence in our theoretical understanding of Deneb's extended atmosphere, the data set contains a signature we did not anticipate: *asymmetry*. An expanding atmosphere model fits well both the high V^2 data and the peak of the second lobe, but fails to match the data in the 1st null.

Figure 3(b) shows that the 1st null data and the 2nd lobe data were obtained at position angles (PA: east of north) in the range $20°$ to $80°$, while nearly all the 1st lobe data falls outside this range. Fitting only the V^2 data between $20°$ and $80°$ yields a reasonably good fit and an angular diameter $\simeq 3\%$ smaller than at the other PAs. Note that the 2nd lobe data are well fit by both diameter values. This is because the peak of the 2nd lobe is relatively insensitive to small diameter shifts. Follow-up observations are scheduled with CHARA/FLUOR to confirm the variation of visibility with position angle for a fixed projected baseline. In retrospect, such an asymmetry is not implausible (D. Peterson, private communication). Adopting a Roche model [26] to describe the asymmetry, with stellar parameters appropriate for Deneb [21]: $R = 180\ R_\odot$, $\log(g) = 1.1$ to 1.3, and $v \sin i = 25$ km s^{-1} with $i = 90°$, the star reaches 30-35% of the angular breakup velocity and has an equatorial radius to polar radius ratio in the range 1.014 to 1.019.

3 Rigel: A Larger Diameter and Limb Darkening

Like Deneb, B-type (B8 Ia) supergiant Rigel (β Ori = HR 1713 = HD 34085) serves as a benchmark for the studies of more distant BA-supergiants in the Milky Way, Local Group, and beyond. Rigel is approximately four times closer to the Sun than Deneb, so its location in the theoretical HR diagram is significantly better constrained relative to Deneb. A precise angular diameter for Rigel helps to further constrain both its effective temperature and its physical radius, important factors in the Wind Momentum-Luminosity Relationship.

There are conflicting measurements of Rigel's angular diameter in the literature. The first interferometric measurements of Rigel, at projected baselines: 9.9 m, 10.0 m, 19.7 m, 23.0 m (all at $\simeq$440 nm), from the Intensity Interferometer [27] yield $\theta_{LD} = 2.55 \pm 0.05$ mas, while three recent measurements, with projected baselines from 63 to 64 m at H-band, from VLTI/IONIC [28] yield $\theta_{LD} = 2.8 \pm 0.1$ mas. This disagreement was not previously recognized because the VLTI/IONIC diameter was not compared to the Intensity Interferometer diameter, but to an angular diameter *estimate* (2.77 ± 0.03 mas) from the first CHARM catalog [29]. This estimate has since been dropped from the updated catalog [30]. We have confirmed with both hydrostatic and expanding model atmospheres that the limb-darkening correction adopted for 440 nm, 1.049, is reasonable. Seeking another solution to this discrepancy, we refit the Intensity Interferometer data [27] and confirm the published mean uniform disk angular diameter. We find, however, that this value is heavily biased by the 19.7 m data point. When weighting the data inversely by the variance, the 19.7 m point carries 9.4 times the weight of the 23.0 m point. Removing the 19.7 m point from the fit yields $\theta_{LD} = 2.69 \pm 0.11$ mas, consistent with the VLTI/IONIC diameter.

Seeking to confirm and improve on the Intensity Interferometer diameter, and to obtain the first 2nd lobe measurements of Rigel, we used the FLUOR beam combiner and CHARA Array baselines E2-W2 and E1-W1 to obtain 4 data points in 2003 and 9 data points in 2004 [23]. Figure 4 shows that our preliminary angular diameter fits are consistent with the VLTI/IONIC H-band angular diameter. The angular diameter is indeed $\simeq$2.75 mas, not

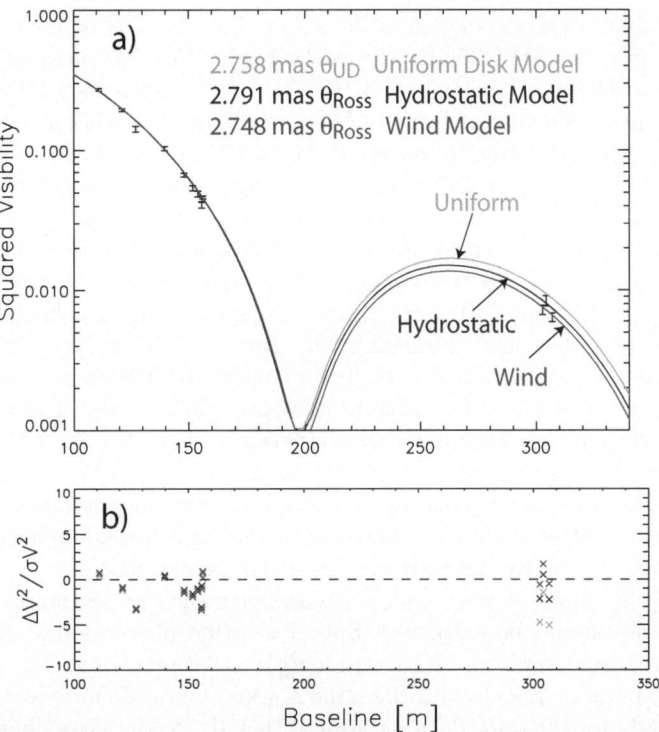

Fig. 4. CHARA/FLUOR squared visibility data for Rigel (B8 Ia) in the 1st and 2nd lobes as a function of projected baseline. (a) All the data plotted on a logarithmic scale along with models from Table 1. The 2nd lobe data are marginally more consistent with the wind model. (b) Deviations from the best fit models.

2.55±0.05 mas. The effective temperature, radius, and hence the angular diameter, must be carefully defined in an extended atmosphere [31]. We adopt a Rosseland radius [21, 32] and note that different models yield significantly different radii: models "J" and "K" differ by 0.04 mas, while the formal uncertainties are 0.01 mas or smaller. We planned our 2nd lobe observations using the Intensity Interferometer diameter. As a result, our 300-meter observations probed the far 2nd lobe, missing the peak where the models are most easily distinguished. Still, the wind model is marginally more consistent with the 2nd lobe observations than the hydrostatic model. Further precise sampling of the 2nd lobe is needed.

4 Limb Darkening: Getting Warmer

Until quite recently, with the exception of Sirius (A1 V) [33], direct (2nd lobe) interferometric limb-darkening measurements have been limited to a small number of stars cooler than the sun: α Cas (K0 III) α Arietis (K2 III) [34], Arcturus (K1.5 III)[35], Betelgeuse (M1 Iab) [36, 37], γ Sagittae (M0 III), V416 Lac (M4 III), BY Boo (M4.5 III) [38], ψ Phoenicis (M4 III) [32]. The 2nd lobe measurements of Deneb (A2 Ia) and Rigel (B8 Ia) presented here and recent 2nd lobe measurements of Altair (A7 V) [39] have substantially increased the number of direct limb-darkening measurements of stars hotter than the Sun. Analyses are now moving beyond just confirming the presence of limb darkening to testing predictions of complex physical phenomena in stellar atmospheres: gravity darkening (from rapid rotation), convection, and mass-loss. In this sense, the field of limb-darkening studies is heating up!

Tables of stellar limb darkening [40], from the linear law with one coefficient to analytic formulae with four or more coefficients, are established by fitting center-to-limb intensity profiles provided by model atmospheres. These coefficients are tabulated as a function of the effective temperature, surface gravity, chemical abundance, and microturbulence. We find that these stellar parameters alone generally do not sufficiently characterize stellar limb darkening to better than a few percent. In the blue, for example, Procyon's limb darkening clearly depends on the specific treatment of convection assumed in the model. For extended atmospheres such as M giants, spherical atmospheres should be employed. Spherical atmosphere models require an additional parameter beyond T_{eff} and $\log(g)$, either mass or radius. Spherical and plane-parallel models with the same T_{eff} and $\log(g)$ do not yield the same angular radii for the same data set, even if they fit the data equally well, as in the case of ψ Phe [32] and as we have shown for Rigel. Finally, for early-type supergiants the limb-darkening profile is sensitive to these stars' mass-loss rates, as demonstrated by the comparison of 2nd lobe measurements and expanding model atmospheres for Deneb. So, we are learning that standard limb-darkening predictions may be unreliable, high-precision interferometric observations (in the near-IR *and* visible, and in the 2nd lobe) provide an opportunity to see where classical models fail, to test better models, and to contribute to the development of the second century of limb-darkening science.

Acknowledgments JPA would like to thank the SOC for their kind invitation. Thanks to D. Mozurkewich and R. Kurucz for supplying the Mark III data and ATLAS12 models, respectively. Thanks to P.J. Goldfinger for assistance with the CHARA observations. This work was in part performed under contract with the Jet Propulsion Laboratory (JPL) funded by NASA through the Michelson Fellowship Program. The Mark III was funded by the Office of Naval Research and the Oceanographer of the Navy. The CHARA Array is operated by the Center for High Angular Resolution Astronomy with support from Georgia State University and the National Science Foundation, the Keck Foundation and the Packard Foundation.

References

1. D. Mozurkewich *et al.* Angular Diameters of Stars from the Mark III Optical Interferometer. *AJ*, 126:2502–2520, 2003.
2. P. Kervella, F. Thévenin, P. Morel, G. Berthomieu, P. Bordé, and J. Provost. The diameter and evolutionary state of Procyon A. Multi-technique modeling using asteroseismic and interferometric constraints. *A&A*, 413:251–256, 2004.
3. F. Abe *et al.* Probing the atmosphere of a solar-like star by galactic microlensing at high magnification. *A&A*, 411:L493–L496, 2003.
4. T. M. Girard *et al.* A Redetermination of the Mass of Procyon. *AJ*, 119:2428–2436, 2000.
5. C. Allende Prieto, M. Asplund, R. J. G. López, and D. L. Lambert. Signatures of Convection in the Spectrum of Procyon: Fundamental Parameters and Iron Abundance. *ApJ*, 567:544–565, 2002.
6. K. Schwarzschild. On the equilibrium of the sun's atmosphere. In D. H. Menzel, editor, *Selected Papers on the Transfer of Radiation*. New York: Dover, 1966. *Über das Gleichgewicht der Sonnenatmosphäre"* Nachrichten von der Königlichen Gesellschaft der Wissenschaften zu Göttingen. Math.-phys. Kalsse (1906) 295, 41.
7. M. Asplund, N. Grevesse, A. J. Sauval, C. Allende Prieto, and D. Kiselman. Line formation in solar granulation. IV. [O I], O I and OH lines and the photospheric O abundance. *A&A*, 417:751–768, 2004.
8. J. B. Hearnshaw. *The Analysis of Starlight: One Hundred and Fifty Years of Astronomical Spectroscopy*. Cambridge: Cambridge University Press, 1990.
9. G. Müller. *Die Photometrie der Gestirne*. Leipzig: Wilhelm Engelmann, 1897.
10. J. B. Hearnshaw. *The Measurement of Starlight, Two Centuries of Astronomical Photometry*. Cambridge: Cambridge University Press, 1996.
11. C. H. Payne. *Stellar Atmospheres*. Number 1 in Harvard Observatory Monographs. Cambridge: Harvard Observatory, 1925.
12. J. P. Aufdenberg, H.-G. Ludwig, and P. Kervella. On the Limb Darkening, Spectral Energy Distribution, and Temperature Structure fo Procyon. *ApJ*, 633, 424–439, 2005.
13. F. Castelli, R. G. Gratton, and R. L. Kurucz. Notes on the convection in the ATLAS9 model atmospheres. *A&A*, 318:841–869, 1997.
14. H. Neckel and D. Labs. Solar limb darkening 1986-1990 ($\lambda\lambda$ 303 to 1099nm). *Solar Physics*, 153:91–114, 1994.
15. M. Asplund, Å. Nordlund, and R. Trampedach. Confrontation of Stellar Surface Convection Simulations with Stellar Spectroscopy. In *ASP Conf. Ser. 173: Stellar Structure: Theory and Tests of Convective Energy Transport*, page 221, 1999.
16. M. Shao, M. M. Colavita, B. E. Hines, D. H. Staelin, and D. J. Hutter. The Mark III stellar interferometer. *A&A*, 193:357–371, 1988.
17. P. Kervella *et al.* VINCI, the VLTI commissioning instrument: status after one year of operations at Paranal. In *Proceedings of the SPIE, Volume 4838*, pages 858–869, 2003.
18. P. H. Hauschildt, F. Allard, J. Ferguson, E. Baron, and D. R. Alexander. The NEXTGEN Model Atmosphere Grid. II. *ApJ*, 525:871, 1999.
19. B. Freytag, M. Steffen, and B. Dorch. Spots on the surface of Betelgeuse – Results from new 3D stellar convection models. *AN*, 323:213–219, 2002.

20. U. Heiter *et al.* New grids of ATLAS9 atmospheres I: Influence of convection treatments on model structure and on observable quantities. *A&A*, 392:619–636, 2002.

21. J. P. Aufdenberg *et al.* The Spectral Energy Distribution and Mass-loss Rate of the A-type Supergiant Deneb. *ApJ*, 570:344–368, 2002.

22. R. P. Kudritzki *et al.* The Wind Momentum-Luminosity Relationship of Galactic A- and B-Supergiants. *A&A*, 350:970, 1999.

23. J. P. Aufdenberg *et al.* CHARA/FLUOR Observations of B- and A-type Supergiants Rigel and Deneb. in preparation.

24. T. A. ten Brummelaar *et al.* First Results from the CHARA Array. II. A Description of the Instrument. *ApJ*, 628:453, 2005.

25. V. Coudé du Foresto *et al.* FLUOR fibered beam combiner at the CHARA array. In *Proceedings of the SPIE, Volume 4838*, pages 280–285, 2003.

26. S. R. Cranmer and S. P. Owocki. The effect of oblateness and gravity darkening on the radiation driving in winds from rapidly rotating B stars. *ApJ*, 440:308–321, 1995.

27. R. Hanbury Brown, J. Davis, and L. R. Allen. The angular diameters of 32 stars. *MNRAS*, 167:121, 1974.

28. J. B LeBouquin *et al.* First observations with an H-band integrated optics beam combiner at the VLTI. *A&A*, 424:719–726, 2004.

29. A. Richichi and I. Percheron. CHARM: A Catalog of High Angular Resolution Measurements. *A&A*, 386:492–503, 2002.

30. A. Richichi, I. Percheron, and M. Khristoforova. CHARM2: An updated Catalog of High Angular Resolution Measurements. *A&A*, 431:773–777, 2005.

31. B. Baschek, M. Scholz, and R. Wehrse. The parameters R and Teff in stellar models and observations. *A&A*, 246:374, 1991.

32. M. Wittkowski, J. P. Aufdenberg, and P. Kervella. Tests of stellar model atmospheres by optical interferometry. VLTI/VINCI limb-darkening measurements of the M4 giant ψ Phe. *A&A*, 413:711–723, 2004.

33. R. Hanbury Brown, J. Davis, R. J. W. Lake, and R. J. Thompson. The Effects of Limb Darkening on Measurements of Angular Size with an Intensity Interferometer. *MNRAS*, 167:475, 1974.

34. A. R. Hajian *et al.* Direct Confirmation of Stellar Limb Darkening with the Navy Prototype Optical Interferometer. *ApJ*, 496:484, 1998.

35. A. Quirrenbach *et al.* Angular Diameter and Limb Darkening of Arcturus. *A&A*, 312:160, 1996.

36. D. Burns *et al.* The surface structure and limb-darkening profile of Betelgeuse. *MNRAS*, 290:L11, 1997.

37. G. Perrin *et al.* Interferometric observations of the supergiant stars α Orionis and α Herculis with FLUOR at IOTA. *A&A*, 418:675–685, 2004.

38. M. Wittkowski *et al.* Direct Multi-Wavelength Limb-Darkening Measurements of Three Late-Type Giants with the Navy Prototype Optical Interferometer. *A&A*, 377:981, 2001.

39. N. Ohishi, T. E. Nordgren, and D. J. Hutter. Asymmetric Surface Brightness Distribution of Altair Observed with the Navy Prototype Optical Interferometer. *ApJ*, 612:463–471, 2004.

40. A. Claret. A New Non-linear Limb-darkening Law for LTE Stellar Atmosphere Models. *A&A*, 363:1081–1190, 2000.

Cepheid Distances from Interferometry

P. Kervella[1], N. Nardetto[2], D. Bersier[3], D. Mourard[2], P. Fouqué[4], and
V. Coudé du Foresto[1]

[1] LESIA, UMR 8109, Observatoire de Paris-Meudon, 5, place Jules Janssen,
    F-92195 Meudon Cedex, France, pierre.kervella@obspm.fr
[2] GEMINI, UMR 6203, Observatoire de la Côte d'Azur, Avenue Copernic,
    F-06130 Grasse, France
[3] Space Telescope Science Institute, 3700 San Martin Drive, Baltimore, MD
    21218, USA
[4] Observatoire Midi-Pyrénées, UMR 5572, 14, av. Edouard Belin, F-31400
    Toulouse, France

Summary. Long baseline interferometry is now able to resolve the pulsational
changes of the angular diameters of a significant number of Cepheids in the solar neighborhood. This allows us to apply a new version of the Baade-Wesselink
method to measure their distances, for which we do not need to estimate the star's
temperature. Using this method and angular diameter measurements from the VLT
Interferometer, we derived the distances to four nearby Cepheids. For three additional stars, we obtained average values of their angular diameters and we estimated their distances from previously published values of their linear sizes. Based
on these new measurements and already existing data, we derived new calibrations
of the Period-Luminosity and Period-Radius relations. Additionally, we obtained
high precision surface brightness-color relations based solely on interferometric angular diameter measurements on Cepheids and $BVRIJHK$ magnitudes. We finally
discuss the prospects of the direct distance measurements of Cepheids.

1 Introduction

For almost a century, Cepheids have occupied a central role in distance determinations. This is thanks to the existence of the Period-Luminosity relation
$M = a \log P + b$ which relates the logarithm of the variability period of
a Cepheid to its absolute mean magnitude. These stars became even more
important since the *HST Key Project* on the extragalactic distance scale
has totally relied on Cepheids for the calibration of distance indicators to
reach cosmologically significant distances. In other words, if the calibration
of the Cepheid P–L relation is wrong, the whole extragalactic distance scale
is wrong.

There are various ways to calibrate the P–L relation. The avenue chosen by the *HST Key Project* was to assume a distance to the Large Magellanic Cloud (LMC), thereby adopting a zero point of the distance scale,
but the LMC distance is currently the weak link in the extragalactic distance scale ladder. Another avenue is to determine the zero point of the

Period-Luminosity relation with Galactic Cepheids, using for instance parallax measurements, Cepheids in clusters, or through the Baade-Wesselink (BW) method (see e.g. [3]). We describe in these proceedings our recent work with the VLTI, that aims at improving the calibration of the Period-Radius (P–R), Period-Luminosity (P–L) and surface brightness-color (SB) relations through the combination of spectroscopic and interferometric observations of bright Galactic Cepheids.

2 Existing Galactic Cepheid distances

Historicaly, a number of methods have been applied to the determination of the distances of nearby Galactic Cepheids (Table 1).

Despite their large apparent brightness, Cepheids are located at large distances, and ESA's Hipparcos satellite could only obtain parallaxes with a poor precision. If we exclude the peculiar first overtone pulsator Polaris, the closest Cepheid is δ Cep. Benedict et al. [2] obtained an excellent parallax measurement using the Fine Guidance Sensor of the *Hubble Space Telescope*, corresponding to a distance of 273 ± 11 pc. It is expected that the parallaxes to ten Cepheids will soon be measured with the same instrument.

The distance to Cepheids in open clusters can be inferred from several secondary techniques to determine the distance to the cluster itself (main sequence fitting, eclipsing binaries,...). This is a useful technique to measure the distances of Cepheids located relatively far away, but the systematic uncertainties (of the order of 0.2 mag) limit its final accuracy for the calibration of the P–L relation.

The classical BW method ([1, 26]) is based on a combination of photometry and spectroscopy. The photometric part is used to estimate the angular diameter of the star over its pulsation based on its color, using SB relations. Our interferometric calibration of these relations is presented in Sect. 5.3.

The latest addition to the arsenal of techniques to measure Cepheid distances is the interferometric version of the BW method. The difference with the classical version is that the angular diameter is not derived from the SB, but measured directly. This method is applicable only to the nearest Cepheids, for which we can measure the amplitude of the angular diameter variation (see Sect. 3 for details). Even for the nearest Cepheids, this demands an extremely high resolving power, that long baseline interferometry is the only technique to provide.

3 The interferometric Baade-Wesselink (IBW) method

3.1 Principle

The basic principle of the BW method is to compare the linear and angular size variation of a pulsating star, in order to derive its distance through

Table 1. Distance determinations to Galactic Cepheids from different methods.

Method	Cepheids	References
Hipparcos parallaxes (low precision)	247	Lanoix et al. [15]
HST-FGS parallax of δ Cep	1	Benedict et al. [2]
Cepheids in open clusters	31	Tammann et al. [24]
Classical Baade-Wesselink	34	Gieren et al. [6]
Interferometric Baade-Wesselink	5	Kervella et al. [10], Lane et al. [14]

a simple division. This method is a well-established way to determine the luminosity and radius of a pulsating star. The two quantities required to apply this technique are the radius variation curve and the angular diameter variation.

On one hand, the linear size variation can be obtained by high resolution spectroscopy, through the integration of the radial velocity curve obtained by monitoring the Doppler shift of the spectral lines present in the spectrum (Fig. 1). A difficulty in this process is that the measured wavelength shifts are integrated values over the full stellar disk. To convert them into a pulsation velocity, i.e. a physical displacement of the photosphere at the center of the disk, we have to multiply it by a projection factor p that encompasses the sphericity of the star and the structure of its atmosphere (limb darkening,...). Unfortunately, the p-factor is still uncertain at a level of a few percents but classical high resolution spectroscopy coupled with the dispersed fringes mode of the AMBER instrument ([20]) are expected to bring strong constraints on this important factor.

On the other hand, the angular size is difficult to estimate directly. Until recently, the only method to estimate the angular size was through the surface brightness of the star. With the advent of powerful infrared long baseline interferometers, it is now possible to resolve spatially the star itself, and thus measure directly its photospheric angular diameter. An uncertainty at a level of about 1% remains on the limb darkening of these stars, that is currently taken from static atmosphere models.

3.2 Limitations of the IBW method

Though the IBW technique is a powerful method, it still presents some of the limitations associated with its classical BW couterpart. Firstly, the proper integration of the radial velocity $v_r(\phi)$ requires an excellent sampling in order to obtain a realistic curve for the radius variation $R(\phi)$. An undersampled radial velocity curve, or an inadequate interpolation technique, can easily bias the radius curve amplitude. Secondly, the currently assumed value of the p-factor could be biased. This parameter is very difficult to measure directly, but numerical modeling ([18]) has shown that it can change significantly depending on which method is used to estimate the radial velocity v_r from

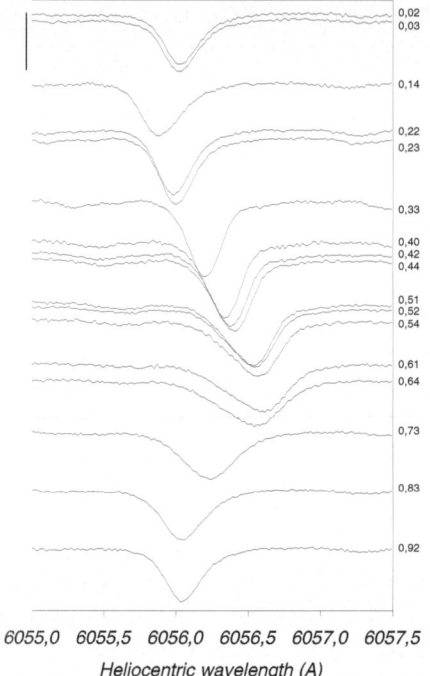

0,02
0,03
0,14
0,22
0,23
0,33
0,40
0,42
0,44
0,51
0,52
0,54
0,61
0,64
0,73
0,83
0,92

6055,0   6055,5   6056,0   6056,5   6057,0   6057,5

Heliocentric wavelength (A)

Fig. 1. Illustration of the Doppler shift of the Fe I line at 6056 Å (rest frame wavelength) of the Cepheid β Dor, for different phases (HARPS data).

the spectra. Recent interferometric measurements obtained with the CHARA array on δ Cep, combined with the parallax from the HST-FGS instrument, have allowed to measure directly for the first time the projection factor ([17]).

The limb darkening correction that is used to compute the photospheric angular size of the Cepheids observed by interferometry is also a potential source of uncertainty. It has been suspected for some time ([21]) that the limb darkening of a Cepheid could depart from that of a stable star with the same physical properties ($T_{\mathrm{eff}}, \log g, ...$), due to the compression of the atmosphere and the presence of shock waves. The application of the LD corrections of stable supergiants at each phase of the pulsation could introduce a bias on the estimate of the photospheric size. On the positive side, good models exist for stable stars, and dynamical modeling is in progress for Cepheids (e.g. [16], [18]). In addition, current interferometers are operating in the infrared H and K bands, which tends to mitigate the sensitivity to this effect, as the LD is relatively small as these wavelengths. Moreover, the IBW technique makes use of differential measurements over the pulsation, and part of the systematic error due to the LD is expected to cancel out.

The fact that most Cepheids (up to 80%, [23]) are binary or multiple stars could create a bias on both the interferometric and photometric data at a few percents level. However, the fact that current interferometers are operating in the infrared reduces the sensitivity to this source of bias, as the companions are usually hot dwarfs that radiate essentially in the blue. The contrast in the H and K bands is thus significantly larger than in the visible, and their contribution can generally be neglected.

4 VINCI observations

The capabilities of the VLTI for the observation of nearby Cepheids are outstanding, as it provides long baselines (up to 202 m) and thus a high resolving power. Though they are supergiant stars, the Cepheids are generally very small objects in terms of angular size. A consequence is that the limit on the number of interferometrically resolvable Cepheids is not set by the size of the light collectors, but by the baseline length. From photometry only, several hundred Cepheids can produce interferometric fringes using the VLTI Auxiliary Telescopes (ATs, 1.8 m aperture). However, in order to measure accurately their size, one needs to resolve their disk to a sufficient level, and this reduces the total number of accessible Cepheids to about 40.

Considering the usual constraints in terms of sky coverage, limiting magnitude and accessible resolution, we selected seven bright Cepheids observable from Paranal: X Sgr, η Aql, W Sgr, β Dor, ζ Gem, Y Oph and ℓ Car. The periods of these stars cover a wide range, from 7 to 35.5 days, an important advantage to properly constrain the P–R and P–L relations. Using the IBW method, we derived the distances to η Aql, W Sgr, β Dor and ℓ Car. For the remaining three objects of our sample, X Sgr, ζ Gem and Y Oph, we obtained average values of their angular diameters, and we applied a hybrid method to derive their distances, based on published values of their linear diameters. Fig. 2 shows the angular diameter curve and the fitted radius curve of ℓ Car ($P = 35.5$ days), that constrains its distance to a relative precision better than 5%. A discussion of these data can be found in [10] and [9].

For our observations, the beams from the two VLTI Test Siderostats (0.35 m aperture) or the two Unit Telescopes UT1 and UT3 were recombined coherently in VINCI ([11]). We used a regular K band filter ($\lambda = 2.0 - 2.4 \mu m$) that gives an effective observation wavelength of $2.18 \mu m$ for the effective temperature of typical Cepheids. Three VLTI baselines were used for this program: E0-G1, B3-M0 and UT1-UT3.

In total, we obtained 69 individual angular diameter measurements, for a total of more than 100 hours of telescope time (2 hours with the UTs), spread over 68 nights ([10]). One of the key advantages of VINCI is to use single-mode fibers to filter out the perturbations induced by the turbulent atmosphere. The resulting interferograms are practically free of atmospheric corruption, except the piston mode (differential longitudinal delay of the

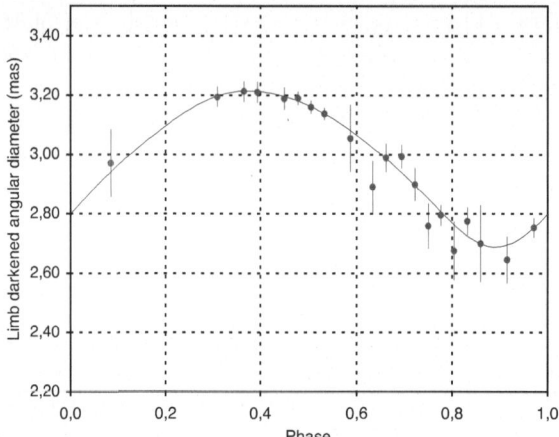

Fig. 2. Angular diameter measurements of the Cepheid ℓ Car ($P = 35.6$ days) obtained with VINCI (dots), and the adjusted diameter curve integrated from radial velocity data (solid curve).

wavefront between the two apertures) that tends to smear the fringes and affect their visibility. But this residual can be brought down to a very low level by using a fast acquisition rate.

5 P–L, P–R and surface brightness-color relations

5.1 Period-Luminosity

The Cepheid P–L relation is the basis of the extragalactic distance scale, but its calibration is still uncertain at a $\Delta M = \pm 0.10$ mag level. Until now, the classical BW method (where one combines photometry and radial velocity data) was used to obtain the distance and radius of a Cepheid. A recent application of this method to individual stars can be found for instance in [25]. Our sample is currently too limited to allow a robust determination of the P–L relation, defined as $M_\lambda = \alpha_\lambda(\log P - 1) + \beta_\lambda$ that would include both the slope and the $\log P = 1$ reference point β_λ. However, if we suppose that the slope is known a priori from the literature, we can still derive a precise calibration. We have considered for our fit the P–L slope measured on LMC Cepheids from [6].

For the V band, we obtain $\beta_V = -4.209 \pm 0.075$ ([7]). The positions of the Cepheids on the P–L diagram are shown on Fig. 3. Our calibrations differ from [6] by $\Delta\beta_V = +0.14$ mag, corresponding to $+1.8\sigma$. The sample is dominated by the high precision ℓ Car and δ Cep measurements. When these two stars are removed from the fit, the difference with [6] is slightly increased, up to $+0.30$ mag, though the distance in σ units is reduced ($+1.5$). From this agreement, ℓ Car and δ Cep do not appear to be systematically different from the other Cepheids of our sample. It is difficult to conclude firmly to a significant discrepancy between [6] and our results, as our sample is currently too limited to exclude a small-statistics bias. However, if we assume an intrinsic dispersion of the P–L relation $\sigma_{PL} \simeq 0.1$ mag, as suggested by [6], then our results point toward a slight underestimation of the absolute magnitudes of Cepheids by these authors. On the other hand, we obtain precisely the same $\log P = 1$ reference point value in V as [15], using parallaxes from Hipparcos. The excellent agreement between these two fully independent calibrations of the P–L relation is remarkable.

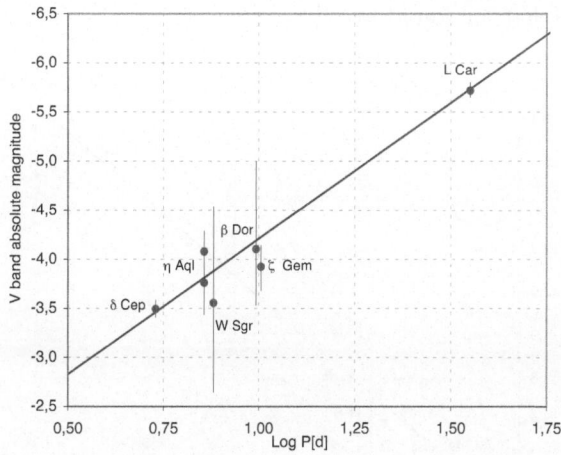

Fig. 3. Positions of the Cepheids measured using the IBW method in the Period-Luminosity diagram (V band), (dots) and best fit P–L relation assuming the slope from [6] (solid line).

5.2 Period-Radius

The Period-Radius relation (P–R) is an important constraint for the Cepheid models (see e.g. [4]). It takes the form of the linear expression $\log R = a \log P + b$. In order to calibrate this relation, we need to estimate directly the linear radii of a set of Cepheids. To complement the VINCI sample of seven Cepheids, we added the measurements of δ Cep, ζ Gem and η Aql obtained previously by other interferometers. We have applied two methods to determine the radii of the Cepheids of our sample: the IBW method, and a combination of the average angular diameter and trigonometric parallax. While the first provides directly the average linear radius and distance, we need to use trigonometric parallaxes to derive the radii of the Cepheids for which the pulsation is not detected. For these stars, we applied the Hipparcos distance, except for δ Cep, for which we considered the recent parallax measurement by [2].

Fig. 4 shows the distribution of the measured diameters on the P–R diagram. When we choose to consider a constant slope of $a = 0.750 \pm 0.024$, as found by [6], we derive a zero point of $b = 1.105 \pm 0.029$ ([7]). As a comparison, [6] obtained a value of $b = 1.075 \pm 0.007$, only -1.6σ away from our result. Fitting simultaneously both the slope and the zero point to our data set, we obtain $a = 0.767 \pm 0.009$ and $b = 1.091 \pm 0.011$. These values are only $\Delta a = +0.7\sigma$ and $\Delta b = +1.2\sigma$ away from [6]. Considering the limited size of our sample, the agreement is satisfactory. On the other hand, the slopes derived from some numerical models are significantly different, such as the slope $a = 0.661 \pm 0.006$ found by [4].

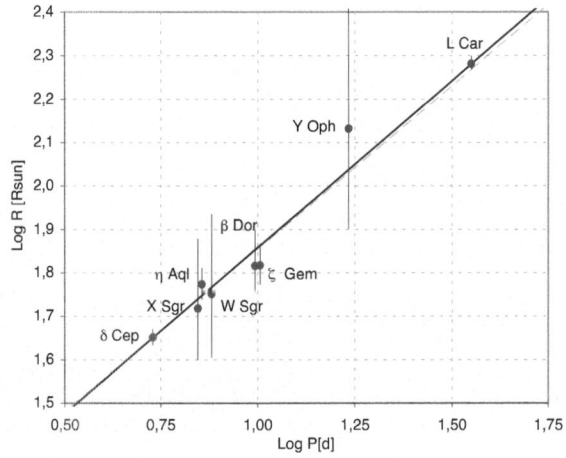

Fig. 4. Period-Radius relation deduced from the interferometric measurements of Cepheids. The solid line corresponds to the simultaneous fit of both the zero point and slope, while the dashed line assumes the slope from [6], fitting only the zero point.

5.3 Surface brightness-color relations

The SB relations link the emerging flux per unit solid angle of a light-emitting body to its color, or effective temperature. Intuitively, the conservation of the SB can easily be understood as both the solid angle subtended by a star and its apparent brightness are decreasing with the square of its distance. In theory, for a perfect blackbody emission, their ratio is therefore a constant for a given effective temperature. In practice, the stars are not perfect blackbodies, and we have to rely on a color index (i.e. the difference between magnitudes in two photometric bands) as a tracer of the effective temperature. The SB relations are of considerable astrophysical interest for Cepheids, as a well-defined relation between a particular color index and the surface brightness can provide accurate predictions of their angular diameters. When combined with the radius curve, integrated from spectroscopic radial velocity measurements, they give access to the distance of the Cepheid through the classical BW method. This method has been applied recently to Cepheids in the SMC ([22]), i.e. at a far greater distance than what can be achieved by the IBW method. But the accuracy that can be achieved on the distance estimate is conditioned for a large part by our knowledge of the SB relations.

When considering a perfect blackbody curve, any color can in principle be used to obtain the SB, but in practice, the linearity of the correspondance between $\log T_{\mathrm{eff}}$ and color depends on the chosen wavelength bands. The surface brightness F_λ is given by the following expression taken from [5]: $F_\lambda = 4.2207 - 0.1 m_\lambda - 0.5 \log \theta_{\mathrm{LD}}$ where θ_{LD} is the limb darkened angular diameter (i.e. the angular size of the stellar photosphere). To fit the individual measurements, we used a linear function of the stellar color indices, expressed in magnitudes (logarithmic scale), using for example the following expression: $F_V(V - K) = a(V - K)_0 + b$.

We assembled a list of 145 individual interferometric measurements of Cepheids, related to nine stars. For each measurement epoch, we estimated the $BVRIJHK$ magnitudes based on Fourier interpolations of the available photometric data from the literature and corrected them for the interstellar extinction. The resulting $F_V(V - K)$ relation fit is presented in Fig. 5. The other relations based on the V band surface brightness F_V and $BIJHK$ colours are plotted on Fig. 6. The smallest residual dispersions are obtained for the infrared based colors, for instance: $F_V = -0.1336 \pm 0.0008(V - K) + 3.9530 \pm 0.0006$ ([8]). The intrinsic dispersion is undetectable at the current level of precision of our measurements, and could be as low as 1%.

6 Prospects on direct Cepheid distances

The HST-FGS trigonometric parallax programme has already provided the parallax of δ Cep to 4% ([2]). It is expected that it will soon result in $\simeq 10$ additional Cepheid parallaxes, with a 10% accuracy. Their combination will theoretically give a final uncertainty of ± 0.03 mag on the P–L zero point. On the interferometric side, several important facilities are now being commisioned or starting routine operation (AMBER/VLTI, FLUOR/CHARA, MIRC/CHARA, SUSI...). We can reasonably expect important results from these instruments starting in 2006. They have the capability to measure the 40 Cepheids accessible to the IBW method, with a distance accuracy better than 5% for 20 of them. This should result in a ± 0.01 mag precision on the

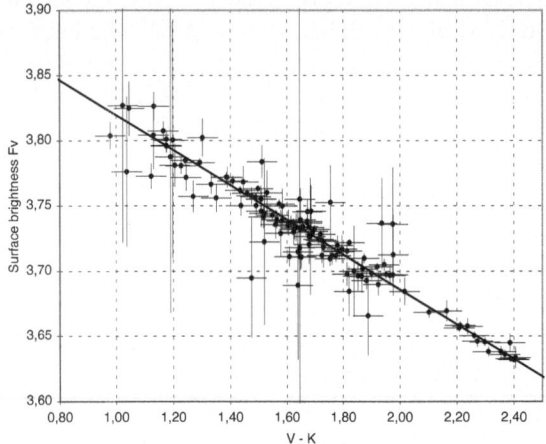

Fig. 5. Surface brightness-color diagram $F_V(V - K)$ for the Cepheids measured by interferometry. The 145 individual interferometric measurements are displayed as solid dots, and the best fit linear model as a solid line.

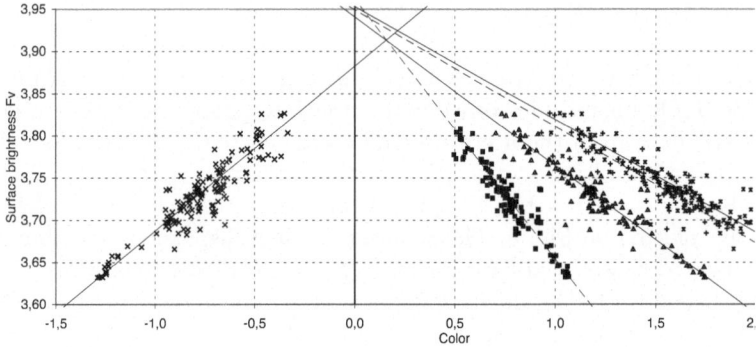

Fig. 6. Overview of the surface brightness-color relations in the V band, based on the $(V - B)$, $(V - I)$, $(V - J)$, $(V - H)$ and $(V - K)$ colors (from left to right).

P–L zero point. One should note that the complementarity with the HST-FGS parallax programme is excellent: for the 10 nearest Cepheids, we will have both the IBW and trigonometric distances available. This will allow to calibrate directly the projection factor for these stars ([17]), and thus waive the uncertainty on this important factor. Further in the future, the GAIA mission ([19]) will measure the trigonometric parallax to an accuracy of $5\,\mu$as up to $V = 13$. For δ Cep, the relative accuracy on the parallax will reach 0.1%! One should note that this is a good match to the interferometric accuracy obtained by Mérand et al. ([17]) on this star. GAIA will also measure directly the distances to LMC/SMC Cepheids to a precision of $\simeq 20\%$. The mission is expected to fly in 2011, and the final catalogue should be released in 2018.

7 An ideal interferometer for Cepheids ?

Operated in the J band on a 200m baseline, an interferometer can resolve 40 Cepheids up to the point that it can measure accurately their pulsation amplitude and apply the IBW method. This is equivalent to AMBER with the VLTI Auxilliary Telescopes ([12, 13]) on the longest baseline of the VLTI. These accessible Cepheids are selected to present fringe visibilities below $V \simeq 90\%$ (angular diameters $\theta \geq 0.3\,\mathrm{mas}$). The apparent magnitude of these targets is $m_J \leq 6.3$, well within the accessible range of AMBER with the ATs. Considering the same 200 m baseline, but this time in the V band, 100 Cepheids are accessible to the IBW method ($\theta \geq 0.15\,\mathrm{mas}$, d up to 10 kpc, $m_V \leq 9.0$), a significant increase compared to AMBER in the J band. This comparison highlights the need for a higher angular resolution to achieve the full potential of 2 m class telescopes for interferometry. The question remains of which of the two factors to improve: shorter wavelength (visible) or longer baseline at infrared wavelength? In the framework of the IBW method, kilometric baselines at infrared wavelengths are probably more favorable due to the lower uncertainty on the LD correction compared to the visible. Moreover, in this wavelength domain, there is no need for adaptive optics with the VLTI ATs, which is an operational advantage. On the other hand, shorter wavelengths are particularly interesting to study the LD properties of Cepheids, at the cost of dedicated adaptive optics systems. A combination of visible and infrared wavelengths on long baselines would naturally be the most powerful configuration, at LD measurements in the visible would allow to waive completely the uncertainty on the LD correction in the infrared.

8 Conclusion

Different methods can be applied to measure Cepheid distances. For $d \leq 500\,\mathrm{pc}$ ($\simeq 10$ stars), trigonometric parallax is clearly the method of choice, as it is purely geometrical. Up to $d \simeq 3000\,\mathrm{pc}$, the IBW method with the AMBER instrument is expected to provide precise distances for 40 stars. The accuracy of the IBW distance estimates will likely be limited to a few percents by the LD model (for the interferometric part) and the p-factor (for the spectroscopic part). In order to access Cepheids in the Magellanic Clouds, interferometrically calibrated SB relations are well suited and will provide accurate distances. The next step of the interferometric study of Cepheids is to measure the limb darkening of ℓ Car as a function of phase, and for a range of wavelengths. This is important in order to better constrain the numerical models of Cepheid atmospheres. Overall, the IBW method has its limitations, but it is still an excellent way of estimating middle-range Cepheid distances, in particular as the quality of the interferometric data is improving rapidly ([17]). Regarding the Cepheid distance scale, it is clear that the GAIA mission will solve the question of the calibration of the P–L relation, but only 12+ years from now!

References

1. Baade, W.: Astron. Nachr. **228**, 359 (1926)
2. Benedict, G. F., McArthur, B. E., Fredrick, L. W., et al.: AJ **123**, 473 (2002)
3. Bersier, D., Burki, G., & Kurucz, R. L.: A&A **320**, 228 (1997)
4. Bono, G., Caputo, F. & Marconi, M.: ApJL **497**, 43 (1998)
5. Fouqué, P. & Gieren, W. P.: A&A **320**, 799 (1997)
6. Gieren, W. P., Fouqué, P. & Gómez, M.: ApJ **496**, 17 (1998)
7. Kervella, P., Bersier, D., Mourard, D., Nardetto, N. & Coudé du Foresto, V.: A&A **423**, 327 (2004)
8. Kervella, P., Bersier, D., Mourard, D., et al.: A&A **428**, 587 (2004)
9. Kervella, P., Fouqué, P., Storm, J., et al.: ApJ **604**, 113 (2004)
10. Kervella, P., Nardetto, N., Bersier, D., Mourard, D. & Coudé du Foresto, V.: A&A **416**, 941 (2004)
11. Kervella, P., Ségransan D., & Coudé du Foresto, V.: A&A **425**, 1161 (2004)
12. Koehler, B., Flebus, C., Dierickx, P., et al.: ESO Messenger **110**, 21 (2002)
13. Koehler, B., Kraus, M., Moresmau, J.-M., et al.: SPIE **5491**, 600 (2004)
14. Lane, B. F., Creech-Eakman, M. & Nordgren, T. E.: ApJ **573**, 330 (2002)
15. Lanoix, P., Paturel, G. & Garnier, R.: MNRAS **308**, 969 (1999)
16. Marengo, M., Sasselov, D. D., Karovska, M. & Papaliolios, C.: ApJ **567**, 1131 (2002)
17. Mérand, A., Kervella, P., Coudé du Foresto, V., et al.: A&A **438**, L9 (2005)
18. Nardetto, N., Fokin, A., Mourard, D., et al.: A&A **428**, 131 (2004)
19. Perryman, M.A.C., de Boer, K., Gilmore, G., et al.: A&A **369**, 339 (2001)
20. Petrov, R., Malbet, F., Richichi, A., et al.: SPIE **4006**, 68 (2000)
21. Sasselov, D. D. & Karovska M.: ApJ **432**, 367 (1994)
22. Storm, J., Carney, B. W., Gieren, W. P., et al.: A&A **415**, 531 (2004)
23. Szabados, L.: IBVS **5394**, 1 (2003), *http://www.konkoly.hu/CEP/intro.html*
24. Tammann, G. A., Sandage, A. & Reindl, B.: A&A **404**, 423 (2003)
25. Taylor, M. M. & Booth A. J.: MNRAS **298**, 594 (1998)
26. Wesselink, A.: Bull. Astron. Inst. Netherlands **10**, 91 (1946)

The *K*-Band Intensity Profile of R Leonis Probed by VLTI/VINCI

D. Fedele[1,2], M. Wittkowski[1], F. Paresce[1], M. Scholz[3,4], P. R. Wood[5], and S. Ciroi[2]

[1] European Southern Observatory, Garching bei München, Germany
dfedele@eso.org
[2] Dipartimento di Astronomia, Università di Padova, Italy
[3] Institut für Theoretische Astrophysik der Universität Heidelberg, Germany
[4] School of Physics, University of Sydney, Australia
[5] Research School for Astronomy and Astrophysics, Australian National University, Canberra, Australia

Summary. We present near-infrared *K*-band interferometric measurements of the Mira star R Leonis obtained in April 2001 and January 2002 with VLTI/VINCI. The April 2001 measurements indicate a center-to-limb intensity variation (CLV) that is clearly different from a uniform disk (UD) intensity profile. We show that these measured visibility values are consistent with predictions from recent self-excited dynamic Mira model atmospheres. We derived high-precision Rosseland diameters for the two epochs and, together with literature estimates of the distance and the bolometric flux, we find linear radii of $350^{+50}_{-40}\,R_\odot$ and $320^{+50}_{-40}\,R_\odot$ and effective temperatures of $2930 \pm 270\,\mathrm{K}$ and $3080 \pm 310\,\mathrm{K}$, respectively.

1 Introduction

Mira stars are cool, low-mass, pulsating variables located on the asympotic giant branch of the Hertzsprung-Russel diagram and that exhibit a conspicous mass-loss. Because of the low temperatures, molecules are present in their extended atmospheres, and dust is formed at larger distances from the star. In this paper, we present a comparison of near-infrared *K*-band VLTI/VINCI interferometric observations of R Leo with predictions by self-excited dynamic Mira model atmosphere ([10, 23, 13, 14]).

R Leo is an oxygen-rich Mira star with spectral type M6e-M8IIIe-M9.5e, a period of 310 days, a *V* magnitude of 4.4-11.3 ([8]), and a mass-loss rate of $\sim 1 \times 10^{-7}\,M_\odot/\mathrm{yr}$ ([3, 9]). [20] measured a relatively low dust emission coefficient of 0.23, i.e. the ratio of the total emission of the dust to the total emission of the star in the mid-infrared. We use a parallax value of $8.81 \pm 1.00\,\mathrm{mas}$ as given by [26], which is the weighted average of the values by [5] and [19]. [27] derived a mean bolometric magnitude $m_{\mathrm{bol}}=0.65$ with total (peak-to-peak) amplitude $\Delta m_{\mathrm{bol}}=0.63$.

2 Observations

The R Leo interferometric data were obtained on 1 & 3 April 2001 (JD = 2452003, stellar phase $\phi_{\mathrm{vis}} = 0.08$) and on 20 January 2002 (JD = 2452295, $\phi_{\mathrm{vis}} = 1.02$) with the ESO Very Large Telescope Interferometer (VLTI) equipped with the K-band commissioning instrument VINCI ([6]). The VLTI test siderostats were used on stations E0 and G0 forming an unprojected ground baseline length of 16 m. The calibration of the visibility values was performed as described in [25], using a weighted average of all transfer function values obtained during the night. In order to derive effective temperatures from the measured angular radius and the bolometric flux we use the mean bolometric magnitude and its amplitude given by [27].

3 Comparison with models

Our measured R Leo squared visibility amplitudes are shown in Fig. 1 together with a typical model prediction for each epoch based on the P model series by [10, 23, 13, 14]. The P model series is complete self-excited dynamic model atmospheres based on fundamental mode pulsation. For the details of the model calculations, we refer [10]. The parent star of the here considered P series has solar metallicity, luminosity $L/L_{\odot}$=3470, period 332 days, mass $M/M_{\odot}$=1.0, radius $R/R_{\odot}$=241 ([10]). In order to characterize the angular diameter of the fitted CLV, any well-defined reference radius of the model CLV can be used, such as the Rosseland radius or the 1.04 μm continuum radius. Physically most meaningful may be a true continuum radius, such as the 1.04 μm radius, which is not affected by time variable molecular contamination (see [10], [15, 13, 14]). In the following, the Rosseland radius is mainly used as reference quantity, as is usual in the literature. The Rosseland angular diameters we found for the two epochs are

Θ_{Ross}(April 2001, phase 0.08) = 28.5 ± 0.4 mas and
Θ_{Ross}(January 2002, phase 1.02) = 26.2 ± 0.8 mas.

The Rosseland angular diameter at the variability phase closer to the maximum (1.02) is smaller by $\sim 8\,\%$ than that at variability phase 0.08. This is consistent with pulsation models (see, e.g. [13] and [14]). Together with the adopted values for π and f_{bol}, these angular diameters for April 2001 and January 2002 correspond to linear radii of $350^{+50}_{-40}\,R_{\odot}$ and $320^{+50}_{-40}\,R_{\odot}$, and to effective temperatures of 2930 ± 270 K and 3080 ± 310 K, respectively.

4 Conclusions

We have compared VLTI/VINCI observations of the Mira star R Leonis to recent self-excited dynamic models. We find that these model CLVs for the

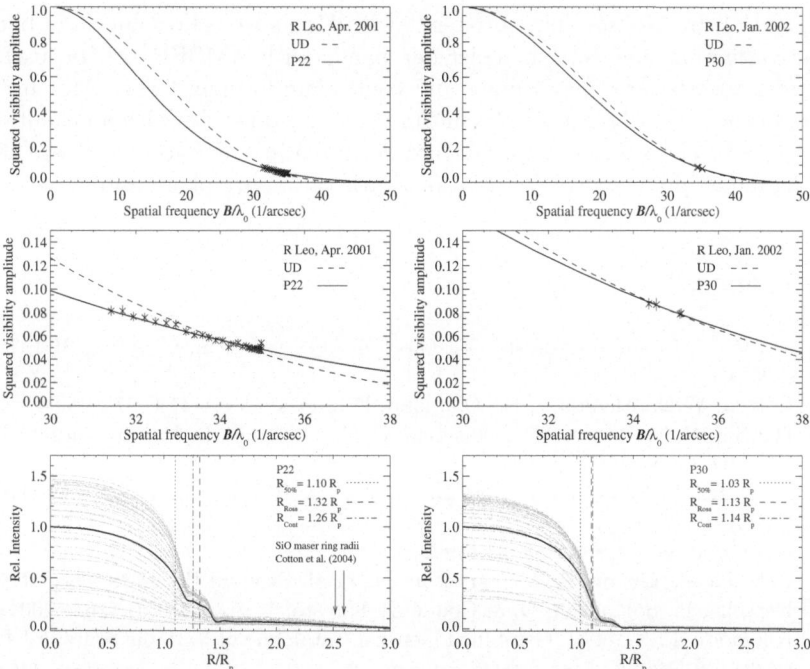

Fig. 1. (Top) Measured R Leo squared visibility amplitudes obtained in April 2001 and January 2002, together with the well fitting P22 and P30 models predictions. For comparison, the UD curve is shown as well. (Bottom) CLV prediction from the P22 and P30 model corresponding to the model visibility curves above. The thin lines denote the monochromatic CLVs while the thick line denotes the CLV averaged over the VINCI sensitivity function. Indicated are also the Rosseland radius, the $1.04\mu m$ true continuum radius, as well as the radius at which the filter-averaged CLV drops by 50%. The mean SiO maser ring radii measured by [2] close in time to our April 2001 data are indicated by the arrows.

phases of our observations are consistent with our measurements. The correspondence of our obtained linear radii with model radii of the fundamental mode pulsation models used is in agreement with the general recent conclusions that Mira stars pulsate in fundamental mode. Two recent works ([28] and [17]) show accordance with predictions by the P-model series and clear deviation from UD profile for post-maximum observations of o Cet and R Leo. It is remarkable that the two stars appear to show similar CLVs, in agreement with the P model series, while their lightcurves and variability amplitudes are different. These findings increase our confidence in these dynamic Mira star models, which are often used to transform broad-band filter-specific UD diameters into more meaningful Rosseland or continuum diameters. More detailed observations are desirable in the future in order to better constrain the models. Such observations should probe the CLV at a larger range of spatial frequencies. In addition, measurements with high spectral resolution in both true continuum and certain molecular bands with AMBER will be useful in order to separate line-forming and continuum-forming layers. Moreover, monitoring of the observed CLVs in time over several cycles with a resolution of $\sim 10\%$ of the variability period are desirable in order to investigate the strong model-predicted CLV variations with variability phase and cycle.

References

1. Bordé, P., Coudé du Foresto, V., Chagnon, G., & Perrin, G. A&A, **393**, 183 (2002)
2. Cotton, W.D., Menesson, B., Diamond, P.J., et al. A&A **414**, 275 (2004)
3. Danchi, W. C., Bester, M., Degiacomi, C. G., Greenhill, L. J., & Townes, C. H. AJ, **107**, 1469 (1994)
4. Dyck, H. M., Benson, J. A., van Belle, G. T., & Ridgway, S. T. AJ, **111**, 1705 (1996)
5. Gatewood, G. PASP, **104**, 23 (1992)
6. Kervella, P., Gitton, P., & Ségransan, D., et al. Proc. SPIE, **4838**, 858 (2003)
7. Kervella, P., Ségransan, D. & Coudé du Foresto, V. A&A, **425**, 1161 (2004)
8. Kholopov, P. N., et al. Combined General Catalogue of Variable Stars, 4.1 Ed (II/214A).(1998)
9. Knapp, G. R., Young, K., Lee, E., & Jorissen, A. ApJS, **117**, 209 (1998)
10. Hofmann, K.-H., Scholz, M., & Wood, P.R.A&A, **339**, 846 (1998)
11. Hofmann, K.-H., Balega, Y., Scholz, M., & Weigelt, G. A&A, **376**, 518 (2001)
12. Hofmann, K.-H., Beckmann, U., Blöcker, T., et al. New A, **7**, 9 (2002)
13. Ireland, M. J., Scholz, M., Wood, P. R., MNRAS, **352**, 318 (2004)
14. Ireland, M. J., Scholz, M., Tuthill, P. G., & Wood. P. R., MNRAS, **355**, 444 (2005)
15. Jacob, A. P. & Scholz, M. MNRAS, **336**, 1377 (2002)
16. Mennesson, B., Perrin, G., Chagnon, G., et al. ApJ, **579**, 446 (2002)
17. Perrin, G., Coudé du Foresto, V., Ridgway, S.T., et al. A&A, **345**, 221 (1999)
18. Perrin, G., Ridgway, S.T., Mennesson, B., et al. A&A, **426**, 279 (2004)
19. Perryman, M. A. C. & ESA 1997, The Hipparcos and Tycho catalogues, Publisher: Noordwijk, Netherlands: ESA Publications Division, 1997, Series: ESA SP Series vol no: 1200, ISBN: 9290923997
20. Sloan, G. C. & Price, S. D. ApJS, **119**, 141 (1998)
21. Tej, A., Chandrasekhar, T., Ashok, N. M., et al. AJ, **117**, 1857 (1999)
22. Tej, A., Lançon, A., & Scholz, M. A&A, **401**, 347 (2003)
23. Tej, A., Lançon, A., Scholz, M., & Wood, P. R. A&A, **412**, 481 (2003)
24. van Belle, G. T., Thompson, R. R., & Creech-Eakman, M. J. AJ, **124**, 1706 (2002)
25. Wittkowski, M., Aufdenberg, J. P., & Kervella, P. A&A, **413**, 711 (2004)
26. Whitelock, P. & Feast, M.: MNRAS **319**, 759 (2000)
27. Whitelock, P., Marang, F., & Feast, M. MNRAS, **319**, 728 (2000)
28. Woodruff, H. C., Eberhardt, M., Driebe, T., et al A&A, **421**, 703 (2004)

Cepheids Observations Using CHARA/FLUOR : α UMi and δ Cep

Antoine Mérand[1], Pierre Kervella[1], Vincent Coudé du Foresto[1],
Stephen T. Ridgway[1,2,3], Jason Aufdenberg[2], Theo ten Brummelaar[3],
David Berger[3], Judit Sturmann[3], Lazlo Sturmann[3], Nils Turner[3], and
Harold A. McAlister[3]

[1] LESIA, UMR8109, Observatoire de Paris, 5 place Jules Janssen, 92195 Meudon,
France antoine.merand@obspm.fr
[2] National Optical Astronomical Observatory, Tucson, AZ, USA
[3] Center for High Angular Resolution Astronomy
Georgia State University, Atlanta, GA, USA

Summary. Over the past decade, stellar interferometry emerged as a powerful tool
to measure distances to Cepheids using the pulsation parallax method, also called
the Baade-Wesselink (BW) method. However, as interferometers gain in baseline
length and precision, the assumptions on which the interferometric BW method relies have to be refined: the projection factor (to convert radial velocity in pulsation
velocity) and the center to limb darkening (to properly interpret interferometric
data) have to be studied theoretically and/or observationally. We report here interferometric observations of two classical Cepheids, using the FLUOR beam combiner
[1] installed at the CHARA Array [2]. α UMi was observed in both the first and the
second lobes of the visibility curve, enabling us to potentially measure its center
to limb darkening. We measured with high precision the angular diameter variation for δ Cep. Combined with the distance to this star from HST/FGS [3], these
measurements provide directly the projection factor.

1 Cepheid distance measurements using the Interferometric Baade-Wesselink Method

This method consists of observing a pulsating star, with good phase coverage,
using both spectroscopy and interferometry. Then, one compares the radial
velocity (from the spectroscopy) of the pulsating atmosphere on the one hand,
and the stellar angular diameter (from interferometry) on the other hand to
get the distance using the following relation:

$$\theta(T) - \theta(0) = -2\frac{p}{d}\int_0^T (V_{\mathrm{rad.}}(t) - \gamma)dt \qquad (1)$$

where θ is the angular diameter, p is the projection factor, d is the distance,
$V_{\mathrm{rad.}}$ is the radial velocity as measured by the spectrograph and γ is the
systematic radial velocity of the object. This method relies on two major
assumptions:

– Spectroscopists have to convert the absorption line displacement (radial velocity) into photospheric displacement. The ratio between the two is called the projection factor, because it accounts mainly for the sphericity of the star;
– Interferometrists have to convert the visibility measurements into angular diameters, which requires a correction for center to limb darkening.

The two parameters of the method, the projection factor, or p-factor, and the center to limb darkening, are usually obtained from stellar models. Reaching the precision where the uncertainty on these quantities can not be neglected, two approaches are possible to break the current model-dependent limit. One possibility is to push further the refinement in the modeling of Cepheids, by coupling hydrodynamic and radiative transfer (e.g. [4, 5, 6]). The other possibility is to directly measure these two parameters.

We report here the observations of two Cepheids, α UMi and δ Cep with CHARA/FLUOR. α UMi has been observed to derive the center to limb darkening, whereas the BW method has been applied to δ Cep, for which the distance is known with high precision [3], in order to derive observationally the p-factor.

2 α UMi and its visibility profile

Sensing the stellar center to limb darkening can be done if the star is very resolved by a stellar interferometer, that means going in the so-called second lobe of the visibility profile. This requires larger baselines than the ones used to sense the diameter variations. Current facilities are limited in baseline, so the number of stars for which such a study can be done is limited. Cepheids, because of the rarity, tend to be far from us and have small angular diameters. The longest baselines are required to reach their second lobe. The only currently available combination of Cepheid and interferometric facility for the second lobe measurment in the northern hemisphere is α UMi as observed by the CHARA Array. We present here observations of α UMi obtained with the FLUOR beam combiner hosted by the CHARA Array.

α UMi is a complex object: not only it is a Population I Cepheid, it is also a spectroscopic binary. Observing it with a stellar interferometer is interesting because both the pulsation and the companion could possibly be detected. Based on other observational evidence [7], it is clear that the companion is currently beyond detection : the magnitude difference between α UMi and its companion is believed to be 6.5 in the K band. Our sensitivity threshold is of the order of the precision of a single visibility measurement, namely 1%. Concerning the pulsation, according to the latest radial velocity measurements, the amplitude in terms of radius variation is of the order of 0.5%, again, beyond our sensitivity limit. This is why we consider these two effects not to have been detected in our observations and we will consequently model α UMi as a single, stable projected disk.

Our data set consists of 61 individual visibility measurements obtained at 3 different baselines (59m, 145m and 245m projected baseline). Because α UMi is located at the north celestial pole, the projected baseline does not vary in length, only in projection angle. The smaller baseline (59m) reaches the middle of the first lobe, the medium (145m) reaches near the minimum while the largest reaches the second lobe (245m).

Although a Cepheid, α UMi is close in many aspects to a non-pulsating yellow super-giant star [7]. We thus compared the visibility data points we obtained on α UMi with prediction of hydrostatic models of a star with same effective temperature, surface gravity and chemical composition [8]. Doing so, the reduced χ^2 improves from 14 (uniform disk) to 3 (limb darkened disk) but is still not satisfactory (Fig. 1). We fitted a generic analytical center to limb darkening profile to the data. Using the power law center to limb darkening profile [9] ($I(\mu) = \mu^a lpha$, $\alpha = 0$ corresponds to a udinofrm disk), the disk is a lot more darkened as expected ($\alpha \approx 0.25$ instead of $\alpha \approx 0.14$), the model does not reproduce well the visibility measurement in the first lobe (B$\approx$59 m).

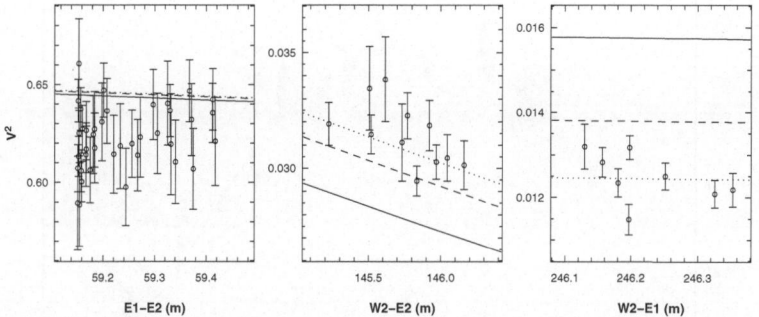

Fig. 1. Fit of α UMi data, neglecting the companion and the pulsation (see text). Squared visibilities as a function of baselines (named after the telescope pairs: E1-E2, W2-E2, W2-E1). Solid line is a uniform disk fit, dashed line is the expected limb darkening from an hydrostatic model for a yellow supergiant, dotted line is an *ad hoc* limb darkening obtained by adjusting a one-parameter analytical profile.

As seen in Fig. 1, the visibilities are systematically lower than the models at the 59m baseline. This is because the visibility profile is not sensitive to limb darkening at low spatial resolution. In order to reproduce the α UMi visibility profile, one has to invoke something that will result in visibility changes a low resolution, namely something significantly larger than the star itself. We are currently investigating this possibility.

3 The pulsation of δ Cep

δ Cep, a 5.36 day period population I Cepheid, is one the brightest classical Cepheids in the northern hemisphere, and has been observed and studied extensively in the past decades. High precision radial velocity measurements are available from the literature (e.g. [10]). Although δ Cep is one of the closest Cepheids, its parallax was measured with a great uncertainty with HIPPARCOS. Recently, a more precise measurement with the HST/FGS interferometer was obtained: 273 ± 11 pc [3]. It is thus possible to estimate the projection factor for this star. We obtained high precision angular diameter measurements using the 313m and 245m baselines of the CHARA Array, linked to the FLUOR beam combiner. Using cross correlation radial velocities [10], our distance determination using the BW method has a precision of 2%, assuming a perfectly known p-factor. On the other hand, using the known distance from HST/FGS, we can directly measure p. Our estimate of p is 1.27 ± 0.06 (Fig. 2), the error bar accounting for all the source of uncertainty: radial velocities, interferometric measurements and HST/FGS distance. The later one is the major contributor with 0.05 out of 0.06.

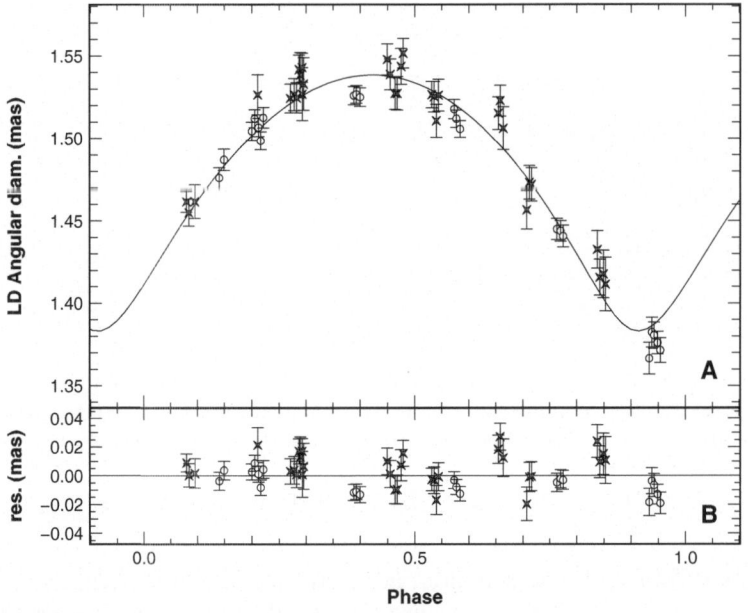

Fig. 2. δ Cep pulsation. Limb darkened diameter as a function of pulsation phase (A). Circles correspond to the 313m baseline, whereas crosses correspond to the 245m baseline. The solid line is not a fit but results from the integration of the radial velocity from Bersier *et al.* (1994) [10], using HST/FGS distance, 1.475 mas (milli-arc second) for the average angular diameter and 1.27 for the projection factor. (B) residuals.

Until now, the p factor usually adopted for the BW method is 1.36 [11]. Attempts to fully model the hydrodynamics of the pulsation and the radiative tranfer have been made. The computed p-factors vary from 1.27 [6] to 1.47 [4]. Our estimation appears to be in the lower range.

4 Conclusion

We have presented preliminary results of high precision long baseline interferometry observations of Cepheids in the framework of the distance determination using the Baade-Wesselink method. α UMi shows a unexpected visibility profile, not explained by either hydrostatic limb darkening models or ad hoc LD profiles. It is more likely that this star is surrounded by an over-resolved component, hence much larger than the Cepheid itself, accounting for a few percent of the flux of the star. δ Cep's angular diameter was observed with high precision. By fixing the distance to the value determined geometrically by HST/FGS, we have directly determined the projection factor and found a slightly smaller p than usually adopted: 1.27 ± 0.06 instead of 1.36.

References

1. Coudé du Foresto *et al.* In *Proceedings of the SPIE, Volume 4838.*, pages 280, 2003.
2. T.A. ten Brummelaar *et al.* *ApJ*, 628:453, 2005.
3. G.F. Benedict *et al.* *Aj*, 124:1695, 2002.
4. C. N. Sabbey, D. D. Sasselov, and M. S. Fieldus *et al.* *ApJ*, 446:250, 1995.
5. M. Marengo, M. Karovska, and D. D. Sasselov *et al.* *ApJ*, 589:968, 2003.
6. N. Nardetto, A. Fokin, and D. Mourard *et al.* *A&A*, 428:131, 2004.
7. N.R. Evans, D.D. Sasselov, and C.I. Short. *ApJ*, 567:1121, 2002.
8. A. Claret. *A&A*, 363:1081, 2000.
9. D. Hestroffer. *A&A*, 327:199, 1997.
10. D. Bersier, G. Burki, M. Mayor, and A. Duquennoy. *A&A*, 108:25, 1994.
11. G. Burki *et al.* *A&A*, 109:258, 1982.

Cepheid Observations with the Sydney University Stellar Interferometer: ℓ Carinae and β Doradus

J. Davis, M. J. Ireland, A. P. Jacob, J. R. North, S. M. Owens, J. G. Robertson, W. J. Tango, and P. G. Tuthill

School of Physics, University of Sydney, NSW 2006, Australia
davis@physics.usyd.edu.au

Summary. Observations of the southern Cepheids ℓ Car and β Dor to yield the mean angular diameters and angular pulsation amplitudes have been made with the Sydney University Stellar Interferometer (SUSI) at a wavelength of 700 nm. The results of a preliminary analysis are compared with those obtained with the VLTI at 2.2 μm and excellent agreement between the results from the two instruments is found for ℓ Car but there are significant differences for β Dor.

1 Introduction

One of the prime programmes for which the Sydney University Stellar Interferometer (SUSI) [2, 3] was developed was the measurement of the mean diameters and angular pulsation amplitudes of Cepheids but, in its initial configuration, it lacked the required sensitivity. The recent commissioning of a red beam-combination system in SUSI has resulted in a significant increase in sensitivity. In this paper we report SUSI measurements of the mean angular diameters and angular pulsation amplitudes of ℓ Car and β Dor, two of the seven Cepheids measured with the VLTI by Kervella et al. [5]. A preliminary analysis of the observational data has been carried out and a comparison is made with the VLTI results.

2 Observations

Observations were made on 26 nights for ℓ Car. A 40 m baseline was used for all the observations except for one night when a 30 m baseline was used. For β Dor, observations were made on 22 nights—2 nights with a 60 m baseline and the remaining 20 nights with a 40 m baseline. The projected baseline was less than the fixed ground baseline given above due to the southerly declinations of the Cepheids and calibrators—ℓ Car and β Dor are both at a declination of approximately -62.5 deg and the projected baseline is ∼0.85 times the ground baseline at transit. Observations were generally restricted to within ±2 hours of the meridian.

The detection system employed the fringe-scanning technique with matched interference filters in the output beams from the beam-combiner. The spectral passband of the filters was 80 nm centred on 700 nm.

Calibrators were selected as close in the sky as possible to each Cepheid with the additional requirement of being minimally resolved. The limiting magnitude of SUSI at 700 nm is ~+5 which limited the choice of calibrators and compromise was necessary. The calibrators used are listed in Table 1.

Table 1. Calibrators used for the observations of ℓ Car and β Dor. m_{700} is the estimated magnitude at 700 nm, θ_{UD} the adopted uniform disk angular diameter at 700 nm and Ω the angular distance of the calibrator from the target Cepheid.

HR	Star	Spectral Type	m_{700}	θ_{UD} (mas)	Ω (deg)	Target
2020	β Pic	A5 V	3.8	0.72	11.5	β Dor
2550	α Pic	A7 IV	3.1	1.01	8.6	β Dor
3685	β Car	A2 IV	1.6	1.59	7.9	l Car
3699	ι Car	A8 Ib	2.0	1.54	4.7	l Car
4114	s Car	F2 II	3.5	0.91	6.5	l Car

Observations of calibrators and Cepheids were alternated in each observing session so that every Cepheid observation was bracketed by observations of calibrators. Each observation contained a set of 1000 scans each 140 μm long and made up of 1024 by 0.2 ms samples. Each scan set was followed by smaller sets of photometric and dark scans. One complete observation of a Cepheid bracketed by calibrators took a total of ~18 minutes.

3 Analysis

The initial analysis of the fringe scans for both Cepheids and calibrators was carried out in a software "pipeline" which outputs the raw and calibrated correlation (equal to visibility^2), projected baseline, hour angle, fluxes etc. for each set of scans. The output file from the pipeline was imported into an Excel spreadsheet for examination of the data and for further analysis to compute mean values for the calibrated correlation, hour angle and projected baseline.

The uniform-disk angular diameter (θ_{UD}) for each night was determined by fitting the equation

$$\theta_{UD} = \left| \frac{2J_1(x)}{x} \right|^2 \tag{1}$$

to the mean value for the calibrated correlation assuming that the correlation at zero baseline was unity. In equation (1) $J_1(x)$ is a Bessel function and

$x = \pi.b.\theta_{\mathrm{UD}}/\lambda$ where b is the projected baseline and λ is the wavelength of the observation.

The uniform-disk angular diameters have been converted to limb-darkened angular diameters (θ_{LD}) using limb-darkening correction factors (ρ) from the tabulation computed for Kurucz model atmospheres by Davis, Tango & Booth [1]. The values of ρ are tabulated as a function of λ, T_{eff}, log g, and [M/H]. Values for T_{eff} and log g, as a function of pulsation phase, have been taken from Taylor [8] and we have adopted the same values for [M/H] as Kervella et al. [5]. The limb-darkened angular diameters are plotted as a function of phase in Fig. 1 (a) for ℓ Car and in Fig. 2 (a) for β Dor.

The relationship at phase ϕ_i between the limb-darkened angular diameter $\theta_{\mathrm{LD}}(\phi_i)$ in mas and the radial displacement of the Cepheid surface $\Delta R(\phi_i)$ in solar radii is given by

$$\theta_{\mathrm{LD}}(\phi_i) = \overline{\theta}_{\mathrm{LD}} + 9.305 \left(\frac{\Delta R(\phi_i)}{d} \right) \qquad (2)$$

where $\overline{\theta}_{\mathrm{LD}}$ is the average limb-darkened angular diameter in mas and d is the distance in pc.

The spectroscopically determined radial displacements of the surfaces of the two Cepheids as a function of phase have been published by Taylor et al. [6] for ℓ Car and by Taylor & Booth [7] for β Dor and these have been adopted for combination with the angular diameter measurements.

The uncertainties in $\Delta R(\phi_i)$ are small compared with the uncertainties in the observed values of $\theta_{\mathrm{LD}}(\phi_i)$ and a linear least squares fit has been made with equation (2) to determine $\overline{\theta}_{\mathrm{LD}}$ and d for both Cepheids. The results are discussed in the next two sections.

4 ℓ Car

The resulting curve from fitting equation (2) to the ℓ Car data is plotted in Fig. 1 (a) with the SUSI angular diameter values.

Kervella et al. [4, 5] used a different source for LD corrections and different combinations of radial velocity data for the radial displacements of the surfaces of ℓ Car and β Dor. In order for a direct comparison to be made between the SUSI and VLTI results we have taken the uniform-disk angular diameters from Kervella et al. [5] and applied LD corrections from the Davis et al. [1] tabulation and used the Taylor et al. [6] radial displacements in equation (2) exactly as for the SUSI analysis. The plot of VLTI angular diameters and fitted curve as a function of phase is shown in Fig: 1 (b). The agreement between the SUSI 700 nm and VLTI 2.2 μm results is excellent and in Fig. 1 (c) the SUSI and VLTI angular diameter values are plotted together with the curve fitted to the combined data. This third plot emphasizes the excellent agreement between the results from the two instruments for ℓ Car.

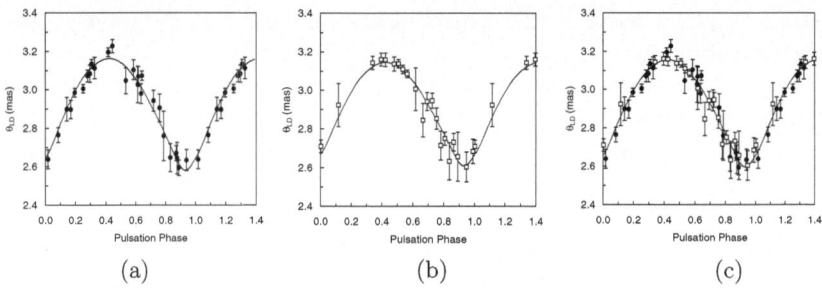

Fig. 1. Limb-darkened angular diameters of ℓ Car as a function of pulsation phase: (a) SUSI results at 700 nm; (b) VLTI results at 2.2 μm; (c) SUSI and VLTI results combined. Further details are given in the text.

Table 2. The SUSI and VLTI rows contain the results of fits using equation (2) as described in the text. The Combined row contains the results of the fit to the combined SUSI and VLTI data. Note that the uncertainties are underestimated as no allowance has been made for systematic uncertainties due, for example, to the angular diameters adopted for the calibrators.

Data Source	ℓ Car		β Dor	
	$\overline{\theta}_{LD}$ (mas)	d (pc)	$\overline{\theta}_{LD}$ (mas)	d (pc)
SUSI	2.873 ± 0.009	525 ± 24	1.744 ± 0.018	265 ± 50
VLTI	2.878 ± 0.011	568 ± 33	1.882 ± 0.020	400 ± 130
Combined	2.876 ± 0.004	545 ± 11	–	–
Hipparcos	–	460 ± 100	–	320 ± 60

5 β Dor

The observational data from SUSI and from the VLTI for β Dor have been analysed following exactly the same procedure as for ℓ Car. The radial displacements for β Dor have been taken from Taylor & Booth [7]. Figure 2 (a) shows the SUSI results and Fig. 2 (b) the VLTI results. It is clear that there are significant differences between the SUSI and VLTI results. The SUSI results give a smaller mean angular diameter and a larger angular diameter pulsation amplitude. At this stage we do not have an explanation and know of no reason for doubting the SUSI results.

6 Summary

Excellent agreement has been demonstrated between the SUSI and VLTI results for ℓ Car but there appear to be significant differences between the mean angular diameter and the amplitude of angular diameter changes for

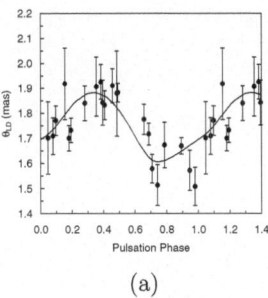
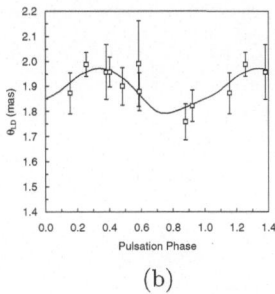

(a)   (b)

Fig. 2. Limb-darkened angular diameters of β Dor as a function of pulsation phase: (a) SUSI results at 700 nm; (b) VLTI results at 2.2 μm. Further details are given in the text.

β Dor between the SUSI and VLTI results. The preliminary nature of the analysis of the SUSI data is emphasized. Refinement of the calibration of the observational data, the limb-darkening corrections, and the radial displacement data is in progress. We stress that our treatment of the VLTI results was made solely for direct comparison with the SUSI results and should not be used as an alternative to the results published by Kervella et al. [4, 5].

Acknowledgements The SUSI programme is funded jointly by the Australian Research Council and the University of Sydney. MI acknowledges the support of an Australian Postgraduate Award, JN the support of a University Postgraduate Award, and AJ and SMO the support of Dennison Postgraduate Awards.

References

1. J. Davis, W.J. Tango, A.J. Booth: MNRAS **318**, 387 (2000)
2. J. Davis, W.J. Tango, A.J. Booth, T.A. ten Brummelaar, R.A. Minard and S.M. Owens: MNRAS **303**, 773 (1999a)
3. J. Davis, W.J. Tango, A.J. Booth, E.D. Thorvaldson and J. Giovannis: MNRAS **303**, 783 (1999b)
4. P. Kervella, P. Fouqué, J. Storm, W.P. Gieren, D. Bersier, D. Mourard, N. Nardetto and V. Coudé du Foresto: ApJ **604**, L113 (2004)
5. P. Kervella, N. Nardetto, D. Bersier, D. Mourard and V. Coudé du Foresto: A&A **416**, 941 (2004)
6. M.M. Taylor, M.D. Albrow, A.J. Booth and P.L. Cottrell: MNRAS **292**, 662 (1997)
7. M.M. Taylor and A.J. Booth: MNRAS **298**, 594 (1998)
8. M.M. Taylor: Analysis of Cepheid spectra. PhD thesis, University of Sydney, Sydney (1999)

The Circumstellar Environment of Evolved Stars as seen by VLTI/MIDI

K. Ohnaka[1], J. Bergeat[2], T. Driebe[1], U. Graser[3], K.-H. Hofmann[1],
R. Köhler[3], Ch. Leinert[3], B. Lopez[4], F. Malbet[5], S. Morel[6], F. Paresce[7],
G. Perrin[8], Th. Preibisch[1], A. Richichi[7], D. Schertl[1], M. Schöller[6], H. Sol[8],
G. Weigelt[1], and M. Wittkowski[7]

[1] Max-Planck-Institut für Radioastronomie, Bonn, Germany
    kohnaka@mpifr-bonn.mpg.de
[2] Observatoire de Lyon, St.-Genis-Laval, France
[3] Max-Planck-Institut für Astronomie, Heidelberg, Germany
[4] Observatoire de la Côte d'Azur, Nice, France
[5] Laboratoire d'Astrophysique, Observatoire de Grenoble, Grenoble, France
[6] European Southern Observatory, Santiago, Chile
[7] European Southern Observatory, Garching, Germany
[8] Observatoire de Paris-Meudon, Meudon, France

Summary. We present the results of the first mid-infrared interferometric observations of the Mira variable RR Sco with the VLTI/MIDI, together with K-band observations using VLTI/VINCI. The uniform-disk diameter was found to be 18 mas between 8 and 10 μm, while it gradually increases at wavelengths longer than 10 μm to reach 24 mas at 13 μm. These uniform-disk diameters in the mid-infrared are significantly larger than the K-band uniform-disk diameter of 10.2 ± 0.5 mas measured using VLTI/VINCI, three weeks after the MIDI observations. Our model calculations show that optically thick emission from a warm molecular envelope consisting of H_2O and SiO can cause the apparent mid-infrared diameter to be much larger than the continuum diameter, and this can explain the mid-infrared angular sizes roughly twice as large as that measured in the K band. The observed increase of the uniform-disk diameter longward of 10 μm can be explained by an optically thin dust shell consisting of corundum and silicate grains.

1 Introduction

Mass loss in asymptotic giant branch (AGB) stars is believed to play an important role in the chemical evolution of the Galaxy, since nuclear processed material is dredged up to the surface and finally returned to the interstellar space via mass loss. However, the mass loss mechanism in AGB stars is not yet fully understood. In particular, the understanding of the region where material is accelerated is meager. In order to better understand the mass loss phenomenon in Mira-type AGB stars, it is crucial to obtain a comprehensive picture of the region where mass outflows are expected to be initiated, that is, the region between the top of the photosphere and the inner edge of the expanding dust shell. Mid-infrared interferometry provides a unique opportunity to probe the circumstellar environment of Mira variables. Since a large

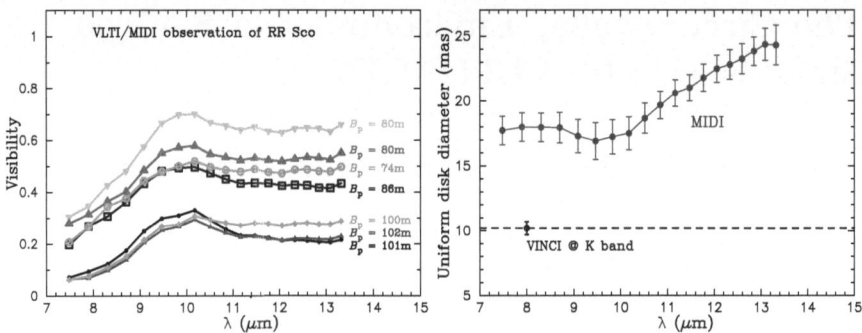

Fig. 1. Left: Visibility as a function of wavelength for different projected baseline lengths ranging from 74 m to 102 m. All curves show a similar shape: a gradually increasing part shortward of 10 μm and a roughly constant part longward of 10 μm. The errors of the calibrated visibilities are typically 10–15%, but the error bars are omitted in this panel for the sake of clarity. **Right:** Uniform-disk diameter as a function of wavelength. The diameters are derived from uniform-disk fits using all seven visibility points at each wavelength. The K-band uniform-disk diameter measured with VINCI is also plotted.

fraction of mid-infrared photons originate in regions cooler than the photosphere, mid-infrared interferometry is well suited for studying the outer atmosphere and the circumstellar dust shell, where complicated, mutually coupled physical and chemical processes take place, finally leading to the onset of mass outflows.

The MIDI instrument at VLTI has a particularly great potential with its spectro-interferometric capability, which enables us to directly observe the spatial structures of molecular and dust formation regions. We present the first spectrally dispersed N-band interferometric observations of the Mira variable RR Sco with VLTI/MIDI and, in addition, K-broadband observations with VLTI/VINCI.

2 MIDI and VINCI observations

RR Sco was observed with MIDI at variability phase 0.6 on three consecutive nights in June 2003 within the framework of the Science Demonstration Time (SDT) program. A prism with a spectral resolution of $\lambda/\Delta\lambda \simeq 30$ was used to obtain spectrally dispersed fringes. In total, seven observations were carried out using the 102 m baseline between the telescopes UT1 and UT3. Due to projection effects, the projected baseline lengths range between 74 and 102 m. For the data reduction, we used the MIDI software package based on the power spectrum analysis (Leinert et al. [4]). The calibrated visibilities of RR Sco are plotted in Fig. 1a. We also fitted all seven visibility points at each wavelength with a uniform disk and the derived uniform-disk diameters are plotted in Fig. 1b as a function of wavelength.

In addition to the MIDI measurements, we used a total of five K-band VLTI/VINCI observations of RR Sco which are publicly available in the ESO archive. These VINCI observations were carried out on 2003 July 10 and 2003 July 11, roughly three weeks after the MIDI SDT observations. The two VLTI siderostats on stations E0 and G0 were used, forming a baseline length of 16 m. We fitted the observed visibility points with a uniform disk, as described in Wittkowski et al. [10]. The fit results in a uniform-disk diameter $d_{UD} = 10.2\pm0.5$ mas which is significantly smaller than the diameters derived with MIDI in the mid-infrared (see Fig. 1b).

3 Modeling of the observed N-band and K-band visibilities

In the present work, we attempt to interpret the observed visibility using a simple model of the warm molecular envelope and an optically thin dust shell. We approximate the star with a blackbody of 3000 K and a radius $R_\star$. The star is surrounded by a warm molecular envelope consisting of H_2O and SiO gas with a constant temperature and density, extending to R_{mol}. The inner radius of the molecular envelope is set to be equal to $R_\star$. The input parameters of our model are the outer radius and the temperature of the molecular envelope (R_{mol} and T_{mol}, respectively) as well as the column densities of H_2O and SiO (N_{H_2O} and N_{SiO}, respectively) in the radial direction.

We estimate the stellar radius $R_\star$ from the K-band uniform-disk diameter measured with VINCI. The analysis of VINCI data on the prototypical Mira o Cet by Woodruff et al. [11] shows that the uniform-disk diameter of o Cet at phase 0.4 is 33.3 mas, while the continuum diameter at 1.04 μm at the same phase is 29.5 mas, which is 13% smaller than the uniform-disk diameter. We apply this conversion factor to the K-band uniform-disk diameter of 10.2 mas obtained with VINCI. This results in a 1.04 μm continuum diameter of 9.0 mas, which corresponds to a radius of 4.5 mas. We adopt this 1.04 μm continuum radius as the angular radius corresponding to the stellar radius $R_\star$.

We first calculate the line opacity due to H_2O and SiO in the wavelength range between 8 and 13 μm. We adopt a Gaussian line profile with a FWHM of 5 km s^{-1}, which represents the thermal and (micro-)turbulent velocities in the atmosphere of RR Sco, and assume that the molecular gas is in local thermodynamical equilibrium. The line list of H_2O was taken from the HITEMP database (Rothman [8]), while the line list of the fundamental bands of ^{28}SiO was generated from the Dunham coefficients given by Lovas et al. [5] and the dipole moment matrix elements derived by Tipping & Chackerian [9].

We calculate the contribution of the optically thin dust shell based on a simple, spherical model consisting of a mixture of silicate and corundum dust.

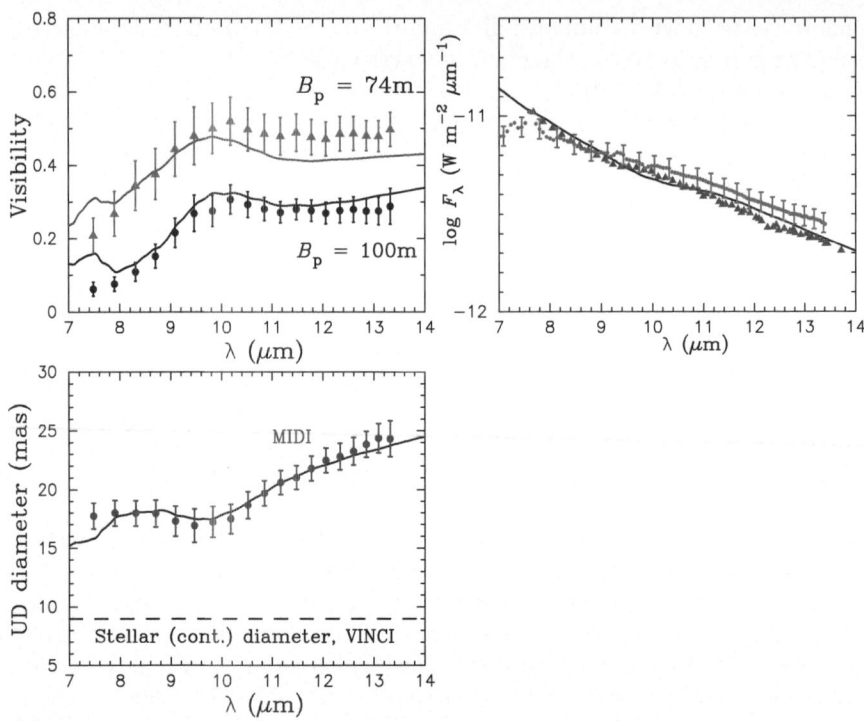

Fig. 2. Upper left: The filled circles and triangles represent the visibilities observed with projected baseline lengths of 99.9 m and 73.7 m, respectively, while the corresponding predicted visibilities are represented with the solid lines. **Lower left**: The filled circles represent the observed uniform-disk diameters, while the solid line represents those predicted for a projected baseline length of 100 m. The dashed line represents the continuum angular diameter estimated from the VINCI observations. **Upper right**: The filled circles represent the calibrated MIDI spectrum of RR Sco obtained from the observations on 2003 Jun 14, while the filled triangles represent the IRAS LRS. The solid line represents the spectrum predicted by the best-fit model.

We calculate the opacity of silicate and corundum in the Mie theory, assuming a single grain size, $a = 0.1$ μm, and using the code published by Bohren & Huffman [1]. We add this dust shell to the warm H_2O +SiO envelope model discussed above, and the total flux and the intensity profile are the sum of the contributions of these two components. The details of the model used in the present work are described in Ohnaka et al. [7].

Figure 2 shows a comparison of the observed visibility, uniform-disk diameter, and spectrum with those predicted by the best-fit model. The parameters of the warm H_2O+SiO envelope are $T_{mol} = 1400 \pm 100$ K, $R_{mol} = 2.3 \pm 0.2$ $R_\star$, $N_{H_2O} = 3 \times 10^{21}$ cm^{-2}, and $N_{SiO} = 1 \times 10^{20}$ cm^{-2}, respectively. The uncertainties of the H_2O and SiO column densities are roughly a factor of 2 and 10, respectively. The optical depth of the dust shell is derived to be 0.025 ± 0.01 at 10 μm, which translates into 0.23 in the visual. The inner boundary of the dust shell is found to be 7.5 $R_\star$ with a dust temperature of 700 ± 200 K. We find that the observed spectrum and visibilities can be best reproduced with a dust mixture of 20% silicate and 80% corundum. Figure 2 shows that the observed visibility and uniform-disk diameter are well reproduced from 8 to 13 μm. The predicted spectrum is also in good agreement with the observed MIDI spectrum. The temperatures and column densities of H_2O and SiO gas derived here are in agreement with those previously derived for other oxygen-rich Mira variables by Yamamura et al. [12] and Matsuura et al. [6]. It is also worth noting that the radius of the molecular layer derived here, 2.3 $R_\star$, is consistent with the result obtained by Cotton et al. [2], who found that the SiO maser toward Mira variables originates in the region of 1.7–2.8 $R_\star$.

4 Concluding remarks

The present analysis demonstrates that spectro-interferometry in the mid-infrared is a powerful tool to study the warm molecular envelope, whose properties have mostly been derived from spectroscopic observations up to now. We have shown that the measurement of the wavelength dependence of the angular size over molecular as well as dust spectral features in the mid-infrared provides direct information on the geometrical extension of the warm molecular envelope, and such information, combined with spectral data, can help put more constraints on physical properties of the warm molecular envelope.

References

1. Bohren, C. F., & Huffman, D. R.: Absorption and Scattering of Light by Small Particles, Wiley, New York (1983)
2. Cotton, W. D., Mennesson, B., Diamond, P. J., et al.: A&A, **275**, 288 (2004)
3. Koike, C., Kaito, C., Yamamoto, T., et al.: Icarus, **114**, 203 (1995)
4. Leinert, Ch., van Boekel, R., Waters, L. B. F. M., et al.: A&A, **423**, 537 (2004)
5. Lovas, F. J., Maki, A. G., & Olson, W. B.: J. Mol. Spectros., **87**, 449 (1981)
6. Matsuura, M., Yamamura, I., Cami, J., Onaka, T., & Murakami, H.: A&A, **383**, 972 (2002)
7. Ohnaka, K., Bergeat, J., Driebe, T., et al.: A&A, **429**, 1057 (2005)
8. Rothman, L. S.: HITEMP CD-ROM (Andover: ONTAR Co.) (1997)
9. Tipping, R. H., & Chackerian, C., Jr.: J. Mol. Spectros., **88**, 352 (1981)
10. Wittkowski, M., Aufdenberg, J., & Kervella, P. : A&A, **413**, 711 (2004)
11. Woodruff, C., Eberhardt, M., Driebe, T., et al.: A&A, **421**, 703 (2004)
12. Yamamura, I., de Jong, T., & Cami, J.: A&A, **348**, L55 (1999)

Part II

Science: Stars — circumstellar matter, IR objects

Evolved Stars: Interferometer Baby Food or Staple Diet?

Peter Tuthill

School of Physics, Sydney University, NSW 2006, Australia
p.tuthill@physics.usyd.edu.au

Summary. With their extreme red and infrared luminosities and large apparent diameters, evolved stars have nurtured generations of interferometers (beginning with Michelson's work on Betelgeuse) with unique science programs at attainable resolutions. Furthermore, the inflated photosphere and circumstellar material associated with dying stars presents complex targets with asymmetric structure on many scales encoding a wealth of poorly-understood astrophysics. A brief review the major past milestones and future prospects for interferometry's contribution to studies of circumstellar matter in evolved stars is presented.

1 Introduction

The first 2-element optical/infrared interferometer observations of evolved star envelopes were performed more than 30 years ago. The present generation of optical interferometers are now poised to transform this once pioneering field into a reliable, routine calibrated tool for a wide range of astrophysics. It seems therefore an opportune moment to reflect on past milestones and potential future developments in the study of evolved stars by optical interferometry.

Firstly, it is important to define the terms of reference of this review. There are a bewildering variety of techniques, often interconnected and used in various multiple combinations, all of which fall under the general heading of optical/infrared interferometry. Among the most productive has been Speckle interferometry, originally implemented as a one-dimensional slit-scanning technique, which in recent decades has gained much momentum from the advent of large format two-dimensional array cameras capable of rapid frame readout. Lunar occultations, which utilize an eclipse by the limb of the moon to perform an interferometric measurement, will not be discussed in any detail here. The development of separate-element interferometry as applied to studies of evolved envelopes has been slower to gain traction, although a number of very significant results have been produced. In passing, there will be some discussion of applicability of various techniques, and of various wavelength regimes.

It is important to point out that the astrophysics which might come under the general heading of "circumstellar material" or indeed of "evolved stars" is a sprawling, and at times ill-defined set of very disparate physical environments. In particular, there is a very blurred boundary for many late-type evolved stars between the upper layers of the atmosphere and the circumstellar environment. Interferometry of the stars' photosphere and extended atmosphere, for which extensive recent progress has been reported, will not be covered here. For the case of Miras, a very thorough and comprehensive literature review of high resolution measurements has been recently published[40]. For a wider perspective of circumstellar envelopes and evolved stars, the reader is referred to the review of Habing[14] (now perhaps a little dated), with recent progress found in topical conference proceedings[37].

We proceed, in the following sections, with a brief historical introduction to high angular resolution studies of circumstellar material in evolved systems, and discuss some of the issues facing present-day observational campaigns with the newly-commissioned generation of optical/IR interferometers.

2 Historical overview: interferometric studies of circumstellar material

2.1 Early Michelson Interferometry

Interferometric observations of circumstellar dust shells were pioneered by the Arizona group utilizing a masking arrangement on the primary mirror of a large telescope. Initial experiments at $5\,\mu m$ used a single-element InSb detector[23] (subsequently upgraded to a Ge bolometer), with fringe encoding in the temporal domain achieved by active pathlength modulation.

Pursuing these techniques, this group eventually collected high angular resolution data on a number of prototypical evolved stars (e.g. α Ori, VY CMa, IRC +10216, Mira) spanning the near to mid infrared spectral regions [24, 25, 26]. A number of interesting phenomena, such as departures from spherical symmetry, time evolution, pulsation-related changes and differing forms of dust shell types were all reported in this body of work.

2.2 Heterodyne Interferometry: the Berkeley Infared Spatial Interferometer

Under the direction of Prof. Charles Townes, a heterodyne interferometry experiment was performed using the McMath auxiliary solar telescopes at Kitt Peak which had two apertures separated on a baseline of 5.5 m. A CO_2 laser operating near $10\mu m$ was used to provide a local oscillator signal, down-converting the fringe signal from the interferometer to radio frequencies, after which correlation and signal detection proceeded much as for a typical radio or millimeter-wave interferometer. Studies of evolved stellar envelopes were undertaken [42, 43, 44], which together with the contemporaneous work of the

Arizona group (with whom they also shared a number of common targets), laid the fundamental observational groundwork for the field. Dust shells were found to fall into distinct classes, with both hot/close and cold/detached morphologies. Changes in time for Mira-type variables were interpreted in the context of stellar pulsation.

Unlike the Arizona group, however, the success of the prototype hetero-dyne interferometer was used by Townes to motivate the construction of a far more ambitious, dedicated, large-aperture interferometer with moving elements to be located on Mt. Wilson, California. This Berkeley Infrared Spatial Interferometer (ISI) [16] saw its first science results on IRC +10216 in 1990 [5], showing that dust shells do indeed respond to the changes in luminosity of the parent star through its pulsational cycle.

Among the most significant papers to come from this project was a big compilation of dust shell observations for 13 late-type stars [6] spanning a variety of different classes of object (supergiants, Miras, carbon stars). This work set the stage for a continual stream of robust, rigorous studies, usually focusing on individual objects, appearing over the last 15 years. The global picture of mass-loss emerging from this body of work is quite complex, and significantly at odds with initial theoretical expectations of a smooth, spheri-cally symmetric, continuous or reliably episodic process. Dramatic departures from simple spherical outflows were reported in many systems [21, 47, 30]. Proper motions as shells of material were carried out with the stellar wind were observed at milli-arcsecond scales [30, 8].

The first major upgrade to the instrumental capability, an RF Filterbank [32], permitting simultaneous extremely high spectral resolution ($\lambda/\delta\lambda = 10^5$) together with high spatial resolution, was completed in the late 1990's. This allowed unique studies of the absorption features in molecules such as NH_3 and SiH_4 [33, 34]. By observing a number of different lines, the molecular stratification of these molecules was determined for VY CMa and IRC +10216, giving a unique window on the chemical evolution of the wind and formation layer of important components as it moves outward with the flow.

A second major upgrade to the ISI, entailing installation of a third tele-scope to yield a 3-baseline closure-phase array [17, 9] is now fully commis-sioned and expecting first science this year.

2.3 Speckle Interferometry

Although Speckle Interferometry was put on a formal footing in the early 1970's by the work of Labeyrie, and had been used even before that in obser-vations particularly of binary stars, it was not until 1979 (a few years after the Arizona and Berkeley groups) that the first infrared observations of re-solved dust shells began appearing from a number of French workers[13, 41]. These early workers invariably used different variants of the technique of slit-scanning, in which a single-pixel detector (typically InSb for the infrared) was used to sample the two-dimensional speckle cloud by scanning the slit across, thus encoding the high resolution information as a temporal signal.

As the pace of the Speckle techniques picked up in the 1980's, a number of significant innovations were implemented. In addition to improvements to the hardware, the post-processing of the data became far more sophisticated with the development of methods to retrieve some estimates of source phase, thus giving a stronger handle on asymmetry. Furthermore, data began to be interpreted in terms of physical models based on numerically computed radiative transfer (albeit in very simple dust shell geometries), rather than just brightness distributions. This leads immediately to more concrete astronomy yielding optical depths, densities and mass-loss rates.

Apart from the French group already mentioned above, among the most productive workers in this early phase of Speckle was Dyck, whose study including 16 late-type stars[10] remains the largest sample of resolved shells in a single paper. However, there were growing numbers of workers in this field, and a complete summary of even the most significant groups is beyond the scope of this review. One-dimensional speckle was able to confirm the accumulating evidence for significant asymmetries, and changes in the form of the dust shell with time and with observing wavelength.

This momentum was not carried forward into the 1990's, possibly due to the expectation that the advent of adaptive optics would soon make Speckle interferometry obsolete. However, a number of technical considerations have led to a small renaissance in Speckle, with the few groups pursuing these techniques at present able to dominate the study of evolved dust shells in the near-IR. First and foremost, adaptive optics has so far notably failed to live up to expectations in delivering a robust, diffraction limited PSF, in which structure at the core can be reliably deconvolved to yield dust shell sizes and morphologies. Secondly, fast-readout two-dimensional array cameras capable of recording the entire speckle cloud and obtaining large volumes of data to average out seeing noise are now standard. Third and finally, a new generation of large (10 m) aperture telescopes have become available, giving a factor of 2-3 times more spatial resolution.

The predominant Speckle group is in Bonn led by G. Weigelt, utilizing a modified bispectral analysis technique capable of full image recovery (called "Speckle Masking", not to be confused with *aperture masking* described below which uses a physical mask to block regions of the telescope pupil). Results on a wide range of astrophysical targets have been reported, with perhaps the distinguishing features of this "second generation" speckle interferometry being the superb two-dimensional image quality recovered in arbitrary complex systems to quite high dynamic ranges. Generally, the imaging efforts have also been matched by extensive astrophysical interpretation and modelling, and many of the systems studied by this group are now representative of the state of the art in our knowledge of evolved star envelopes [3, 4, 19, 51, 29, 48, 39].

2.4 Aperture Masking Interferometry

In a modern context, aperture masking interferometry has a lot in common with Speckle (requirements for the telescope and camera are very similar, and data processing also has much overlap), although we treat it here as a separate technique. The basic idea dates back over 130 years to the first stellar interferometry suggestions of Fizeau, in which a mask is placed over a telescope pupil to form a sparse-aperture interferometer. After a very long hiatus since the pioneering work of Michelson in the early 20th century, aperture masking was resurrected by Prof John Baldwin in Cambridge in the mid 1980's. In the study of evolved stellar envelopes in the near-IR, the most active group originates from Berkeley and utilizes the 10 m Keck telescope giving the highest resolution full images of complex targets yet recovered in this waveband.

Exquisite detail has been revealed in many prototypical systems, with sufficient resolution to observe flows, plumes and disks at scales of a few tens of milli-arcseconds, and to track proper motions of only a few milli-arcseconds per year. Supergiants, AGB and post-AGB stars, Carbon Stars and OH/IR stars have all been studied [31, 45, 35, 8, 22, 49], with those of particular merit in the understanding of the final stages of the mass-loss process being the transition objects at the point of evolving into planetary nebulae.

2.5 IRC +10216 – an example case study

Rather than attempting to summarize the major findings of the interferometric imaging campaigns targeting dozens of different stars within a handful of different classes, we select here the famous prototype infrared-bright carbon star, IRC +10216. Discussion of past and present studies of this object can tell us much about what has been achieved, and also about gaps in our present capability.

Due to its large apparent size and high infrared luminosity, this object is among the most studied of all interferometry targets. Thus many first demonstrations of principle for new observational methodologies (positive detection of dust shell asymmetry [26, 11]; variations in dust condensation radius with luminosity phase[5]) were pioneered on this star.

Here we focus on more recent results, in which full imaging has been obtained in the near-IR. Among the most satisfying outcomes of recent work on this object is that three entirely independent groups, the Bonn[50] and Berkeley[45] groups (mentioned above), and also a group from Cambridge[18], have arrived at near-identical images of this system from observations at three separate telescopes. Example images, from the highest resolution Keck study, are given in Figure 1. Where data exist at a contemporaneous epoch, it is possible to smooth these images slightly to the resolution of the other telescope (e.g. the 6 m Russian SAO in the case of [50]) and the agreement is near perfect.

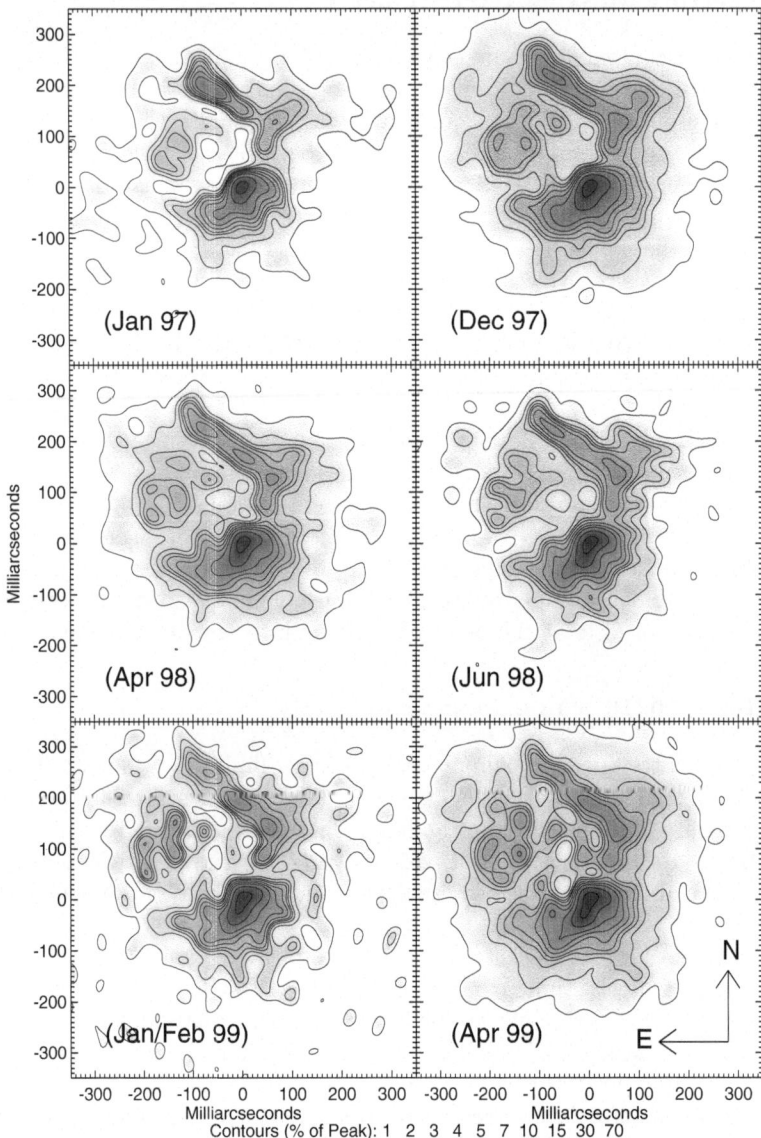

Fig. 1. Images of IRC +10216 in the K-band taken from the work of Tuthill et al.[45], showing the time-evolution of structure over a 2-year period.

With time-evolution studies now approaching a good fraction of a decade in duration, it has been possible to follow proper motions of clumps of material as they condense and are accelerated within the circumstellar outflow. This motion can already be seen clearly in the time-sequence of images presented in Figure 1. Such in-situ studies revealing the dynamics and kinematics of the dust as it condenses and moves is surely a key to unraveling the remaining outstanding problems in mass-loss in evolved stars.

However, despite these dramatic advances in imaging capability, deep and fundamental uncertainties remain. An example concerns the ambitious series of modelling papers from the Bonn group[27, 28] which aim to reproduce much of the existent photometric and interferometric data on IRC +10216. The resultant model, which contains a complex series of nested shells with varying density, dust properties, and symmetry, provides a reasonable fit to most of the observations [27]. However, in this model the central star itself is not the brightest clump visible, but lies behind considerable intervening opacity in the northern regions. A dramatic enhancement in this opacity[28] is then required to explain subsequent epochs in which the feature thought to be the star disappears entirely[28]. Problems with this model have been pointed out by other groups[49], who favor the interpretation that the dominant bright knot is indeed the star (as also favored by Lunar occultation studies[38]). The purpose of this discussion here is not to take any sides in this controversy, but to use it to illustrate that even with extensive, high precision imaging covering a number of wavebands/epochs, these stellar systems are complex enough that uncertainty over something as fundamental as the location of the central star can persist.

3 Present and Future prospects

In the early 1980's, when the present generation of long-baseline optical interferometers were being built, studies of circumstellar matter in evolved systems featured highly in their science proposals and justifications. However, with the notable exception of the Berkeley ISI, most of these interferometers have not been used in the study of circumstellar shells. (A second exception is the VLTI which has already a number of significant publications in this field, and the potential to do much more, particularly with MIDI. However, as these results are covered in much more detail elsewhere in this conference proceedings, I will not discuss them further here.)

There are a number of reasons why present-generation interferometers have largely overlooked these targets. Many are very faint in the optical (which can lead to difficulties for visible acquisition/guiding system). Their large apparent sizes and complex morphologies argue for only modest baseline lengths in the near-IR ($\sim$30 m), yet highly filled Fourier coverage. This corner of observational parameter space is at present better addressed by techniques like aperture masking, and for many modern near-IR interferometers the best (largest) targets are over-resolved and impossible to observe. When we also consider the requirement for full imaging often with high dynamic range, we can see that far from being easy "commissioning science" targets, evolved dust shells present a significant observational challenge to long-baseline optical interferometry. However, there are observational strategies being employed to mitigate these problems, a few of which are discussed below.

3.1 Multi-Resolution Studies

A lesson learned long ago by radio astronomers is that to image regions containing structure on a number of spatial scales, a corresponding set of observations at different resolutions is required. In the context of evolved star envelopes, astrophysical systems vary from cases where light from the central star completely dominates (optically-bright supergiants and Miras) to cases where the central star cannot be directly observed at all (OH/IR objects and deeply embedded carbon stars). For all objects, and most particularly for intermediate cases where star and shell both contribute, unraveling the flux contributions from components whose size may differ by orders of magnitude is not straightforward.

One way to approach this problem is to observe the object with different instruments covering the same waveband. First of all, a wide-field can be imaged (with or without adaptive optics) to search for very extended, low surface-brightness flux. An example of the power of combining AO (yielding high dynamic range, wide field-of-view), together with interferometer (high angular resolution) data is illustrated by recent studies of VY CMa [31] which found a dramatic extended curved plume several arcseconds long. This structure may originate from a one-sided jet-like flow seen within the inner hundred milli-arcseconds in high-resolution imaging.

One of the most extensive studies of circumstellar shells in a broad range of evolved targets to date[36] entailed combined synthesis imaging of a sample of targets using data from both the Keck aperture masking experiment (baselines out to 10 m) and the IOTA interferometer (baselines out to 38 m). For a number of dusty evolved systems, it was possible to obtain crucial information on the size and flux contribution from the central star from the long baseline IOTA data. Armed with this knowledge of the stellar component, the resolved structure in the extended shell sampled by the masking could be imaged without ambiguity. A number of these dust shells are depicted in Figure 2. In particular, the form of the dust distribution around stars in the lower panels, VY CMa and NML Cyg, were dramatically modified over earlier images generated without the extra constraints from IOTA.

3.2 Differential Methods

Many of the most representative types of circumstellar dust shell are difficult to study for reasons of dynamic range. Although the circumstellar material is

Gallery of Dust Shells

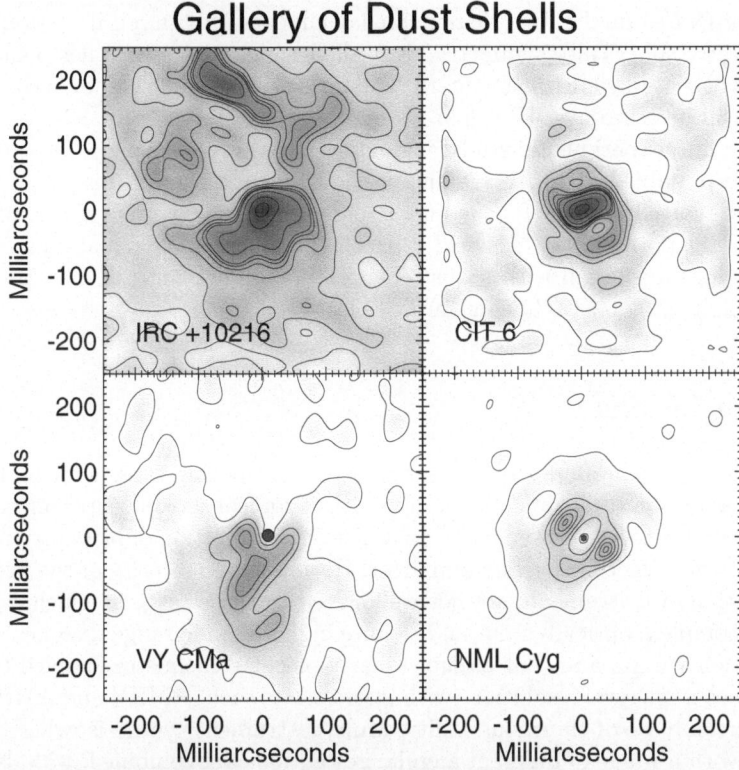

Contours (% of Peak): 0.1 0.5 1 2 3 4 5 10 30 70

Fig. 2. A gallery of images of evolved stellar systems recovered from Keck aperture masking data. The images in the lower two panels have had extra information added to the mapping procedure obtained from long-baseline observations with the IOTA interferometer, giving constraints on the flux and size of the central star.

not inherently faint, the problem (in common with the planet-hunting community) lies in detecting it against the glare, which in this case arises from of some of the most luminous stars in the galaxy. Although a number of approaches to overcome this difficulty are presently under investigation (e.g. Nulling), we will here turn our attention to the *differential methods* in which isolation against the noise processes is obtained by contrasting two measurements which should have similar instrumental and systematic variations, but differing signals due to the stellar source.

One example is a differential phase observation, where the change in the center-of-light caused by two objects of different spectrum is exploited with a dual-channel interferometer measurement. Among the most promising candidates for progress in the study of evolved dust shells is the technique of Optical Interferometric Polarimetry[12], in which comparison is made between signals at orthogonal polarizations. This is ideal for studying the scattered light processes in the dusty circumstellar environment. Even if we consider a perfectly spherically symmetric dust shell which yields zero net polarized flux, when studied with an interferometer baseline sufficient to resolve it, there will be a strong OIP signal.

The first experimental study demonstrating this promising technique has just been published[20], revealing previously unsuspected dust structures around evolved objects R Car and RR Sco. The paper gives a fascinating taste of the potential for this field, with the finding of a quasi-stationary (non outflowing) dust shell with a dust chemistry enabling survival at extremely close radii (previously thought impossible).

4 Conclusions

The dust shells around evolved stars present fascinating targets for high angular resolution study, and encode a wealth of topical astrophysics concerning poorly understood phenomena of mass loss, the formation of planetary nebulae and the galactic recycling of matter. However the majority of the present generation of instruments are not well suited to these targets, which can be large, complex, optically faint and require high dynamic range imaging. High resolution studies aimed at circumstellar molecular chemistry enabled by simultaneous spectral resolution (with instruments such as MIDI and AMBER) will certainly be of enormous utility. Multi-instrument synthesis, where data taken with a range of different angular resolutions are combined, will also be crucial. We also encourage creative approaches, such as OIP, to solving the dynamic range problem in revealing faint material against the high-luminosity primary stars.

References

1. Bester, M., Danchi, W. C., Degiacomi, C. G., Townes, C. H., & Geballe, T. R. *Astrophysical Journal Lett.* 367, L27 (1991)
2. Bester, M., Danchi, W. C., Hale, D., Townes, C. H., Degiacomi, C. G., Mekarnia, D., & Geballe, T. R. *Astrophysical Journal* 463, 336 (1996)
3. Blöcker, T., Balega, Y., Hofmann, K.-H., Lichtenthäler, J., Osterbart, R., & Weigelt, G. *Astron. Astrophys.* 348, 805, (1999)
4. Blöcker, T., Balega, Y., Hofmann, K.-H., & Weigelt, G. *Astron. Astrophys.* 369, 142, (2001)
5. Danchi, W. C., Bester, M., Degiacomi, C. G., McCullough, P. R., & Townes, C. H. *Astrophysical Journal Lett.* 359, L59 (1990)
6. Danchi, W. C., Bester, M., Degiacomi, C. G., Greenhill, L. J., & Townes, C. H. *Astronomical Journal* 107, 1469 (1994)
7. Danchi, W. C., Bester, M., Greenhill, L. J., Degiacomi, C. G., Geis, N., Hale, D., Lopez, B., & Townes, C. H. *Astrophys. & Space Sci.* 224, 447 (1995)

8. Danchi, W. C., Green, W. H., Hale, D. D. S., McElroy, K., Monnier, J. D., Tuthill, P. G., & Townes, C. H. *Astrophysical Journal* 555, 405, (2001)
9. Danchi, W. C., Townes, C. H., Fitelson, W., Hale, D. D. S., Monnier, J. D., Tevosjan, S., & Weiner, J. *Proc. SPIE* 4838, 33, (2003)
10. Dyck, H. M., Zuckerman, B., Leinert, C., & Beckwith, S. *Astrophysical Journal* 287, 801, (1984)
11. Dyck, H. M., Howell, R. R., Zuckerman, B., & Beckwith, S. *Publ. Astr. Soc. Pac.* 99, 99, (1987)
12. Elias, N. M. *Astrophysical Journal* 549, 647, (2001)
13. Foy, R., Chelli, A., Lena, P., & Sibille, F. *Astron. Astrophys.* 79, L5, (1979)
14. Habing, H. J. *Astronomy and Astrophysics Review* 7, 97 (1996)
15. Hale, D. D. S., et al. *Astrophysical Journal* 490, 407, (1997)
16. Hale, D. D. S., et al. *Astrophysical Journal* 537, 998, (2000)
17. Hale, D. D. S., Fitelson, W., Monnier, J. D., Weiner, J., & Townes, C. H. *Proc. SPIE* 4838, 387, (2003)
18. Haniff, C. A., & Buscher, D. F. *Astron. Astrophys.* 334, L5, (1998)
19. Hofmann, K.-H., Balega, Y., Blöcker, T., & Weigelt, G. *Astron. Astrophys.* 379, 529, (2001)
20. Ireland M.J., Tuthill P.G., Davis J., and Tango W. *Mon. Not. R. astr. Soc.*, 361, 337 (2005)
21. Lopez, B., et al. *Astrophysical Journal* 488, 807, (1997)
22. Lopez, B., Tuthill, P. G., Danchi, W. C., Monnier, J. D., & Niccolini, G. *Astron. Astrophys.* 377, 90, (2001)
23. McCarthy, D. W., & Low, F. J. *Astropysical Journal Lett.*, 202, L37 (1975)
24. McCarthy, D. W., Low, F. J., & Howell, R. *Astrophysical Journal Lett.*, 214, L85 (1977)
25. McCarthy, D. W., Howell, R., & Low, F. J. *Astrophysical Journal Lett.*, 223, L113 (1978)
26. McCarthy, D. W., Howell, R., & Low, F. J. *Astrophysical Journal Lett.*, 235, L27 (1980)
27. Men'shchikov, A. B., Balega, Y., Blöcker, T., Osterbart, R., & Weigelt, G. *Astron. Astrophys.* 368, 497, (2001)
28. Men'shchikov, A. B., Hofmann, K.-H., & Weigelt, G. *Astron. Astrophys.* 392, 921, (2002)
29. Men'shchikov, A. B., Schertl, D., Tuthill, P. G., Weigelt, G., & Yungelson, L. R. *Astron. Astrophys.* 393, 867, (2002)
30. Monnier, J. D., et al. *Astrophysical Journal* 481, 420, (1997)
31. Monnier, J. D., Tuthill, P. G., Lopez, B., Cruzalebes, P., Danchi, W. C., & Haniff, C. A. *Astrophysical Journal* 512, 351, (1999)
32. Monnier, J. D., Fitelson, W., Danchi, W. C., & Townes, C. H. *Astrophysical Journal Suppl.* 129, 421, (2000)
33. Monnier, J. D., Danchi, W. C., Hale, D. S., Lipman, E. A., Tuthill, P. G., & Townes, C. H. *Astrophysical Journal* 543, 861, (2000)
34. Monnier, J. D., Danchi, W. C., Hale, D. S., Tuthill, P. G., & Townes, C. H. *Astrophysical Journal* 543, 868, (2000)
35. Monnier, J. D., Tuthill, P. G., & Danchi, W. C. *Astrophysical Journal* 545, 957, (2000)
36. Monnier, J. D., et al. *Astrophysical Journal* 605, 436, (2004)

37. Y. Nakada, M. Honma, M. Seki (eds.) "Mass-losing pulsating stars and their circumstellar matter", Workshop, May 13-16, 2002, Sendai, Japan, *Astrophysics and Space Science Library, Vol. 283, Dordrecht, Kluwer* (2003)

38. Richichi, A., Chandrasekhar, T., & Leinert, C. *New Astronomy* 8, 507, (2003)

39. Riechers, D., Balega, Y., Driebe, T., Hofmann, K.-H., Men'shchikov, A. B., & Weigelt, G. *Astron. Astrophys.* 424, 165, (2004)

40. Scholz, M., *Proceedings, SPIE*, 4838, 163 (2003)

41. Sibille, F., Chelli, A., & Lena, P. *Astron. Astrophys.* 79, 315 (1979)

42. Sutton, E. C., Storey, J. W. V., Betz, A. L., Townes, C. H., & Spears, D. L. *Astrophysical Journal Lett.*, 217, L97, (1977)

43. Sutton, E. C., Storey, J. W. V., Townes, C. H., & Spears, D. L. *Astropysical Journal Lett.*, 224, L123, (1978),

44. Sutton, E. C., Betz, A. L., Storey, J. W. V., & Spears, D. L. *Astropysical Journal Lett.*, 230, L105, (1979)

45. Tuthill, P. G., Monnier, J. D., Danchi, W. C., & Lopez, B. *Astrophysical Journal* 543, 284, (2000)

46. Tuthill, P. G., Monnier, J. D., Danchi, W. C., Wishnow, E. H., & Haniff, C. A. *Publ. Astr. Soc. Pac.* 112, 555, (2000)

47. Tuthill, P. G., Danchi, W. C., Hale, D. S., Monnier, J. D., & Townes, C. H. *Astrophysical Journal* 534, 907, (2000)

48. Tuthill, P. G., Men'shchikov, A. B., Schertl, D., Monnier, J. D., Danchi, W. C., & Weigelt, G. *Astron. Astrophys.* 389, 889, (2002)

49. Tuthill, P. G., Monnier, J. D., & Danchi, W. C. *Astrophysical Journal* 624, 352, (2005)

50. Weigelt, G., Balega, Y., Bloecker, T., Fleischer, A. J., Osterbart, R., & Winters, J. M. *Astron. Astrophys.*, 333, L51, (1998)

51. Weigelt, G., Balega, Y. Y., Blöcker, T., Hofmann, K.-H., Men'shchikov, A. B., & Winters, J. M. *Astron. Astrophys.* 392, 131, (2002)

Eta Car through the Eyes of Interferometers

O. Chesneau[1], R. van Boekel[2], T. Herbst[2], P. Kervella[3], M. Min[4],
L.B.F.M. Waters[4], Ch. Leinert[2], R. Petrov[5], and G. Weigelt[6]

[1] Observatoire de la Côte d'Azur-CNRS-UMR 6203, Dept. Gemini, Avenue
Copernic, 06130 Grasse, France Olivier.Chesneau@obs-azur.fr
[2] Max-Planck-Institut für Astronomie, Königstuhl 17, 69117 Heidelberg, Germany
[3] LESIA, CNRS-UMR 8109, Observatoire de Paris-Meudon, 5 place Jules
Janssen, 92195 Meudon Cedex, France
[4] Sterrenkundig Instituut 'Anton Pannekoek', Kruislaan 403, 1098 SJ
Amsterdam, The Netherlands
[5] Université de Nice-Sophia Antipolis, Parc Valrose, 06108 Nice, France

Summary. The core of the nebula surrounding Eta Carinae has recently been observed with VLT/NACO, VLTI/VINCI, VLTI/MIDI and VLTI/AMBER in order to spatially *and* spectrally constrain the warm dusty environment and the central object. Narrow-band images at 3.74 μm and 4.05 μm reveal the structured butterfly-shaped dusty environment close to the central star with an unprecedented spatial resolution of about 60 mas. VINCI has resolved the present-day stellar wind of Eta Carinae on a scale of several stellar radii owing to the spatial resolution of the order of 5 mas ($\sim$ 11 AU). The VINCI observations show that the object is elongated with a de-projected axis ratio of approximately 1.5. Moreover the major axis is aligned with that of the large bipolar nebula that was ejected in the 19th century. Fringes have also been obtained in the Mid-IR with MIDI using baselines of 75m. A peak of correlated flux of 100 Jy is detected 0.3" south-east from the photocenter of the nebula at 8.7 μm is detected. This correlated flux is partly attributed to the central object but it is worth noting that at these wavelengths, virtually all the 0.5" x 0.5" central area can generate detectable fringes witnessing the large clumping of the dusty ejecta. These observations provide an upper limit for the SED of the central source from 3.8 μm to 13.5 μm and constrain some parameters of the stellar wind which can be compared to Hillier's model. Lastly, we present the great potential of the AMBER instrument to study the numerous near-IR emissive lines from the star and its close vicinity. In particular, we discuss its ability to detect and follow the faint companion.

1 Introduction

Eta Carinae is one of the best studied but least understood massive stars in our galaxy [8]. Eta Car is classified as a Luminous Blue Variable (LBV); a short lasting phase of hot star evolution characterized by strong stellar winds and instabilities leading to possible giant eruptions. Eta Car offers a unique opportunity to observe the consequences of such a giant events with the two historical eruptions in the 1840s and 1890s.

The large bipolar nebula surrounding the central object, known as the "Homunculus", was formed during the first one, while some fainter and less extended structures were attributed to the lesser 1890 event[1]. Despite numerous observations and theoretical studies, the cause of the two outbursts remains unknown. Currently, the Homunculus lobes span a bit less than 20" on the sky (or 45000 AU at the system distance of 2.3 kpc) and are largely responsible for the huge infrared luminosity of the system. The central source is therefore deeply embedded and suffers from high extinction due to the nebular dust.

Improved spatial resolution observations have often been the key for the progress in our understanding of this emblematic embedded object. The central source has been studied by speckle interferometry techniques in visible light, which revealed a complex knotty structure [31, 14]. Originally, three remarkably compact objects between 0.1" and 0.3" northwest of the star were isolated (the so-called BCD *Weigelt blobs*, the blob A being the star itself). Other similar but fainter objects have since been detected [32, 7]. They are found to be surprisingly bright ejecta moving at low speeds ($\sim$50 km.s^{-1}). They belong to the equatorial regions close to the star; their separation from the star is typically 800 AU.

Eta Car has been systematically observed with Hubble Space Telescope (HST) instruments ([7], GHRS; [21, 19], WFPC2/NICMOS; [27, 28], ACS/HRC among many others) and in particular the Space Telescope Imaging Spectrograph (STIS) since the beginning of 1998 [18, 26, 30, 16]. With 0.1 arcsec angular resolution and a spectral resolving power of 5000, the central point-like source can be studied more or less independently from the extended nebulosity and the nebular structures can be dissected (cf. contribution from T. Gull in these proceedings). The central source could be separated from the Weigelt blobs allowing a careful study of the central object by Hillier et al. [17]. The mostly reflective nebula also allowed an indirect study by STIS of the stellar wind from several points of view at different latitudes in the nebulae by means of reflected P Cygni absorption in Balmer lines [26]. The authors convincingly prove the asphericity of the wind, suggesting an enhanced polar wind mass-loss.

Eta Car was observed with the Infrared Space Observatory (ISO, [20]). The ISO spectra indicated that a much larger amount of matter should be present around Eta Car in the form of cold dust than previously estimated. Observations with higher spatial resolution by Smith et al. [23, 25] showed a complex but organized dusty structure within the three inner arcseconds. They showed that the dust content in the vicinity of the star is relatively limited and claimed that the two polar lobes should contain the large mass of relatively cool dust necessary to explain the ISO observations (but see [10]). The mechanism required to form the two gigantic lobes of gas and dust remains poorly understood.

---

[1] although by that time the dust absorption was such that the opacity around the central object was such that the amplitude of this event is not really constrained

One of the main limitations of HST observations is probably that its instruments are mostly restricted to the optical domain. The Homunculus is a dusty nebula dominated by reflected starlight at optical wavelengths and even in K band, the scattered light from the central object is far from being negligible. This makes it problematic to really look through all this material without being affected by the diffuse light. Lower extinction in the IR allows us to look inside the Homunculus to study embedded structures, provided that the spatial resolution is sufficient to study them.

In the following, we present the recent results obtained by the impressive gain in spatial resolution provided by the VLT with the NACO instrument (Sect.1) and the VLTI with the VINCI (Sect.2) and MIDI (Sect.3) instruments. In Sec.4 we present the potential of AMBER observations for detecting and observing the companion of Eta Car.

2 NACO observations: the inner dusty nebula

The interferometric observations presented in the following sections were complemented with broad- and narrow-band observations taken with the NAOS/CONICA (NACO) imager installed on VLT UT4 (Kueyen), equipped with an adaptive optics (AO) system. These observations are described extensively in van Boekel et al. [1] and Chesneau et al. [2].

The diffraction limit (defined by the Point Spread Function, PSF of the telescope) of a 8 meter telescope at 3.8 μm is about 100 mas. At this wavelength, the NACO adaptive optics sufficiently corrects the atmospheric seeing, routinely providing a Strehl ratio approaching 0.5. A careful deconvolution procedure can improve the spatial resolution to about 50-80 mas, i.e. close to the diffraction limit of K band images usually obtained with lower strehl performances and more affected by scattered light (see below). These NACO images represent the highest-resolution images of Eta Car in the K and L bands presently available (shown in the Figure 1). They resolve much of the sub-arcsecond structures.

The NACO observations offer the opportunity to bridge the gap between existing observations and the interferometric data obtained with very high resolution but sparse UV coverage. They are also a great help for interferometric observations employing single mode fibers for which the field of view (FOV) is strongly limited.

Dust plays a key role in the study of Eta Car. It intervenes in every observation as strong and patchy extinction is frequently invoked as an important process in explaining the photometric variability of Eta Car. However, the exact nature and location of dust formation/destruction sites has never been observed. In the close vicinity of the central object, dust is still present in large quantities and even the K band images are contaminated by the intense scattered light which decreases strongly only in L band.

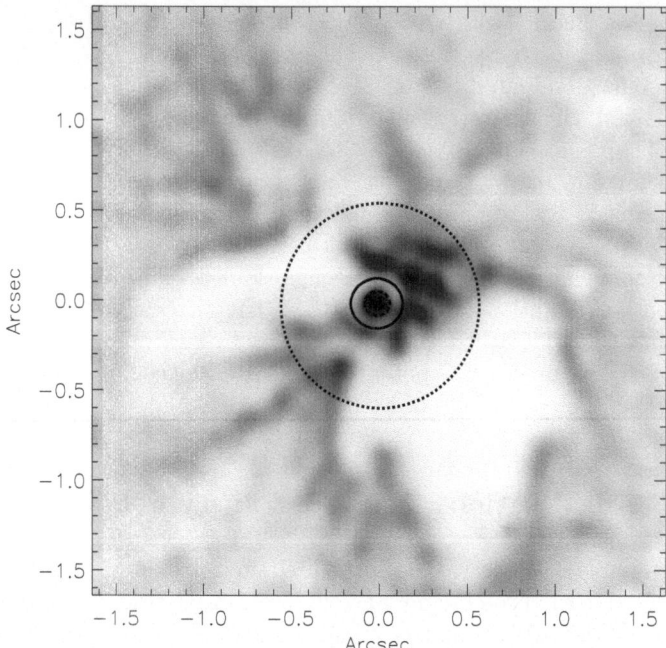

Fig. 1. NACO Pfγ deconvolved image. The resolution achieved is of the order of 60 mas. To reduce the contrast, the image $I^{1/4}$ is shown. The FOV of the different interferometric instruments of the VLTI with UTs and ATs are shown: 2.2μm with ATs and 10μm with UTs in solid line, 2.2μm with UTs in dashed line, 10μm with ATs in dotted line.

A 3.8 μm narrow-band deconvolved image is shown is Fig. 1. This image illustrates the difficulty of observing such a complex and extended object with long baseline interferometers. The key difficulty is that all spatial structures at the scale of one Airy disk contribute to the interferometric signal, especially the single mode AMBER and VINCI interferograms. AMBER and VINCI are sensitive to the 1-20 mas structures at the center of the FOV and are "contaminated" by everything inside a complex 100-150 mas patch with the UTs and inside a 400-600 mas patch with ATs at 2.2 μm. This means that the visibility and phase can be contaminated by the closest regions of the "Weigelt complex". If we want to be fully able to interpret AMBER data, we would have to accurately control the pointing and do mosaicing to explore the inner 30-150 mas region. The AMBER capability to disentangle between narrow (from blobs) and broad (from central star) emission lines owing to its high spectral resolution (R=10000) should be used to provide the necessary astrometric information for assessing the pointing quality (see Sect. 5).

For MIDI observations, the situation is even worse since in this case the dust emission flux is several times larger than that of the central star and virtually all the nebula is bright. The airy patterns of UTs and ATs delimit a region of 0.25 and 1.25 arcsec respectively. One avantage though, is the availability of a 3" FOV to simultaneously get fringes from extended regions; i.e., not to be restricted to the FOV of a single Airy disk.

3 VINCI observations: the evidence for rotation

Rotation is an intrinsic property of all stars, which definitely cannot be neglected in the case of early spectral types. The most obvious consequence is the geometrical deformation that results in a radius larger at the equator than at the poles. Another well established effect, known as gravity darkening or the von Zeipel effect, is that both the surface gravity and emitted flux decrease from the poles to the equator. Although well studied in the literature, such effects of rotation have rarely been directly tested against observations.

It must be pointed out that in the case of Eta Car, the wind density is such that the true photosphere is not visible. Therefore, the consequences of rotation can only be indirectly observed through their effects on the dense wind of this star. Until the indirect observations of Smith et al. [26], the common thought was that centrifugal forces favor mass-loss in the equatorial plane. We will see in the following that interferometry is the most appropriate tool to detect directly the asymmetry if the central source.

The VINCI observations of Eta Car are described in van Boekel et al. [1]. The two 35 cm siderostats and the instrument VINCI were used to obtain interferometric measurements at baselines ranging from 8 to 62 m in length. The observations were carried out in the first half of 2002 in four different nights, and again in early 2003. The baselines used, have a ground length of 8, 16, 24, and 66 m respectively. In particular observations with the 24 m baseline cover a wide range of projected baseline orientations.

VINCI observations provide information on the K continuum from the central object, the flux from emission lines being limited to less than a few percent of the total flux in this band. As mentioned in the previous section, the extended FOV of siderostats includes, in addition to the central source (representing 57% of the observed flux), the regions from the 'Weigelt complex' within the dotted curve in Fig.1. AMBER should be used to evaluate potential measurement biases.

VLTI/VINCI observations clearly resolve this central object; its size can now be measured to be 5 mas at 2 μm corresponding to 10 AU at the distance of Eta Car. This is much larger than the stellar photosphere so that we must be observing an optically thick wind. The radiation is dominated by free-free emission and electron scattering; the emerging spatial intensity profile is determined by the mass-loss rate and the wind clumping factor. The intensity profile measure with the VLTI breaks the degeneracy between these two parameters in previous modeling efforts; mass loss rate and clumping factor can be derived separately from the combination of HST/STIS spectroscopy and

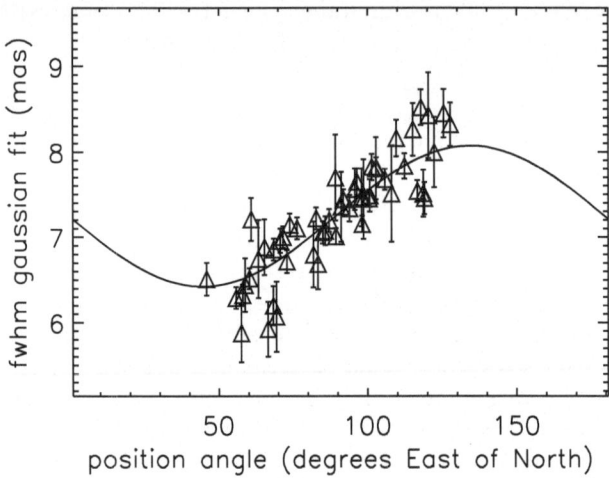

Fig. 2. Variation of FWHM fitted to the visibilities measured with VINCI as a function of projected orientation of the 24 m baseline. The solid line gives the best fit to the measurements, assuming a 2D Gaussian shape of the source at each projected baseline orientation. The amplitude of the size variations gives a ratio of major to minor axis of 1.25±0.05. The major axis has a position angle of 134°±7° East of North.

the interferometric data. These observations are consistent with the presence of a star which has an ionized, moderately clumpy stellar wind with a mass loss rate of about 1.6×10^{-3} $M_\odot$ yr^{-1}. This star-plus-wind spherical model, developed by Hillier et al. [17], is also consistent with the HST STIS observations of the central object.

A second important conclusion from the VLTI data is that the central object is not spherically symmetric, the star is elongated with a de-projected axis ratio of about 1.5. Moreover, its major axis is aligned with that of the Homunculus. These VLTI observations provide a direct measurement of the wind geometry proposed by Smith et al. [26]. The alignment on all scales means that the 1840 outburst looks like a scaled-up version of the present-day wind, and that this wind is stronger along the poles than in the equatorial plane. As Dwarkadas & Owocki [12] showed, the radiation pressure in these massive stars is stronger in the polar regions because of the von Zeipel effect. The wind is primarily controlled by this phenomenon and not by the local gravity.

4 MIDI observations: dusty clumps, everywhere!

The MIDI recombiner attached to the VLTI is the only instrument that is able to provide sufficient spatial and spectral resolution in the mid-infrared to disentangle the central components in the Eta Car system from the dusty environment. The Hillier model suggests a flux level of 200-300 Jy at 10 μm and 10-15 mas diameter of the star plus wind at this wavelength.

Eta Car was observed with MIDI with the UT1 and the UT3 telescopes during commissioning and guaranteed time observations. These observations are reported in Chesneau et al. [2]. Virtually all the capabilities of MIDI were used during these 4h observations: single-dish imaging during acquisition, spatially resolved long slit spectroscopy, undispersed mapping of the correlation pattern and, finally, dispersed fringes. All data were taken within the small, but indeed of great interest, 3" FOV of the instrument.

The spatial distribution of the fringes detected by MIDI with the 8.7 μm filter is shown in Fig.3. The peak of the fringes is localized at the position of the star itself but an extended halo is also visible in the Weigelt complex about 0.4-0.6" northwest from the star. This is the confirmation that highly compressed material emitting strongly at 8.7 μm exists in this region. The fringes at the location of the Weigelt blobs are definitely more extended than a single PSF FWHM at 8.7 μm (220 mas).

This implies that in the equatorial Weigelt region a fraction of the dust is embedded in clumps with a typical size smaller than 10-20 mas (25-50 AU) within a total extent of about 1000 AU. Nevertheless, this correlated flux represents only a few percent of the total flux at these locations. It must be pointed out that only a few scans with fringes have been recorded during this commissioning measurement and the lowest detectable fringe signal visible in Fig. 2 is about 20 Jy. With more integrated frames it should be possible to see that all the Weigelt complex indeed can generate fringes.

Dispersed fringes were also obtained which reveal a correlated flux of about 100 Jy situated 0.3" south-east of the photocenter of the nebula at 8.7 μm, which corresponds with the location of the star as seen in NACO images. This correlated flux is partly attributed to the central object, and these observations together with the VINCI ones provide an upper limit for the SED of the central source from 2.2 μm to 13.5 μm (Fig. 4).

The 74m baselines were roughly perpendicular to the main axis of the nebula and the putative rotation axis of the prolate star itself. This means that the baselines were oriented perpendicular to the main stellar axis, in which direction the star is smaller, corresponding to a maximum correlated flux. Hence, the MIDI measurements can be considered as an upper limit of the correlated flux observable from the star. Despite large error bars, the MIDI correlated flux is obviously below the model predictions. We also note that the MIDI data were acquired at the periastron passage of the faint

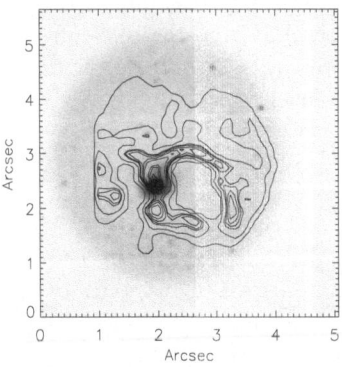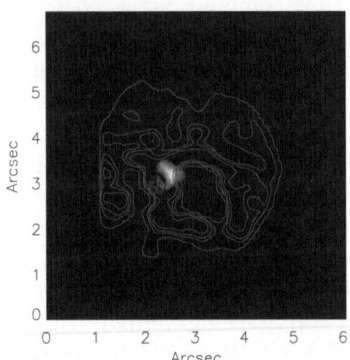

Fig. 3. Left, the figure shows the signal fluctuations within the MIDI acquisition FOV. The external regions are dominated by the detector noise and the internal regions by the tunnel and sky background fluctuations. The signal from the fringes is strong and centered on the position of the star as seen in the Fig.1 image. The contour plot represents the contours of the deconvolved MIDI 8.7 μm acquisition image. Right, the noise pattern and the fringe pattern from a calibrator have been subtracted from the previous figure in order to show the extended fringe signal. The vertical orientation is approximately parallel to the bipolar nebula (PA=138°).

companion (still undetected), and no emission lines in the extracted spectra are visible compared to the model[2].

Nine spectrally dispersed observations from the nebula itself at PA=318 degree, i.e. in the direction of the bipolar nebula within the MIDI field of view of 3" were also extracted. A large amount of corundum (Al_2O_3) is discovered, peaking at 0.6-1.2" south-east from the star, whereas the dust content of the Weigelt blobs is dominated by silicates. These observations are extensively discussed in Chesneau et al. [2].

The guaranteed time observations presented here were intended to judge the feasibility of MIDI observations of Eta Car. The observations, conducted with UTs only, demonstrated the interest of such a scientific program but also the difficulty to extract the central star signal from the dust one. The results were obtained with only one baseline and at periastron passage of the companion (see following section). In particular, a possible explanation of the low correlated flux observed, compared with the model of Hillier, could

---

[2] The whole Eta Car spectrum is dominated by strong emission lines which disappear during a short time at periastron passage. Despite the low spectral resolution of MIDI (R=30 with the prism), we can clearly say that these lines were absent for these observations. The emission lines are for instance well detectable in HD316285, a twin of Eta Car from the spectral point of view, with the same spectral resolution.

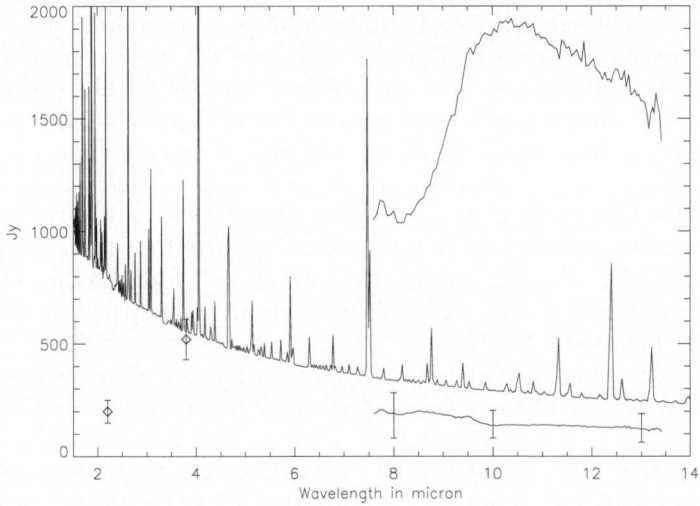

Fig. 4. Spectral energy distribution from Hillier's model compared with the photometry obtained from an airy disk centered on the star (upper curve) and the extracted correlated flux corrected from the expected visibities of the model (lower cuver). The near-IR photometry in the K and L band are also indicated.

be the presence of the extended wind-collision zone between the central star and the fainter companion, commonly thought to be the place of an intense dust formation. The garanteed time with UTs has been completed and the star is now being offered for open time observations. It would be of great interest, now that the orbital phase is near apastron, to undertake a more ambitious program with UTs taking advantage of the fact that AMBER is ready to perform contemporary observations. Some test observations are also planned to check the feasibility of a large observing program with ATs. Their large FOV could prevent the extraction of pertinent information for such a complex object but they also offer a wealth of baselines that could provide a good uv coverage.

5 AMBER observations: revealing the binary?

The original detection of a 5.52 year period in Eta Carinae in the spectroscopic and near-infrared photometric data of Damineli [4] has been confirmed by later observations [6, 9, 33]. The existence, mass, and orbit of a companion and its possible impact on the behavior of the primary are still strongly disputed (e.g. [9, 3, 15, 22, 11]).

The deduced parameters of the orbit from X-ray observations by Corcoran et al. [3] imply that the periastron and apastron distances are roughly 1.5

AU and almost 30 AU, respectively (see also [22]). At closest approach, the secondary is well embedded, deep within the dense wind of the B-star primary.

The first observations of Eta Car with AMBER appear promising (see Petrov et al., this volume). Already, absolute and differential visibilities, differential and closure phases were obtained at medium (R=1500) and high (R=10000) spectral resolution. Even with this impressive amount of information, the spatial interpretation of this complex (and time variable) object remains limited by the poor uv coverage. We advocate a spectroscopic approach based on the study of the numerous near-infrared lines formed at different spatial locations, viewing the central star through different angles and experiencing different excitation mechanisms (cf. in particular Smith et al. 2002b, Fig.14 and Fig.15).

Near-infrared emission lines are unique diagnostics of the geometry and kinematics the wind of Eta Car and for studying in its close vicinity. The infrared spectrum is a strong function of the position in Eta Car's nebula, with molecular hydrogen and [FeII] tracing more easily cold or collisionally excited material formed in the circumstellar outside the inner 1" region [24, 29]. The HeIλ10830 line, variable with the orbital cycle ([4] and emitted from some equatorial ejecta [24] is of particular interest. A. Damineli has been monitoring HeIλ10830, Paγ and Paδ over the past several years and showed that this line is very sensitive to the orbital motion of the companion [4, 5], see also the web site of A. Damineli http://www.etacarinae.iag.usp.br/).

Smith [24] was able to marginally resolve the emission from the Weigelt blobs, allowing their combined spectrum to be separated from the central star at IR wavelengths for the first time. During the first AMBER observations, it was observed that the narrow emission lines are offset from the star's position by 0.2 to 0.4 arcsec, while the broad lines from the star wind appear when the telescope is accurately pointed. Thus, it should be possible to separate these blobs from the star by extracting segments along the slit on either side of the star's position. To the NW the emission is dominated by the blobs, and to the SE the star dominates.

The seach for any indices that are able to constrain the companion is now being intensively conducted. By studying the high and low state of excitation from the emission lines of Weigelt blobs, Verner et al. [30] found consistent results with an O supergiant or a Wolf- Rayet (W-R) star. Falceta-Goncalves et al. [13] recently studied the accumulation of material in the wind-wind collision zone in an attempt to conciliate X-rays and optical observations. They also considered the formation rate of dust in the collision zone and came to interesting conclusions, complementary to those in the MIDI and NACO observations by Chesneau et al. [2]. Their Fig.3, showing the expected tail of gas and dust created in the collision zone, illustrates the potential complexity of the object, which will be soon intensively observed with AMBER.

AMBER is particularly well suited to study the wind-wind collision zone and should detect it more easily than the companion itself: the emitting zone is large (i.e. well resolved), the wind-wind collision zone might be bright near periastron, and the radial velocities encountered for this kind of phenomenon can probe further the potentially complex geometry. The simultaneous use of three telescopes is indeed the best advantage of these AMBER observations. The observations of the famous WR-O binary system γ^2 Velorum (cf. Petrov et al., this volume) show the potential of AMBER for such a study and the preliminary results are very promising.

References

1. van Boekel, R., Kervella, P., Schöller,M, et al., A&A, **410**, L37 (2003)
2. Chesneau,O., Min, M., Herbst, T. et al., A&A, **435**, 1043 (2005)
3. Corcoran, M.F., Ishibashi, K., Swank, J.H., Petre, R., ApJ **547**, 1034 (2001)
4. Damineli, A., ApJ Letters, **460**, L49, (1996)
5. Damineli, A., Conti, P.S., Lopes, D.F., New Astronomy, 2, 107 (1997)
6. Damineli, A., Kaufer, A., Wolf, B., Stahl, et al. ApJ Letters, **528**, L101 (2000)
7. Davidson K., Ebbets, D., Morse, J.A. , AJ, **113** 335 (1997)
8. Davidson K. & Humphreys, R.M., Annu. Rev. Astron. Astrophys., **35**, 1 (1997)
9. Davidson, K., Ishibashi, K., Gull, T.R. et al., ApJ **530**, L107 (2000)
10. de Koter, A., Min, M., van Boekel, R., Chesneau, O., ASPC, **332**, 319 (2005)
11. Duncan, R.A. & White, S.M., MNRAS, **338**, 425 (2003)
12. Dwarkadas, V.V. & Owocki, S.P., ApJ, **581**, 1337 (2002)
13. Falceta-Goncalves, D., Jatenco-Pereira, V., Abraham, Z., MNRAS, **357**, 895 (2005)
14. Falcke, H., Davidson, K., Hofmann, K.-H., Weigelt, G., A&A, **306**, L17 (1996)
15. Feast, M., Whitelock, P., Marang, F., MNRAS, **322**, 741, (2001)
16. Gull, T.R., Vieira, G. and Bruhweiler, F., ApJ, **620**, 442 (2005)
17. Hillier, D.J., Davidson, K., Ishibashi, K., Gull, T., ApJ, **553**, 837 (2001)
18. Ishibashi, K., Gull, T.R., Davidson, K. et al., ApJ, **125**, 3222 (2003)
19. King, N.L., Nota, A., Walsh, J.R. et al. ApJ, **581**, 285 (2002)
20. Morris, P.W., Waters, L.B.F.M., Barlow, M.J. et al., Nature, **402**, 502 (1999)
21. Morse, J.A., Davidson, K., Bally, J. et al., AJ, **116**, 2443 (1998)
22. Pittard, J.M., Corcoran, M.F., A&A, **383**, 636 (2002)
23. Smith, N., Gehrz, R.D., Hinz, P.M. et al., ApJ, **567**, L77 (2002a)
24. Smith, N., MNRAS, **337**, 1252 (2002b)
25. Smith, N., Gehrz, R.D., Hinz, P.M. et al., AJ, **125**, 1458 (2003a)
26. Smith, N., Davidson, K., Gull, T.R. et al., ApJ, **586**, 432 (2003b)
27. Smith, N., Morse, J.A., Gull, T.R., et al., ApJ, **605**, 405 (2004a)
28. Smith, N., Morse, J.A., Collins, N.R. and Gull, T.R., ApJ, **610**, L105 (2004b)
29. Smith, N., MNRAS, **357**, 1330 (2005)
30. Verner, E., Bruhweiler, F., Gull, T.R., ApJ, **624**, 973 (2005)
31. Weigelt, G. & Ebersberger, J., A&A, **163**, L5 (1986)
32. Weigelt, G. Albrecht, R., Barbieri, C. et al., RMxA&A, Ser. Conf., **2**, 11 (1995)
33. Whitelock, P.A., Feast, M.W., Marang, F. et al., MNRAS, **352**, 447 (2004)

The Ejecta of Eta Carinae: What we have Learned from Space Telescope Imaging Spectrograph and the Ultraviolet Echelle Spectrograph

Theodore R. Gull

ExoPlanets and Stellar Astrophysics, Code 667, Exploration of the Universe Division, NASA/Goddard Space Flight Center, Greenbelt, MD, 20771, USA
Theodore.R.Gull@NASA.GOV

Summary. Between 1997.0 and 2004.3, a series of observations were accomplished with the Hubble Space Telescope Imaging Spectrograph (HST/STIS) and from 2002.9 to 2005.3 with the Very Large Telescope/Ultraviolet Echelle Spectrograph (VLT/UVES). Coordinated observations were also done with RXTE, CHANDRA, FUSE and several ground-based telescopes. Much new information has been obtained about this Luminous Blue Variable (LBV) which appears to be a binary star system with at least one component being in very late stages of CNO-burning. These observations are summarized with intent on suggesting future observations with the VLT and VLT(I) especially during the upcoming apastron (2006.25) and periastron (2009.0).

1 Introduction

Eta Carinae is a Rosetta Stone for understanding how massive stars lose mass as they approach the end of the CNO cycle. In the past, astronomers have touted it as the prototype for Luminous Blue Variables - those very massive stars in very late stages of evolution [6]. In the 1840's, the star was observed to rival Sirius in apparent magnitude, then faded. A secondary brightening occurred in the 1890's with subsequent fading. Today the star is slowly gaining in apparent brightness. Surrounding Eta Carinae is a nested pair of hourglass structures, known as the Homunculus [7] with the internal Little Homunculus [13], expanding outward consistent with ejection in the 1840's and the 1890's. In recent years, a spectroscopic 5.54-year period was noticed indicating that Eta Carinae is a binary system [4]. At least one member appears to be at a critical stage of evolution, the end of the CNO-processing cycle. The ejected material, which has had little opportunity to encounter and to mix with the precursor wind or the interstellar material, is rich in nitrogen and helium, yet poor in oxygen and carbon [25, 26, 1]. Nebular emission is dominated by lines of neutral and singly-ionized iron peak elements [30].

While observations are consistent with a massive binary system, we have yet to directly see spectral evidence of a companion star. Obtaining direct evidence of the binary star system is a major goal that may be possible with the VLTI.

This discussion describes observational results from HST/STIS, VLT/ UVES, RXTE, CHANDRA and FUSE to date and projects what might be accomplished with the VLTI with current and future instrumentation.

2 Recent Observations

From early 1997 to summer 2004, HST/STIS imaging spectroscopy provided a unique capability to observe complex extended structures from 1175 to 10100Å [14, 29]. The 10"×18" angular dimensions of the Homunculus on the sky and its nebular brightness, plus the brightness of Eta Carinae permitted full spectral coverage of the central source and brighter portions of the nebulosity in a reasonable number of HST orbits. A series of observations were done with the CCD moderate dispersions extending from 1640 to 10100Å using a 52"×0.1" aperture with R≈8000 and 0.05"/CCD pixel beginning 1998.0 through 2003.25. From late 2001 to 2003.25, limited observations were done in the 1175-2370Å region at R≈30,000 and the 2390 to 3160Å region at R≈109,000. Spatial sampling with the MAMA detectors permitted 0.0125"/HIRES pixel, yielding angular resolution approaching 0.030". STIS apertures(0.09"×0.2", 0.2"×0.2" and 0.3"×0.2") were used to acquire high dispersion spectral and spatial information of the bright nebular components close to the central source.

Confirmation of the predicted 1998.0 minimum came through X-ray monitoring with RXTE led by M. Corcoran [2, 3]. Several observing programs led by K. Davidson and T. Gull were accomplished with STIS leading up to an HST Treasury activity designed to follow the changes of the central source and response by the ejecta across the predicted 2003.5 minimum. Corcoran and colleagues acquired X-ray spectroscopy at several phases with CHANDRA. Far ultraviolet observations with FUSE before and after the minimum were done by T. Gull and collaborators.

The spatial structure, as recorded by HST/WFPC2 and HST/ACS/HRC, and the radial velocity of the Homunculus indicate that the hourglass structure is tilted out of the sky plane by ≈45 degrees [7] and implies that the central source is likewise tilted. The VLTI observations at 2 microns [23] indicate that the stellar wind is shaped like a prolate spheroid (American football) oriented along the same axis. In coordination with the HST Treasury proposal, K. Weis and colleagues used the VLT/UVES to monitor the polar spectrum scattered by the flat portion of the foreground lobe and the spectrum of Eta Carinae seen directly. As there is about a three week delay in travel time, changes in the polar spectrum would be expected to arrive about three weeks later [28, 21].

A binary system consisting of a very massive B-star and a less massive O or WN stellar companion in a highly elliptical orbit appears to explain the periodic behavior of the system. The primary star, likely $\approx$ 100 M⊙ and temperature, $T_e \approx$15,000K, supports a massive wind: 10^{-4} M⊙yr^{-1} at 500 to 600 kms^{-1}. The companion is thought to be $\approx$ 30 - 40 M⊙ with $T_e \approx$35,000K with a wind of 10^{-5} at 3000 kms−1 terminal velocity. The wind-wind interaction of these two stars explain the spectral properties of the CHANDRA X-ray spectra [18]. Changes in nebular response are thought to be due to Lyman radiation being trapped by the very extended wind of the primary during the periastron portion of the orbit. With the drop of Lyman radiation, the relatively dense ejecta relax in ionization and excitation for a few months, but then returns to a more highly excited level when the Lyman radiation again breaks through the primary extended wind.

3 Outer Ejecta: Homunculus and Little Homunculus

The orientation of the Homunculus is such that we view Eta Carinae through the walls of the foreground lobes of the hourglass-shaped, nested Homunculus and Little Homunculus. While our vision of the central region is obscured, the line of sight provides an opportunity to study the ejecta thrown out over the past two centuries. The spatially resolved STIS echelle spectra reveal multiple absorbing velocity components in line of sight [9, 10]. Bright nebular-scattered starlight back-illuminates the wall. We are able to see velocity shifts that correlate with the spatial curvature of the Homunculus and the Little Homunculus. Excitation, ionization and temperature indicate that the Little Homunculus, being ionized, has velocities around −146 kms^{-1} and the Homunculus, being neutral, has velocities ranging from −386 to −587 kms^{-1}. Over thirty velocity components have been identified [10]. Absorption lines of Fe II, Cr II, Ni II, Ti II, V II, Fe I, and other ions have been identified. Given the quasi-periodic spacing in velocity, likely these components are the remnants of the 5.52-year modulated wind from the central source. In contrast to interstellar medium absorptions that originate from ground energy levels, these absorptions are from multiple iron-peak elements in neutral and singly-ionized state arising from numerous metastable levels.

The shell structure of the Little Homunculus is seen in emission lines of H I, [N II] and other lines typically seen in relatively dense H II regions. By contrast the Homunculus is seen primarily as a dust-scattering shell with a spectrum of the central source red-shifted. A thin skin just internal to the Homunculus is seen in forbidden emission lines of singly-ionized iron, nickel, etc. Thus the Homunculus is largely neutral but with metals ionized by Balmer continuum, not Lyman continuum.

Two well isolated components, -146 and -513 kms^{-1}, have been well characterized [9] with T_e =7000 and 760 K during the broad maximum and densities around 10^6cm^{-3}. During the few month long minimum of 2003.5, the -146 kms^{-1} component dropped in temperature as reflected by the populations of the metastable energy levels of Fe II. The Ti II population levels seen in the

-513 kms^{-1} component did not change substantually. However, in the STIS FUV spectral region nearly 1000 H$_2$ absorption lines, identified to be present in the -513 kms^{-1} ejecta during the broad maximum, fade or disappear during the minimum [17]. For both velocity systems and for intermediate velocity systems, the source of excitation appears to be the ultraviolet radiation from the central source. During the broad maximum, Lyman radiation from the hot secondary star excites and ionizes the Little Homunculus; UV radiation longward of 912Å penetrates beyond the Little Homunculus to ionize the metals on the interior of the Homunculus and to photo-excite/destroy the H$_2$ in the outer shell. When the hot companion approaches periastron, the cavity blown by the companion is enveloped by the much more extensive, massive primary wind and the Lyman radiation is trapped. For a few short months the Lyman radiation-dependent structures drop in ionization and excitation, then return to the maximum state.

In the disk region, located between the lobes of the Homunculus hourglass, a very peculiar 'neutral' emission region persists. We call this structure the strontium filament as the first lines identified of this structure were [Sr II] emission lines near Hα [31]. Subsequent studies [11] identified over 500 emission lines, some of the strongest being from Ti II, Fe I, Fe II and V II. No hydrogen, nitrogen or oxygen emission lines were found. Hence this is not the typical ionized hydrogen emission region, but an example of a neutral emission region excited by photons with energies below 13.6 eV. Singly ionized titanium is like the proverbial canary in the coal mine: the absence of Ti II in a partially ionized gas indicates an abundance of photons above 13.58 eV, just shortward of the ionization potential of H I; the presence of Ti II indicates insufficient Lyman continuum to completely ionize the gas. This is well known from ISM studies. Most of the metals seen as singly ionized ions here are normally combined with oxygen and precipitated onto dust grains. Metals such vanadium and titaniaum require oxygen to form oxides that build dust grains. As noted by Verner et al [25, 26], oxygen and carbon appear to be about one percent solar abundances. Bautista et al [1] note that Ti/Ni is 100x solar in the strontium filament, consistent with underabundance of oxygen both through nuclear processing and chemical activity.

4 The Central Region

The broad stellar line profiles of Eta Carinae change with the 5.52-year period and the ejecta respond by the appearance and disappearance of highly excited nebular emission lines. Damineli et al [5] noted that the high excitation narrow lines of [Ne III], [Ar III] and broad lines of He I decreased or disappeared altogether during the several month long minimum. The speckle interferometry of Weigelt and Ebersberger [27] revealed at least three nebular components, known as the Weigelt Blobs B, C and D. Located to the northwest at angular distances of 0.1 to 0.25" from the stellar source, these condensations prove to be extremely bright, dense partially ionized structures. Recent imagery with HST/ACS/HRI in the 2200 and 3300Å spectral regions revealed additional condensations much like beads on a necklace encircling the central source at a distance of 0.1 to 0.25" [20]. Only three condensations to the northwest are ionized. The rest appear to be either reflection nebulosities or clumps in the resolved wind of the primary.

Models of the Weigelt condensations B and D ([25, 26] first led to a quantitative estimate that oxygen and carbon are underabundant while nitrogen and helium are overabundant. Meynet and Maeder [16] have modeled evolution of massive stars. The highly convective envelope leads to overabundance of nitrogen at the expense of carbon and oxygen for stars with 100 M⊙. Hence the estimate abundances of these condensations are consistent with ejecta of a massive star in the late stages of CNO processing.

STIS observations throughout the visible and UV indicate that the stellar wind of the primary is resolved extending out to about 0.4", which at the estimate distance of 2300 parsecs projects to about 1000 AU radius. Hillier et al [12] have compared stellar wind models to the STIS observations and find that indeed the primary wind extends over distances comparable to these scales. J. Pittard (private communication) has provided simple graphics of the wind-wind interface for this binary system. As the companion plunges down past the primary during periastron, the bow-shock cavity becomes highly distorted and then is cutoff by the wrapping around of the much more massive primary wind. Line profiles recorded by the STIS with high spatial resolution are being analyzed to determine if evidence of the wind interactions can be detected.

The binary model [3] suggests the companion is in an orbit with major axis of about 30 AU and ellipticity of 0.9 (M. Corcoran, private communication). Likely the orbit lies in the plane of the disk which is tilted at 45 degrees to the sky and projects in the general direction between Weigelt condensations C and D. This places the major axis at about 30 degrees from the sky plane, so the maximum projected separation at apastron is about 20 AU or 8 milliarcseconds separation.

4.1 Recent Temporal Variations

The close monitoring of Eta Carinae with RXTE, HST/STIS, HST/ACS, FUSE and VLT/UVES is providing much new insight on the predictability of changes by the central source and the nebulosity. Unfortunately at the time of this conference, STIS is no longer operational and FUSE is recovering from loss of a momentum wheel and a gyro. Continued monitoring with VLT/UVES has become even more important not only because of its high spectral resolution, but because the star and nebular structure appears to be changing. Comparisons of the overlap region (3060 to 3160Å) between the STIS NUV echelle observations (R=109,000 and spatial resolution ≈ 0.030")

with the VLT/UVES (R=80,000 and spatial resolution ≈0.7 to 1.5") indicated that from late 2002 to early 2004, a very significant amount of 'stellar
continuum' was originating from nebular structure (and the extended wind
component) surrounding the central source. Deep Ti II, V II and other nebular absorption lines originating from the -513 kms^{-1} ejecta seen in the STIS
E230H spectra were broadened and shallow due to velocity shear across the
wall of the Outer Homunculus. The latest observations with VLT/UVEs (K.
Weis, PI) indicate that UVES sees these lines nearly as sharp and deeply as
STIS had previously. HST/ACS/HRC observations in December 2004 confirm that the nebular-scattered light has dropped significantly. Along with
the trend since 1998 of the central source brightening by at least three-fold,
we suspect that the dust is clearing out of the immediate vicinity of the
central source.

5 What VLT and VLTI could do

The 5.54-year spectroscopic periodicity [4], the X-ray spectral modeling[18],
the nebular emission line modeling [25, 26] are consistent with the central
source being a massive binary system. Near infrared continuum [23] observations with the VLTI indicate the wind of the central source is extended
towards the polar regions of the Homunculus. Is there a way to detect the
secondary star? Thus far searches in the visible and ultraviolet spectral regions have yielded no direct indication of the secondary. If the companion is
a WN star, then the very bright, broad emission of He I 4686Å and 1640Å
would be expected to be seen in the spectrum. The FUV is so riddled with
absorption complexes from the circumstellar ejecta (Homunculus and Little
Homunculus) that the 1640Å line is not identifiable. Emission at 4686Å has
been identified [22, 8, 15], but appears only months before the X-ray peak
and spectroscopic minimum. Likely this emission, if He II emission, is originating from the colliding wind region, or bowshock, not the companion star.
No other He II emission line has been identified. Lines of He I, with structure
are present in the spectra, but are awaiting further analysis. Absorption lines
of C IV do not appear to be present in the STIS FUV spectra, but the primary stellar emission, with the foreground nebular absorptions, confuse the
spectrum. In the FUSE spectra, in addition to the ejecta absorptions, huge
absorptions from nebular and interstellar H_2 obscure the spectrum. Hillier
et al [12] point out that the stellar flux shortward of 1500Å must be originating from the companion, but the foreground absorptions simply block out
any information on the character of the star. While likely a mid-O star in
energy distribution, we simply cannot see the companion star clearly enough
to demonstrate its unique properties in the UV or the visible portions of the
spectrum.

High spatial and reasonable spectral resolution in the near infrared offers the strong possibility of detecting the companion directly. Continuum

observations by van Boekel et al. [23], while model-dependent, are interpreted to show a prolate spheroidal wind structure. However, these observations were taken over an extended period of the orbit. The spatial relationship of the two stars is changing within the orbital period, and the wind structure is likely to be changing at some level. Given that Hillier et al [12] demonstrate that the wind structure extends out to 500 AU (0.2"), changes of the wind structure could be distorting the observations. More complete sampling of the UV plane will be necessary over intervals of a few weeks to conclusively show the wind structure at any given portion of the orbit.

Broad, P-Cygni emission lines of H I, He I, Fe II are noticeable in the near infrared spectra published by N Smith [19]. Observations with VLTI/AMBER focused on these line profiles show promise to reveal the companion star. Most intriguing is the He I 1.083 and 2.06μ P-Cygni lines. The original discovery of the 5.52-year period by Damineli was by spectrophotometry of the He I 1.083μ line. The continuum-normalized line flux shows a broad maximum for five years, then drops with the X-ray drop. Curiously the line flux has been dropping relative to the continuum, but the explanation here is that the central source is getting brighter with dust clearing. Comparison of the UV spectra of Weigelt D observed by STIS in March 1998 to September 2003 (5.5 years apart and so in the same spectroscopic phase) indicates that Weigelt D has not changed significantly. Rather the star has brightened relative to the nebula. Likely the He I 1.083μ emission is originating not from the central source but from the nebulosity immediately surrounding the source.

One explanation may be that the He I emission is originating from the wind-wind collisional region, as the 4686Å emission is thought to originate -- if the 4686Å emission is indeed originating from He II. Varricatt, Williams and Ashok [24] studied WR140 in the near infrared and found that the 1.083μ emission in that system could be explained in terms of nebular emission originating from the denser wind piled up against the bowshock in that system. Is this the source of He I emission in the Eta Carinae system?

Application of VLTI/AMBER to the He I 1.083 and 2.06μ line profiles with resolving power of 1500 or 20,000 likely would resolve whether the He I emission is originating from the secondary star, or from the wind-wind interaction between the primary and secondary. Likely the H I and Fe II emissions may show wind structure of the primary, but could include emissions and changes in structure in the wind-wind interaction. Based upon STIS spectra the Fe II broad features have velocities that imply they originate from the more extensive, denser primary wind, not the secondary wind. The H I profile may or may not differentiate between the wind-wind and the stellar component origins.

Examination of the simple toy model produced by Pittard (private communication) indicates that the two stellar winds produce an undistorted bowshock only near apastron. As the companion approaches the primary, the massive winds interact heavily leading to distortion of the bowshock and then wraps around the primary. During the minimum, the distortion leads to a complete disconnection of the secondary wind from its outer structure and the Lyman radiation becomes trapped. Hence the ionized gas in the direction of the Weigelt condensations is no longer supported by the Lyman flux of the secondary and recombines. A month to three months after periastron, the secondary has plowed through much of the primary wind structure and then begins to build up a new bowshock structure with an ionized path in the direction of the Weigelt condensation. The structure then begins to rebuild an ionized region.

6 Conclusions

Based upon STIS, UVES, FUSE, RXTE and CHANDRA observations, a picture is emerging on the structure of the ejecta around Eta Carinae and the central source. From STIS we know that there are two hourglass structures surrounding Eta Carinae, one nested within the other; the outer being mostly neutral gas and the inner being an ionzed bipolar shell. A large disk structure is located between the two lobes with much neutral gas including singly-ionized metals. Close to the central source are a number of bright condensations associable with the disk, but possibly immersed in the outer regions of the combined stellar winds. STIS and FUSE observations indicate that a mid-O or WN star must be a companion to the massive B-star seen as the extended stellar source. CHANDRA spectroscopy and STIS nebular spectroscopy are explainable in terms of the winds and Lyman flux from a hot secondary companion. UVES extends the STIS observations both in spectral coverage and now, with the loss of STIS, with temporal monitoring at least in the visible spectral region.

The real need is high spatially-resolved imaging with moderate to high spectral resolution. The ideal situation would be those capabilities from the ultraviolet to the infrared, allowing us to build careful diagnostics to gain temperature, density with velocity eventually leading to a three dimensional image of the interactions. Multiple lines of interest lie in the visible region especially from 4400 to 7400Å, but active optics and interferometry currently are limited to longward of one micron. In that case the combination of VLTI with AMBER is the currently best option. An integral spectroscopic combination of VLTI with UVES fed by fiber optics is of great interest in the study of Eta Carinae. The central source and ejecta change rapidly within the 5.5-year period, but also linearly with time. Combining observations taken at at quite different times leads to a confusing picture. Complete sampling in narrow intervals relative to the orbit (equal area in equal time!) needs to be done.

The VLTI with AMBER shows great promise in building the needed three dimensional structure of the interacting binary system within Eta Carinae.

Acknowledgements These studies were accomplished through the Space Telescope Science Institute and with funding from the STIS GTO resources and STScI projects 9420 and 9973, plus FUSE funding through projects D007. The author thanks the STIS GTO team, the HST Eta Carinae Treasury Team and especially the Eta Carinae Lunch Bunch that meets weekly at Goddard Space Flight Center to discuss this intriguing object.

References

1. M. Bautista, H. Hartman, T. R. Gull and K. Lodders: MNRAS **370**, 1191 (2006)
2. M. Corcoran, J. Swank, R. Petre et al IAUC6842 (1998)
3. M. Corcoran, *http : //lheawww.gsfc.nasa.gov/users/corcoran/eta_car*
4. A. Damineli: ApJ **460**, L49 (1996)
5. A. Damineli, O. Stahl, A. Kaufer et al: A&AS **133**, 299 (1998)
6. K. Davidson, R. Humphreys: ARA&A **35**,1 (1996)
7. K. Davidson, N. Smith, T.R.Gull et al: AJ **121**, 1569 (2001)
8. T. R. Gull: ASPC **332**, 281 (2005)
9. T. R. Gull, G. Vieira, F. Bruhweiler et al: ApJ **620**,442 (2005)
10. T. R. Gull, G. Vieira and K. Nielsen: ApJS **163**,173 (2005)
11. H. Hartman, T. R. Gull, S. Johansson and N. Smith: A&A **419**, 215 (2004)
12. J. Hillier, T. Gull, K. Nielsen et al. ApJ, **642**, 1098 (2005)
13. K. Ishibashi, T.R. Gull, K. Davidson et al: AJ **125**, 3222 (2003)
14. R. Kimble, B. Woodgate, C. Bowers et al: ApJ **492**, L83 (1998)
15. J. C. Martin, K. Davidson, R. Humphreys et al. ApJ, **640**, 474 (2006)
16. G. Meynet and A. Maeder: A&A **404**, 975 (2003)
17. K. Nielsen, T. R. Gull and G. Vieira Kober : ApJS **157**, 138 (2005)
18. J. Pittard and M. Corcoran: A&A **383**, 636 (2002)
19. N. Smith: MNRAS **337**, 1252 (2002)
20. N. Smith, J. Morse, N. Collilns and T. R. Gull: ApJ **610**, L105 (2004)
21. O. Stahl, K. Weis, D. Bomans et al: A&A **435** 303 (2005)
22. J. E. Steiner and A. Damineli: ApJ **612**, L133 (2004)
23. R. van Boekel, P. Kervella, M. Sholler et al: A&A **410**, L37 (2003)
24. W.P. Varricatt, P. M. Williams and N. M. Ashok: MNRAS **351**, 1307 (2004)
25. E. Verner, T.R. Gull, F. Bruhweiler et al: ApJ **581**, 59 (2002)
26. E. Verner, F. Bruhweiler and T. R. Gull: ApJ **624**, 973 (2005)
27. G. Weigelt and J. Ebersberger: A&A **163**, L5 (1986)
28. K. Weis, O. Stahl, D. Bomans et al: AJ **129**, 169 (2005)
29. B. Woodgate, R. Kimble, C. Bowers et al: PASP **110**, 1183 (1998)
30. T. Zethson: Hubble Space Telescope spectroscopy of Eta Carinae and Chi Lupi. PhD Thesis, Univerisity of Lund, Lund, Sweden (2001)
31. T. Zethson, T.R. Gull, H. Hartman et al: AJ **122**, 322 (2001)

First AMBER/VLTI Observations of Hot Massive Stars

R. G. Petrov[1], F. Millour[1,2], O. Chesneau[3], G. Weigelt[4], D. Bonneau[3],
Ph. Stee[3], S. Kraus[4], D. Mourard[3], A. Meilland[3], M. Vannier[5], F. Malbet[2],
F. Lisi[6], P. Antonelli[3], P. Kern[2], U. Beckmann[4], S. Lagarde[3], K. Perraut[2],
S. Gennari[5], E. Le Coarer[2], Th. Driebe[4], M. Accardo[5], S. Robbe-Dubois[1],
K. Ohnaka[4], S. Busoni[6], A. Roussel[3], G. Zins[2], J. Behrend[4], D. Ferruzi[5],
Y. Bresson[3], G. Duvert[2], E. Nussbaum[4], A. Marconi[5], Ph. Feautrier[2],
M. Dugué[3], A. Chelli[2], E. Tatulli[2], M. Heininger[4], A. Delboulbe[2],
S. Bonhomme[3], D. Schertl[4], L. Testi[6], Ph. Mathias[3], J.-L. Monin[2],
L. Gluck[2], K. H. Hofmann[4], P. Salinari[6], P. Puget[2], J.-M. Clausse[3],
D. Fraix-Burnet[2], R. Foy[7], and A. Isella[6]

[1] LUAN, Université de Nice - Sophia Antipolis. Parc Valrose, F-06108 Nice Cedex
2, France. petrov@unice.fr
[2] LAOG, Université Joseph Fourier, BP 53, F-38041 Grenoble Cedex 9, France
[3] Observatoire de la Côte d'Azur,BP 4229, F-06304 Nice Cedex 4, France
[4] Max Planck Institute für Radioastronomie, D-53121 Bonn, Germany
[5] European Southern Observatory, Casilla 19001, Santiago, Chile
[6] Osservatorio Astrofisico di Arcetri, Istituto Nazionale di Astrofisica, Largo E.
Fermi 5, I-50125 Firenze, Italy
[7] CRAL, Observatoire de Lyon, F-69561 Saint Genis Laval Cedex, France

Summary. AMBER is the first near infrared focal instrument of the VLTI. It combines three telescopes and produces spectrally resolved interferometric measures. This paper discusses some preliminary results of the first scientific observations of AMBER with three Unit Telescopes at medium (1500) and high (12000) spectral resolution. We derive a first set of constraints on the structure of the circumstellar material around the Wolf Rayet γ^2 Velorum and the LBV η Carinae.

1 Introduction

A feature common to many hot and massive stars is a complex circumstellar envelope revealed by strong emission lines in the spectrum and excesses in the continuum Spectral Energy Distribution (SED). The classical Be star κ Canis Majoris, the brightest Wolf Rayet γ^2 Velorum and the more than famous Luminous Blue Variable η Carinae were among the first medium spectral resolution AMBER Guaranteed Time Targets.

These stars belong to fairly different classes, with, to make it short, specific evolution stages, extremely different mass loss rates (respectively about 10^{-9}, 10^{-5} and $10^{-3} M_\odot/y$), different envelope densities, opacities and chemical compositions.

However, they raise variants of common questions. What are the exact mass loss rates, since their computation requires a model of the envelope geometry? Can we discriminate from the contribution of dust and of free-free emission of gas to the continuum spectrum? Are they mechanisms allowing dust to be present closer to the star than the expected sublimation radius? What are the relative contributions of radiation pressure and stellar rotation to the production and shaping of the envelope? Can we confirm that the radiation pressures show very strong variations with latitude, in particular with the recently renewed importance of the Von Zeipel effect increasing the apparent gravitation and radiation pressure near the poles of fast rotators? Are the stars close enough to critical velocity for this to be the main explanation for mass loss and even for variability through a rotational instability mechanism? Do we have to look into stellar activity producing local perturbations of the velocity field and/or of the radiation pressure? Are there other eruption mechanisms? Is the envelope completely shaped by the emission mechanisms (wind, rotation, kinetic momentum transfer) or is it severely perturbed by other sources such as binarity, which is of course decisive for γ^2 Velorum but also suspected to be important for η Carinae. In the influence of the companion on the circumstellar material what is the part of gravitation and this of the companions own stellar wind and can we see a wind-wind interaction zone? It is now quite clear that spherically symmetric models are outdated but can we still hold on central symmetric ones, eventually moderately perturbed?

It seems that a full answer to these questions would require full images of the targets in the continuum as well as in many narrow spectral channels in many different emissions lines, allowing to derive intensity maps and velocity fields at different optical depths. In principle, AMBER is able to produce such color images in the near infrared thanks to its three telescopes beam combiner feeding a medium (R=1500) and high (R=12000) resolution spectrograph. They are the goal of long term programs of AMBER which will require a very large number of observations with the Auxiliary 1.8 meter telescopes. In the meantime, we decided that these objects are an excellent test case for one of the main bets behind the conception of the "only three telescopes but ambitious spectrograph" AMBER instrument. Since such most studied candidates are already fairly constrained by multi wavelength (from X to far IR) spectro-photometry, high resolution spectroscopy, larger scale imaging, polarimetry and often interferometry without spectral resolution, then a small number of interferometric measures simultaneously in a large number of spectral channels, should allow decisive breakthroughs. What we have measured is very little information compared to full color images but it also multiplies purely spectroscopic information by at least a factor 8.

In the following, we give a first insight of our very preliminary understanding of the first AMBER measures made on these stars.

2 AMBER observations

AMBER is a three beams near infrared VLTI focal instrument producing dispersed fringes with spectral resolutions 35, 1500 and 12000 [1]. This paper refers to medium and high spectral resolution observations made in the K band. Figure 1 displays an example of the individual image detected by the AMBER detector, with left to right the photometric 1 and 2 channels, then the interferometric one and the photometry of the third beam, all dispersed in the vertical direction. One can see a three telescopes fringe pattern in the interferometric channel and the Br_γ emission line crossing all spectra horizontally. The figure also describes the work channel and the reference channel. AMBER measures the stellar spectra, the absolute visibility in each channel, the differential visibility which makes sense even when the absolute visibility is poorly calibrated, the differential phase and the closure phase. These quantities can be interpreted in an unique manner only if we have a good u-v coverage. However, it can be remembered that the visibility is related to the angular scale in λ/B_i units in the direction of the baseline B_i. The differential phase gives an idea of the position of the object photocenter in the direction of the baseline B_i in the spectral channel λ. This is particularly true when the phase is smaller than 1 radian. The closure phase is a measure of the asymmetry of the source: central symmetric or unresolved sources have a zero closure phase. These generalities can be wrong in an infinity of particular cases in which they are wrong, but they still are quite useful for a first interpretation and initial orientation of the model fitting.

For bright sources we currently can guarantee an absolute visibility accuracy ranging between 0.03 and 0.06 and differential visibility and phase smaller than 0.01 (visibility units and radians respectively). The closure phase is usually substantially more noisy than the added individual phases, because the probability to have simultaneously three good fringe patterns is quite low in the currently vibration dominated VLTI.

3 The Wolf Rayet star γ^2 Velorum

We observed the brightest Wolf Rayet γ^2 Velorum in a set of spectral windows in the K band from 1.95 to 2.17 μm. This star have been very extensively observed by spectroscopy, polarimetry and intensity interferometry [3] and is known to be a SB2 spectroscopic binary system (WC8+O7.5III) with period 78.53 days, $(a1 + a2)\sin i = 164.10^6$ Km, eccentricity =0.326. The inclination i=63 ± 5deg is fairly tightly constrained by intensity interferometry and polarimetry [2]. The Hipparcos distance is 258 ± 40 pc and there is a not completely closed controversy with pre-Hipparcos spectro-photometric estimations which were around 400 pc. After constraining again the orbit and the distance by a new interferometric measurement the main goal was to find out where is the circumstellar material in the system. The individual stars should be too small to be resolved (less than 0.5 mas) even if the WR seems larger

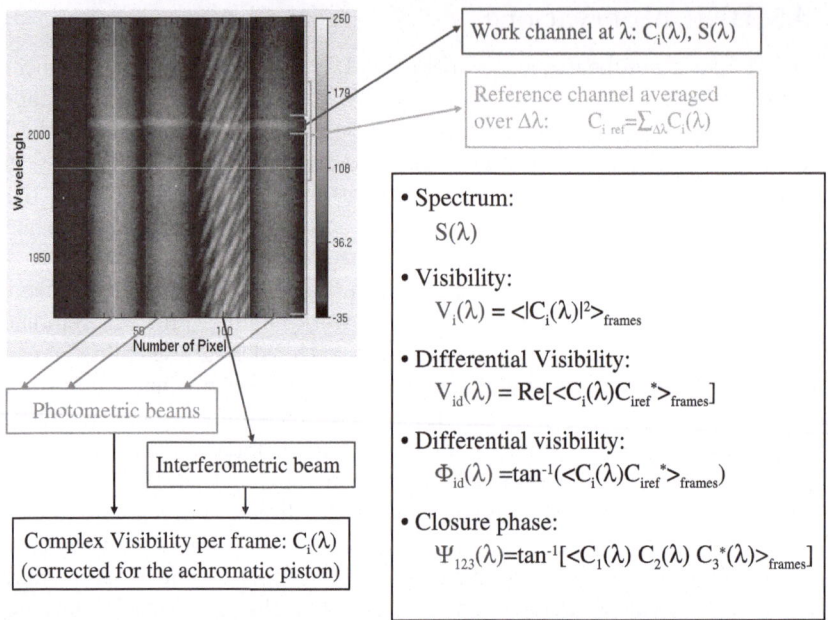

Fig. 1. AMBER typical bright star, three telescopes, medium spectral resolution image and AMBER measures.

because of the optically thick wind. However, we can expect some amount of dust in the system even if SED fits indicate that this should not yield more than 10% of the total flux. We also expect some gas concentrated in the zone where the radiation pressure of the two stars are comparable. This wind-wind zone can contribute to the emission lines but also to the continuum through a free-free emission. The results are displayed if figure 2 with, clockwise from the upper left corner: the spectrum, showing many carbon and helium emission lines, the closure phase, the differential phases and the visibilities as a function of wavelength. The main spectral features are indicated in the spectrum, which is dominated by two strong emission groups at 2.079 m (C IV) and 2.11 (He I) both strongly blended. The closure phase is of the order of 1 radian in the "continuum" channels and shows very strong variations in the emission lines. So does the differential phases, which are set to a zero average in each observing window by the definition of the reference channel, but which very strongly vary in the lines. To fit the measurements, we are performing the following steps:

- Try to find the binary parameters in "continuum" channels. This failed even with increased error bars. A solution is to introduce a faint Resolved Component (RC), contributing to 15% of the total flux Then we have a very good fit.

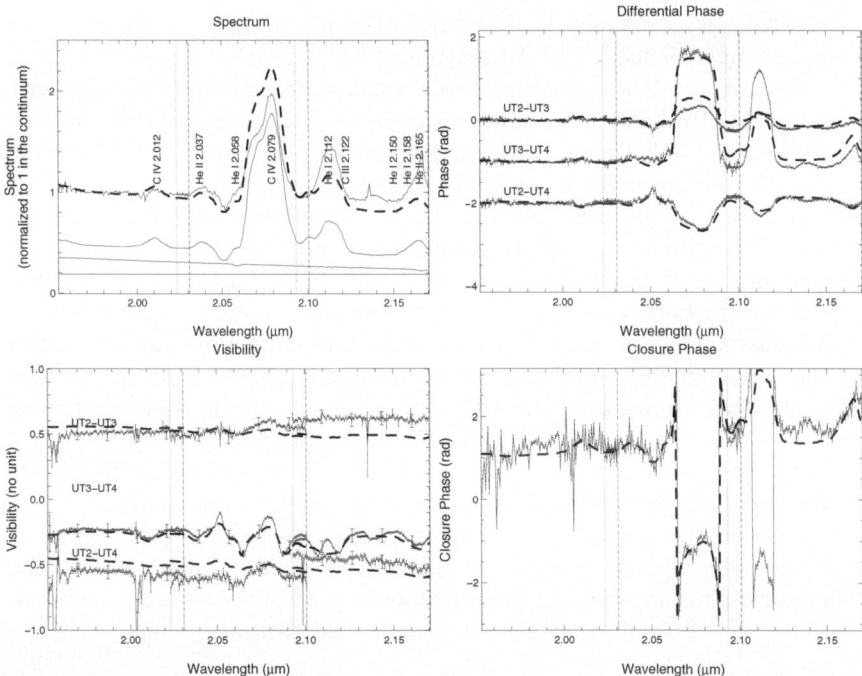

Fig. 2. AMBER observations of $\gamma^2 Velorum$. The vertical lines show the limit of the observing spectral windows. The spectrum displays from top to bottom: the calibrated normalized spectrum (full line), the best fit (dashed line), the WC8 component model spectrum, the O7III component model spectrum and the Resolved Component flat contribution. In the visibility figure we have V (top) for UT2-UT3, V-0.5 (middle) for UT3-UT4 and V-1 (bottom) for UT2-UT4. The small error bars are for differential visibility accuracy and the large ones are from the external calibration uncertainty. For the differential phases we have $\Delta\phi$ for UT2-UT3, $\Delta\phi-1$ for UT3-UT4 and $\Delta\phi-2$ for UT3-UT4. In all cases, the full line represents measures or inputs to the model and the dashed line is for the best fit with a binary.

- Use models of the component spectra, derived from spectroscopic observations [4], to further constrain the binary parameters. This is what is represented in figure 2 and yields the parameter which are discussed below. Remarkably, we find similar binary and RC parameters in the continuum and in the lines.
- Improve the fit by changing the WR spectrum. Actually we will derive the best possible WR spectrum knowing the parameters of the binary. This step is currently under progress and improves the fit of the closure and differential phases. From the observation of figure 2, it already appears that the phases are well fitted where the WR spectrum reconstruction is coherent, for example around the C IV 2.079 μm line, and wrong where the intensity of the WR line is incorrect as it can be seen in the spectrum near 2.112 μm. However, it is also clear that a change of the component

spectra cannot explain the sharp variation in visibility which can be seen for example for the UT3-UT4 baseline at 2.053, 2.081, 2.108 and 2.12 μm,. The strong visibility variations reveal small scale structures, possibly with large radial velocities since they seem to be in the wings of the C IV and He I lines. We believe that this is the signature of circumstellar material concentrated near the wind-wind collision zone.

– Constrain the geometry and kinematics of the wind-wind zone by fitting the visibility, closure and differential phase residuals left after the best binary fit. These residuals are 3 to 5 times larger than the differential error bars. First, we will try to introduce in each spectral channel a Gaussian "cloud" whose size, intensity and position might materialize the corresponding isovelocity wind-wind region. Second, we will use the supposedly known geometry of the wind region around the axis joining the two stars to have a fit with less parameters.

– From the geometry of the wind region, we can derive its contribution to the continuum spectrum. Then, we will discriminate dust and gas contributions to the continuum and repeat all the process.

– Eventually, all this relatively simple mix of geometrical constraints on physical parameters might yield a global model of the system, whose parameters will again be evaluated from the measures.

Table 1. Parameters of the γ^2 Vel binary system

separation ρ .	position angle .	flux ratio .	resolved component .
3.6 ± 0.03 mas	165 ± 5 deg	0.6 ± 0.1	0.14 ± 0.04

The best global fit is obtained for the binary and resolved (RC) component parameters given in table 1. This has to be compared with the $\rho = 5mas$ and $\theta = 161deg$ expected from the spectroscopic binary and the Hipparcos distance. The new value of the separation can be explained by an error on the inclination $\sin i$ and/or on the parallax. Since $\sin i$ seems much better constrained, our results yields a parallax of $348 \pm 50pc$ which is compatible both with the Hipparcos estimate (we are at 2.2 σ) and with the pre-Hipparcos spectro-photometry.

4 The Luminous Blue Variable η Carinae

The Luminous Blue Variable η Carinae, most luminous star known in our Galaxy, is one of the best studied and maybe less understood massive stars. It is the source of the spectacular Homunculus nebulae, produced by a maybe 10 solar masses outburst in 1840. A second massive outburst occurred in 1890 and produced a second nebulae which seems to be an embedded smaller replica of the first one. The reason for the outburst remains unknown but many attempts are being made to connect the shapes of the present day stellar wind and the nebulae general geometry. After indications from the HST Imaging Spectrograph, the first VLTI observations with VINCI strongly indicated that the stellar wind appears elongated in the nebulae axial direction which is also believed to be the stellar rotation axis. This is compatible with a radiation pressure substantially increased near the poles by a Von Zeipel effect. Our goal was to extend the VINCI observations by adding spectral resolution to the interferometric measures in order to constrain the velocity field.

One of the main difficulties in observing η Car with single mode fiber instruments such as AMBER or VINCI is that the fibers collect information from an extended patch of the sky. In the case of VINCI siderostats, an array of about 1.4 arc second contribute to the interferogram. Images from the NACO adaptive optics (resolution 50 to 100 mas) have been used to find out what fraction of the collected flux can actually contribute to the fringes (i.e. is produced in an array smaller than 10 to 50 mas). This allowed to estimate that the central source contains 57% of the flux. When it was assumed that the remaining 43% are completely resolved by the interferometer, the resulting visibilities show a smooth variation with the position angle of the baseline. This is interpreted as a present day wind shape elongated exactly along the axis of the nebulae, with a ratio of 1.25 between the projected major and minor axes.

The first result of the AMBER 3 UT measurements is that closure phase is zero (within the 0.05 radians accuracy) in the continuum for all observing times, covering a fairly large range of hour angles. This is a good confirmation that at least in the continuum the object is well represented by a central symmetric structure such as the Gaussian ellipsoid.

The UTs inject in AMBER fibers the light coming from about 70 mas. We first assumed that AMBER signal would be completely dominated by the central stellar wind. Figure 3a shows the AMBER measures (triangles), each visibility point converted in the FWHM of a Gaussian. The dashed line shows the Van Boekel et al. Vinci fit [5]. The AMBER points are not compatible with the VINCI fit and show strong variations of Gaussian FWHM with small variations of position angle. If we assume that AMBER data has been contaminated by a fraction of light for an interferometrically resolved source, then it is possible to eliminate the strong variations of FWHM with PA. The remarkable point is that we then obtain a structure elongated exactly in the same direction and with the almost the same major to minor axis ratio than from VINCI. Figure 3b shows the AMBER measures corrected assuming that the flux ratio between the resolved and central structures is 0.45 (i.e. the resolved structure contributes to 31% of the total flux instead of 43% in the VINCI case). The dotted curve shows the VINCI fit, scaled down by a factor 1.33. The easiest way to explain this difference in apparent size is to challenge the VINCI estimation of the contribution of non resolved structures to the total flux, since we now know that a fraction of the flux in one UT Airy disk comes from structures non resolved by the interferometer. The values in figure 3b are indeed compatible with VINCI measurement where 47% instead of 43% of the flux contribute to the unresolved structure. This would slightly change the major to minor axis ratio but not the position angle of the structure.

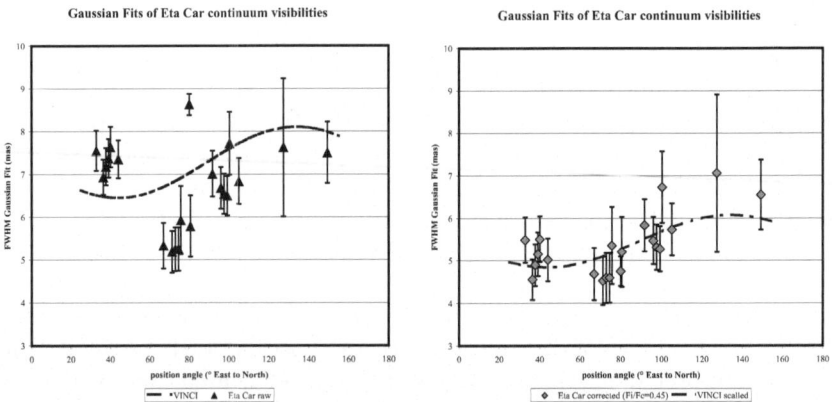

Fig. 3. AMBER observations of η Carinae in the continuum. FWHM of the Gaussian fit as a function of the position angle of the projected baseline. Left: from raw visibility (triangles) and comparison with the Van Boekel et al. fit of VINCI data (dashed curve). Right: from visibility corrected assuming that 31% of the flux comes from an interferometrically fully resolved component (diamonds). The dashed curve shows the VINCI data fit scaled down by a factor 1.33 or assuming that in VINCI data the resolved component contributes to 47% in the total flux instead of 43% in the original VINCI fit.

The analysis of this continuum data confirms the importance and the difficulty of separating the contributions from different elements in the single mode field. A full solution would be to have interferometric observations with some field or at least to be able to have an efficient mosaicing strategy. Maybe the new infrared image sensor IRIS can be used for that purpose. The spectral information will help refining this task. We already know that the "resolved component" discussed here does not show sharp spectral features, unlike for example the Weigelt blobs which have been shown to generate narrow band features when we are pointing about 200 mas away from the brightest spot. A more detailed discussion of the shape of the complex visibility through the line is quite premature and would make this paper even longer. A preliminary result is that the differential phase through the spectral lines is basically flat but for local spikes corresponding to photocenter displacements substantially smaller than the object size. This seems in contradiction with an optically thick wind with a surface on which the different equal velocity zones are clearly separated. In fact the data seem to show similar ovoid structures in all spectral channels, the key change being the optical depth.

5 Conclusion

At an early stage in the reduction and interpretation of the first AMBER data on bright massive hot stars, we have tried to illustrate our strategy for model fitting in spite of an extremely limited u-v coverage. We want to use as much as possible the pre existing information and to find new features where our measures as a function of λ differ significantly from the best model fitted in continuum data. For $\gamma^2 Vel$ we are able to show that the interferometric signal is dominated by the binary system but that it is necessary to include an unresolved component with a spectra almost flat over the K band. However we also detect smaller scale structures in the system that are a good candidate for a signature of the wind-wind collision zone. The modeling of this zone will also allow to constrain further the nature of the spectral continuum component. For ηCar, the situation is made more complex by the necessity to evaluate quite accurately the contribution of the larger scale structure to the flux collected by the fibers. The analysis of the AMBER data confirms the VINCI observation of a structure elongated in the direction of Homonculus nebulae, but it also shows that it is necessary to revise the VINCI evaluation of larger scale structure contribution. The consequences are quite important since a fairly limited variation ($<5\%$) of the contribution can change the estimated size by about 30%. Next, we will try to combine AMBER measurements with spectrally resolved NACO+PF observations, to have a map as accurate as possible of the different scales of structures and will further analyze the differential and closure phases which are much less sensitive to large scale underlying structures

Acknowledgements: The authors deeply acknowledge the AMBER consortium members, the staff of the associated Institutes[8] and the ESO/VLTI team who permitted to obtain these results.

References

1. R.G. Petrov et al: Interferometry for Optical Astronomy II. Edited by Wesley A. Traub. Proceedings of the SPIE, Volume 4838, pp. 924-933 (2003).
2. W. Schmutz et al:Astronomy and Astrophysics, v.328, p.219-228 (1997)
3. R. Hanbury Brown et al: Mont. Not. R. Astr. Soc. **148**, 103-117 (1970)
4. L. Dessart et al: Mon. Not. R. Astron. Soc. **315**, 407-422 (2000)
5. R. van Boekel et al: Astron. Astrophys. **410**, L37-L40 (2003)
6. O. Chesneau et al: these proceedings (2005)

---

[8] See list of consortium members and associates at:
http://amber.obs.ujf-grenoble.fr/article.php3?id_article=45

Mineralogy of Circumstellar Dust

L.B.F.M. Waters[1,2] and Ch. Leinert[3]

[1] Astronomical Institute "Anton Pannekoek", University of Amsterdam, Kruislaan 403, NL-1098 SJ Amsterdam, The Netherlands
rensw@science.uva.nl
[2] Instituut voor Sterrenkunde, Katholieke Universiteit Leuven, Celestijnenlaan 200B, B-3001 Leuven, Belgium
[3] Max Planck Institut Für Astronomie, Königstuhl 17, D-69117 Heidelberg, Germany leinert@mpia.de

Summary. The composition and spatial distribution of circumstellar dust in young and old stars can be used as a probe of the physical and chemical processes that take place. Such studies require a modest spectral resolution in combination with high spatial resolution and large collecting area, now for the first time provided by the new generation of optical interferometers. This review discusses some early results from the VLT Interferometer in the area of circumstellar dust.

1 Introduction

Both in their infancy as well as at the end of their life, stars are surrounded by large amounts of gas and dust. While in young stars this material originates from the interstellar cloud which formed the star, in old stars the material is from a stellar wind. This difference in origin is naturally reflected in its composition: in young stars interstellar material is observed, as it is modified by the process of star- and planet formation. In old stars, material from the stellar photosphere is blown into space, and its composition is determined by the chemical evolution of the underlying star. In both cases however it is crucial to observe this circumstellar matter at very high angular resolution, in order to determine its spatial distribution, and to learn more about the physical and chemical processes that take place: near the star strong gradients in composition of the gas and dust are expected to occur. This review focuses on the *circumstellar dust* in young and old stars (such as Asymptotic Giant Branch (AGB) stars), and how it can be used as a diagnostic for planet formation (in young stars) and dust formation from gas-phase condensation (in old stars).

2 Infrared diagnostics of dust

The infrared spectral region offers a rich diagnostic of the chemical composition, lattice structure and (within certain limits) size and shape of dust

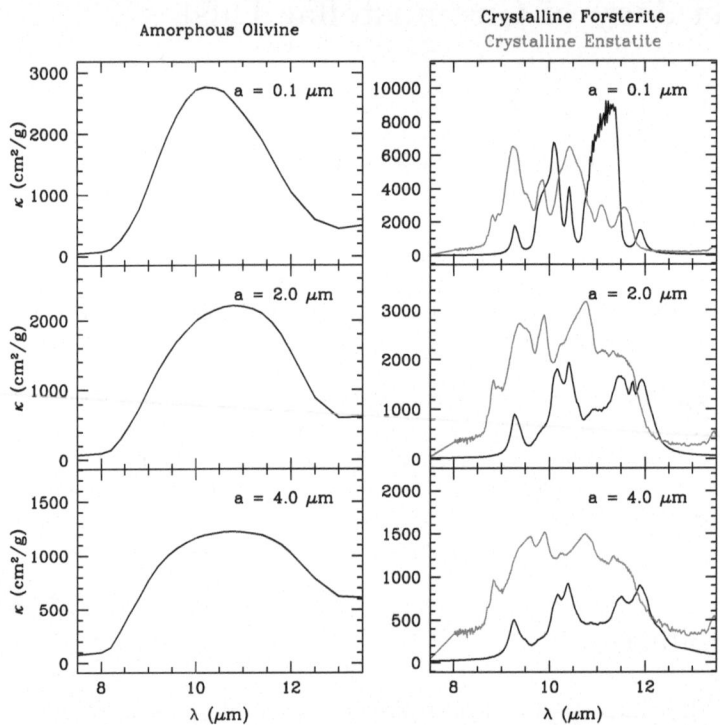

Fig. 1. Mass absorption spectra of amorphous (left column) and crystalline (right column) silicates with several grain sizes (see [18]).

particles. This is because strong vibrational resonances of abundant dust-forming molecules are in the (mid-) infrared. Table 1 lists some of the most important resonances in circumstellar dust with their identification, that can be detected using ground-based IR instrumentation. There are also dust components that lack resonances, which makes these more difficult to quantify, notably amorphous carbon and metallic Fe. Based on meteoritic evidence [1], amorphous carbon must be a dominant dust species produced in C-rich AGB outflows, and is also abundantly present in interstellar space. Fits to the spectra of O-rich AGB stars suggest that metallic Fe is present at modest abundance [15].

The strength and shape of infrared dust resonances is sensitive to grain size (and shape). Figure 1 shows the mass absorption at mid-infrared wavelengths of two crystalline dust species, forsterite and enstatite, and amorphous olivine ($FeMgSiO_4$), for a range of dust grain size, and assuming a distribution of hollow spheres to represent a distribution of irregularly shaped particles [18]; similar calculations using different grain models were published by [10]. The resonances react differently to changes in grain shape because of

Table 1. Circumstellar dust species in the 10 and 20 μm atmospheric windows

λ (μm)	species	location[1]
9.7, 18	amorph. silicates	Y,O
10.6, 11.3, 19.5, 23.5	forsterite Mg_2SiO_4	Y,O
9.2	enstatite $MgSiO_3$	Y,O
8.6, 20.1	SiO_2	Y
23	FeS	Y,O
13	spinel? $AlMg_2O_4$	O
11	alumina	O
19.5	$Fe_xMg_{1-x}O$	O
7.7, 8.6, 11.3, 12.7	PAH	Y,O
11.3	SiC	O
20.3	TiC?	O

[1] Y = young stars, O = old stars (Asymptotic Giant Branch (AGB) post-AGB, Supergiant)

differences in the complex refractive index of the grain material. In general, resonances will weaken with respect to the adjacent continuum as grain size increases, and the peak position will also shift, mostly to longer wavelengths. In practice this implies that, for instance, silicate grains larger than a few microns in size will be difficult to detect because of their weak resonances.

3 Dust in proto-planetary disks

The dust in proto-planetary disks originates from interstellar space, but may have gone through considerable changes in the molecular cloud, and/or while being incorporated into the accretion disk. Spectroscopy of interstellar dust shows evidence for only a few species, of which amorphous Fe-Mg silicates and amorphous carbon are the most abundant by mass. Apart from these, Polycyclic Aromatic Hydrocarbons (PAHs) are ubiquitous. Many processes occur in the proto-planetary disk that may alter this simple composition (fig 2). At very low temperatures, that occur far from the star and in the disk mid-plane, most molecules will be frozen onto the grains causing a very interesting ice chemistry. These ices slowly evaporate when coming closer to the star, meanwhile releasing new molecular species formed in the ice. At modest temperatures, above about 400 K, reactions between Fe and H_2S may result in the formation of FeS. In the innermost disk regions, the dust temperature may rise above 1000 K. Above this temperature the silicate lattice will convert from amorphous to the energetically more favourable crystalline form (annealing); At roughly similar temperatures, amorphous carbon becomes graphitic. The fate of Fe is not well known. During annealing the Fe

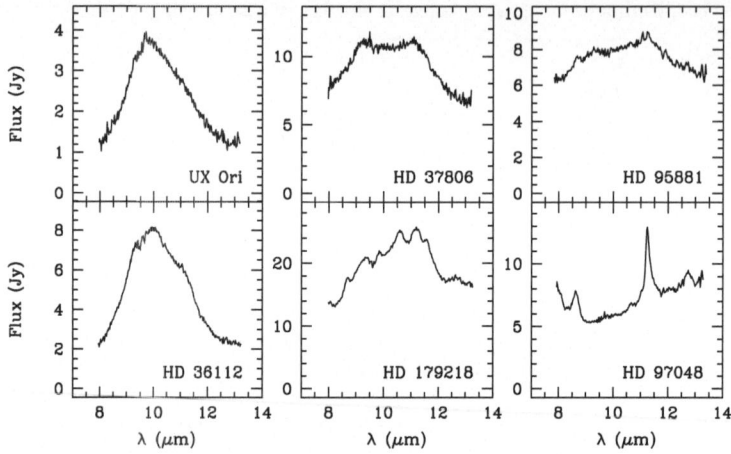

Fig. 2. Spatially unresolved 10 μm spectra of some Herbig Ae/Be stars [29, 28]. Note the wide range in shape and strength of the silicate bands, due to changes in grain size and composition in the upper disk atmosphere. In HD97048, no silicates are seen, the spectrum is dominated by PAH emission.

may be lost from the Fe-Mg-rich amorphous silicate (creating a Mg-rich crystal), or it may stay in the solid (producing an Fe-rich olivine). At 1500 K, silicates will evaporate, and an equilibrium chemistry will allow the condensation of Mg-rich crystalline forsterite, and at slightly lower temperatures the conversion of forsterite into enstatite and SiO_2. Radial mixing will cause these altered materials to be transported away from their formation sites to larger distance from the star, causing a chemically very inhomogeneous dust composition. In addition, local processes may also be important in causing chemical alteration of the dust, by means of shock heating, lightning, and parent body processing.

While these chemical alterations take place, grain-grain collisions are mostly at low velocities because the grains are well coupled to the gas. These collisions will likely be constructive resulting in the slow aggregation of grains to larger structures. Settling will cause large grains to disappear from the disk surface and sink towards the mid-plane. However, the same mechanism which causes radial mixing, will also mix vertically on timescales much shorter than those for radial mixing. It is thus expected that some fraction of the mid-plane grain population, presumably the micron-sized and smaller grains, are mixed up to the surface.

The dust in the disk surface layer is important because these are the grains that absorb stellar optical and ultraviolet photons used to heat the disk, determining the vertical scale-height of the disk as a function of distance. When small grains are abundant, dust opacities will be high and the disk will be easily heated. Self-consistent disk models in hydrostatic equilibrium (e.g.[5, 9]) show that flaring disks are a natural solution of the equations governing disk structure. Disk models with an inner dust-free zone [9] predict that their inner rim is puffed-up and that this inner rim can cast a shadow over part of the outer disk region. With decreasing dust opacity due to grain aggregation, the disk scale-height may drop so much that the entire outer disk remains in the shadow of the inner rim: a self-shadowed, flat disk geometry results [8]. These changes in disk structure will feed back to the dust aggregation, since this is a density-sensitive process.

3.1 Early results using the VLTI

As described above, disk models predict a strong radial and vertical gradient of the nature of dust in proto-planetary disks, from highly processed and aggregated in the inner disk regions, to much more pristine in the outer disk regions. In addition, strong vertical gradients exist. Planet formation may locally have a strong effect on grain properties. These gradients occur on spatial scales of $\approx$ 1 AU, which corresponds to about 10 milliarcsec in the nearest star forming regions. The VLT Interferometer is ideal to study such spatial scales, in particular the MIDI instrument [16] which operates in the 10 μm atmospheric window. Below we summarize some early VLTI results on proto-planetary disks. We also refer to the review by Dutrey (this volume) on MIDI results.

[17] and [27] studied the spatial structure and composition of dust in the innermost regions of seven Herbig Ae/Be stars. These are intermediate mass pre-main-sequence stars surrounded by a passively heated proto-planetary disk. The age of the stars is in the range 1-10 Myrs [30]. In Fig. 3 we show the results for two stars, HD 142527 and HD 163296. The visibility curve shows a weak drop between 8 and $\approx$ 10 μm, and is flat between 10 and 13.5 μm. Such global behaviour is expected for [9] models, although the effect is predicted to be stronger than observed. The high visibility at 8 μm is caused by the inner rim flux contribution, while at longer wavelengths the outer regions dominate more.

The 10 μm spectra of the spatially unresolved disk, of the inner disk (correlated spectrum) and of the outer disk (total minus inner) of HD 142527 and HD 163296 are also shown in fig. 3. The large difference between inner and outer disk spectra of both stars is due to a strong gradient in crystallinity of the small (less than a few μm) silicate grain component in the upper disk. This shows that a large reservoir of crystalline silicates is present in the inner disk. The two stars show a large difference in the degree of crystallinity: in HD 142527 the entire inner disk ($<$ 2-3 AU) is crystalline, while in HD 163296 about 40 per cent of the grains in that region is crystalline. Model calculations indicate that the MIDI observations are consistent with the innermost, T $>$ 1000 K grains being fully crystalline in HD 163296 and the T $<$ 1000 K grains being mostly amorphous: the grains are crystalline *where they must* and amorphous *where they can* (Meijer et al, in preparation). The case of

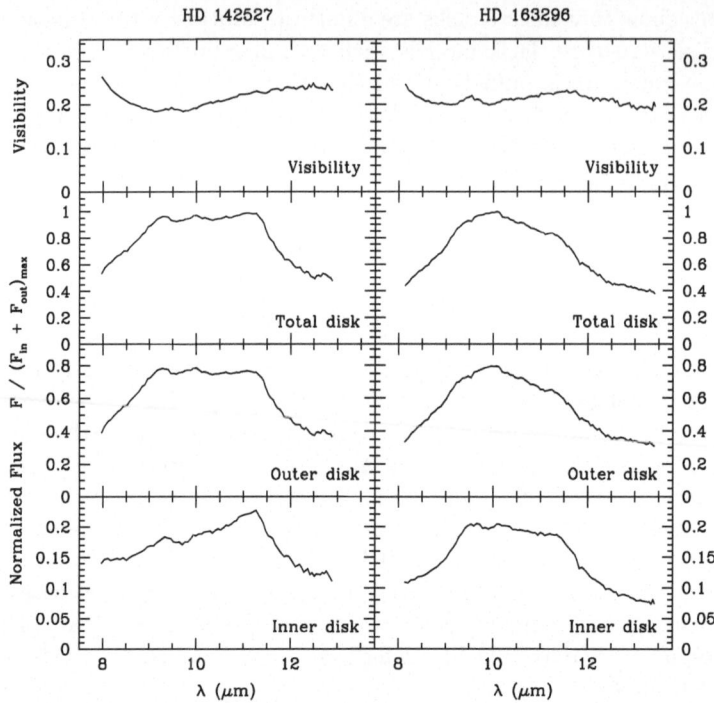

Fig. 3. MIDI observations of two Herbig Ae/Be stars. Shown are the visibility curve, the total (unresolved) 10 μm spectrum, the correlated flux (inner disk) and the total minus correlated flux (outer disk). See [17] and [27].

HD 142527 is different: here substantial parts of the outer disk must also be crystalline. This could be due to radial mixing or local processes. Interestingly, [27] find evidence for a gradient in grain mineralogy of the crystals in HD 142527, with forsterite more abundant in the innermost regions, and roughly equal amounts of forsterite and enstatite further out. Such a gradient is qualitatively predicted by chemical equilibrium disk models[11].

The disk surrounding 51 Oph seems to have a different structure compared to that of most other Herbig Ae/Be stars. The 2-8 μm spectrum is dominated by emission from molecules as CO and H_2O [31] and the presence of CO first overtone emission indicates that very hot CO gas is present close to the star [25]. The MIDI observations of 51 Oph suggest a very compact source, difficult to account for using DDN type models. Gil et al. (these proceedings) fit the infrared spectrum and MIDI observations using a small, geometrically flat disk.

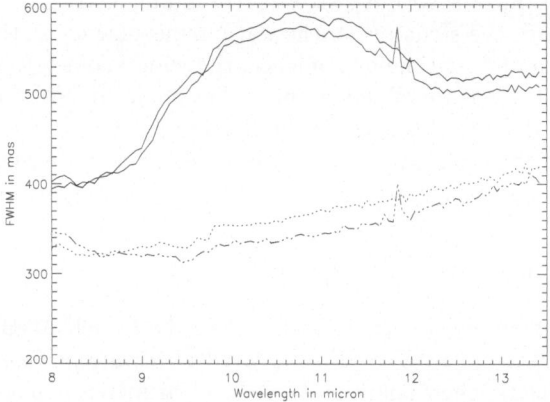

Fig. 4. Upper curves: size of the dust envelope of OH26.5+0.6 measured using the MIDI prism on a single UT telescope. The object is largest at the maximum of the dust optical depth, near 10 μm [4]. Lower curves: same for point source calibrators.

4 Dust in evolved stars

Cool evolved stars, both Asymptotic Giant Branch (AGB) stars and Red Supergiants (RSG) are characterized by a dense, extended layer of (molecular) gas which surrounds the stellar photosphere. This extended warm molecular layer is caused by stellar pulsations with periods between $\approx$ 100 and 1500 days. The physical and chemical conditions in the inner molecular envelope are favorable for the formation of dust; the chemical composition of the stellar photosphere of course determines the nature of the dust that can form: O-rich (e.g. oxides, silicates) for mira's and OH/IR stars (C/O < 1) and C-rich (e.g. amorphous carbon, SiC) for carbon stars (C/O > 1). The newly formed dust absorbs stellar photons and momentum, which causes the material (gas and dust) to flow away from the star at velocities between 5 and 30 km/s. For mass loss rates in excess of about 10^{-5} M$_\odot$/yr the dust column density is high enough for the photosphere to be obscured at short wavelengths.

Mid-infrared spectroscopy of dusty, oxygen-rich AGB stars show a dust composition which varies strongly between stars, possibly related to the dust mass loss rate (wind density) and/or the pulsation properties of the underlying star. Stars with little circumstellar dust are dominated by simple oxides, while the amount of amorphous silicates can be very small [23, 24, 22, 21, 3]. With increasing amount of dust, amorphous silicates become the dominant species, and for obscured (OH/IR) stars crystalline silicates are also observed [33]. Carbon stars show less obvious trends, with amorphous carbon and SiC always present, and MgS for mass loss rates above about 10^{-6} M$_\odot$/yr.

Since stellar winds of late type stars strongly influence their evolution, the physics of their winds merits detailed studies. High spatial resolution observations of the dust forming layers surrounding AGB stars are important to constrain both the structure of the warm molecular layer, the dust formation mechanism and its time dependence, the wind geometry, and the effects of binarity on the nature of the circumstellar environment. Ground-braking work has been done in this area by the ISI interferometer [26, 12]. These ISI results allowed for the first time to determine stellar radii near 10 μm, and the dust formation region around AGB stars and supergiants (e.g. [6, 2]).

4.1 AGB stars

[20] studied the circumstellar environment of the O-rich AGB star RR Sco, combining both near IR (VINCI) and MIDI observations (see also Ohnaka, these proceedings). Their analysis showed a clear increase in stellar diameter between 2 and 10 μm, which can be explained by a warm molecular layer of about 2 stellar radii, containing H_2O and SiO. The inner radius of the dust shell is located at ≈ 7 R_* and probably consists of Al_2O_3.

The OH/IR star OH26.5+0.6 is a prototype AGB star in the so-called "superwind" phase; this phase is characterized by excessive surface mass loss which may terminate the AGB. The most recent high mass loss phase of OH26.5+0.6 started some 200 yrs ago [14]. Since virtually all proto-planetary nebulae are non-symmetric, the question arises when these global asymmetries develop. The MIDI instrument provided both single telescope as well as interferometric observations of OH26.5+0.6 [4], and both set constraints on the wind geometry. [4] find that the object is resolved in the single telescope images, with a clear asymmetry. Furthermore, using the prism of MIDI the size of the star was found to vary with wavelength, being large in the centre of the 10 μm silicate feature (fig 4). [4] conclude that this wavelength dependent size of the envelope is difficult to understand if the dust shell extends to very close to the star. Models with an inner dust-free region of 20-30 stellar radii can fit the observations. The reason why such a large inner dust-free cavity should exist are unclear. Interestingly, MIDI detected no fringes on the 100 meter baseline, i.e. the central star was not detected.

4.2 Binary post-AGB stars

The evolution and circumstellar environment of binaries of which one star is on the AGB is often very different from that of single stars. Close binaries probably merge, while wider systems can survive the AGB period, albeit with strongly modified orbital parameters. Mass lost during the early AGB period can be stored in the system, either surrounding the companion or the entire system. In this way long-lived reservoirs of material are formed; the AGB star may evolve to a C-rich object, causing both O-rich and C-rich dust to be present. The dust in these reservoirs will undergo the same processing as in the disks surrounding young stars (for a recent review see [32]). The

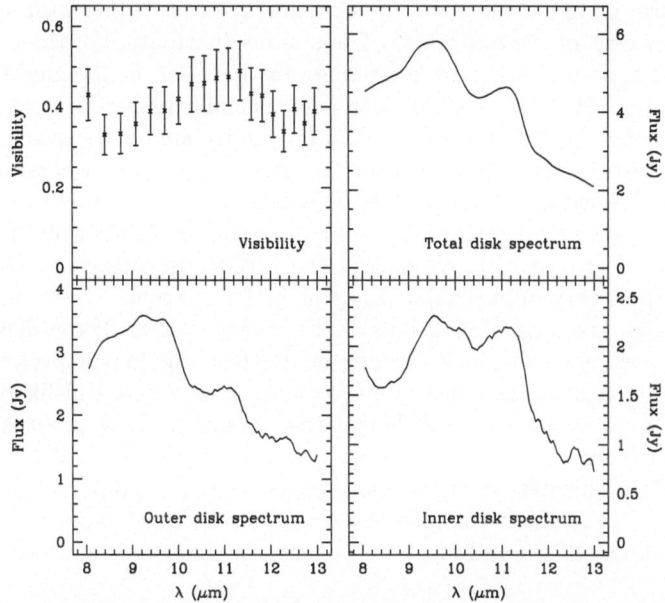

Fig. 5. MIDI observations of the binary post-AGB star HD52961 [7]

difference between young and old stars is that in the former *most* dust is primitive, interstellar, while in evolved stars *all* dust is freshly condensed.

The infrared spectra of post-AGB binaries with circumbinary disks show a wide temperature distribution, with the hottest dust close to the dust sublimation temperature of about 1500 K (De Ruyter et al, submitted to A&A). The millimeter wave continuum has a spectral slope consistent with the presence of large (i.e. mm-sized), cold grains. The Red Rectangle, the most famous example of this class of objects, has a circum-binary disk with grains that could be of cm size [13]. The (cold) dust in these disks is oxygen-rich, and often highly crystalline (e.g. [19]). The spatial distribution of the dust holds important clues concerning the formation history of the disk, and the nature of the dust processing. As in the Herbig stars, the spatial scales of these systems are well suited for the VLTI.

Deroo et al. [7] observed the binary post-AGB stars HD 52961 and SX Cen with MIDI. While SX Can was found to be unresolved, HD 52961 is well resolved on $\approx$ 40 m baselines (fig. 5). Uniform disk fits to the visibility give a size of 35 milli-arcsec at 8 μm, increasing to about 55 milli-arcsec at 13 μm. The spatially unresolved 8-13 μm spectrum shows a weak silicate feature which clear evidence for crystalline silicates. The correlated flux spectrum has a much different shape, with much more prominent contributions from forsterite and enstatite. Clearly the crystalline silicates in HD 52961 are concentrated in the innermost disk regions. This is confirmed by fits to the spectra, giving $\approx$ 80 per cent crystallinity of the inner disk spectrum, and 20 per cent of the outer disk. The spatial distribution of the crystals in HD 52961 is at first sight rather similar to that seen in the Herbig Ae/Be stars! We note that the outer disk shows a peculiar mineralogy, with an unusually strong band near 9.5 μm, possibly due to large SiO_2 grains.

The question arises what causes this strong gradient in crystallinity. If the grains formed in a "normal" AGB wind, one would expect the bulk to be amorphous when they enter the disk. Thermal annealing could then crystallize the innermost disk, explaining the MIDI observations of HD 52961. However, it is very unlikely that the star had a "normal" dusty wind when on the AGB, given the dimensions of the binary system. If the disk is built from captured gas, with dust forming in the disk, one may expect all grains to be highly crystalline, since grains forming in the disk will likely remain at high temperatures for much longer than grains forming in an AGB outflow. Indeed, the (unresolved) mid-IR spectrum of SX Cen indicates that 80 percent of the silicates are crystalline, suggesting a much different spatial distribution of these materials. More observations of similar objects are needed to establish the origin and formation history of these very interesting disks.

5 Conclusions

The early results from the VLT Interferometer show the tremendous potential of spectrally resolved high angular resolution observations. This holds great promise for the near future, when the VLTI and other interferometers as the Keck Interferometer will come to full development. Instruments such as MIDI and AMBER are setting the stage for new generations of instruments, allowing unprecedented sensitivity for the highest spatial resolution imaging.

Acknowledgements It is a pleasure to thank the MIDI team and ESO staff for their essential contributions to make VLTI work. Thanks also to Roy van Boekel, Michiel Min, Pieter Deroo and Hans van Winckel for their important input to this review.

References

1. Bernatowicz, T. J., Cowsik, R., Gibbons, P. C., et al. 1996, ApJ, 472, 760
2. Bester, M., Danchi, W. C., Hale, D., et al. 1996, ApJ, 463, 336
3. Cami, J. 2002, Ph.D. Thesis
4. Chesneau, O., Verhoelst, T., Lopez, B., et al. 2005, A&A, 435, 563
5. Chiang, E. I. & Goldreich, P. 1997, ApJ, 490, 368
6. Danchi, W. C., Bester, M., Degiacomi, C. G., McCullough, P. R., & Townes, C. H. 1990, ApJ, 359, L59
7. Deroo, P., et al. 2006, A&A, 450, 181
8. Dullemond, C. P. & Dominik, C. 2004, A&A, 417, 159

9. Dullemond, C. P., Dominik, C., & Natta, A. 2001, ApJ, 560, 957
10. Fabian, D., Henning, T., Jäger, C., et al. 2001, A&A, 378, 228
11. Gail, H.-P. 2004, A&A, 413, 571
12. Hale, D. D. S., Bester, M., Danchi, W. C., et al. 2000, ApJ, 537, 998
13. Jura, M., Turner, J., & Balm, S. P. 1997, ApJ, 474, 741
14. Justtanont, K., Skinner, C. J., Tielens, A. G. G. M., Meixner, M., & Baas, F. 1996, ApJ, 456, 337
15. Kemper, F., de Koter, A., Waters, L. B. F. M., Bouwman, J., & Tielens, A. G. G. M. 2002, A&A, 384, 585
16. Leinert, C., Graser, U., Waters, L. B. F. M., et al. 2003, in Interferometry for Optical Astronomy II. Edited by Wesley A. Traub. Proceedings of the SPIE, Volume 4838, pp. 893-904 (2003)., 893–904
17. Leinert, C., van Boekel, R., Waters, L. B. F. M., et al. 2004, A&A, 423, 537
18. Min, M., Hovenier, J. W., & de Koter, A. 2003, A&A, 404, 35
19. Molster, F. J., Yamamura, I., Waters, L. B. F., et al. 2001, A&A, 366, 923
20. Ohnaka, K., Bergeat, J., Driebe, T., et al. 2005, A&A, 429, 1057
21. Posch, T., Kerschbaum, F., Mutschke, H., Dorschner, J., & Jäger, C. 2002, A&A, 393, L7
22. Posch, T., Kerschbaum, F., Mutschke, H., et al. 1999, A&A, 352, 609
23. Sloan, G. C. & Price, S. D. 1995, ApJ, 451, 758
24. Sloan, G. C. & Price, S. D. 1998, ApJS, 119, 141
25. Thi, W.-F., van Dalen, B., Bik, A., & Waters, L. B. F. M. 2005, A&A, 430, L61
26. Townes, C. H., Bester, M., Danchi, W. C., et al. 1998, in Proc. SPIE Vol. 3350, p. 908-932, Astronomical Interferometry, Robert D. Reasenberg; Ed., 908–932
27. van Boekel, R., Min, M., Leinert, C., et al. 2004, Nature, 432, 479
28. van Boekel, R., Min, M., Waters, L. B. F. M., et al. 2005, A&A, 437, 189
29. van Boekel, R., Waters, L. B. F. M., Dominik, C., et al. 2003, A&A, 400, L21
30. van den Ancker, M. E., de Winter, D., & Tjin A Djie, H. R. E. 1998, A&A, 330, 145
31. van den Ancker, M. E., Meeus, G., Cami, J., Waters, L. B. F. M., & Waelkens, C. 2001, A&A, 369, L17
32. van Winckel, H. 2003, ARA&A, 41, 391
33. Waters, L. B. F. M., Molster, F. J., de Jong, T., et al. 1996, A&A, 315, L361

First Evidence for a Spatially Resolved Disk Structure around the Herbig Ae Star R CrA

S. Correia[1], R. Köhler[2], G. Meeus[1], and H. Zinnecker[1]

[1] Astrophysikalisches Institut Potsdam, An der Sternwarte 16, D-14482 Potsdam, Germany scorreia@aip.de, hzinnecker@aip.de, gwen@aip.de
[2] Sterrewacht Leiden, Niels Bohrweg 2, NL-2333 CA Leiden, The Netherlands koehler@strw.leidenuniv.nl

Summary. We present mid-infrared interferometric observations of the Herbig Ae star R CrA obtained with MIDI at the VLTI using several projected baselines in the UT2-UT3 configuration. The observations show resolved circumstellar emission in the wavelength range 8-13 μm on a ~ 6 AU scale with a non-symmetric intensity distribution, providing support for an inclined disk geometry. Visibilities are best fitted using a uniform ring model with outer radius in the range 6-10 AU, in the wavelength range 8-13 micron. The inclination of the ring with respect to the plane of the sky is found to be $\sim 45°$, consistent with the 40° suggested from near-infrared imaging polarimetry (Clark et al. 2000, MNRAS, 319, 337).

1 Introduction

The presence of circumstellar disks around the intermediate mass ($M \lesssim 5 M_\odot$) Herbig Ae stars is supported by a large body of observational evidence [1, 2]. More specifically, while the observed spectral energy distribution (SED) of such stars can be explained by both a disk-like distribution of material (e.g. [3]) and other geometries like envelopes (e.g. [4]), clear evidence for circumstellar disks comes from resolved flattened structures observed by interferometry at millimeter, near-IR and recently also mid-IR wavelengths (e.g. [5, 6, 7]).

R CrA is a bright (100 $L_\odot$) young Herbig A5e star, located at the center of a small cluster (The Coronet [8, 9]) at 130 pc [10]. Several characteristics indicate the presence of a circumstellar disk around R CrA : a flat mid-IR to far-IR/mm SED [3] ([2]), a broad silicate emission feature [12], a UX Ori Type [13], a high degree (8%) of optical linear polarisation [14] , the possible association with a extended molecular outflow [15, 16] as well as with several Herbig-Haro systems [17], and a near-infrared reflection nebulosity whose resolved spatial polarization is consistent with a bipolar outflow being truncated by an evacuated spherical cavity [20].

---

[3] although most of the mm-excess is actually from the nearby embedded infrared source IRS7 [11] and source confusion in the large IRAS beams might be an issue.

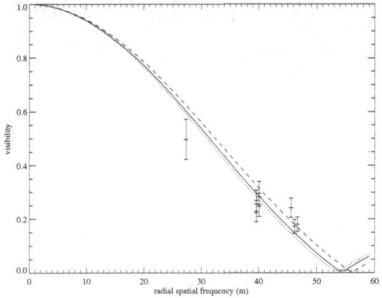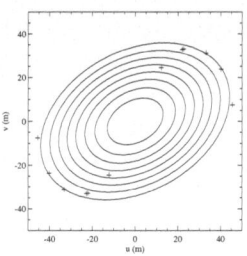

Fig. 1. Left panel: Wavelength averaged visibilities as a function of uv-radius and best-fit face-on uniform ring brightness distribution (R_{out}=6.3 AU$^{+0.1}_{-0.2}$). Dashed and dotted lines are respectively the inner and outer boundary values of R_{out}. Right panel: Contour plot of the best-fit inclined uniform ring model whose parameters are listed in Table 1. The contour levels are for visibilities of 0.1, 0.2, 0.3, 0.4, 0.5, 0.6, 0.7, 0.8, 0.9. East is up, north is right.

2 Observations and Calibration

R CrA has been observed with MIDI at the VLTI between 2004 July 8 and 30. We obtained 6 sets of spectrally dispersed (Prism, R=30) visibilities in the spectral range 8-13 μm with the 47 m long northeast UT2-UT3 baseline. Projected baseline lengths and PAs are 27.4 m, 46.6 m, 46.2 m, 45.6 m, 40.1 m, 39.6 m and 63°.6, 30°.5, 9°.4, 43°.4, 55°.6, 56°.1, respectively. Each data set has been reduced with the MIA package [18]. The raw visibilities were calibrated by observing the source HD 173484, whose adopted angular size is 3.35 ±0.31 mas (MIDI calibrator list). We evaluated the error on the visibilities by comparing the instrumental visibility obtained with different calibrators observed during the same night. This leads to a typical relative uncertainty of 15%, a value we adopt in the following. The acquisition images from both telescopes show that R CrA is unresolved at 8.7 μm, i.e. at a resolution limit of λ/D=0″.27, corresponding to ∼ 30 AU at 130 pc.

3 Evidence for an inclined disk structure

3.1 First approach

We compared the observed visibilities with those derived from a simple geometric disk model chosen to be a uniform ring brightness distribution. The inner radius was fixed to a dust sublimation radius of 0.3 AU, typical for a Herbig AeBe star of that luminosity [19]. In a first step, only the wavelength averaged visibilities were fitted. We will see later that this rough approximation can be justified by the relative flatness of the visibilities observed on R

CrA in the 8-13 μm wavelength range. As a start, and for comparison purposes, we took a ring with zero inclination (face-on), so that the outer radius R_{out} was the only free parameter.

Left panel of Fig. 1 shows the visiblities plotted as a function of uv-radius together with the best-fit model. For the latter, R_{out}=6.3 AU $^{+0.1}_{-0.2}$ and the reduced χ^2=2.4. Two conclusions can be drawn from this preliminary analysis. First, the mid-IR emission of R CrA is clearly resolved into a structure of ~ 6 AU radius. Second, this structure is asymmetric which hints at a inclined disk geometry. Including the inclination in the model leads to a better fit, with a reduced χ^2=1.1. The best-fit model is shown in contour plot in the right panel of Fig. 1 together with the uv-coverage of our observations. The fitted parameters are R_{out}=8.6 AU$^{+6.4}_{-1.9}$, an inclination of i=47° $^{+21}_{-22}$ with respect to the plane of the sky, and a semimajor axis position angle[4] PA=152° $^{+24}_{-14}$.

3.2 Disk structure and orientation

Spectrally-dispersed visibilities as measured by MIDI are likely to provide additional information, because of the extension of uv-coverage. However, one should keep in mind that there is in principle a degeneracy between spatial brightness distribution (morphology) and dust spectral emission features. A fortunate observation, in this respect, is that the ISO-SWS spectra are rather featureless apart from a broad 9.7 μm amorphous silicate feature, i.e. no prominent PAH feature at 8.6 μm and no 11 μm complex feature [12]. This allows us to derive meaningful results by fitting our simple uniform ring model to the observed spectrally-dispersed visibilities. Fig. 2 shows the set of observed spectrally-dispersed visibilities together with the best-fit face-on and inclined ring models. Given the possibility that the observed visibilities may be due to the presence of a close companion[5], we additionally fit a binary model to the complete data set. The fitted parameters as a function of wavelength are presented in Fig. 3, while their wavelength-averaged values can be found in Table 1. The quoted errors correspond to a variation of reduced χ^2 of unity. While we can rule out with a high degree of confidence a binary model, the best agreement with these mid-IR visibility data is found for an inclined uniform ring model.

Fitting a uniform ring model to the total data set for each spectral channel as we did leads to a monotonic variation of R_{out} with wavelength (Fig. 3). Smaller outer radii are found at shorter wavelengths. This explains the fact that emission closer to the star arises from hotter dust and could in principle be used to derive the temperature profile of the circumstellar disk.

The inclination of the ring with respect to the plane of the sky is 44° $^{+8}_{-17}$. This is fully consistent with the inclination of the plane (40°) perpendicular to the symmetry axis of the bipolar reflection nebula as suggested from NIR

---

[4] oriented east of north.

[5] R CrA is flagged in Hipparcos as "stochastic binary" and "suspected nonsingle"

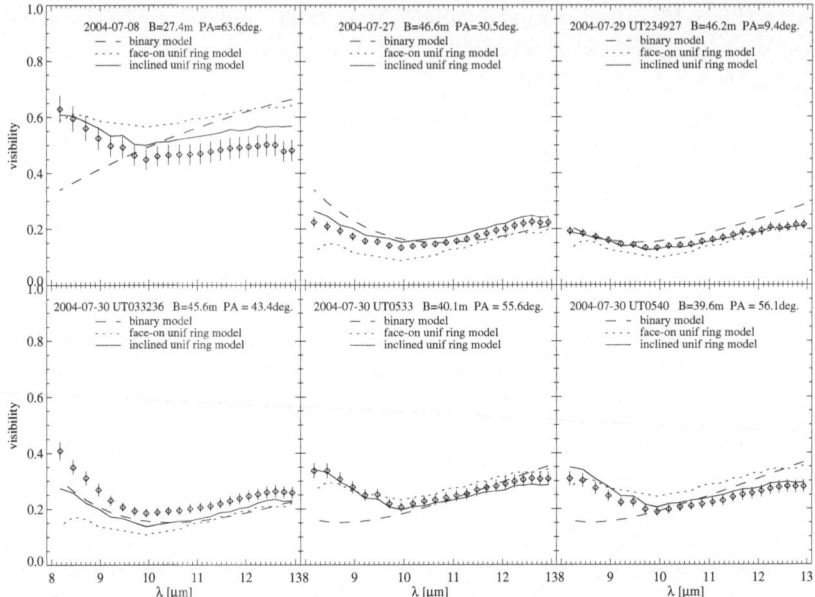

Fig. 2. Spectrally-dispersed visibilities between with 8-13 μm (diamonds) with the best-fit models (binary, face-on and inclined uniform rings).

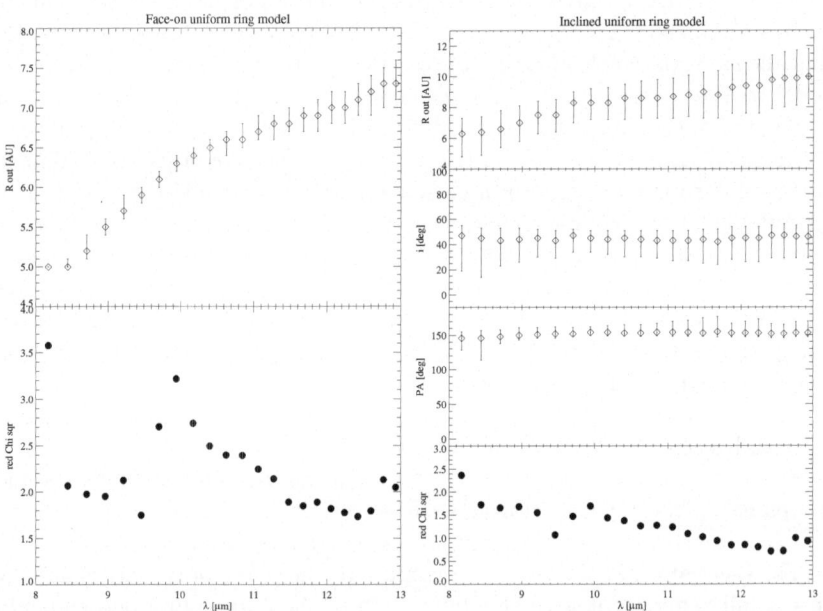

Fig. 3. Best fit parameters as a function of wavelength for respectively a face-on (left panel) and inclined (right panel) uniform ring model.

Table 1. Best-fit models based on spectrally-dispersed visibilities. Uniform ring parameters are wavelength-averaged values. The binary model separation, PA and brightness ratio are quoted in the R_{out}, PA and i columns, respectively.

	reduced χ^2	R_{out} (AU)	i (°)	PA (°)
Binary	2.4	$3.1_{-0.2}^{+0.2}$	$0.44_{-0.04}^{+0.05}$	36_{-10}^{+10}
Face-on uniform ring	2.2±0.7	$6.43_{-0.15}^{+0.15}$	...	...
Inclined uniform ring	1.3±0.6	$8.48_{-1.47}^{+1.26}$	44_{-17}^{+8}	152_{-7}^{+14}

imaging polarimetry [20]. However, the derived position angle of the disk symmetry axis ($62°_{-7}^{+14}$) is only marginally consistent with R CrA as the driving source for the extended NE-SW molecular outflow at PA$\sim 30°$ [15, 16]. The same conclusion holds for the direction of the closest HH objects with respect to R CrA [17] (HH 96: $\sim 38°$, HH 97W: $\sim 34°$, HH 98: $\sim 33°$, HH 100: $\sim 34°$, HH 99A: $\sim 51°$, HH 104A: $\sim 95°$).

Acknowledgments We are grateful to B. Wilking for his help. GM is supported by DFG grant ME 2061/3-1.

References

1. Natta, A., Grinin, V., Mannings, V., Protostars and Planets IV, p. 559 (2000)
2. Hillenbrand, L.A., Strom, S.E., Vrba, F.J. et al., ApJ, 397, 613 (1992)
3. Dullemond, C.P., Dominik, C. and Natta, A., ApJ, 560, 957 (2001)
4. Hartmann, L., Kenyon, S.J., Calvet, N., ApJ, 407, 219 (1993)
5. Mannings, V. & Sargent, A.I., ApJ, 490, 792 (1997)
6. Eisner, J.A., Lane, B.F., Hillenbrand, L.A., ApJ, 613, 1049 (2004)
7. Leinert, Ch., Van Boekel, R., Waters, L.B.F.M. et al., A&A 423, 537 (2004)
8. Taylor, K.N.R. & Storey, J.W.V., MNRAS, 209, 5 (1984)
9. Wilking, B., McCaughrean, M., Burton, M.G. et al., AJ, 114, 5 (1997)
10. de Zeeus, P.T., Hoogerwerf, R., de Bruijne, J.H.J., AJ, 117, 354 (1999)
11. Choi, M. & Tatematsu, K., ApJ, 600, L55 (2004)
12. Acke, B. & Van den Ancker, M.E., A&A 426, 151 (2004)
13. Dullemond, C.P., Van den Ancker, M.E., Acke, B. et al., ApJ 594, L47 (2003)
14. Bastien, P., ApJ, 317, 231(1987)
15. Levreault, R.M., ApJSS, 67, 283 (1988)
16. Anderson, I.M., Harju, J., Knee, L.B.G., A&A, 321, 575 (1997)
17. Wang, H., Mundt, R., Henning, T. et al., ApJ, 617, 1191 (2004)
18. Köhler, R. and Jaffe, W.: "MIA+EWS, the software for MIDI data reduction".
19. Monnier, J.D., Millan-Gabet, R., Billmeier, R. et al., ApJ, 624, 832 (2005)
20. Clark, S., McCall, A., Chrysostomou, A. et al., MNRAS, 319, 337 (2000)

The Chaotic Winds of AGB Stars: Observation Meets Theory

Peter Woitke and Andreas Quirrenbach

Sterrewacht Leiden, P.O. Box 9513, 2300 RA Leiden, The Netherlands
woitke@strw.leidenuniv.nl

Summary. Spherically symmetric (1D) and axisymmetric (2D) dynamical simulations for dust-driven winds of carbon stars are presented which aim at a qualified interpretation of interferometric image and visibility data.

1 Introduction

With the new interferometric instruments AMBER and MIDI mounted on the VLT, the innermost dust formation and wind acceleration zones of Asymptotic Giant Branch (AGB) stars become spatially resolvable in the thermal infrared. The measured visibilities reveal new details about the spatial dust distribution around these stars, with the ability to detect wind asymmetries and possibly cloud-like inhomogeneities [1, 2, 3]. In combination with likewise detailed, multi-dimensional models, new constraints for the long-standing astrophysical problem of AGB star wind generation can be provided.

2 The Model

We have developed new multi-dimensional models for dust-driven winds of AGB stars, which include hydrodynamics (based on the FLASH-solver [4]) with radiation pressure on dust grains, thermodynamics with radiative cooling, equilibrium chemistry, time-dependent dust formation [5] and grey radiative transfer [9]. The models can be run in spherical symmetry (1D) or in axisymmetry (2D). As inner boundary condition, we use a hydrostatic stellar atmosphere *without pulsation*. Except for the ability to calculate solutions in 2D, the model is similar to [6] and [7].

The calculated model structures provide the input for a-posteriori frequency-dependent Monte Carlo radiative transfer calculations [8] in order to predict spectral energy distributions, monochromatic images and visibilities.

3 Results

The resultant wind solutions for carbon stars, here with common parameters $L_\star = 10^4 \, L_\odot$, $M_\star = 1 \, M_\odot$ and $T_\star = 2500 \, \mathrm{K}$, fall into three different classes: *stationary, oscillating* and *chaotic winds.*

Figure 1 shows an example for the stationary wind solutions. These winds result to be spherically symmetric (even with 2D treatment), to have small mass loss rates $\dot{M} \approx 10^{-7} M_\odot/\mathrm{yr}$, small outflow velocities $v_\infty < 10 \, \mathrm{km/s}$ and small dust-to-gas ratios. The images are dominated by stellar photons. The visibilities are featured by a relatively broad central bump on top of the stellar signal, which reflects the slightly larger dust envelope.

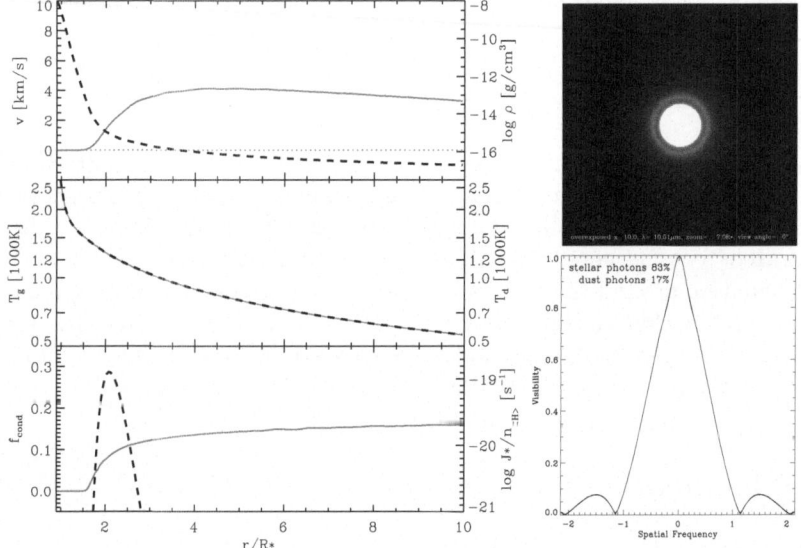

Fig. 1. *Stationary 1D Model* for C/O = 1.8. **Left:** model snap shot: v velocity (full), ρ mass density (dashed); T_g gas temperature (full), T_d dust temperature (dashed); f_cond degree of condensation (full), $J_\star$ nucleation rate (dashed). **Upper Right:** Calculated image at $10\mu\mathrm{m}$ of the innermost $7 \, R_\star$, here overexposed by a factor of 10 to make visible the tenuous dusty wind. **Lower Right:** Calculated visibilities at $10\mu\mathrm{m}$. The spatial frequency is given in relative units, such that the first minimum of the stellar signal is at 1.22. Note the broad bump on top of the stellar signal originating from the dust envelope.

Figure 2 shows an example for the 1D oscillating wind solutions. Due to an instability called "external κ-mechanism" [6], dust does not form continuously, but event-like in long time intervals (often longer than typical stellar pulsation periods). These wind solutions have typically large $\dot{M} \approx 5 \times 10^{-6} M_\odot/\mathrm{yr}$, large $v_\infty \approx 25 \, \mathrm{km/s}$ and large dust-to-gas ratios. The images are dominated by thermal dust emission. The star can be massively obscured. Visibilities have narrow central bumps which are "breathing"

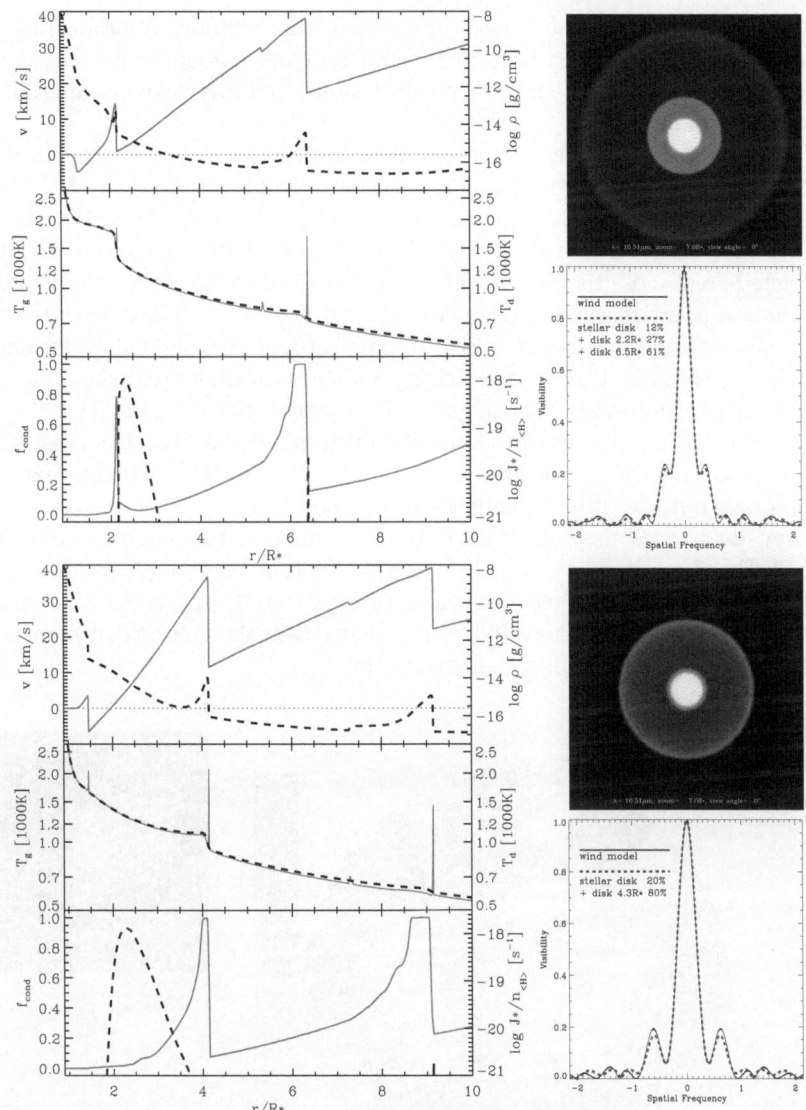

Fig. 2. *Oscillatory 1D Wind Model* for C/O = 2.1. **Upper part:** Just after the formation of a new dust shell. **Lower part:** Same model, but 0.8 years later, after the new shell has moved to about 4 stellar radii. See text for more details.

in time. Simple two-uniformly-bright-disk or three-uniformly-bright-disk fits, respectively (see dashed lines in Fig. 2), provide reasonable fits to the visibilities calculated from the image (black lines) and give correct values of the dust shell radii.

Figure 3 shows a new type of solution obtained by axisymmetric (2D) models. These models reveal a much more complicate picture of the dust and wind formation as compared to the spherically symmetric (1D) models. Excited by instabilities, dust formation takes place from time to time in restricted areas close to the star. These clouds are then accelerated outward by radiation pressure, while less dust-containing matter is falling back towards the star at other places. A highly dynamical and turbulent dust formation zone is created in this way, which again leads to inhomogeneous dust production. Further away from the star, flow instabilities (e.g. Rayleigh-Taylor) have time to modify and to shape the outward moving cap-like dust structures. The spectral appearance is generally similar to the 1D oscillating wind solutions, but the time-dependence is less significant, because the creation of a new dust cloud has less effect than the formation of a complete new dust shell. By comparison of visibility data at different baseline orientations, one can easily distinguish between winds of type 2 and type 3. At high spatial frequencies, unusual signals may occur that reflect the presence of small-scale structures in the spatial dust distribution.

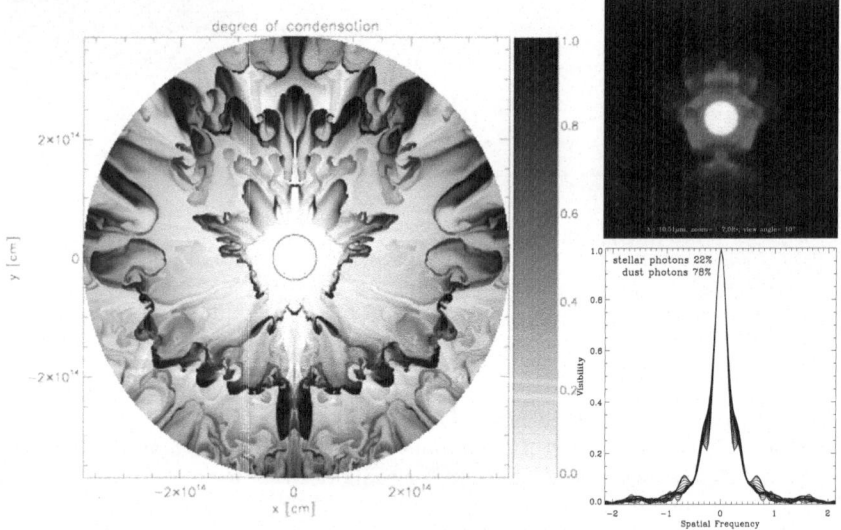

Fig. 3. *Axisymmetric (2D) Chaotic Wind Model* for C/O = 1.9, showing the degree of condensation ($\propto$ dust-to-gas ratio) within the innermost $10\,R_\star$. The inner circle marks the stellar radius. Image and visibilities are calculated for an inclination angle between the equator plane and the observer's direction of $10°$. Visibilities are overplotted for 50 different baseline orientations.

4 Conclusions

New 2D models for AGB star winds have been presented which predict that carbon stars should produce either a smooth stationary wind at small mass loss rates or dust clouds and shells at high mass loss rates in more or less regular, long time intervals (> 600 days). At high mass loss rates, symmetry breaking occurs naturally in these models[1] due to various instabilities, even without stellar pulsation.

The MIDI-interferometer is well-suited to study the morphology of dusty AGB star winds. All three proposed types of winds (stationary / oscillating / chaotic) can be distinguished. The radial extensions of the innermost 1-2 dust shells can be directly measured[2]. Monitoring is required to reveal the proper site of the dust formation and the dynamics of dust shells.

A true collaboration between observation and theory is required to further explore the nature of dust-driven winds. Can the "smoke signals" from these stars be related to a turbulent dust formation zone?

Acknowledgements This work is part of the ASTROHYDRO3D initiative supported by the NWO Computational Physics programme, grant 614.031.017. The software used in this work was in part developed by the DOE-supported ASCI/Alliance Center for Astrophysical Thermonuclear Flashes at the University of Chicago.

References

1. Tuthill P., Monnier J., Danchi W., Lopez B., ApJ **543**, 284 (2000)
2. Weigelt G., Balega Y.Y., Blöcker T., Hofmann K.-H., Men'shchikov A.B., Winters J.M., A&A **392**, 131 (2002)
3. Monnier J.D., Millan-Gabet R., Tuthill P.G., et al., ApJ **605**, 436 (2004)
4. Fryxell B., Olson K., Ricker P., Timmes F.X., Zingale M., et al., ApJ **131**, 273 (2000)
5. Gail H.-P., Sedlmayr E., A&A **206**, 153 (1988)
6. Fleischer A. J., Gauger A., Sedlmayr E., A&A **297**, 543 (1995)
7. Höfner S., Feuchtinger M. U., Dorfi E.A., A&A **297**, 815 (1995)
8. Niccolini G., Woitke P., Lopez B., A&A **399**, 703 (2003)
9. Woitke P., A&A 452, 537 (2006)

---

[1] see movies at http://www.leidenuniv.nl/~woitke
[2] Dust shell radius ≠ dust formation radius!

Observations of 51 Ophiuchi with MIDI at the VLTI

C. Gil[1], F. Malbet[2], M. Schöller[1], O. Chesneau[3], and Ch. Leinert[4]

[1] European Southern Observatory, Casilla 19001, Santiago 19, Chile
cgil@eso.org, mschoell@eso.org
[2] Laboratoire d'Astrophysique de l'Observatoire de Grenoble, BP 53 38041
Grenoble Cedex 9, France Fabien.Malbet@obs.ujf-grenoble.fr
[3] Observatoire de la Côte d'Azur, CNRS UMR 6203, Avenue Copernic, Grasse,
France Olivier.Chesneau@obs-azur.fr
[4] Max-Planck-Institut fur Astronomie, Knigstuhl 17, 69117 Heidelberg, Germany
leinert@mpia-hd.mpg.de

Summary. We present interferometric observations of the Be star 51 Ophiuchi. These observations were obtained during the science demonstration phase of the MIDI instrument at the VLTI in June 2003. The baseline UT1-UT3 was used, corresponding to 102 m. It is currently known that this object presents a circumstellar dust and gas disk that shows a very different composition from other Herbig Ae disks. The nature of the 51 Oph system is still a mystery to be solved. Observations with MIDI at the VLTI allowed us to reach high-angular resolution (20 mas). We have several uv points that allowed us to constrain the disk model. We have modeled 51 Oph visibilities and were able to constrain the size and geometry of the 51 Oph circumstellar disk.

1 Introduction

51 Ophiuchi is a Be (B9.5IIIe) star, located at 131 pc with a rotational velocity $vsini = 267 \pm 5$ km/s [1]. 51 Oph shows a large infrared excess, that was first noticed by [12]. This infrared excess was explained as being originated by the circumstellar dust around the star.

The 51 Oph circumstellar gas and dust origin is still unknown. Since its infrared excess was noticed, this object has been observed in many different wavelengths, from the infrared to the ultraviolet. Grady and Silvis [4] found the first evidence for the presence of dust by detecting a 10 μm silicate feature in emission. 51 Oph has been compared to β Pic due to the presence of a gas envelope with a variable column density, high density gas accreting to the star and the fact that this gas is also collisionally ionized. There was no detection of an extended disk at 18 μm, which suggested that the hot dust must be located in the close proximity of the star [5]. Van den Ancker et al. [10] analyzed 51 Oph ISO archive data and suggested a few different scenarios for the nature of this system. The authors have studied the composition of the circumstellar gas and detected spectral features due to hot gas-phase molecules not typical of Ae/Be stars. They have suggested that 51 Oph might not be a young system but a highly evolved one instead. Nevertheless, the presence of all the circumstellar material can not be explained if the system is really an evolved one. The other scenario includes the presence of a companion in order to explain the hot gas observed. A more exotic explanation consists of assuming that the detected material is a result of a recent collision of two gas-rich planets or the accretion of a solid body as the star increases its size at the end of its main-sequence life [10].

Roberge et al. [9] observed 51 Oph circumstellar disk with the Far Ultraviolet Spectroscopic Explorer and found that the composition of the infalling gas is highly nonsolar. However, all studies suggest that the circumstellar matter is most likely in the form of a Keplerian disk rather than in a spherical shell. In 2004 the first MIDI observations of Herbig Ae/Be stars were published by Leinert et al. [6]. The authors have studied a sample of 7 Herbig Ae/Be stars but could not fit a Dullemond et al. [3] type disk model to the 51 Oph data like they did for the other stars. In this paper, we have used the data obtained by Leinert et al. [6] on 51 Oph and constrained the size and geometry of the 51 Oph circumstellar disk by using a standard disk model. In section 2 the observations and data reduction are presented. The results are introduced and discussed in sections 3 and 4, and summarized in section 5.

2 Observations and Data Reduction

51 Oph was observed during MIDI Science Demonstration Time (SDT) at the VLTI on June 15th and 16th 2003. MIDI is a 2 beam combiner that operates in the mid-infrared, 8 to 13 μm. The baseline used was the UT1-UT3 (102m), with a maximum full spatial resolution of 20 mas at 10 μm. The observations log is presented in Table 1. We have obtained 5 visibility points, that were dispersed at a resolution of 30 in the N band.

We have reduced the 51 Oph data using the IDL software written for MIDI. The data reduction is described in detail in Leinert et al. [6]. The data analysis and modeling was done with the interpreted programming language yorick.

Table 1. Log of the observations

UT Date	Projected Baseline (m)	P.A. (°)	Hour Angle	Calibrators
15/Jun/03	101.2	23	3	HD 168454, HD 168454, HD 167618
15/Jun/03	101.4	38	7	HD 168454, HD 168454, HD 167618
15/Jun/03	85.6	45	8	HD 168454, HD 168454, HD 167618
16/Jun/03	98.8	-7	0	HD 165135, HD 152786, HD 165135
16/Jun/03	99.6	14	2	HD 165135, HD 152786, HD 165135

Table 2. Disk model parameters for 51 Oph.

Parameter	Best fit Value
Distance	131 pc
$R_\star$	$7\ R_\odot$
$M_\star$	$3.8\ M_\odot$
T_{eff}	10000 K
Av	0.15
Accretion rate	$7.10^{-5}\ M_\odot/\text{yr}$
Disk outer radius	7 AU
Disk inner radius	0.55 AU
Inclination	88°
Position Angle	78°

3 Results

We have modeled the circumstellar disk by using a flat disk model with a temperature law proportional to $r^{-3/4}$ [8]. The disk radial structure was derived from [7]. In this model, the disk is considered as infinitely thin and the flux is calculated through the sum of all the rings of radius r and width dr.

In Table 2 one can find a detailed list of the parameters used to fit the standard disk model to our data. The values for the stellar mass ($M_\star$), effective temperature (T_{eff}) and atmospheric extinction (Av) were all taken from the literature [10]. We have found a good fit for the data with a disk nearly edge-on, inclination of 88°, position angle of 78° and an accretion rate of $7.10^{-5}\ M_\odot/\text{yr}$.

For a radius of $7\ R_\odot$, we have found a good fit for the spectral energy distribution (SED), with an inner disk radius of 0.55 AU and an outer disk radius of 7 AU (see Fig. 1). In this figure one can see the SED for 51 Oph well reproduced by a standard disk model. The photometry points used in these figure were taken from [12].

In Figure 2 on the left, we can see the visibility model for the 51 Oph circumstellar disk and the uv-track for our measurements. The 5 visibility points are marked with crosses over the uv-track. In this figure north is up and west is left, so we note that the position angle is obtained after a Fourier transform of the visibility model (-12°+90°=78°). We have also studied the visibilities behavior against projected baselines and hour angle. In Figure 2 on the right, we have plotted the visibilities as a function of hour angle. For both cases we have found a good fit for all the visibility points by using the disk model parameters in Table 2.

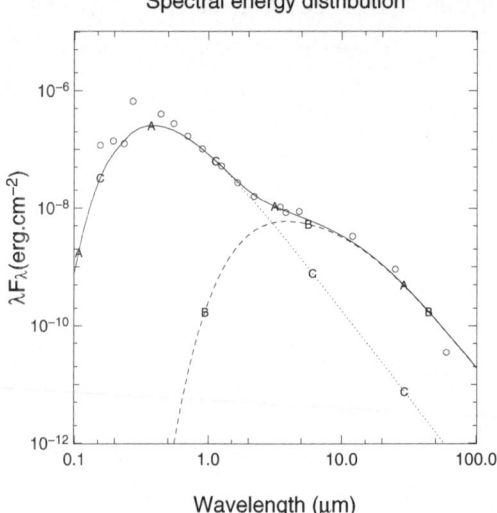

Fig. 1. Spectral energy distribution for 51 Oph. The circles correspond to photometry measurements compiled from the literature. The dotted line is the stellar
contribution, the dashed line is the disk contribution and the solid line corresponds
to the total energy distribution (star+disk). Our standard disk model successfully
reproduces the 51 Oph SED.

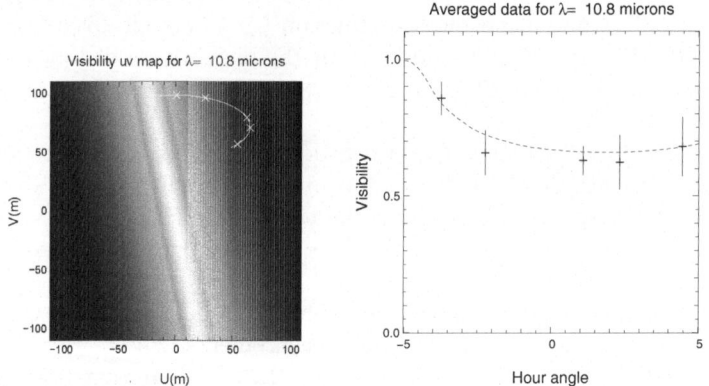

Fig. 2. Left: 51 Oph disk visibility model. The five visibility points (marked as
crosses on the uv-track) were obtained in 2 different nights. Right: Visibilities as a
function of hour angle. Best fit for a standard disk with an inclination of 88° and
a position angle of 78°.

4 Discussion

The MIDI observations are compatible with a flat circumstellar disk, inclined of 88° with a position angle of 78°. These results are compatible with the recent results obtained by Thi et al. [11], who have found a high column density of CO rotating in a disk in the inner astronomical unit of 51 Oph, seen nearly edge-on. We have determined a stellar radius of 7 $R_\odot$, a disk inner radius of 0.55 AU, an outer radius of 7 AU, a disk inclination angle of about 88° and a disk position angle of about 78°. Due to the nature of the object, our interpretation relies critically on one visibility point and thus drew some criticism. Four of the visibility points are nearly identical, while the fifth one is significantly higher. This higher visibility point was obtained with the lowest projected baseline at 85.6m. There is consequently a potential degeneracy of information between a base length effect and an angle effect. In order to secure our data set, we plan to obtain a few extra visibility points with MIDI.

5 Conclusions

-   51 Oph was observed for the first time in the mid-infrared at high-angular resolution.
-   We have modeled 51 Oph visibilities and were able to constrain the size and geometry of the 51 Oph circumstellar disk.
-   The best fit to the data corresponds to a flat circumstellar disk, inclined of 88° with a position angle of 78°.
-   We will need more visibility measurements in order to secure our data set and confirm the results obtained.

References

1. Dunkin, S. K., Barlow, M. J., & Ryan, S. G. 1997, MNRAS, 290, 165
2. Fajardo-Acosta, S. B., Telesco, C. M., & Knacke, R. F. 1993, ApJ, 417, L33
3. Dullemond, C. P., Dominik, C., & Natta, A. 2001, ApJ, 560, 957
4. Grady, C. A., & Silvis, J. M. S. 1993, ApJ, 402, L61
5. Jayawardhana, R., Fisher, R. S., Telesco, C. M., Piña, R. K., Barrado y Navascués, D., Hartmann, L. W., & Fazio, G. G. 2001, AJ, 122, 2047
6. Leinert, C., et al. 2004, A&A, 423, 537
7. Lynden-Bell, D., & Pringle, J. E. 1974, MNRAS, 168, 603
8. Malbet, F., & Bertout, C. 1995, A&AS, 113, 369
9. Roberge, A., Feldman, P. D., Lecavelier des Etangs, A., Vidal-Madjar, A., Deleuil, M., Bouret, J.-C., Ferlet, R., & Moos, H. W. 2002, ApJ, 568, 343
10. van den Ancker, M. E., Meeus, G., Cami, J., Waters, L. B. F. M., & Waelkens, C. 2001, A&A, 369, L17
11. Thi, W.-F., van Dalen, B., Bik, A., & Waters, L. B. F. M. 2005, A&A, 430, L61
12. Waters, L. B. F. M., Cote, J., & Geballe, T. R. 1988, A&A, 203, 348

Mid-Infrared Spectrally-Dispersed Visibilities of Massive Stars Observed with the MIDI Instrument on the VLTI

D. J. Wallace[1], J. Rajagopal[1], R. Barry[1], L. J. Richardson[1], B. Lopez[2], O. Chesneau[2] and W. C. Danchi[1]

[1] Exoplanets and Stellar Astrophysics, NASA/GSFC, Greenbelt, MD 20771
    Debra.Wallace@gsfc.nasa.gov, Jayadev.Rajagopal@gsfc.nasa.gov,
    Richard.Barry@gsfc.nasa.gov, William.C.Danchi@nasa.gov
[2] Observatoire de la Cote D'Azur Bruno.Lopez@obs-nice.fr,
    Olivier,Chesneau@obs-nice.fr

Summary. The mechanism driving dust production in massive stars remains somewhat mysterious. However, recent aperture-masking and interferometric observations of late-type WC Wolf-Rayet (WR) stars strongly support the theory that dust formation in these objects is a result of colliding winds in binaries. Consistent with this theory, there is also evidence that suggests the prototypical Luminous Blue Variable (LBV) star, Eta Carinae, is a binary. To explore and quantify this possible explanation, we have conducted a high resolution interferometric survey of late-type massive stars utilizing the VLTI, Keck, and IOTA interferometers. We present here the motivation for this study as well as the first results from the MIDI instrument on the VLTI. (Details of the Keck Interferometer and IOTA interferometer observations are discussed in this workshop by Rajagopal et al.). Our VLTI study is aimed primarily at resolving and characterizing the dust around the WC9 star WR 85a and the LBV WR 122, both dust-producing but at different phases of massive star evolution. The spectrally-dispersed visibilities obtained with the MIDI observations will provide the first steps towards answering many outstanding issues in our understanding of this critical phase of massive star evolution.

1 Introduction

Massive star evolution depends upon rotational speed, metallicity, and mass. The most massive stars ($M_{initial} \geq 20M_\odot$) will likely evolve from an initial O-type star through an LBV state (and possibly a brief Red SuperGiant (RSG) phase) to end their lives as a WR star. LBVs are extremely luminous, unstable, evolved supergiants near the Eddington limit that undergo irregular bursts of mass ejections ranging from $10^{-5} - 10^{-2} M_\odot yr^{-1}$ [1]. This is a short-lived phase where the star exhibits dramatic photometric and spectroscopic variations over many timescales. The star is generally deficient in C and O while overabundant in N suggesting the end of hydrogen burning. Most LBVs are surrounded by dusty ring-like nebulae and exhibit elliptical or bi-polar axi-symmetry. The WR phase is identified with the onset of core helium burning and is sub-divided into two phases. The first evolutionary phase is the WN-phase which exhibits strong spectroscopic lines of He and N as a result of CNO-cycle burning. The later WC-phase is associated with strong lines of He and C which are believed to result from the triple alpha process. Strong, dense winds and extreme mass loss from the bare stellar core are symptomatic of this stage.

Dust formation is strongly associated with the LBV phase and in late-type WC stars, but noticeably absent in the WN phase. In LBV stars this dust is consistent with large metallic and silicate grains or C-rich grains such as Polycyclic Aromatic Hydrocarbons (PAHs) [2]. In the WC-type stars, the dust takes the form of small carbon grains. For either case, the dust formation requires low temperatures and high densities. For the WR stars, there are two models proposed to explain the formation of dust. A single star model maintains that grain growth is possible due to the shielding effects of the clumps in the WR stellar wind which allows for lower temperatures nearer to the star where sufficient high densities for grain nucleation and condensation may be found [3]. The binary model [4, 5] suggests that the high densities and lower temperatures can be achieved in a binary wind collision zone at small stellar separations ($\approx$ few AU). This model has been proven experimentally for both WR 104 [6] and WR 98a [7] as Keck aperture-masking interferometric observations have resolved each WC9-type star into a spiral "pinwheel"-shaped nebula with an apparent rotation period of $\approx$ 1 year. These data strongly suggest that the stars are short-period WC+OB binaries continuously forming dust in a colliding-wind interface. In conjunction with the strong evidence for the LBV Eta Carinae's binary state suggests that dust production in massive stars could be evidence of a binary wind collision region.

2 Binarity in Massive Stars

Are all dust-producing, late-type massive stars exclusively binary? To probe the evidence of multiplicity in late-type massive stars, we must first examine the evidence for multiplicity in young massive stars. Mason et al. [8] find the binary distribution of O-type stars is bimodal in log P. They suggest that this is a selection effect due to the difficulty in discovering those systems with periods in the range of years to decades. The authors believe that with advances in instrumentation (notably interferometry) more systems will be resolved to reveal a true binary fraction of O-type stars close to unity.

This hypothesis is consistent with new Hubble Space Telescope (HST) Fine Guidance Sensor (FGS) observations of O-type stars. Nelan et al. [9] find that of 23 OB stars observed in the Carina Nebula (NGC 3372) 5 are newly resolved binaries with separations ranging from $0\rlap{.}''015$ to $0\rlap{.}''352$ (37 to 880 AU at a distance of 2.5 kpc). This is slightly less than the 32% (accounting for projection effects, inclination angles, limited magnitude differences between binary components, and the FGS detection threshold) expected if the prediction of Mason et al. [8] is correct, but well within statistical errors given the small sample size. The authors are extending their survey and increasing their sample size to lower statistical errors and further explore this interesting result.

Given that WR and LBV type stars evolve from the O stars, does this imply that all LBVs and WRs should also be multiple? In 2001, van der Hucht [10] estimated a conservative binary frequency of 38% for the entire Galactic WR population. The inclusion of apparently single WR stars showing evidence for dust emission (assuming that the dust formation is indicative of a binary wind collision zone) increases this percentage to 44%. Even if we find that the assumption of binarity due to dust emission is valid, this is not consistent with the current information on O-type stars — the WR star predecessors. LBVs are even harder to resolve as binary due to their normal extended dusty shells. Nota et al. [11] suggest that a unified picture of these objects implies a common formation mechanism which could involve a binary companion. Eta Carinae has recently been established as a probable binary, but others of this class are less well determined.

3 Keck and IOTA Interferometer Observations of Wolf-Rayet Stars

As part of a larger effort to quantify the binarity of massive stars, we used the Keck Interferometer (KI) in V^2 mode to study a sample of 8 WR stars. Two of these early observations (WR 106 and WR 113) failed due to technical difficulties. Of the 6 remaining stars, WR 137 and WR 140 were resolved as binaries. WR 137 is resolved as a binary for the first time in these observations with a minimum separation of $\approx$ 2.6 mas. WR 140 was resolved previously by Monnier et al. [12] using the IOTA interferometer. WR 134, WR 135, WR 136, and WR 148 were not resolved by KI. Follow-up observations are being performed with the IOTA interferometer. Details of these observations can be found in the contribution by Rajagopal et al. in these proceedings.

4 VLTI Observations of WR 95 and WR 122

To complement and expand upon these KI observations, we received time to study 2 WR stars (WR 95 and WR 106) and 2 LBV-WR transitionary objects (WR 31b and WR 122) with the MIDI instrument on the Very Large Telescope Interferometer (VLTI). Both WR 95 and WR 106 are WC9-type stars that show evidence of dust emission, but lack other indications of binarity. WR 31b, better known as AG Car, is a LBV-type object that forms dust continuously in its stellar wind and is surrounded by an elongated, bipolar dust nebula. Paresce & Nota [13] suggest that, like Eta Carinae, the bipolar structure is most naturally explained by outflowing material from a small, secondary component in a close binary system and the precession of its associated accretion disk. WR 122, also known as NaST1, is a similiar object also surrounded by an elliptical, dusty nebula. We present here our observations of WR 95 and WR 122.

Our visibility curve for WR 95 is presented in Fig. 1 which shows a relatively linear increase in N-band visibility with wavelength. The visibility curves are very similar despite the difference in projected baseline lengths. This implies slightly different sizes at the two position angles, and could be an indication of a flattened disk geometry. This agrees well with KI aperture masking observations of another dusty WC9 star, WR 104 [6].

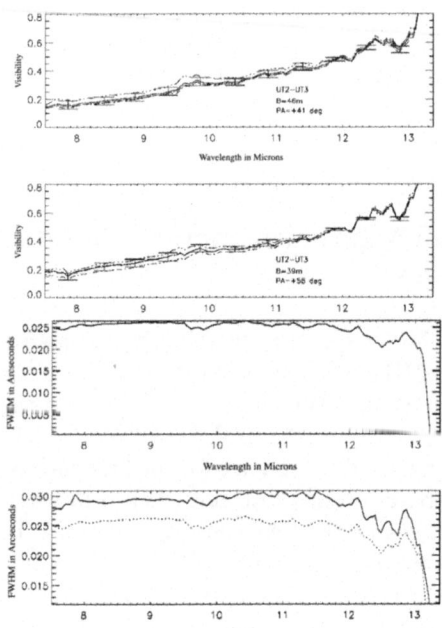

Fig. 1. WR 95 VLTI MIDI visibility function and gaussian fit

The visibility curve of WR 122, presented in Fig. 2, suggests a quite different geometry. The fairly flat visibility as a function of wavelength implies a monotonic increase in the size of the emitting region. This suggests a large dust distribution with extended cooler regions. The simple gaussian fit (shown in Figure 2) is consistent with an expanding shell geometry formed by continuous dust production. Clearly, for both objects, more work is needed to analyze and interpret these observations.

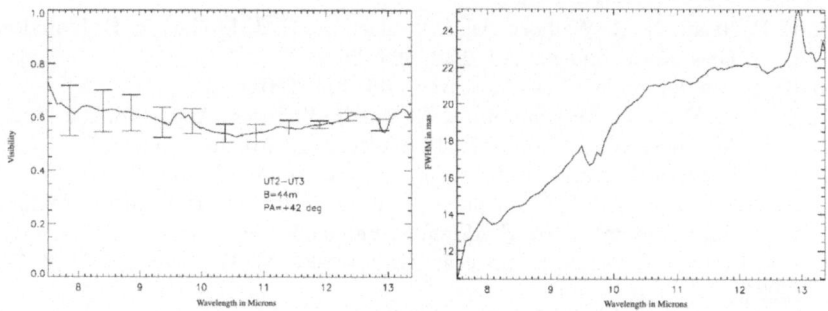

Fig. 2. WR 122 VLTI MIDI visibility function and gaussian fit

5 Conclusions

Increasing evidence demonstrates that massive stars are likely to form in pairs or groups. A survey of O-type stars with the HST FGS is underway to establish the binary percentage of young massive stars. An equally inclusive study is needed to determine the fate of more evolved massive stars. Many of these objects are enshrouded by dust, necessitating the use of ground-based interferometers sensitive at near-infrared and mid-infrared wavelengths. Not only will these observations increase our knowledge of massive star evolution at current epochs, but these observations are crucial to the proper interpretation of stellar evolution and dust production in the early Universe.

Our VLTI MIDI, KI, and IOTA observations of LBV and WR stars form the beginnings of such a survey. These observations reveal the profound differences in dust emission and distribution between WR and LBV stars. Planned future observations at longer baselines may detect the binary systems directly.

References

1. S. N. Shore, B. Altner, & I. Waxin: ApJ **112**, 2744 (1996)
2. R. H. M. Voors, L. B. R. M. Waters, A. de Koter, J. Bouwman, P. W. Morris, M. J. Barlow, R. J. Sylvester, N. R. Trams, & H. J. G. L. M. Lamers: A&A, **356**, 501 (2000)
3. I. Cherchneff, Y. H. Le Teuff, P. M. Williams, & A. G. G. M. Tielens: A&A, **357**, 572 (2000)
4. P. M. Williams, K. A. van der Hucht, D. R. Florkowski, A. M. T. Pollock, & W. M. Wamsteker: Dust Formation in the Wind of HD 193793. In: *Proceedings of the 122nd Symposium of the IAU: Circumstellar Matter*, ed by I. Appenzeller, C. Jordon (Dordrecht: Reidel), 453
5. V. V. Usov: MNRAS **252**, 49 (1991)
6. P. G. Tuthill, J. D. Monnier, & W. C. Danchi: Nature **398**, 487 (1999)
7. J. D. Monnier,, P. G. Tuthill, & W. C. Danchi: ApJL **525**, 97 (1999)
8. B. D. Mason, D. R. Gies, W. I. Hartkopf, W. G. Bagnuolo, Jr., T. ten Brummelaar, & H. A. McAlister: AJ **115**, 847 (1998)

9. E. P. Nelan, N. R. Walborn, D. J. Wallace, A. F. J. Moffat, R. B. Makidon, D. R. Gies, & N. Panagia: AJ **128**, 323 (2004)

10. K. A. van der Hucht: New Astr. Rev. **45**, 135 (2001)

11. A. Nota, M. Livio, M. Clampin, & R. Schulte-Ladbeck: ApJ **448**, 796 (1995)

12. J. D. Monnier, W. A. Traub, F. P. Schloerb, R. Milan-Gabet, J.-P. Berger, E. Pedretti, N. P. Carleton, S. Kraus, M. Lacasse, M. Brewer, S. Ragland, A. Aheam, C. Coldwell, P. Haguenauer, P. Kern, P. Labeve, L. Lagny, F. Malbet, D. Malin, P. Maymounkov, S. Morel, C. Papaliolios, K. Perraut, M. Pearlman, I. L. Porro, I. Schanen, K. Souccar, G. Torres, & G. Wallace: ApJ **448**, 788 (2004)

13. F. Paresce & A. Nota: ApJ **341**, L83 (1989)

The B[e] Star Hen 3-1191 Resolved with MIDI

R. Lachaume, Th. Preibisch, and Th. Driebe

Max-Planck-Institut für Radioastronomie, auf dem Hügel 69, D-53121 Bonn

Summary. We report on spectrally dispersed, spatially resolved VLTI/MIDI observations of the B[e] star Hen 3-1191. The N-band visibilities give an equivalent disc diameter increasing from 24 mas at 8 μm to 36 mas at 13 μm, with a slight silicate feature. We explain the spectral energy distribution and the N-band visibility continuum with an ad-hoc circumstellar disc model with a hot, puffed-up inner rim. However, it does not help us in determining Hen 3-1191's still unknown nature. High resolution visible spectroscopy and/or an accurated determination of its distance are the likely keys to this determination.

1 Introduction

Hen 3-1191 ($\alpha = 16^{\rm h}27^{\rm m}15''$, $\delta = -48°39'27''$, $V = 13.7$), a.k.a. WRAY 15-1484, features strong Balmer emission, permitted and forbidden emission lines, and a strong mid-IR excess. It is therefore member of the B[e] spectral class as (re)defined by [10]. Though B[e] properties hint towards similar physical conditions of the circumstellar (CS) material, the authors conclude that B[e] stars form a highly heterogen group classified into supergiants, young stellar objects (YSO), compact planetary nebulae (cPN), and symbiotic stars. Hen 3-1191 cannot be unambiguously classified into one of these groups because its distance, thus its luminosity, is unknown. It so far was reported as a cPN [13, 17], a post-AGB object [4], a YSO [5, 16], or a symbiotic object see [5].

[5] started modelling work on Hen 3-1191 and noticed that its spectral energy distribution (SED) could be explained in the range 0.4–10 μm by an early B star photosphere surrounded by dust emitting at a temperature of 950 K. The IRAS measurements from 12 to 100 μm are however in excess compared to this model and their slope hint towards the additional presence of an disc or envelope. [7] successfully used a radiative transfer simulation of Hen 3-1191, that they supposed wrapped in a spherical CS shell. This model accounts for the visible and IR SED with the constraint of two polar cavities of relatively small radial extent. No detailed simulation of the spectroscopic data has been carried out so far; studies mainly focus on the presence and width of particular emission lines.

The advent of optical interferometry with spatial resolutions of less than 10 mas and high spectral resolution permits to geometrically probe the inner

Table 1. MIDI observation log

night	time (UT)	B_p (m)	θ_p (°)	$V(8.3\,\mu m)$	$V(10.2\,\mu m)$	$V(12.9\,\mu m)$
2004-04-09[a]	06:53	45.6	33	0.56 ± 0.06	0.51 ± 0.06	0.60 ± 0.06
2004-04-11[b]	05:38	46.3	22	0.45 ± 0.06	0.43 ± 0.06	0.53 ± 0.06
	09:05	41.2	54	0.63 ± 0.06	0.62 ± 0.06	0.70 ± 0.06

[a] Calibrators: HD 107446, 129456, 139127, 152885, 161892, 169916
[b] Calibrators: HD 81797, 107446, 129456, 139997, 152885, 168723, 176411

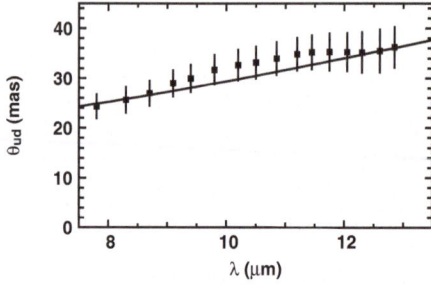

Fig. 1. Equivalent uniform disc diameter of Hen 3-1191 averaged over baselines vs. wavelength. The markers and error bars represent values derived from the observation and the dashed line the estimated continuum value.

parts of stellar environments. With only a few visibilities – giving an equivalent diameter – it is possible to disentangle SED models and rule out hypotheses as shown in recent papers for YSOs [11, 12, 9]. In this paper we present observations of Hen 3-1191 in the N-band obtained with the Very Large Telescope Interferometer (VLTI) and interpret the IR excess and spatial extension as the signature of a CS disc.

2 Observation and data reduction

Hen 3-1191 was observed with the UT2-UT3 baseline (47 m NE) of the VLTI. Three MIDI visibilities with a spectral resolution ≈ 25 at $\lambda\lambda$ 8–13 μm have been obtained at different hour angles, thus, spatial frequencies. The data reduction has been performed with the MPIA MIDI reduction software and the system visibility derived from all the calibrators observed in the same night. The detailed spectrum of the visibility is reported on Fig. 3. Figure 1 displays the equivalent disc diameter of Hen 3-1191 derived from the visibilities. The object is clearly resolved with a equivalent uniform disc diameter increasing from 24 ± 3 mas at 8 μm to 36 ± 4 mas at 13 μm. The increase of diameter is steeper from 8 to 10 μm than from 10 to 13 μm, which we interpret as a spectral feature due to the silicates with an excess of 10–15% in size over the continuum – represented as a dashed line in the figure.

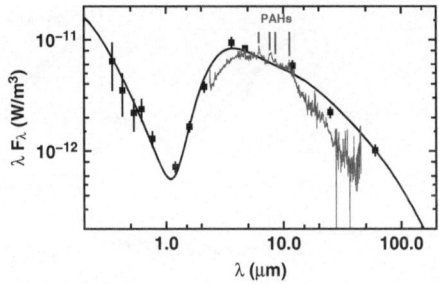

Fig. 2. SED of Hen 3-1191. Marquers with error bars: photometry. Gray line: ISO spectroscopy. Black line: best model fit of the photometric and interferometric data.

The photometric data, represented in Fig. 2, are taken from various sources: JP11 catalogue in the visible, [5, 3] in the near-IR, and IRAS in the far-IR. The reduced and calibrated spectroscopic data (same figure) have been taken from the ISO archive. Aromatic infrared bands (PAHs) are identified at 6.2, 7.7, 8.6 and 11.3 μm.

3 Modelling of the circumstellar environment

Our first attempts to model data with a CS shell with polar cavities after [7] failed at reproducing the visibilities by far while predicting the right order of magnitude for the IR fluxes. We present an alternate scenario of an accretion disc, that explains both visibilities and SED.

Current models of CS discs around intermediate-mass stars [6] predict the presence of a hot, puffed-up inner rim emitting in the near-IR while the rest of the disc may be illuminated at a grazing angle or shadowed and re-emits in the thermal IR. Our model of Hen 3-1191 is an ad hoc simplification of such models. It comprises three geometrically simple elements locally emitting like blackbodies: a sphere (star) with temperature $T_\star$ and radius $R_\star$; a ring (inner rim of the disc) of height $h_{\rm rim}$, radius radius $R_{\rm in}$, temperature $T_{\rm rim}$, inclination i, and position angle θ; flat disc (rest of the disc) of inner radius $R_{\rm in}$, inclination i, position angle θ, and a temperature profile $T(r) \propto T_{\rm in}(r/R_{\rm in})^{-q}$. The distance $d = 4\,{\rm kpc}$ and the stellar mass $M_\star = 14\,M_\odot$ (main-sequence early B star) are assumed. The inner rim is heated by irradiation, so its temperature is linked to that of the star by $T_{\rm rim} = \epsilon^{-1/4}(R_\star/R_{\rm in})^{-1/2}T_\star$, where ϵ is the ratio of the ring opacity for its own radiation to its opacity for the stellar radiation [2].

We performed a best-fit model of the photometric and interferometric data. The parameters of this model are given in Table 2, and the comparison between model and data are displayed in Figs. 2 & 3. The parameters derived for the accretion disc are rather surprising. First, the inner rim has an unusually low temperature of about 1030 K, below the dust sublimation

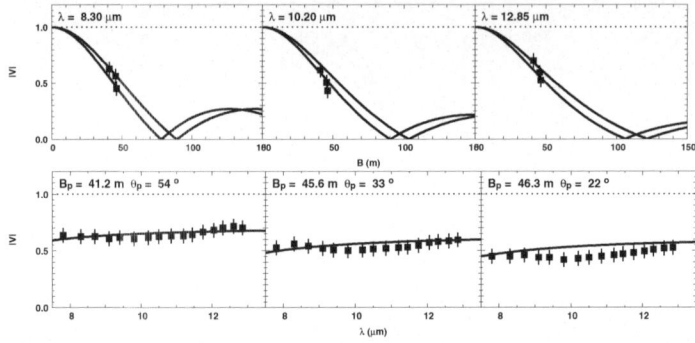

Fig. 3. Visibility of Hen 3-1191. Marquers with error bars: observations. Solid line: bost model fit of photometric and interferometric data. *Top:* visibility vs. baseline at 8.3, 10.2, 12.85,μm along the major and minor axes of the inclined disc. *Bottom:* visibility vs. wavelength for each measured spatial frequency.

Table 2. Best-fit model of Hen 3-1191 observations using a disc with hot inner rim.

$T_\star = 25000\,\mathrm{K}$	$R_\star = 6\,R_\odot$	$M_\star = 14\,M_\odot$	$R_{\mathrm{in}} = 30\,\mathrm{AU}$
$h_{\mathrm{rim}} = 2.3\,\mathrm{AU}$	$T_{\mathrm{rim}} = 1030\,\mathrm{K}$	$T_{\mathrm{in}} = 480\,\mathrm{K}$	$q = 0.75$
$i = 30°$	$\theta = -10°$	$\epsilon = 0.25$	$d = 4\,\mathrm{kpc}$

radius at 1400–1500 K observed in typical HAe/Be stars. Second, the remaining parts of the disc are not heated by irradiation, but by viscosity: One one hand, the temperature profile is typical of viscous disc with the exponent $q = 0.75$. On the other hand, a passive disc model e.g. [2] is not able to account for the high temperature $T = 480\,\mathrm{K}$ at 30 AU. Using the standard accretion disc model by [14] we derive from that temperature an accretion rate $\dot{M}/M_\star \approx 1.5 \times 10^{-3}\,\mathrm{yr}^{-1}$. It means that a stellar mass would be accreted in only $\approx 700\,\mathrm{yr}$. It hints towards a short ($\lesssim 100\,\mathrm{yr}$) transitory state similar to an FU Orionis outburst.

4 The nature of Hen 3-1191

The supergiant hypothesis. Early B[e] supergiants have a bolometric luminosity of $\gtrsim 10^5\,L_\odot$ [10] Fig.1, that is an absolute bolometric magnitude $\lesssim -7.8$ and an absolute visual magnitude $\lesssim -5.0$ bolometric correction by [1]. If Hen 3-1191, located in the galactic plane and with dereddened $V \approx 10.5$ ($A_V \approx 3$), were a supergiant it should be located at $\gtrsim 13\,\mathrm{kpc}$ and feature an extinction of at least 10 magnitudes. We therefore conclude that Hen 3-1191 does not belong to the supergiant group.

The Herbig A[e] hypothesis. The SED and visibilities are consistent with massive accretion ($\dot{M}/M_\star = 1.5 \times 10^{-3}\,\mathrm{yr}^{-1}$) in a disc with a large inner hole.

Despite of the unusual picture, some other Herbig Be stars feature similar characteristics e.g. HK Ori A[15], so that we cannot rule out this possibility.

The proto-planetary nebula hypothesis. We have not yet investigated this possibility. The presence of a accretion discs with high accretion – as our data suggest – is conjectured to explain the high polarity in some proto-PN [8].

5 Conclusion & perspective

High resolution spectroscopy in the optical is one of the keys in order to determine the type of the object. We plan near-IR observations with AMBER to constrain the geometry of the innermost parts (the inner disk rim?).

References

1. Bessell, M. S., Castelli, F., & Plez, B. 1998, A&A, 333, 231
2. Chiang, E. I. & Goldreich, P. 1997, ApJ, 490, 368
3. Cutri, R. M., Skrutskie, M. F., van Dyk, S., et al. 2003, VizieR Online Data Catalog, 2246, 0
4. De Winter, D. & Pérez, M. R. 1998, in ASSL Vol. 233: B[e] stars, 269
5. De Winter, D., The, P. S., & Perez, M. R. 1994, in Astronomical Society of the Pacific Conference Series, 413–416
6. Dullemond, C. P., Dominik, C., & Natta, A. 2001, ApJ, 560, 957
7. Elia, D., Strafella, F., Campeggio, L., et al. 2004, ApJ, 601, 1000
8. Frank, A. & Blackman, E. G. 2004, ApJ, 614, 737
9. Lachaume, R., Malbet, F., & Monin, J.-L. 2003, A&A, 400, 185
10. Lamers, H. J. G. L. M., Zickgraf, F., De Winter, D., Houziaux, L., & Zorec, J. 1998, A&A, 340, 117
11. Malbet, F., Berger, J.-P., Colavita, M. M., et al. 1998, ApJ, 507, 149
12. Millan-Gabet, R., Schloerb, F. P., & Traub, W. A. 2001, ApJ, 546, 358
13. Pereira, C. B., Landaberry, S. J. C., & de Araújo, F. X. 2003, A&A, 402, 693
14. Shakura, N. I. & Sunyaev, R. A. 1973, A&A, 24, 337
15. Smith, K. W., Balega, Y. Y., Duschl, W. J., et al. 2005, A&A, 431, 307
16. The, P. S., de Winter, D., & Perez, M. R. 1994, A&A, 104, 315
17. Zickgraf, F.-J. 2003, A&A, 408, 257

Part III

Science: Stars — binaries and multiples

Binaries: A Main Staple of Interferometry

Christian A. Hummel

European Southern Observatory, Casilla 19001, Santiago 19, Chile
chummel@eso.org

Summary. Besides the measurement of stellar diameters, resolving binaries has been another important astrophysical application of interferometry since the beginning. As a model independent probe of stellar masses and distances, binaries continue to play a role as new windows in the near and middle infrared are opened to observe binaries in earlier stages of evolution. In my talk, I will first briefly review the technique and its challenges, then discuss some of the results in view of their astrophysical impact, and finally touch on interferometric imaging of multiple stars.

1 Introduction and outline

In many ways, binaries have been a real cash cow for the young and quite specialized technique of optical and infrared long baseline interferometry. (Though dating back to 1922[20], interferometry did not really take off until the first generation of laser and computer controlled instruments in the eighties.) First, after a single uniform disk, binaries are only the second most simple source geometry, ideal to be constrained even with the sparse data of early interferometry. Second, binaries are of course astrophysically important, providing the only hypothesis-free method to measure stellar masses and distances, as has been demonstrated again recently in the resolution of the Pleiades distance controversy[28]. The high spatial resolution capability of long baseline interferometry finally promised to resolve that majority of the double-lined spectroscopic binaries, which had been out of reach for the predecessor, the speckle technique. That binaries are found to be quite a bit more common than single stars just adds to the importance of observing these objects.

The following sections present the progress in this field from a thematical point of view, which, even if biased by the author's own perception of the progress, should provide a useful review. More comprehensive introductions into this field can always be found elsewhere, e.g. [24, 21].

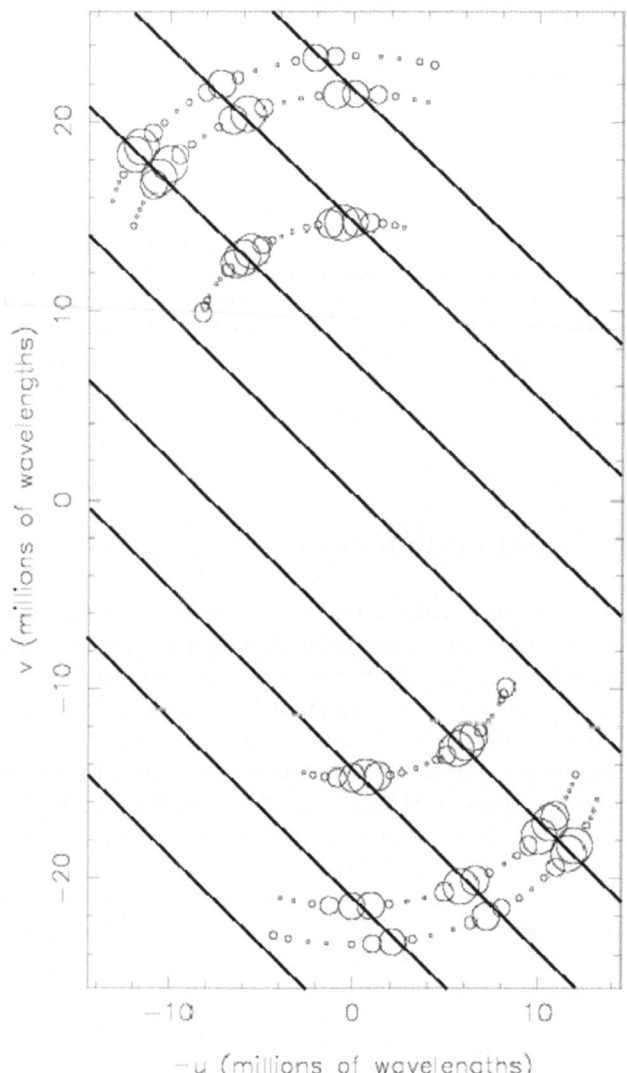

Fig. 1. Observations of Capella with the Mark III stellar interferometer. Measurements in 3 channels cover the aperture as shown. The size of the circles is proportional to the squared visiblity amplitude. The straight lines correpond to the locus of the visiblity maxima predicted by a model for the binary of separation 41.5 mas, and position angle 314.9 degrees.

2 Challenges and achievements

While the basic task of fitting separation and position angle between the two components of the binary to data as shown in Fig.1 is straighforward, one challenge may arise in finding an unambiguous solution if just one wavelength channel was recorded during the night. As one can also learn from Fig.1, the position angle has an ambiguity of 180°, which is only lifted if closure phases are measured by the interferometric array. What can happen to an orbit solution if quadrant errors go undetected can be seen by comparing the results for θ Aquilae between [10] and [13].

As another "reward" for having such a high spatial resolution capability at their disposal, astronomers had to deal with the significant motion of short period binaries during the interval necessary to accumulate enough data using Earth's rotation to change orientation and projected length of the baseline. This challenge was met by fitting orbits directly to the visibility data, or by fitting a reference position with low order approximations for the changes in separation and position angle from a preliminary orbit. It is important to realize that the presentation of the orbits in the literature varies between these two methods, as do the visualization for the benefit of the reader of the constraints afforded by the observations.

3 Encounters with theory

At the time when early long baseline interferometers were starting to resolve spectroscopic binaries, most precise stellar masses were only known for the components of eclipsing double-lined spectroscopic binaries. These stars however are usually on the main-sequence, as there is an observational bias for eclipsing systems to be small in size. Consequently, comparatively little was known about the masses of giants, let alone supergiants. An especially highly priced commodity were the composite spectrum binaries, with one component on the main-sequence, and the other evolved off the main-sequence to become a red giant or supergiant. Isochrones therefore had to fit two widely separated locations in the HR diagram.

The knowledge of stellar masses alone does not do much good to astrophysics, unless accompanied by knowledge of the other parameters predicted by stellar evolution for a star of given mass, chemical composition, and perhaps rotation. Most important of these, the stellar luminosity can be determined from long baseline interferometry quite well due to the comparatively precise, with respect to speckle interferometry that is, photometric calibration of the magnitude difference of the binary components, preferably done in several bands to also determine the colors.

When the precision of the orbits improved to the point that with equally precise radial velocities masses (and luminosities) could be derived with a precision of better than 1%, the stellar evolution models had to be computed specifically for the targets under investigation. That included most importantly the knowledge about the chemical abundances, and we refer the reader to [27] for an excellent example of such a thorough investigation.

4 Footprints in the HR diagram

The vast majority, if not all orbit determinations of double-lined spectroscopic binaries by long-baseline interferometry are shown in Fig.2. The interferometers involved are the Mark III, PTI, NPOI, SUSI, and NII. Not all of these have led to reliable results due to problems in historic radial velocity observations usually related to insecure detections of the secondary spectrum.

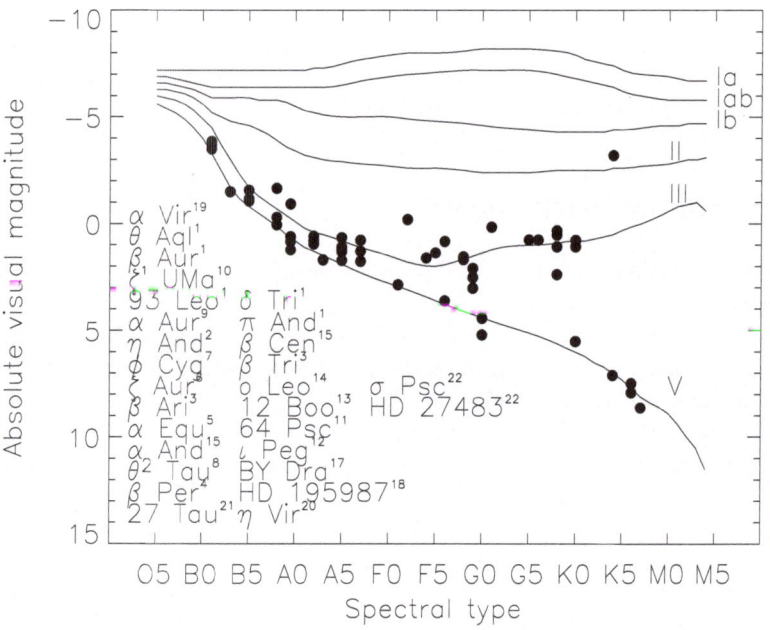

Fig. 2. References: [1]: [13]; [2]: [11]; [3]: [22]; [4]: [23]; [5]: [1]; [6]: [3]; [7]: [2]; [8]: [26]; [9]: [12]; [10]: [14]; [11]: [4]; [12]: [5]; [13]: [7]; [14]: [15]; [15]: [9]; [16]: [25]; [17]: [6]; [18]: [27]; [19]: [8]; [20]: [16]; [21]: [28]; [22]: [18];

5 Advanced modeling

Since there is a significant overlap between orbital elements measured in a visual binary and those of a spectroscopic binary, it didn't take long before researchers (e.g. [14]) combined the corresponding data sets, i.e. visibilities (or relative positions) and radial velocities. An orbit was then fitted to the combined data, overall reducing the number of free parameters used in the end for a complete determination. Some orbits included the semi-amplitudes K_1 and K_2, others the component masses instead. In either case, it is important to ascertain the consistency of the two data sets before combination. As to the relative weighting of the different data sets, they should be chosen via the measurement uncertainties such that individual fits would produce residuals consistent with the uncertainties.

Back to the question on what parameters to include, it is important to select a set where no parameter could be written as a combination of any of the other ones. For example, in the case of a set including the component masses as the parameters of interest for double-lined systems, the parallax is a dependent variable through Kepler's third law and thus cannot be included in the model. (A parallax measurement can be added to the data in order to constrain single-lined spectroscopic binaries.) The uncertainties of the parameters included in the model would take correlations properly into account, while the uncertainty of derived parameters would have to be determined carefully due to the non-linear nature of the model.

6 Multiple systems and the hierarchical model

A third component in the field of view (FOV) poses interesting new challenges for the interferometer. The more common case is a component which contributes "incoherent" flux because it is within the photometric FOV, but outside the interferometric FOV set by the bandwidth of the recorded fringes. In this case, the measured visibilities will be lower by a factor corresponding to the magnitude difference between the star outside the FOV and the ones producing fringes.

Not too much has been achieved in this field yet by long baseline interferometers, despite triples being the next stepping stone in complexity for imaging applications, as well as being astrophysically interesting for their relative orbit orientations.

We can distinguish a few cases which are probably going to be typical for interferometry. The classification depends somewhat on the interferometer baselines. For example, in the Algol class we would have triples where the wide pair is resolved, but not the close pair, while all radial velocity curves are well sampled. We would be able to determine distance and all component masses, but not relative orbit orientations. Then there is the η Virginis class, where all components are resolved, and the close pair is a double-lined binary.

If the systemic velocity of the close pair samples well the wide pair orbit, as in η Virginis, then all desired parameters can be derived ([16]).

Multiple stellar systems are usually hierarchical, meaning that they can be assembled of binaries the components of which can be multiple themselves but are dynamically represented by the total mass at their center of mass. This makes them dynamically stable. When computing such models for specific measurements, sometimes a photo center has to replace the components of the closest unresolved pair in a multiple system.

7 Interferometric imaging

As is well known, imaging is the domain of interferometric arrays of three or more elements combined simultaneously in order to measure the so-called closure phases in addition to the visibility modulus. Without closure phases, images would suffer from mirror symmetry. While imaging has become the standard mode of large radio interferometric arrays, it has so far played a side role in optical long baseline interferometry. Few images have been published, exclusively of binary stars. In fact, these are usually "PR" images, while important image parameters with astrophysical relevance have been extracted from modeling the visibility data. This is unlikely to change anytime soon due to the high demands in terms of data quality and quantity imposed by this hypothesis-free object structure reconstruction technique.

Modeling has the advantage over imaging in that it uses far fewer free parameters, simply because it is "assumed" that we know the structural components of the object. How would we know? Well, this is where imaging would come in handy. But the images would be too crude to allow parameters determined from them. Especially in the case of spectrally resolving interferometers with broad wavelength coverage, one could improve the aperture coverage by combining all visibilities, which would be equivalent to neglecting to first order the wavelength dependence of the structure. Then, after having determined the basic structure, modeling with proper structural elements would take over.

Imaging software for optical interferometry has indeed come a long way. Since this technique produces observables different from the ones produced by radio interferometry, i.e. typically squared visibilities and closure phases versus complex visibilities with amplitude and phase, the standard radio imaging techniques cannot be applied easily. At the meeting of the SPIE in 2004, a contest to image simulated interferometry data produced several packages which take fairly advanced approaches to solving the imaging problem for optical data ([19]).

In Fig.3, an algorithm is shown which has been very successfull in radio interferometric imaging. It has also been applied to optical interferometry. While the steps are explained in the caption, the basic principle, i.e. successively removing the model from the data, lends itself to experiments with more advanced ideas such as combining modeling and imaging by parameterizing image components to account for the wavelength dependence of structure, for example. Assigning photospheric effective temperatures to CLEAN components of a standard interferometric map is such a parametrization[16], enabling the imaging of composite spectrum binaries using broadband aperture synthesis for example.

Difference mapping

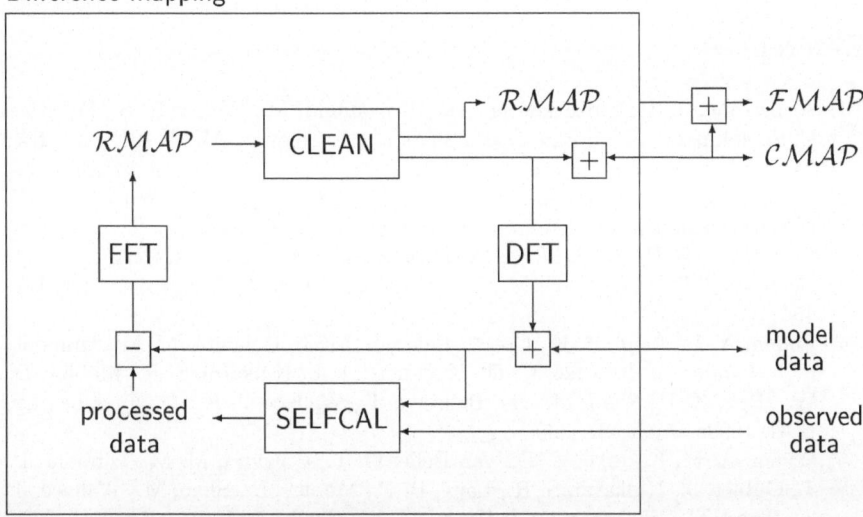

Fig. 3. Flow diagram of the difference mapping algorithm. The iterations start with a default model, most often a point source at the phase center, and the observed visibility data consisting of amplitudes and phases. After a phase self-calibration step, the model visibilities are subtracted, and the result is Fourier-transformed into a map containing the residual flux, i.e. the flux which is not yet accounted for in the model. Additional model components are extracted from the residual map using the CLEAN algorithm, and added to the model. The model visibilities are computed from the model components using a direct Fourier transform, and the cycle continues.

8 Conclusions and future challenges

While there has been an initial tendency to "count butterflies" when selecting easily feasible binaries for the early pioneering long baseline interferometers, current research targets specific astrophysically interesting objects. When this method will be be extended very soon to the faint PMS and perhaps even protostars with the giant interferometers like Keck and VLTI, a lot more very exciting science will result.

Among the future challenges will be the so-called "messy" target, i.e. one which displays complex structure on many scales, which cause complications even if we are interested in parameters of only one compact region. This has to do with the emission filling the entire field of view, which happens with η Carinae for example. Similarly, cluster environments or the galactic center environment will challenge the method, and it would be unfortunate if these were not tackled soon.

References

1. Armstrong, J. T., Mozurkewich, D., Vivekanand, M., Simon, R. S., Denison, C. S., Johnston, K. J., Pan, X.-P., Shao, M., & Colavita, M. M. 1992, AJ, 104, 241
2. Armstrong, J. T., Hummel, C. A., Quirrenbach, A., Buscher, D. F., Mozurkewich, D., Vivekanand, M., Simon, R. S., Denison, C. S., Johnston, K. J., Pan, X.-P., Shao, M., & Colavita, M. M. 1992, AJ, 104, 2217
3. Bennett, P. D., Harper, G. M., Brown, A., & Hummel, C. A. 1996, ApJ, 471, 454
4. Boden, A. F., Lane, B. F., Creech-Eakman, M. J., Colavita, M. M., Dumont, P. J., Gubler, J., Koresko, C. D., Kuchner, M. J., Kulkarni, S. R., Mobley, D. W., ,Pan, X. P., Shao, M., van Belle, G. T., Wallace, J. K., & Oppenheimer, B. R. 1999a, ApJ, 527, 360
5. Boden, A. F., Koresko, C. D., van Belle, G. T., Colavita, M. M., Dumont, P. J., Gubler, J., Kulkarni, S. R., Lane, B. F., Mobley, D., Shao, M., Wallace, J. K., The PTI Collaboration, & Henry, G. W. 1999b, ApJ, 515, 356
6. Boden, A., & Lane, B. 2001, ApJ, 547, 1071
7. Boden, A. F., Torres, G., & Hummel, C. A. 2005, ApJ, 627, 464
8. Herbison-Evans, D., Hanbury Brown, R., Davis, J., & Allen, L. R. 1971, MN-RAS, 151, 161
9. Davis, J., Mendez, A., Seneta, E. B., Tango, W. J., Booth, A. J., Byrne, J. W., Thorvaldson, E. D., Ausseloos, M., Aerts, C., & Uytterhoeven, K. 2005, MNRAS, 356, 1362
10. Hummel, C. A., & Armstrong, J. T. 1991, in High Resolution Imaging by Interferometry II, ed. by J.M. Beckers and F. Merkle (Garching: ESO), P. 697
11. Hummel, C. A., Armstrong, J. T., Quirrenbach, A., Buscher, D.F., Mozurkewich, D., Simon, R. S., Johnston, K. J., 1993, AJ, 106, 2486
12. Hummel, C. A., Armstrong, J. T., Quirrenbach, A., Buscher, D. F., Mozurkewich, D., Elias, II N. M., & Wilson, R. E. 1994, AJ, 107, 1859
13. Hummel, C. A., Armstrong, J. T., Buscher, D. F., Mozurkewich, D., Quirrenbach, A., & Vivekanand, M. 1995, AJ, 110, 376
14. Hummel C. A., Mozurkewich D., Armstrong J. T., Hajian, A. R., Elias N. M. II, & Hutter D. J. 1998, AJ, 116, 2536
15. Hummel, C. A., Carquillat, J.-M., Ginestet, N., Griffin, R. F., Boden, A. F., Hajian, A. R., Mozurkewich, D., & Nordgren, T. E. 2001, AJ, 121, 1623
16. Hummel, C.A., Benson, J.A., Hutter, D.J., Johnston, K.J., Mozurkewich, D., Armstrong, J.T., Hindsley, R.B., Gilbreath, G.C., Rickard, L.J, & White, N.M., 2003, AJ, 125, 2630

17. Hummel, C. A. 2004, in: Spectroscopically and Spatially Resolving the Components of Close Binary Stars, ed. by R.W. Hilditch, H. Hensberge, and K. Pavlovski San Francisco: ASP, p. 13
18. Konacki, M., & Lane, B. F. 2004, ApJ, 610, 443
19. Lawson, P. R., Cotton, W. D., Hummel, C. A., Monnier, J. D., Zhao, M., Young, J. S., Thorsteinsson, H., Meimon, S. C., Mugnier, L., Le Besnerais, G., Thièbaut, E., & Tuthill, P. 2004, in: Interferometry for Optical Astronomy II, ed. by W. Traub (Bellingham: SPIE), 886
20. Merrill, P. W. 1922, ApJ, 56, 40
21. Monnier, J. 2003, Rep. Prog. Phys. 66, 789
22. Pan, X. P., Shao, M., Colavita, M. M., Mozurkewich, D., Simon, R. S., & Johnston, K. J. 1990, ApJ, 356, 641
23. Pan, X., Shao, M., & Colavita, M. M. 1993, ApJ, 413, 129
24. Quirrenbach, A. 2001, Ann. Rev. Astron. Astrophys., 39, 353
25. Ryabchikova, T. A., Malanushenko, V. P., & Adelman, S. J. 1999, A&A, 351, 963
26. Tomkin, J., Xiaopei, P., & McCarthy, J. K. 1995, AJ, 109, 780
27. Torres, G., Boden, A. F., Latham, D. W., Pan, M., & Stefanik, R. P. 2002, AJ, 124, 1716
28. Zwahlen, N., North, P., Debernardi, Y., Eyer, L., Galland, F., Groenewegen, M. A. T., & Hummel, C. A. 2005, A&A, 425, 45

Protoplanetary Disks as seen by Interferometry

Anne Dutrey

L3AB, Observatoire de Bordeaux, 2 rue de l'observatoire, F-33270 Floirac, France
Anne.Dutrey@obs.u-bordeaux1.fr

Summary. Protoplanetary disks encountered around young Pre-Main-Sequence (PMS) stars of low or intermediate masses are now observable by current Near-Infrared (NIR) and Mid-Infrared interferometers. Taking also into account the progress done by millimeter arrays, interferometry is slowly but surely changing the knowledge we have of their physics. In this review, I show how these recent observations help to constrain the physical properties of both the very inner disks where planetary formation should occur and the outer disk where the reservoir of mass is located. I also illustrate this by presenting interferometric studies from a few examples such as AB Auriga.

1 Introduction

Proto-planetary disks orbiting around young stars of a few million years are now routinely observed by interferometers from the radio up to the optical domain. At the distance of the closest star forming regions (D$\sim$ 100 pc), the angular radii of these disks usually in the range $1 - 3''$ or 150-450 pc. However, being heated by the their central stars, their spectral energy distribution (SED) is such that the apparent angular size drops very fast when the observing wavelength goes from the millimeter domain to the optical one. Typically the apparent size of the thermal emission of a hot dust disk at $10\mu m$ would not exceed a few AU or $\sim 0.1''$.

From the observations, one can distinguish three parts in disks. The "very inner disk", located close to its inner radius, is extending up to radius of $\sim 1 - 3$ AU and is only resolved by optical and NIR interferometry. The "inner dust disk" is extending beyond a radius of ~ 30 AU and would somewhat correspond in our system to the inner solar system, located up to the first Kuiper Belt objects. The "outer disk" corresponds to the large dust and gas disk usually traced by CO rotation line emission and observed by millimeter and sub-millimeter interferometers such as the IRAM array, the SMA or OVRO. Fig.1 summarizes the area sampled by current and future large interferometers depending on the wavelength of observations and baseline lengths in a disk orbiting a TTauri star located at $D \sim 150$ pc. Fig.1 clearly shows that a global comprehension of the disk physics can only be achieved by multi-wavelength observations performed at high angular resolution.

Hence, resolving (and modelling) such disks necessarily requires interferometric technics. This is true both for small (very inner disks) and large (outer
disks) scales because only aperture synthesis can today technically achieve
the angular resolution needed.

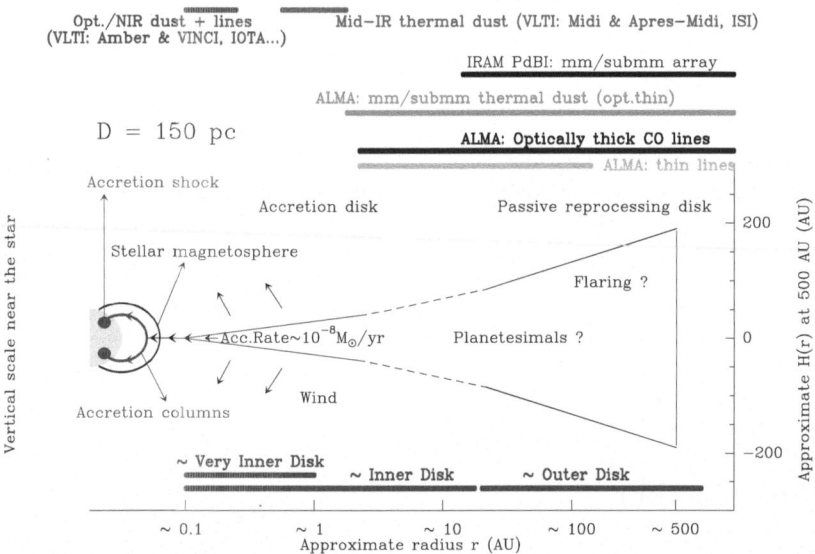

Fig. 1. A cartoon showing the area sampled by interferometers depending on their
wavelengths, sensitivity and angular resolution. Optical/NIR interferometers reveal the disk structure at the inner edge. Millimeter/submillimeter arrays currently
characterize the dust and gas in the outer disk (contrary to ALMA which will also
observe the inner disk) while MID-IR interferometers sample the dust in the inner
disk.

2 Models of disks are currently based on the SED

Disk models were originally based on the analysis of the SED of the dust
disk from the optical down to millimeter domain. SEDs can be reproduced
by models (for a review, see Hartman 1998) where dust disks are either
i) passive: they reprocess the stellar light or ii) active: disk are heated by
viscous dissipation. For viscous or accretion disks, the viscosity ν is usually
expressed by the so-called α parameter linked to ν by $\nu = \alpha c_s H$ where c_s is
the sound speed. The accretion remains subsonic with $\nu/r \sim \alpha c_s H/r << c_s$
and $\alpha \sim 0.01$. In both kinds of models, since there is no vertical flow, the
motions are circular and remain Keplerian ($v(r) = \sqrt{GM_*/r}$, where M_* is the
stellar mass). Hence the disk, in hydrostatic equilibrium, is geometrically thin

with $H \ll r$ where H is the disk scale height. Chiang & Goldreich (1997) have developed a model of passive disk where the optically thin upper layer of the disk is super-heated above the blackbody equilibrium temperature by the stellar light impinging the disk and producing an atmosphere above the disk. Both viscous heating and super-heated layers seem to be however necessary to properly take into account the observed SEDs (D'Alessio et al. 1998, 1999).

More recently, Dullemond and collaborators (Dullemond et al. 2001, hereafter DDN) have introduced a new model which better accounts for the properties of the disk SED at NIR wavelengths. This model was originally developed to explain the SEDs of the Herbig Ae disks around 2-5 μm. In this model of passive irradiated flaring disk, the inner dust disk is truncated by evaporation of the dust. At the inner edge, the inner rim is puffed up and much hotter than the external part of the disk because it is directly illuminated by the central star. The rim can even be high enough to shadow a significant fraction or the totality of the outer disk from the stellar light. This can directly affect the strength of the MID-IR excess and the apparent size of the 10μm emission.

Finally, other models such as those developed by Vinkovic et al. 2006 and including a tenuous and small ($\sim$ 10 AU) dusty halo may also explain the disk SEDs in the NIR domain.

It is clear that the SED itself does not provide enough constraints and that resolved observations (analyzed in the image or the UV planes) are needed to bring new insights on the physics of inner disks. The latter point is illustrated through several recent examples in the next section.

3 Planetary Forming Regions: new constraints provided by NIR and MID interferometry

Resolved quantitative information on the physics of inner disks were first provided by the IOTA interferometric observations of Monnier & Millan-Gabet (2002) who observed several disks around TTauri and Herbig Ae/Be stars. They found that the observed inner disk sizes are proportional to the stellar luminosity, as is expected when the inner dust disk radius is defined by the sublimation radius of the dust. The observed sample of young stars has now been significantly increased and Akeson (2005) reviews in these proceedings the status of these new observations.

MIDI, the interferometer operating at 10μm of the VLT, is now routinely performing observations of Herbig Ae/Be disks. Leinert and collaborators 2004, who have published the first MIDI analysis of Herbig Ae/Be observations, found that the apparent sizes at 10μm of the observed disks are qualitatively consistent with models of flaring disks. Flaring disks which have larger MID-IR excesses show larger sizes at 10μm (group I objects) while self-shadowed (or flat) disks with lower MID-IR excesses have also smaller apparent sizes at 10μm (group II objects). Moreover, from existing fits of the

DDN model derived from the diks SEDs, they calculated the corresponding visibilities for each object. In this process, there was no new fit to take into account the MIDI observations. The results are quoted in Fig.2. Although the fits are in general agreement with the data, they do not reproduce all visibility curves for all sources and more refined models are required to properly interpret the resolved MIDI data.

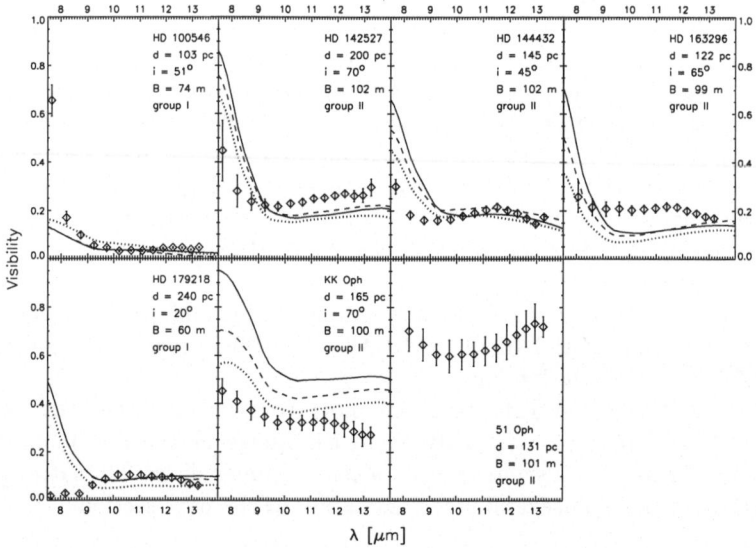

Fig. 2. From Leinert et al. 2004. MIDI visibilities versus wavelength observed in a sample of seven Herbig Ae/Be stars are given with their error bars. For each object, Leinert and collaborators have superimposed the visibility predicted by SED modelling only and coming from the DDN model. They show that the visibility calculated for cuts along the major axis (broken lines), minor axis (solid lines) and pole-on disks (dotted lines).

However, the physics close to the star can be even more complicated than a simple disk model. It is the case when disks and expanding winds or outflows are simultaneously observed. This is demonstrated by the recent observations performed by Malbet and collaborators (see also talk by Malbet in these proceedings). Malbet et al. used AMBER, the NIR interferometer of the VLT, to perform the first observations around the Herbig Be star MWC297. They found (Fig.3) that the visibility in the Brγ line is lower than the visibility of the continuum. This implies that the line emission comes from a more extended region than the continuum emission. This is likely what happens in the standard scenario of a "disk + wind model". The dust disk is poorly resolved by the interferometer while the expanding wind is responsible for the extended Brγ line emission.

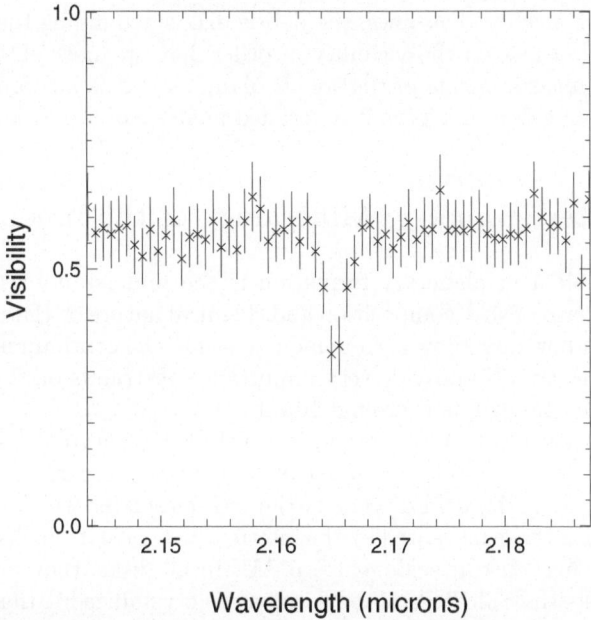

Fig. 3. MWC297 observations performed with AMBER by Malbet et al. 2005. Visibility of the continuum and Brγ line. The visibility of the Brγ line is lower than those of the continuum, suggesting that it comes from a more extended region.

Detailed modelling of interferometric observations is often limited by the small number of visibility points obtained. To avoid this, Malbet et al. 2005 observed the young TTauri star FU Orionis in several campaigns of observations with IOTA, PTI and VINCI. They observed in H and K bands and got about 287 visibility points on baselines extending from 20 up to 110m. They simultaneously modelled all the observed visibilities. They found that the geometry of the circumstellar material is that of a disk and measured its inclination and position angles. They also found that the disk seems better modelled by an active accretion disk with a high accretion rate of $\sim 5C^{18}O10^{-5}M_\odot/yr$. Moreover, they marginally detected an unresolved structure which might be a possible companion. Thanks to the good UV coverage, these observations reveal how complex the close circumstellar material around a young stellar object can be (possible multiplicity) and provide the first quantitative modelling, at the AU-scale, of the surroundings of an YSO.

The recent examples shown above illustrate how interferometric observations are suitable to provide new constraints on the physics of proto-planetary disk which always present high or reasonable IR excess. It is however important to mention that studying by interferometry inner regions of NIR optically thin disks surrounding Vega-like stars is more problematic. In a recent paper, DiFolco et al. 2004 have observed Fomalhaut with VINCI in the K and H bands. At short baselines, the debris disk surrounding the star would correspond to a loss on the visibility of order 3 %, at most. This is enough to affect the measurements of the stellar diameter if it is not properly taken into account but does not permit a detailed analysis of the disk itself.

4 MID interferometry: Mineralogy of the Inner Disk

One key problem in planetary formation is the understanding of the dust evolution in term of size, composition and chemical nature. MIDI observations around 10μm not only allow astronomers to study the continuum emission of dust disks but can also provide very important constraints on the properties of the Silicates bands found around 10μm.

Van Boekel et al. 2004 recently observed three Herbig Ae disks with MIDI using the low spectral resolution mode of ~30. They compare the 10μm spectrum coming from the field of view of the interferometer which contains the whole disk with those measured by the interferometer itself which corresponds to the inner dust disk at scale ~ $1 - 5$ AU. In all disks, they found that in the inner disks the Silicate grains appear more crystallized i) than any other dust observed around YSOs and ii) than in their outer disks. These results suggest that Silicates crystalize very early in the evolution of disks, before the formation of terrestrial planets.

So far, these data bring the first observational constraints on the properties of the Silicates grains at the scale of a few AUs in disks and will provide in the next years the first detailed analysis of inner dust disk mineralogy.

5 Multi-wavelengths interferometry: AB Auriga case

Current millimeter interferometry is less limited than optical interferometry by the UV coverage and mm arrays routinely get images of the environments of YSOs. These images allow astronomers to better take into account the complexity of the star formation processes (e.g. presence of both outflows and disks, existence of multiples stars). Quantitative studies of disks are however usually performed by data analysis inside the UV plan in order to avoid the deconvolution of the image which is a non-linear process (Guilloteau & Dutrey 1998). Current mm arrays provide mostly information on the dust and gas content of the outer disks but, thanks to their heterodyne detection system, they have a high spectroscopic resolution which allows detailed kinematic analysis.

Interferometric studies of the circumstellar material around AB Aurigae performed from the mm to the optical domain reveal the complexity of the environment of a young star and the necessity to understand both the small and large scales in disks.

In many aspects, AB Auriga is taken as the proto-type of the Herbig Ae star. The star is an A0 star of $\sim 10^6$ years. Modelling of the SED shows that the star is a Group I source having a flaring disk (Meeus et al. 2001). It is also surrounded by a large reflection nebula (Grady et al. 1999) extending up to ~ 10000 AU (Fig.4, top). More recently, using the Subaru telescope, Fukagawa et al. 2004 found that at medium scale ($\sim 100 - 500$ AU) the material around the star presents a spiral pattern.

Recent sub-arcseconds images of AB Auriga obtained by Piétu et al. 2005 with the IRAM Plateau de Bure interferometer in the isotopologues of CO, and in continuum at 3 and 1.3 mm reveal that the environment of AB Aur is very different from the proto-planetary disks observed so far with mm arrays. These observations also allow the authors to trace the structure of the circumstellar material in regions where optical and IR mapping is impossible because of the emission from the star itself.

On top of Fig.4, the HST image from Grady et al. 1999 is given in false colors while contours correspond to the thermal dust emission observed at 1.3mm by Piétu et al. (the cross shows the star location). The mm continuum emission is not centrally peaked but is dominated by a bright, asymmetric ("spiral-like") feature at about 140 AU from the central star. Little emission is associated with the star itself.

The molecular emission, shown on bottom of Fig.4, reveals that AB Aur is surrounded by a very extended flattened low mass gaseous structure ("disk") which is also not centrally peaked. Bright molecular emission is also found towards the continuum asymmetry. The large scale molecular structure suggests the AB Aur "disk" is inclined between ~ 25 and 35 degrees. The strong emission from the continuum and molecular asymmetry prevents an accurate determination of the inclination of the inner disk part. Surprisingly, an analysis of the CO line kinematics reveal that the disk rotation is non Keplerian, at the 10σ level.

At small scales, IOTA observations by Millan-Gabet et al. 1999 reveal that the inner disk radius is at ~ 0.3 AU, consistent with expected location of the dust sublimation radius for an A0 star. Using new PTI data and existing IOTA data, Eisner et al. 2004 have fitted several models inside the visibilities, they found that the disk is inclined by $\sim 20°$. Unfortunately, their UV coverage was too poor to provide quantitative model fitting.

It is clear that both at large and medium scale ($\sim 50 - 100$ AU), the circumstellar material presents many departures from a symmetric structure and only a good UV coverage can allow astronomers to get a comprehensive picture of the disk geometry. Inclination angles observed at small and large scales are consistent but the poor UV does not provide enough data to reveal possible asymmetries. Both the spiral-like feature and the departure from purely Keplerian motions in the disk suggest that AB Aur is undergoing a large scale event which might be related to a disturbance (of yet unknown origin) and/or an early phase of star formation in which the Keplerian regime

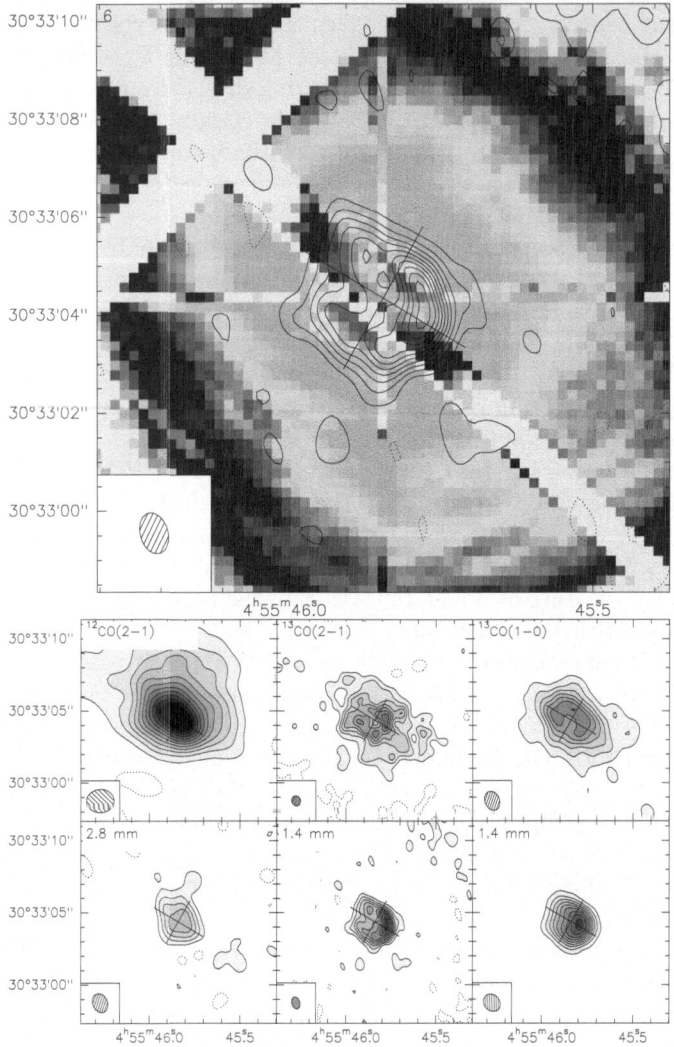

Fig. 4. From Piétu et al. 2005. Top: 1.4 mm continuum data (in contours) super-imposed on the HST image from Grady et al., (1999), in false color. The angular resolution is $0.85 \times 0.59''$ at PA 18°. Bottom: A montage displaying high resolutions image of the continuum emission at 2.8 mm and 1.4 mm, and of the integrated line emission of ^{12}CO J=2→1 ^{13}CO J=2→1 and ^{13}CO J=1→0 transitions. The (lower resolution) emission at 110 GHz is also presented.

is not yet fully established. The latter interpretation is more likely and is reinforced by the fact that the dust observed at mm wavelengths appears less evolved than in most proto-planetary disks. Undergoing analysis of MIDI observations of the Silicates bands at $\sim 10\mu$m also suggests the presence of amorphous Silicates, as it is also observed at larger scales with ISO (Bouwman et al. 2000), in agreement with a relatively young stellar object.

6 Summary

Nowadays, interferometry (from millimeter up to optical wavelengths) reminds the best tool to provide quantitative observations from the large to the small scales in proto-planetary disks. A fundamental step is currently being reached in the knowledge of inner disks thanks to optical, NIR and MID-IR interferometry as it was done about ten years ago by mm arrays for outer disks.

Both MIDI and AMBER results show the importance of having spectroscopic modes to study the composition and distribution of dust and gas. Moreover, all the examples presented in this review emphasize the fundamental importance of the UV coverage. Only a good UV coverage will permit observers to disentangle between the various physical models by providing reliable model fitting.

The example of AB Auriga is very rich. Up to now, no consistent physical model can explain both the observed asymmetric features and the departure from Keplerian rotation. Apparent asymmetries are also predicted by symmetric rim models (e.g. Isella & Natta 2005). Detection of asymmetries and thus measurements of phase closure relations is the necessary next step of optical interferometry.

In summary, a new era is currently opening. New interferometric facilities (including ALMA) will allow astronomers in a near future to get direct insights in the region of proto-planetary disks where planetary formation is thought to occur.

Acknowledgements: I would like to thank Jean-Philippe Berger, Olivier Chesneau, Emmnanuel DiFolco, Stéphane Guilloteau, Fabien Malbet and Vincent Piétu for providing material for this review. I also acknowledge Christoph Leinert and all the MIDI Team for introducing me to the interferometry with MIDI.

References

1. D'Alessio P., et al., 1998, ApJ, 500, 411
2. D'Alessio P., et al., 1999, ApJ, 527, 893
3. Bouwman J., de Koter A., van den Ancker M. E., Waters L. B. F. M., A&A, 2000, 360, 213

4. Chiang E.I. & Goldreich, P., 1997, ApJ, 490, 368
5. DiFolco E. et al., 2004, A&A, 426, 601
6. Dullemond C.P., Dominik, C., Natta A., 2001, ApJ, 560, 957
7. Eisner J.A. et al., 2004, ApJ, 613, 1049
8. Fukagawa M. et al. , 2004, ApJ, 605, L53
9. Grady C. et al., 1999, ApJ, 523, L151
10. Guilloteau S. & Dutrey A., 1998, A&A, 339, 467
11. Hartmann L., 1998, "Accretion Processes in Star Formation", Cambridge University Press (Cambridge Astrophysics series; 32)
12. Isella A., & Natta A., 2005, A&A, 438, 899
13. Malbet F., et al., 2005, A&A, 437, 627
14. Meeus G., et al., 2001, A&A 365, 476
15. Millan-Gabet et al., 1999, ApJ 513, L131
16. Monnier J., Millan-Gabet R., 2002, ApJ, 579, 694
17. Leinert Ch., et al., 2004, A&A, 423, 537
18. Piétu V., Guilloteau G., Dutrey A., 2005, A&A, 443, 945
19. Van Boekel R., et al., 2004, Nature, 432, 479
20. Vincovic D., Ivezic Z., Jurkic R., Elitzur M., 2006, ApJ, 636, 348

Pre-Main Sequence Binaries: The Promise of IR Interferometry

Michal Simon

SUNY-SB, Stony Brook, NY, USA 11794-3800 michal.simon@stonybrook.edu

Summary. Astronomers use theoretical calculations of the evolution of young stars to estimate their masses and ages. Unfortunately, for stars less massive than 1 $M_\odot$, the current theoretical calculations yield discrepant results, and for stars less than ~0:5 $M_\odot$, the calculations are essentially uncalibrated by empirical data. Interferometric observations offer the capability to resolve the orbits of nearby double-lined spectroscopic binaries and thereby measure masses of the components and their distance. The combined observations can also yield reliable estimates of the component effective temperatures and luminosities. A limited test of the age dependence of the tracks will be possible because the components of spectroscopic binaries may be reasonably expected to be coeval. It may also be possible to resolve interferometrically the diameters of nearby pre-main sequence stars which would yield estimates of their relative ages.

1 The Problem

The mass of a star is of fundamental importance because it determines the star's life from birth to death. The usual method to estimate the mass and age of a pre-main sequence (PMS) star, by its location in the H-R diagram relative to theoretical calculations of PMS evolution, is unreliable. Fig. 1 illustrates the scatter of the theoretical calculations with respect to the H-R diagram of a M0 spectral type young star with luminosity 0.5 $L_\odot$. The tracks yield masses and ages discrepant by factors of 2 to 3.

It is also interesting to compare the mass and age distributions that are derived when different theoretical tracks are applied to the same stars in a young cluster. Kenyon and Hartmann [1] in their Fig. 15, plot the H-R diagram for the Taurus cluster using D'Antona and Mazzitelli's [2] tracks and the corresponding age and mass distribution in their Figs. 16 and 17 (left). According to the DM tracks, the median age and mass of the stars are ~ $0.8 \cdot 10^6$ years and ~0.4 $M_\odot$. Fig. 2 shows the same stars, with luminosities and effective temperature as given in [1] plotted on an H-R diagram using the BCAH[3] tracks. The stars appear older and more massive, median age $1.8 \cdot 10^6$ years (Fig. 3) and median mass 0.7 $M_\odot$.

The consequences of the differences among the calculations are that the mass spectrum of the stars produced in a star-forming region, the distribution of masses in binaries, and the region's star-forming history are imprecisely

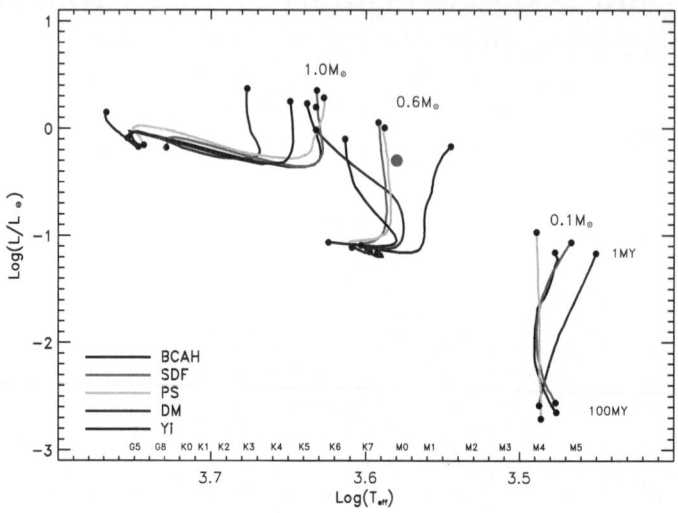

Fig. 1. An H-R diagram showing theoretical evolutionary tracks for PMS stars of mass 1.0, 0.6, and 0.1 $M_{\odot}$ between the ages 10^6 and 10^8 years, as calculated by BCAH [3], Siess et al. [5], Palla and Stahler [6], DM2, and Yi et al. [7]. The spectral type - effective temperature relation is that used by Hildebrand and White [4].

known. These problems also limit our understanding of the origins of planetary systems because the chronology of extrasolar planet formation is uncertain.

The differences arise mostly in the treatments of convection and the initial conditions [2, 3, 4, 5, 6, 7]. Masses good to better than ~10% are required for meaningful tests of the evolutionary tracks. Dynamical methods can reach this precision. Binaries and single stars with extensive circumstellar disks in the nearby star-forming regions (SFRs) provide suitable targets and measurements of a few PMS masses are becoming available (e.g. [8, 9, 10, 11]). However, at this time (April, 2005), masses are known for only 8 stars in the mass range 0.5 to ~1 $M_{\odot}$, and for only one star with mass below 0.5 $M_{\odot}$ (see Hillenbrand and White's Table 1 [9]). This is a serious gap in our knowledge because many stars in SFRs such as Taurus have masses less than 0.5 $M_{\odot}$.

A complete test of the evolutionary tracks requires not only masses of the stars but also their luminosities, effective temperatures, and ages. I will discuss how the observations can provide estimates of these quantities as well.

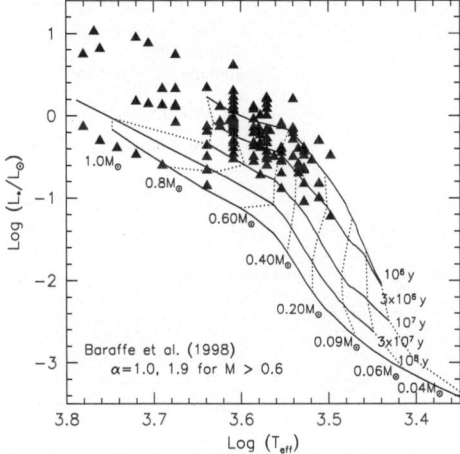

Fig. 2. An H-R diagram for the young stars in the Taurus cluster using stellar data from Kenyon and Hartmann [1] and the BCAH tracks. This figure is to be compared with Kenyon and Hartmann's Fig 15, where the stars are plotted relative to the DM tracks. Here, the stars appear about twice as old and twice as massive.

2 The Promise of IR Interferometry

2.1 Visual and Spectroscopic Binaries

Binaries provide the "gold standard" for dynamical mass measurements of stars. They are traditionally divided into the "Visual Binaries" (VBs) in which the orbit is angularly resolved, and the "Spectroscopic Binaries" in which the velocities of one component (the single-lined binaries, SB1s) or both (the double-lined binaries, SB2s) are measured. Neither the VBs or SBs alone can provide the component masses except in special circumstances. This, however, is possible by observing a binary as a VB as well as a SB2. The virtue of SBs is that the periods of the identifed systems are generally short, less than a few years. Their orbital parameters can therefore be derived fairly quickly. On the other hand, in the nearby SFRs, the angular separations of their components are usually very small, even beyond resolution by adaptive optics techniques at large diameter telescopes. IR interferometry provides the means to resolve some SBs and to map their orbits as VBs.

The orbital parameters provided by a VB are its semi-major axis in angular measure, $a(\prime\prime)$, period P, eccentricity e, and the three parameters describing the binary orientation with respect to our line of sight, the position angle of the line of nodes, Ω, the longitude of the periastron, ω, and most importantly the inclination, i. Nearly all SBs have been discovered in visible light and most are SB1s, apparently because the secondaries have lower

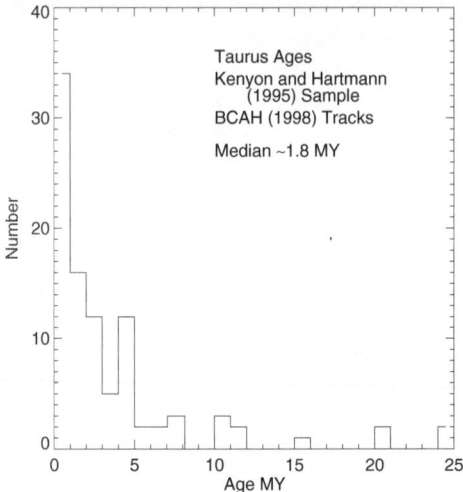

Fig. 3. The age distribution for the stars plotted in Fig. 2. The median age in Kenyon and Hartmann's [1] corresponding Fig 16 (left) is about $0.8 \cdot 10^6$ years.

mass than the primaries, and are therefore fainter in visible light than the primaries, eluding detection. For systems in which the mass of the primary is less than ~ 1 M$_\odot$, the light contrast is much more favorable in the IR. Prato et al. [12] and Mazeh et al. [13] describe the application of high resolution IR spectroscopy to young and old SB1s in order to detect their secondaries and thus to measure the secondary to primary mass ratio, q = M 2 =M 1. In the SB1s, the component masses and inclination are linked in the mass function, $f(m) = (M_2 sini)^3/(M_1 + M_2)^2$. SB2s provide the mass ratio q and $asini$, in physical units. Thus, combining the orbital parameters of a VB, with the parameters of a SB2, yields M_1, M_2, and the distance to the binary. Boden et al. (e.g. [14]) have demonstrated such an analysis for main sequence binaries. Boden et al. and Schaefer et al. apply this technique to the SB2s HD 98800 B and Haro 1-14c in papers presented at this conference.

2.2 Component Luminosities, Effective Temperatures, and Ages

Precision measurement of masses is the necessary first step towards the goal of testing the evolutionary tracks. A full test requires, in addition, measurement of the stellar luminosity, L, effective temperature, T_{eff}, and age.

In our IR spectroscopy of SB2s, we have always found that determination of the spectral types of the primary in the IR at 1.5 μm was consistent with that reported in the visible. The discrepancy is usually no more than one subclass (e.g. [15]). It seems reasonable to assume that the spectral types of the secondaries identifed by the IR measurements are equally reliable and that T_{eff} (determined using a spectral type - T_{eff} relation such as that in [9]) of the components can be determined to the accuracy required. The interferometric and spectroscopic observations measure a flux ratio at the wavelength of observation. Usually several wavelengths are accessible. It is therefore possible to derive approximate spectral energy distributions of the components, and an apportionment of the system luminosity[1]. Boden et al. demonstrate this in their study of HD 98800 B (paper presented at this conference).

Absolute determinations of age are not possible, but close binaries do provide the constraint that their components are the same age. The primary and secondary masses must therefore lie on the same isochrone. Interferometric observations may also enable one to determine relative ages of the stars in a cluster by measuring their diameters. Fig. 4 shows the theoretical diameters of 0.1 and 0.5 $M_\odot$ stars at ages $< 10^7$ years and a distance of 50 pc (see next section), according to the BCAH tracks. Recent diameter measurements with the VLTI and CHARA achieve uncertainties of a few hundreths mas [16, 17]. Measurement of the diameters of PMS stars seems within reach at the longest interferometer baselines and will enable a comparison of their relative ages.

3 Good Targets are Hard to Find

The requirements for a binary to be observable with the current generation of IR interferometers are very stringent. The binary must be bright enough in the near IR (usually K-band) that fringes can be detected and bright enough in the visible (usually V or R-band) for wavefront correction by adaptive optics. The apparent angular extent of the binary orbit must be small enough to lie within the diffraction limit of a single telescope of the interferometer, but large enough that it is resolvable at the projected interferometer baselines. At the Keck Interferometer, operating at K, this range is about 5 to 40 mas. One might also want to complete the observations in a reasonable length of time by requiring that the binary orbital period be 2 years or less. With these limits, the parameter space of binaries at the ~140 pc distance of the star-forming regions in Taurus and Ophiuchus is very restricted. For systems of total mass $M_{tot} = 0.1 M_\odot$, only those with $P \sim 2$ years are accessible and for those with $M_{tot} = 1.0 M_\odot$, those with periods longer than ~0.5 years are observable.

Song and Zuckerman have recently discovered a large number of young stars in nearby moving groups at distances as close as 20 pc (e.g. [18]). The parameter space accessible to interferometric observations is much larger for these closer systems: binaries at a distance of 50 pc with $M_{tot} = 0.1 M_\odot$ are reachable if $P > 0.4$ years and those with $M_{tot} = 1.0 M_\odot$ if $P > 0.1$ years.

---

[1] In the special case of eclipsing SB2s, in which $i \sim \pi = 2$, L and T_{eff} can be determined directly (e.g. Stassun et al. [10]).

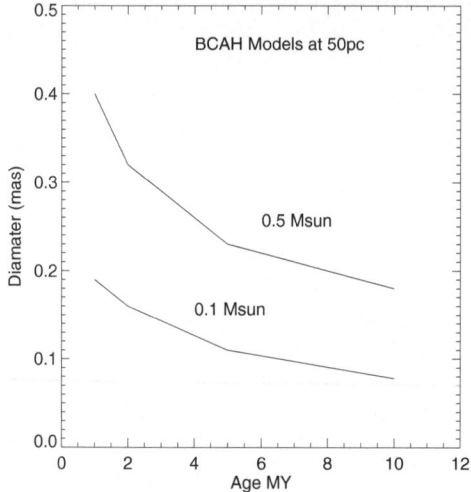

Fig. 4. Stellar diameters as a function of age for PMS stars at 50 pc distance, calculated using the BCAH tracks.

Clearly, it will be very valuable to identify the spectroscopic binaries among these nearby stars in order to select targets suitable for intereferometric observations.

4 Truth in Advertising

Although we think we know how to determine (M,L,T_{eff}) for PMS stars, the complexities of the stars - perhaps because they are heavily spotted or show strong accretion activity - suggest that inevitably some will stand apart from the others and confound the tests. This may already be the case with UZ Tau Ea and NTTS 045251B [4, 15]. In the long run, a statistical approach applied to a large sample of stars, such as that illustrated by Hillenbrand and White [9], may be required to average the individual peculiarities in order to provide an accurate assessment of the theoretical calculations.The interferometric observations I have described represent steps toward that goal.

Acknowledgments By a stroke of misfortune, I was unable to give this paper at the conference. I am very grateful to G. Schaefer for presenting it in my place and for a careful reading of this text version. I thank L. Prato and A. Zolotov for preparing Figs. 1 and 2. This research was supported in part by NSF Grant 02-05427.

References

1. Kenyon, S.J., and Hartmann, L. 1995, ApJS, 101, 117
2. D'Antona F. and Mazzitelli, I. 1997, Mem. S. A. It., 68, 807 (DM)
3. Baraffe, I. et al. 1998, A&A, 337, 403 (BCAH)
4. Baraffe, I., Chabrier, G., and Allard, F. 2003, in *Brown Dwarfs*, ed. E. Martín, p. 43
5. Palla, F. and Stahler, S.W. 1999, ApJ, 525, 772 (PS)
6. Siess, L., Dufour, E, and Forestini, M. 2000, A&A, 358, 593
7. Yi,S.,Kim, Y.-C., and Demarque, P. 2003 ApJS, 144, 259
8. Covino, E. et al. 2001, A&A, 375, 130
9. Hillenbrand, L.A. and White, R.J., 2004, ApJ, 604, 741
10. Stassun, K.G. et al. 2004, ApJS, 151, 357
11. Steffen, A. et al. 2001, AJ, 122, 997
12. Prato, L. et al. 2002, ApJ, 569, 863
13. Mazeh, T., Simon, M., Prato, L., Markus B., and Zucker, S. 2003, ApJ, 599, 1344
14. Boden, A.F., Creech-Eakman, M.J., and Queloz, D. 2000, ApJ, 536, 880
15. Prato, L. et al. 2002, ApJ, 579, L99
16. A. Richichi, A. and Roccatagliata, V. 2005, A&A, 434, 1201
17. McAlister, H.A., et al. 2005, ApJ, 628, 439
18. Zuckerman, B. and Song, I. 2004, ARAA, 42, 685

Observations of Young Stellar Objects with Infrared Interferometry: Recent Results from PTI, KI and IOTA

Rachel Akeson

Michelson Science Center, Caltech MS 100-22, Pasadena, CA USA
rla@ipac.caltech.edu

Summary. Young stellar objects have been one of the favorite targets of infrared interferometers for many years. In this contribution I will briefly review some of the first results and their contributions to the field and then describe some of the recent results from the Keck Interferometer (KI), the Palomar Testbed Interferometer (PTI) and the Infrared-Optical Telescope Array (IOTA). This conference also saw many exciting new results from the VLTI at both near and mid-infrared wavelengths that are covered by other contributions.

1 Introduction

The formation and evolution of circumstellar disks of gas and dust are a well established component of low and intermediate mass star formation. These disks serve as a conduit of material onto the star and also as the reservoir of material from which any potential planetary system may form. To fully characterize the initial conditions of planet formation, we need to detail the density, composition and temperature structure of the disk as a function of radius and also determine how these properties evolve with time.

T Tauri stars are young solar analogs and are in the middle of the evolutionary path from deeply embedded protostar to main sequence star with a residual or debris disk. Herbig Ae/Be stars are the more massive counterparts of the T Tauri stars, which also show disk characteristics and in some cases residual envelope material. FU Orionis objects are a small but remarkable class of young stellar objects (YSO) that display several-magnitude outbursts in visible light. These outbursts are believed to be caused by sudden increases in disk accretion rates around the youngest stars.

Accretion onto the central star from the disk is dependent on and influences the inner disk (<1 AU) properties. Determination of the temperature and density structure of the inner disk is important for understanding the mechanisms which drive outflows and for establishing the initial conditions for planet formation. By studying disks around stars with a range of stellar and disk properties we can begin to characterize the range of disk properties and their relation to stellar properties.

2 KI, PTI and IOTA interferometers

This section briefly describes the three near-infrared interferometers at which the presented results were taken. Full descriptions of the facilities are available in the given references.

The Keck Interferometer (KI) is a direct-detection infrared interferometer which connects the two 10-meter Keck telescopes on an 85-meter baseline. KI is funded by the National Aeronautics and Space Administration and is developed and operated by the Jet Propulsion Laboratory, the W.M. Keck Observatory and the Michelson Science Center. The system includes adaptive optics for each telescope with wave-front sensing in the visible, angle tracking operating at J and H band and fringe tracking at K band. The fringe tracking camera has a 50 mas FWHM field of view with a Gaussian acceptance pattern. KI is described in detail by [7] and references therein.

The Palomar Testbed Interferometer (PTI), described in detail by [6], is a long-baseline, direct detection interferometer which utilizes active fringe tracking in the infrared. Pairwise combination is possible on three baselines: NS (110 meter), NW (85 meter) and SW (85 meters). PTI was developed as a testbed for the Keck Interferometer and science operations are now managed by the Michelson Science Center.

The Infrared Optical Telescope Array (IOTA) is located on Mt. Hopkins, Arizona and is operated by the Smithsonian Astrophysical Observatory at the Harvard Center for Astrophysics [22]. There are 3 movable telescopes on baselines ranging from 5 to 40 meters. Visible light (fiber) and near-infrared (integrated optics) beam combination is currently possible and closure phase observations have recently begun [18].

3 Previous near-infrared interferometry results

The first young stellar object to be observed using this technique was FU Ori [11], followed by Herbig Ae/Be stars [13, 14] and T Tauri stars [1]. The FU Ori results were consistent with accretion disk models, while both the T Tauri and Herbig results found characteristic sizes larger than expected from geometrically flat accretion disk models. More recent observations of Herbigs [9] have found some objects at earlier spectral types which are consistent with accretion disk predictions.

Tuthill et al. [23] used aperture masking to image LkHα101 and found an inner edge for the infrared emission at 3.4 AU, inferring that this position is set by dust sublimation. Theoretical work by Natta et al. [8] and Dullemond et al. [8] explain the discrepancy between near-infrared observations and previous models with an accretion disk that includes a thick inner wall at the dust sublimation radius. Optically thin material may be present within this radius.

4 New results

4.1 General method for observing disks

As none of the new results presented here has sufficient visibility coverage to produce an image, all the visibilities are modeled using geometric and radiative models. See the individual papers for exact details, but in general the procedure to determine the disk size is something like:

1. Get visibility measurements on your favorite target.
2. Determine the non-disk (stellar, scattering, envelope) contribution at the observation wavelength. There are generally two methods: SED fitting, which has issues with variability and non-contemporaneous data, and high resolution spectroscopic veiling measurements, which only exist for a handful of sources in the infrared.
3. Fit your favorite model. If there are enough baselines or enough hour angle coverage, the inclination can be determined, otherwise a face-on geometry is often assumed.

Note that the errors from step 2 are often larger than the errors from step 1.

4.2 T Tauris

Observations at PTI of two T Tauri objects, T Tau N and SU Aur, were presented in [2] and [3] added the sources of DR Tau and RY Tau and extended the observations to three baselines for all objects, allowing the inclination angle to be determined. DR Tau is near the sensitivity limit for PTI and is marginally resolved, while RY Tau and SU Aur are significantly resolved. T Tau N is also resolved by these observations, but the data are difficult to interpret as the companion T Tau S is within the PTI 1 arcsecond field of view and contributes incoherently to the measured visibility. The stellar contribution at 2.2 μm was determined using veiling estimates from the literature and simple geometric models were fit to the data. For an inclined ring model, the radii for SU Aur, DR Tau and RY Tau are 0.27±0.04, 0.11±0.03 and 0.30±0.01 AU respectively.

Additionally, [3] used the Monte Carlo radiative equilibrium technique developed by [5] to model the spectral energy distribution (SED) and infrared visibilities of all sources except T Tau N. The code includes accretion and shock/boundary luminosity and multiple scattering is included in the disk. This technique naturally accounts for the radiative transfer effects and the heating and hydrostatic structure of the inner wall of the disk. In these models, the gas disk extends to within a few stellar radii and the dust disk radius is set at the dust sublimation radius. For SU Aur and RY Tau, the gas within the inner dust disk contributed substantially to the near infrared-emission from the disk (Figure 1). The extended emission in these models (here meaning emission from scales larger than 10 milliarcseconds) was less

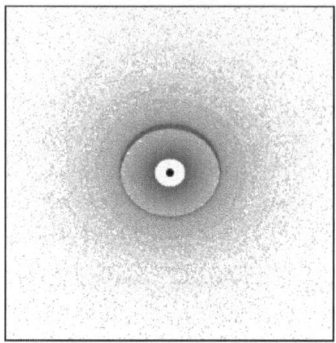

Fig. 1. Radiative transfer model for RY Tau. The flux has been scaled to the 0.15 power to provide better contrast in the image. The image is 12.5 milliarcsec or 1.75 AU across. The model is shown at the position angle used for the calculations, not the best-fit position angle. The outer ring of emission is the inner dust radius and emission within that is due to gas.

than 6% of the total emission for all sources, suggesting that scattered light is not a dominant component for these systems.

Akeson et al. [4] observed an additional seven sources in Taurus with KI. The target sources have a range of mass accretion rates and excess near-infrared emission. All sources show evidence for resolved K band emission, although a few of the sources are marginally consistent with being unresolved. The infrared excess was calculated by fitting stellar photosphere models to optical photometry and estimates of the physical size of the emission region were calculated using simple geometric models for the sources with significant infrared excess. The data required models with a range of dust sublimation temperatures and possibly optical depths to match the measured radii.

Eisner et al. [10] used KI to observe four T Tauris in the rho Ophiuchus star formation region and also obtained nearly simultaneous near-infrared adaptive optics imaging photometry, optical photometry and optical spectroscopy to determine the veiling. This group then estimated the stellar properties, mass accretion rates, and disk corotation radii. The data were compared to both geometrically flat accretion disk and flared disks with puffed-up inner walls. As for the Herbig Ae stars, the data are better fit with models incorporating puffed-up inner walls.

4.3 Herbigs

Monnier et al. [19] used KI to expand the range of Herbig sources measured with infrared interferometry with observations of 14 sources with spectral types from G5 to B0. For all sources except the early B stars, the observations support a simple disk model with an optically-thin central region and a inner dust disk corresponding to dust sublimation temperatures of 1000 to 1500 K. As suggested by previous work [9] the earliest B stars are more consistent with a flat accretion disk.

In Figure 2, I plot the Herbig and T Tauri sources from [19, 10, 4] in measured dust radius vs. luminosity. The left panel includes only the stellar luminosity, while the right panel includes both the stellar and the accretion luminosity, as suggested by [20]. Inclusion of the accretion luminosity has a much greater effect on the T Tauri objects, as the Herbigs generally have very low accretion rates. The dashed lines represent the dust sublimation radius for an optically-thin ring (no backwarming of grains) at sublimation temperatures of 1000 and 1500 K. Although the model is a good fit over several decades of luminosity, the scatter in the T Tauris is considerably larger. Some of this may be due to the larger uncertainties in the stellar component and the intrinsic source variability, but as discussed by [4] there are also trends with accretion rate.

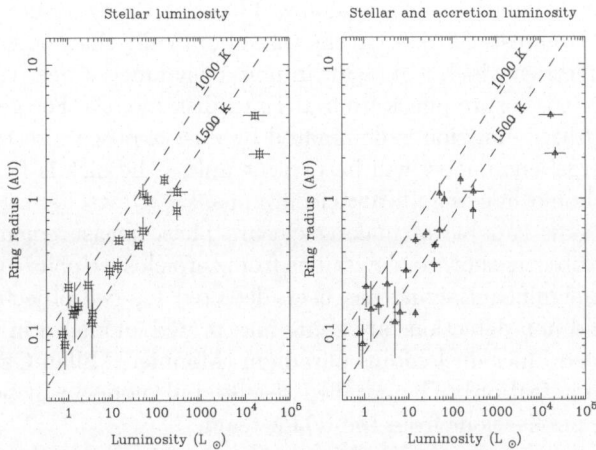

Fig. 2. Dust disk radius for a sample of Herbig and T Tauri objects as measured by KI data plotted against the source luminosity. In the left panel only the stellar luminosity is used, while the right panel includes both the stellar and the accretion luminosity. The dashed lines represent the dust sublimation radius for an optically-thin ring (no backwarming of grains) at sublimation temperatures of 1000 and 1500 K.

4.4 FU Oris

FU Ori objects are young stellar objects in a phase of greatly enhanced accretion. The disk emission dominates the spectral energy distribution at all wavelengths, including the optical. This makes interpretation of infrared interferometry data somewhat simpler as the stellar photosphere contribution can be neglected in the near-infrared. If a disk model is adopted in which the disk material starts at the stellar radius with a viscosity-driven temperature power law of $T \propto r^{-3/4}$, then the only free parameter is the reference disk temperature.

This disk model was sufficient to explain observation of the prototype object, FU Ori by [11]. Recent observations with KI have detected more members of the FU Ori class. Although the observations are limited, the simple one parameter disk model cannot match both the infrared flux and the measured visibilities [15].

4.5 Closure Phase

If the visibility is measured simultaneously on three or more baselines, it is possible to form the closure phase, a quantity which is not corrupted by the atmosphere, but instead reveals the source geometry (see e.g., [17]). Non-axisymmetric distributions of emission will have non-zero closure phase. This can apply to circumstellar disks in the case of optically thick emission, flaring or other geometries which will result in non-axisymmetry. See [12] for examples of predicted closure phase from T Tauri disk models. For geometries in which the infrared emission is dominated by dust exposed directly to stellar heating, a large asymmetry will be present unless the disk is face-on - thus these disks should have significant closure phase.

IOTA is now capable of making closure phase measurements [18]. Although most objects show no deviations from zero closure phase at the IOTA resolution, a significant signal has been detected for two objects. Both the detections and non-detections are being interpreted/modelled in the context of the extended inner dust radius paradigm (Monnier, Millan-Gabet, Berger private communication). The detailed results and models will be presented in upcoming publications from the IOTA team.

5 Conclusions

Infrared interferometry is ideal for observing the inner radii of circumstellar disks around young stellar objects. The results accumulated over the past several years from PTI, IOTA, KI and VLTI are broadly consistent with a model in which the infrared emission is dominated by material at the dust sublimation radius. However there are some exceptions. The early Herbig Be stars are not consistent with this model, but instead are well fit by a geometrically flat accretion disk. Although, the large scatter in the T Tauri sample is complicated by the significant uncertainties in the infrared excess, there may also be correlations with system properties such as accretion luminosity. For the FU Ori systems, observations of additional objects suggest that the simple accretion-dominated disk does not match in all cases.

The next few years will bring even more new results in this field as the increased sensitivity of the large aperture interferometers expands the list of observed sources and the application of techniques like closure phase allows more constraints to be placed on disk properties.

Acknowledgments The author would like to thank Rafael Millan-Gabet, Jean-Phillipe Berger, John Monnier, Josh Eisner and Peter Tuthill for providing slides and results. This work was performed in part at the Michelson Science Center, California Institute of Technology.

References

1. Akeson, R. L., Ciardi, D. R., van Belle, G. T., Creech-Eakman, M. J., and Lada, E. A. 2000, Astrop. J., 543, 313
2. Akeson, R. L., Ciardi, D. R., van Belle, G. T., & Creech-Eakman, M. J. 2002, Astrop. J., 566, 1124
3. Akeson, R. L., Walker, C.H., Wood, K., Esi ner, J.A., Scire, E., Penprase, B., Ciardi, D. R., van Belle, G. T., Whitney, B., and Bjorkman, J.E., 2005, Astrop. J., 622, 440
4. Akeson, R.L. et al 2005, Astrop. J., 635, 1173
5. Bjorkman, J. E. & Wood, K. 2001, Astrop. J., 554, 615
6. Colavita M. M. et al 1999, Astrop. J., 510, 505
7. Colavita, M. M., & Wizinowich, P. L. 2003, Proc. SPIE, 4838, 79
8. Dullemond, C.P., Dominik, C. and Natta, A., 2001, Astrop. J., 560, 957
9. Eisner, J.A., Lane B.F., Hillenbrand, L.A. , Akeson, R.L. & Sargent, A.I., 2004, Astrop. J., 613, 1049
10. Eisner, J.A., Hillenbrand, L.A., White, R.J . Akeson, R.L. & Sargent, A.I., 2005, Astrop. J., 623, 952
11. Malbet, F. et al. 1998, Astrop. J. Letters, 507, L149
12. Malbet, F., Lachaume, R., & Monin, J.-L. 2001, A&A, 379, 515
13. Millan-Gabet, R., Schloerb, F. P., Traub, W. A., Malbet, F., Berger, J. P., & Bregman, J. D. 1999, Astrop. J.l, 513, L131
14. Millan-Gabet, R., Schloerb, F. P., and Traub, W. A, 2001, Astrop. J., 546, 358
15. Millan-Gabet, R., et al. ApJ, 641, 547
16. Monnier, J. D. & Millan-Gabet, R. 2002, Astrop. J., 579, 694
17. Monnier, J. D. 2003, Reports of Progress in Physics, 66, 789
18. Monnier, J. D., et al. 2004, Astrop. J. Letters, 602, L57
19. Monnier, J. D., et al. 2005, Astrop. J., 624, 832
20. Muzerolle, J., Calvet, N., Hartmann, L., & D'Alessio, P. 2003, Astrop. J.l, 597, L149
21. Natta, A., Prusti, T., Neri, R., Wooden, D., Grinin, V. P., & Mannings, V. 2001, A&A, 371, 186
22. Traub, W. A., et al. 2004, Proc. SPIE, 5491, 482
23. Tuthill, P. G., Monnier, J. D., & Danchi, W. C. 2001, Nature, 409, 1012

FU Orionis - The MIDI Perspective

Sascha P. Quanz, Thomas Henning, Christoph Leinert, Thorsten Ratzka,
Sebastian Wolf

Max-Planck-Institute for Astronomy, Königstuhl 17, 69117 Heidelberg, Germany
email: quanz@mpia.de

Summary. We report on recent observations of the pre-main sequence star FU Orionis with VLTI/MIDI. FU Ori was observed at the end of 2004 with three different baselines and a maximum resolution of 24 mas, corresponding to approximately 11 AU at a distance of 460 pc. The object was resolved with all three baselines and visibility curves and spectra from 8 micron to 13 micron were obtained. In addition, the recently discovered companion FU Ori S was visible in the acquisition images from which 8 micron photometry could be derived. The observations will be compared to current models describing the FU Ori system.

1 Introduction

FU Orionis is the prototype of a small, but quite remarkable class of low-mass Young Stellar Objects (YSOs) normally referred to as FU Ori objects (FUORs). For the first members of this class an outburst in optical light of up to 4-6 magnitudes over short timescales, followed by a decrease in luminosity over several years or decades, was observed. Other objects were included in the class as they shared common specific spectroscopic features, e.g., double-peaked line profiles and a spectral type varying with wavelength. For an overview concerning the FU Ori phenomenon we refer to [1]. Most observational data can be explained by the presence of an accretion disk surrounding the young stars (however, see also [2]). A dramatic temporal increase in the accretion rate, where the disk outshines the star by several orders of magnitude, can cause the observed outbursts in luminosity. Several scenarios, possibly triggering such an increased accretion rate, have thus far been proposed. They include (a) interactions of binary or multiple systems where tidal forces disturb the circumstellar disk [3], (b) planet-disk interactions, where thermal instabilities in the disk are caused by the presence of a massive planet [4], or (c) thermal instabilities in the disk alone [5]. Apart from revealing the mechanism leading to the observed outbursts, it is at least as important to find out whether all TTauri stars undergo such epochs of enhanced accretion or whether FUORs are a special class of YSOs. Most observations of classical TTauri stars show that the derived accretion rates might not be able to account for the creation of a low-mass star over reasonable timescales. However, FUOR-phases might provide an elegant solution to this problem as they could speed up the accretion process.

The observations presented in this report were carried out in November and December 2004. FU Ori was observed with MIDI in PRISM mode at projected baselines of 86.3m, 56.7m and 44.8m with position angles of 84.2°, 106.4° and 46.6°, respectively.

2 The Acquisition Image

FU Ori was recently found to be a binary system by Wang et al. [6]. But although theoretically possible, it is very unlikely that the binary component (FU Ori S) triggered the outburst of FU Ori in the late 1930s [7]. Knowing about the existence of the fainter companion the integration time of some MIDI acquisition images was increased in order to derive N-band photometry for both components. In three images FU Ori S was clearly visible (see Figure 1) and aperture photometry could be applied to the observations. The results are summarized in Table 1. Interestingly, FU Ori S shows a relatively higher N-band excess than FU Ori itself. Comparing the de-reddened IR colors of both components to what is found in the literature, it turns out that the NIR and MIR photometry is fairly consistent with Class II YSOs [9]. The

Table 1. Photometric values for the FU Ori system. J-L' are taken from Reipurth & Aspin 2004 [7], N-band values are derived from MIDI 8.7 μm acquisition images.

Component	J [mag]	H [mag]	K' [mag]	L' [mag]	N [mag]
FU Ori	6.30±0.03	5.64±0.05	5.25±0.02	4.18±0.04	2.75±0.19
FU Ori S	10.75±0.23	9.92±0.21	9.15±0.15	8.09±0.16	5.28±0.11

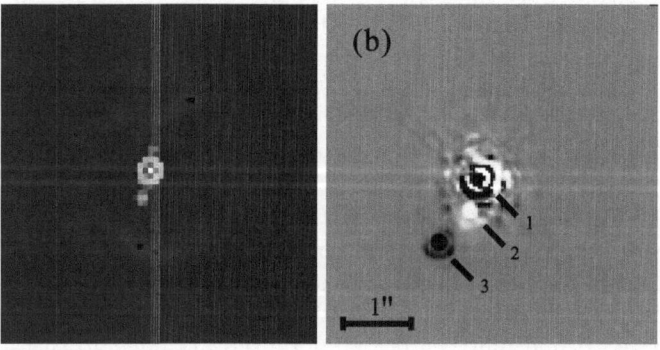

Fig. 1. MIDI acquisition image (left panel) and K-band image obtained by Wang et al. 2004 [6] (right panel). FU Ori S is labelled with a "2". The position angle ($\sim 162.5°$) and the separation ($\sim 0''.484$) to FU Ori agree very well between the N- and K-band images. (North is up, east to the left.)

additional N-band flux of FU Ori S can be explained in terms of differences in the geometry of the assumed circumstellar disks (e.g. larger flaring angle).

3 The Spectrum

In order to derive the total flux of an object, as required for calibrating the dispersed interferometric observations, low-resolution spectra were obtained. However, these spectra themselves contain useful scientific information. Figure 2 shows an averaged MIR spectrum of FU Ori. In contrast to other YSOs no prominent silicate feature is detected. In principle this could mean that there are indeed hardly any silicates in the circumstellar material of FU Ori. However, this appears rather unlikely in the context of YSOs. Another explanation could be that the assumed accretion disk is inclined to a certain extent, so that no $10\mu m$ feature can be seen [10]. For an optically thick accretion disk, however, only a narrow range of inclination angles provides this possibility. Apart from this geometric effect also dust properties can explain the absence of the silicate feature. When the dust particles grow due to coagulation, the contrast between the feature and the continuum will decrease [11]. Thus, it might be the case that silicates surrounding FU Ori have already grown to sizes beyond the micrometer range. In addition to larger grains the vertical structure of an optically thick viscous accretion disk can also account for the observed spectrum. High internal heating and turbulent motion in the disk might prevent the creation of a well-defined vertical temperature gradient required to produce spectral features.

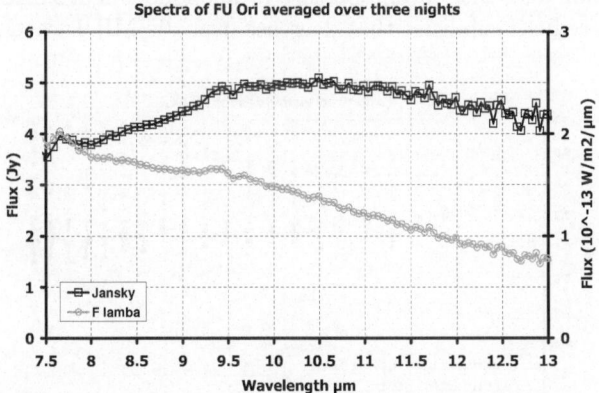

Fig. 2. This spectrum combines data of all three observing nights. The flux is given in F_ν and F_λ and agrees very well to earlier measurements [8]. Neither is an absorption nor a clear emission feature apparent.

4 The Visibility Measurements

In Figure 3 the visibilities measured for FU Ori at three different baselines are shown. The object was clearly resolved during each observation. As expected, for the longest baseline (UT2-UT4, 86m) the lowest visibility is observed ranging from ∼0.8 at 8μm to ∼0.65 at 9.5μm from where it remains almost constant. Whereas the visibility of the shortest baseline (UT2-UT3, 45m) remains almost constant at ∼0.8 over the whole wavelength range, the visibility observed with the intermediate baseline (UT3-UT4, 56.7m) increases from 0.8 at 8μm to almost 1.0 at 13μm. Comparing the results to observations of circumstellar disks around Herbig Ae/Be stars [12] reveals, that those stars as well as the models applied to them normally show a sharp drop in the visibility between 8-10μm from where they remain almost constant. This difference can, however, be expected assuming that Herbig stars are surrounded by extended passive disks whereas FU Ori hosts an optically thick accretion disk. Observations carried out over the past six years using different interferometer (PTI, IOTA and VLTI/VINCI) demonstrated that the NIR visibilities can indeed be fitted with a simple accretion disk model [13]. However, a second model consisting of an accretion disk with an embedded "bright spot" provided an even better fit to those data. Figure 4 depicts MIDI visibilities derived from these two models. The data were kindly provided by R. Lachaume and F. Malbet and are computed for the same baselines as shown in Figure 3. It becomes clear that given the current errors in the visibilities due to calibration uncertainties MIDI is hardly able to distinguish between the two models. Furthermore, both models predict visibilities that are clearly higher than the observed values. And as expected for an accretion disk with a temperature profile following a simple power law the models predict visibilities that decrease over the MIDI wavelength range

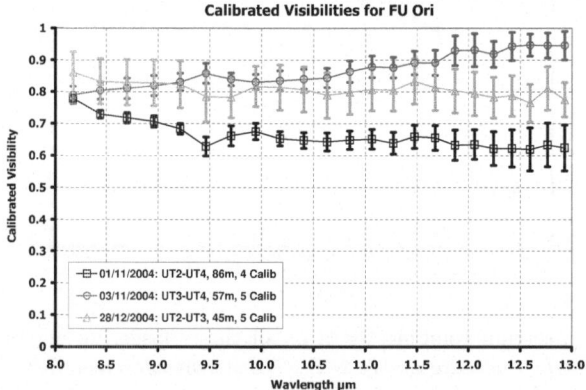

Fig. 3. Visibility measurements of FU Ori for three different baselines. The errors arise by applying different calibrator stars to the interferometric measurements.

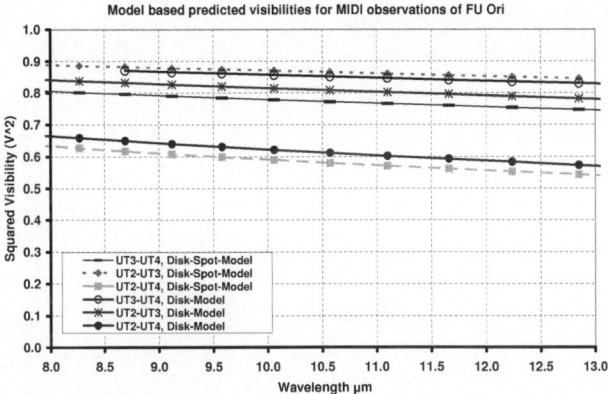

Fig. 4. MIDI squared visibilities computed for the two different models published in [13]. The data were kindly provided by R. Lachaume and F. Malbet.

regardless of baseline. However, this behavior is also not clearly present in our observations.

It can be stated that the current models of FU Ori are not sufficient to explain the interferometric observations with MIDI. A detailed and thorough modelling of FU Ori, based on the spectral energy distribution, the spectrum and the interferometric observations will bring new insights not only in the disk geometry but also in disk structure and dust composition. Different radiative transfer models are currently investigated and applied and the results will be published in the near future.

References

1. Hartmann, L. & Kenyon, S.J.: ARAA **34**, 207 (1996)
2. Herbig, G.H., Petrov, P.P, Duemmler, R.: ApJ **595**, 384 (2003)
3. Bonnell, I. & Bastien, B.: ApJ **401**, L31 (1992)
4. Lodato, G. & Clarke, C.J.: MNRAS **353**, 841 (2004)
5. Bell, K.R., Lin, D.N.C., Hartmann, L.W., Kenyon, S.J.: ApJ **444**, 376 (1995)
6. Wang, H., Apai, D., Henning, Th., Pascucci, I.: ApJ **601**, L83 (2004)
7. Reipurth, B. & Aspin, C.: ApJ **608**, L65 (2004)
8. Hanner, M.S., Brooke, T.Y., Tokunaga, A.T.: ApJ **502**, 871 (1998)
9. Haisch, K.E.Jr., Lada, E.A., Lada, C.J.: AJ **121**, 1512 (2001)
10. Men'shchikov, A.B. & Henning, Th.: A&A **879**, 318 (1997)
11. Bouwman, J., Meeus, G., de Koter, A., Hony, S., Dominik, C., Waters, L.B.F.M.: A&A **375**, 950 (2001)
12. Leinert, Ch., van Boekel, R. et al.: A&A **537**, 423 (2004)
13. Malbet, F., Lachaume, R. et al.: A&A **437**, 627 (2005)

VLTI MIDI Observations of the Herbig Ae Star HR 5999

Thomas Preibisch[1], Thomas Driebe[1], Stefan Kraus[1], Regis Lachaume[1], Roy van Boekel[2], Gerd Weigelt[1]

[1] Max-Planck-Institut für Radioastronomie, Auf dem Hügel 69, D–53121 Bonn, Germany preib, driebe, skraus, lachaume, weigelt@mpifr-bonn.mpg.de
[2] Max-Planck-Institut für Astronomie, Königstuhl 17, D–69117 Heidelberg, Germany boekel@mpia-hd.mpg.de

Summary. We have used long-baseline mid-infrared interferometric observations with MIDI at the VLTI to resolve the circumstellar material around the Herbig Ae star HR 5999 and to provide the first direct measurement of its angular size. The obtained visibilities range between ~ 0.5 and ~ 0.9 at 10μm. This implies that the mid-infrared emission from HR 5999 is clearly resolved, but has a compact structure not much larger than the resolution limit of the MIDI observations. The characteristic size of the emission region depends on the position angles, and ranges between 10 and 25 milli-arcseconds (corresponding to $\sim 2 - 5$ AU) for a uniform-disk model, and $5 - 15$ mas ($\sim 1 - 3$ AU) for a Gauss model (FWHM). We find a dependence of the characteristic size on the projected baseline position angle, which suggests asymmetry of the emission region. To derive constraints on the geometry of the dust distribution, we compare our interferometric measurements to 2D, frequency-dependent radiation transfer simulations of circumstellar disks.

1 Introduction

The Herbig Ae star HR 5999 (= HD 144668 = V856 Sco) is one of the best studied intermediate-mass pre-main sequence stars. Its spectral type is A7 III-IV and its Hipparcos parallax suggests that HR 5999 is located on the far side of the Lupus region at a distance of 208^{+50}_{-30} pc. Comparison of the stellar parameters to pre-main sequence evolution models suggest that HR 5999 is a ~ 0.5 Myr old $\sim 4\,M_\odot$ star. The star shows a strong infrared excess and was detected with IRAS at 12, 25, and 60 μm, as well as a millimeter wavelengths [1]. The infrared excess suggests a total circumstellar mass of the order 0.01 $M_\odot$. The spectral energy distribution of HR 5999 could be fitted with a simple analytical disk model [2].

HR 5999 shows quasi-periodic and irregular photometric and spectro-scopic variability, which has been interpreted as due to instabilities in an optically thick circumstellar accretion disk. A detailed analysis of the UV emission lines in the IUE spectra of HR 5999 [4] revealed gas accreting onto the star with velocities as high as 300 km/sec (see also [5]). The double-peaked Hα and Mg II emission profiles of HR 5999 suggest that the circumstellar disk is seen nearly edge-on [4]. The average mass accretion rate was estimated to be $\sim 7 \times 10^{-7}\, M_\odot/\mathrm{yr}$, but the observed variability of the UV emission clearly suggests that the accretion process happens in a non-steady fashion: during episodic events of clumpy accretion, the accretion rate is much higher than at other times. These instabilities in the accretion rate have similarities to FU Orionis type outbursts.

To summarize, the observations clearly show that HR 5999 is surrounded by circumstellar material, which is probably in the form of a circumstellar accretion disk seen under a relatively high inclination angle (i.e., close to edge-on).

2 Observations and data analysis

The observations discussed in this paper were obtained with the MIDI instrument at the VLTI during several observing runs between April and September 2004. In total, ten measurements were obtained. Data reduction was performed with the MIA software developed at the Max-Planck-Institut für Astronomie, which is described in [6].

Our observations cover a range of projected baseline lengths from 39 m to 102 m and position angles ranging from 15° to 173°, and therefore provide a relatively good coverage of the uv plane. These observations trace the $8 - 13\,\mu$m emission from hot and warm dust in the inner disk regions and can be used to constrain the geometrical structure of the circumstellar material on scales of a few AU.

3 Results

The resulting wavelength-dependent visibilities of HR 5999 are shown in Fig. 1. Although the object is clearly resolved, the relatively high values of the visibilities show that the size of the emission region is only slightly larger than the resolution limit of MIDI (which varies between 4 AU and 10 AU for the baselines used in our observations).

The visibilities are essentially flat, for some of the uv points slightly increasing, over the 8 to 13 μm wavelength range. The measurements obtained at different position angles and baselines produce significantly different visibility values, suggesting that the structure of the emission is not a simple spherical distribution.

In order to derive an initial estimate of the size of the resolved emission region, we fitted the observed visibility values with two simple models, first a uniform disk model, and second, a Gaussian brightness distribution. The resulting diameters range between 10 and 25 milli-arcseconds, corresponding to $\sim 2-5$ AU for the uniform disk model, and $5-15$ mas ($\sim 1-3$ AU) for the Gauss model (FWHM). The Gauss fit diameters as a function of wavelength for the different MIDI observations are shown in Fig. 2. For all data sets, the

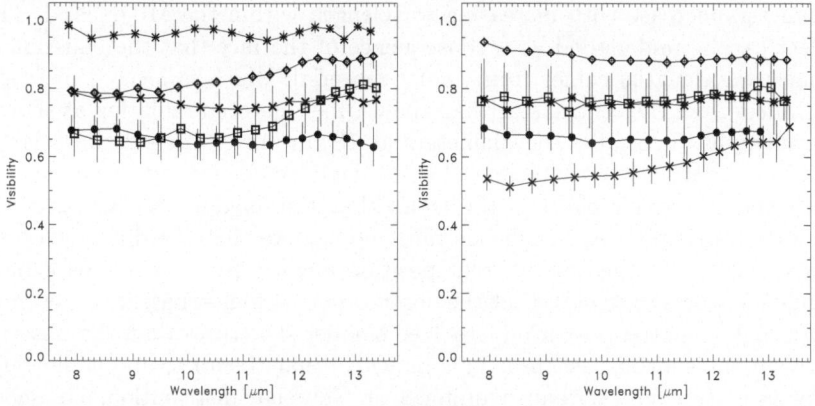

Fig. 1. Visibility as a function of wavelength for the MIDI observations of HR 5999. The left panel shows the data for the observations at baseline-length / position angle of 102 m/15° (dots), 83 m/50° (crosses), 63 m/123° (boxes), 46 m/21° (asterisks), and 46 m/20° (diamonds). The right panel shows the data for the observations at 100 m/30° (dots), 90 m/46° (crosses), 46 m/173° (boxes), 42 m/52° (diamonds), and 39 m/56° (asterisks).

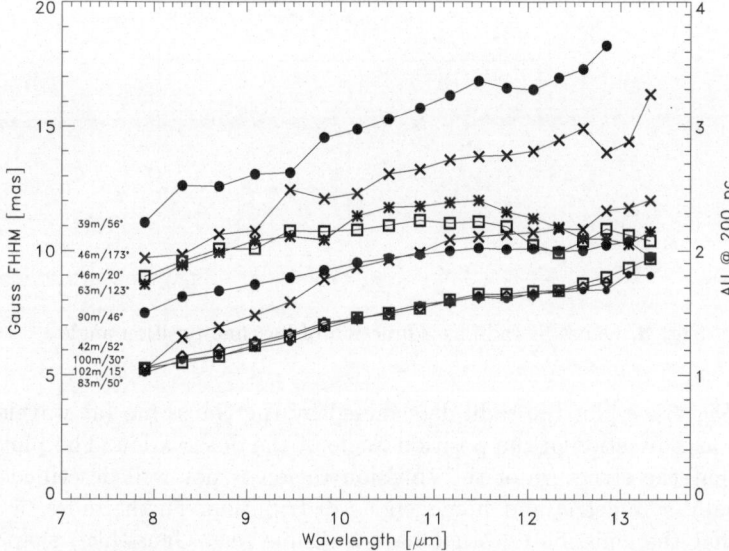

Fig. 2. Gauss distribution FWHM diameters as a function of wavelength for the MIDI observations of HR 5999.

diameters increase with increasing wavelength within the MIDI band. This effect can be understood as a consequence of the fact that the emission at longer wavelengths comes from cooler material, which is located at larger distances from the central star than the warmer material radiating at shorter wavelengths. Typically, the diameters at $13\,\mu$m are $\sim 1.5 - 2$ times larger than at $8\,\mu$m.

Another notable effect is the trend that the longest baselines give the smallest diameters. Although one must not ignore that the different observations were obtained at different position angles (see below), this general tendency seems to suggest that the observations at longer baselines, i.e. those with higher spatial resolution, resolved smaller structures than the observations at the shorter baselines. This indicates that the structure of the emission is more complex than a uniform or Gaussian distribution, i.e. shows sub-structure at different spatial scales.

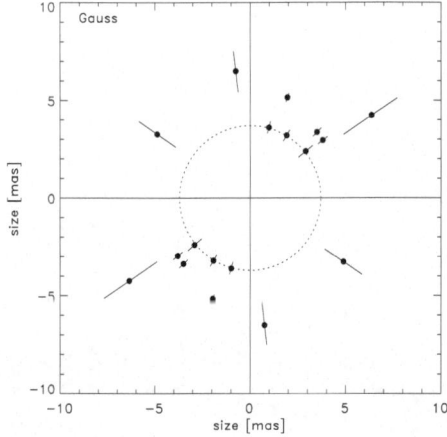

Fig. 3. Gauss fit radii as a function of baseline position angle.

In Fig. 3 we plot the radii determined by the Gauss fits (at wavelength $10\,\mu$m) as a function of the position angle of the observation. The plot suggests that the structure of the emission region is not well described by a spherically symmetric and homogenous distribution. Furthermore, it indicates that the emission does not have a simple (e.g. Gauss-like) shape but a more complicated structure, probably with substructure at different size scales.

4 Discussion

For the first step of the interpretation of our MIDI data it is interesting to compare the observed size of the emission region of HR 5999 to simple theoretical considerations. The minimum size of any circumstellar dust distribution is given by the dust sublimation radius, i.e. the distance from the central star at which the dust grains are heated to their sublimation temperature ($\sim 1500 - 2000$ K). Inside this radius, dust grains cannot survive because they sublimate and their material goes into the gas phase. The dust sublimation radius is a function of the stellar luminosity, the dust sublimation temperature, and the optical properties of the grains [7]. For the stellar parameters of HR 5999 ($L_{bol} = 100\,L_{\odot}$, $T_{eff} = 7925$ K; see [3]), the expected dust sublimation radius for silicate grains is $\sim 0.5 - 0.9$ AU, depending on the size of the grains and the exact value of the sublimation temperature.

The characteristic sizes we find with MIDI for HR 5999, i.e. radii of $\sim 0.75 - 1.5$ AU (computed as $\frac{1}{2}$ of the FWHM for the Gauss fits), are only a few times larger than the expected dust sublimation radius. This shows that the dust distribution around HR 5999 must have a very compact structure. We would like to note that the small measured size is a fully model-independent result, directly related to the rather high values of the measured visibility. This small size seems to be related to the relatively modest (as compared to most Herbig stars) IR excess of HR 5999. It is very interesting to see that HR 5999 fits well into the correlation between the $10\,\mu$m sizes and the slope of the $10 - 25\,\mu$m infrared spectrum derived for a sample of seven Herbig AeBe stars studied with MIDI [6].

The above results show that the MIDI data cannot be explained by simple geometrical brightness distributions. To gain more insight, we currently perform 2D, frequency-dependent radiation transfer simulations to model the distribution of the circumstellar material around HR 5999. From the simulated images we compute the expected visibilities. Comparison to the observed visibilities allow us to derive constraints on the extent, thickness, and orientation of the disk. Since our analysis is still ongoing, the results will be described in a forthcoming publication. A preliminary result is that the data can be explained by a moderately thick Keplerian disk which is truncated at an outer radius of $\sim 2 - 3$ AU, and is seen at a high inclination angle ($\geq 70°$, i.e. close to edge-on).

References

1. Henning, Th., Launhardt, R., Steinacker, J., Thamm, E.: A&A 291, 546 (1994)
2. Hillenbrand, L.A., Strom, S.E., Vrba, F.J., Keene, J.: ApJ 397, 613 (1992)
3. Acke, B. & van den Ancker, M.E.: A&A 426, 151 (2004)
4. Perez, M.R., Grady, C.A., The, P.S.: A&A 274, 381 (1993)
5. Blondel, P.F.C., Talavera, A., Tjin A Djie, H.R.E.: A&A 268, 624 (1993)
6. Leinert, Ch., van Boekel, R., Waters, L.B.F.M., et al.: A&A, 423, 537 (2004)
7. Monnier, J.D., Millan-Gabet, R.: ApJ, 579, 694 (2002)

Disentangling the Wind and the Disk in the Close Surrounding of the Young Stellar Object MWC 297 with AMBER/VLTI

F. Malbet[1], M. Benisty[1], W. J. de Wit[1], S. Kraus[2], A. Meilland[3],
F. Millour[1,4], E. Tatulli[1], J.-P. Berger[1], O. Chesneau[3], K.-H. Hofmann[2],
A. Isella[5], R. Petrov[4], T. Preibisch[2], P. Stee[3], L. Testi[5], G. Weigelt[2], and
the AMBER consortium

[1] Laboratoire d'Astrophysique de Grenoble, UMR 5571 Université Joseph
Fourier/CNRS, BP 53, F-38041 Grenoble Cedex 9, France
[2] Max-Planck-Institut für Radioastronomie, Auf dem Hügel 69, D-53121 Bonn,
Germany
[3] Laboratoire Gemini, UMR 6203 Observatoire de la Côte d'Azur/CNRS, BP
4229, F-06304 Nice Cedex 4, France
[4] Laboratoire Universitaire d'Astrophysique de Nice, UMR 6525 Université de
Nice/CNRS, Parc Valrose, F-06108 Nice cedex 2, France
[5] Osservatorio Astrofisico di Arcetri, Istituto Nazionale di Astrofisica, Largo E.
Fermi 5, I-50125 Firenze, Italy

Summary. The young stellar object MWC 297 is a B1.5Ve star exhibiting strong
hydrogen emission lines. This object has been observed by the AMBER/VLTI instrument in 2-telescope mode in a sub-region of the K spectral band centered around
the Brγ line at 2.1656μm. The object has not only been resolved in the continuum
with a visibility of 0.50 ± 0.10, but also in the Brγ line, where the flux is about twice
larger, with a visibility about twice smaller (0.33 ± 0.06). The continuum emission
is consistent with the expectation of an optically thick thermal emission from dust
in a circumstellar disk. The hydrogen emission can be understood by the emission
of a halo above the disk surface. It can be modelled as a latitudinal-dependant wind
model and it explains the width, the strength and the visibility through the emission lines. The AMBER data associated with a high resolution ISAAC spectrum
constrains the apparent size of the wind but also its kinematics.

1 Introduction

Pre-main sequence stars in the intermediate mass range, called Herbig Ae and
Be stars (HAeBe), are observed to be surrounded by circumstellar material.
It reveals itself by discrete emission lines and by continuous excess emission
in the spectral energy distribution (SED). The spatial distribution of this
material however has been subject to debate, where both geometrically flat
disk models and spherically symmetric envelope models can reproduce the
observed SED.

The geometry of circumstellar material near HAeBe stars seems to differ between the early-type and late-type members of the group, which is not

surprising given the increasing interaction between star and disk at the earlier type stars. For the HAe stars a successful working model exist, while on the other hand, a disk structure near the HBe stars and their intricate star-disk interactions still escape a good understanding. In this study we present high spatial resolution, intermediate spectral resolution interferometric observations with AMBER of the early-type Herbig Be star MWC,297. This star displays a strong emission line spectrum corresponding to a B 1.5Ve spectral type. The rather well determined stellar parameters [1] and its high NIR luminosity render this star the perfect target to investigate in detail the geometry of the circumstellar material near the early type HAeBe stars.

2 Observations

MWC 297 has been observed during the second night of the first commissioning run of the AMBER instrument on the UT2-UT3 baseline of the *Very Large Telescope Interferometer* (VLTI). AMBER is the VLTI beam combiner in the near-infrared [2]. The instrument is based on spatial filtering with fibers and spatial beam combination along one dimension. The interferometric beam is anamorphized perpendicular to the fringe coding in order to be injected into the slit of a spectrograph. MWC 297 has been measured in the [2100,2230nm] spectral range with 1500 spectral resolution.

The results for the line visibilities are relatively consistent with all data reduction methods [3, 4], while this is not the case for the continuum visibilities. Besides the continuum visibilities in the K band has already been measured by other instruments like IOTA and PTI and therefore the important result is the line visibility.

Left part of Fig. 1 shows the variation of the visibility with wavelength. The continuum visibilities correspond to an average of $V_{\mathrm{cont}} = 0.50 \pm 0.10$ and the line visibility to a value of $V_{\mathrm{line}} = 0.33 \pm 0.06$.

MWC 297 was also observed with ISAAC at the ESO VLT UT1 telescope (see right part of Fig. 1) in the short wavelength medium resolution mode ($\mathcal{R} = 8900$) at the Brγ wavelength. Broad-band photometric data were collected from the litterature [1, 5, 6]. Existing interferometric data for MWC 297 consist of IOTA H-band [7] and PTI K-band [8] continuum data.

3 Modeling

We tried to model the large body of interferometric, spectroscopic and photometric data that exists for MWC,297 The modeling is done by applying two different codes, an optically thick disk one and a stellar wind one. The disk code is designed to model the continuum radiation, whereas the stellar wind code reproduces the strong emission lines. Figure 2 represents a sketch of the

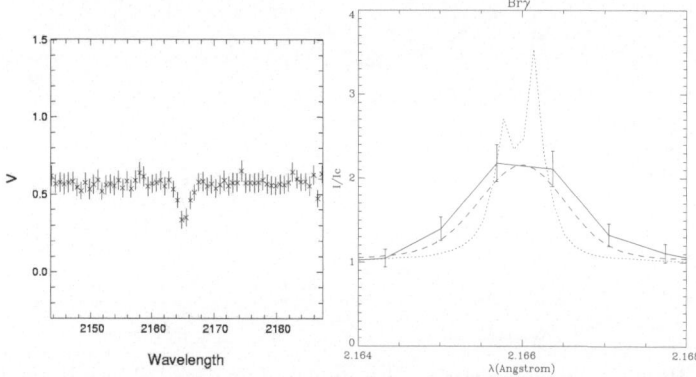

Fig. 1. Left: spectral dependence of the MWC 297 visibilities. Right: comparison of Brγ observed with AMBER (full line) and ISAAC (dotted line). The dashed line corresponds to the ISAAC spectrum convolved at the AMBER spectral resolution.

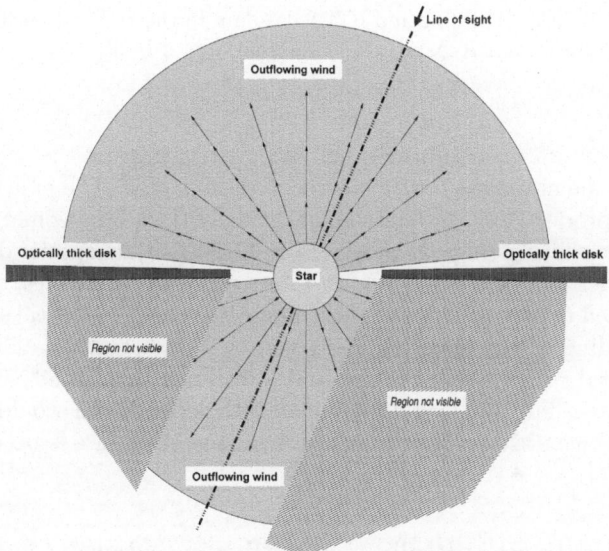

Fig. 2. Sketch of the model including an optically thick disk and an outflowing wind (edge-on view). The receding part of the wind is only partly visible because of the screen made by the optically thick disk.

combined model, where the optically thick disk and the outflowing wind are spatially independent.

3.1 Continuum radiation: optically thick disk

The disk model [9, 10] consists in an axisymmetric radial analytic disk structure which is heated both by stationary accretion and stellar irradiation. The

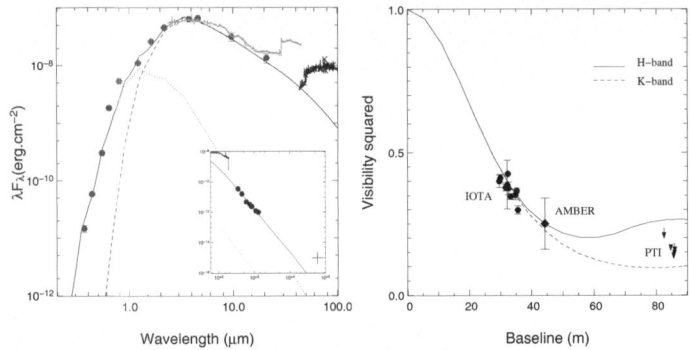

Fig. 3. Result from the optically disk model. *Left panel:* observed and modeled SED for MWC,297. The full dots are the continuum measurements [5], also included are the ISO SWS/LWS spectra. Dotted line is the star, dashed line the accretion disk, and the full line the resulting total flux of the model. *Right panel:* resulting best-fit model radial visibilities compared with AMBER, IOTA and PTI observed continuum visibilities. Full line and IOTA data are in the *H*-band, dashed line and AMBER/PTI are in the *K*-band. PTI values are upper limits.

disk is in hydrostatic equilibrium, non self-gravitating and the accretion flux is following the standard power law for a viscous disk. The emitted continuum flux is produced by the emission of optically thick but geometrically thin black-body radiating rings. It produces an SED, and, its spatial distribution can be Fourier transformed to obtain interferometric visibilities.

We probed the sensitivities of these fits by varying the central star parameters, according to the uncertainties given by Drew et al. [1]. They derived half a spectral subtype uncertainty, and a distance error of 50 pc. The mass accretion rate is far from well determined. If the central star would be of type B2 at a distance of 200 pc, the required mass accretion rate is between 0 and $10^{-6} M_\odot \, \mathrm{yr}^{-1}$.

3.2 Emission lines: optically thin outflowing wind

In our model, the emission lines are produced in a circumstellar gas envelope. In order to model this line profile and the corresponding visibilities, we have used the SIMECA code [11, 12]. The solutions for all stellar latitudes are obtained by introducing a parametrized model constrained by the spectrally resolved interferometric data.

Since the SIMECA code has originally been developed to model the circumstellar environment of classical Be stars, we had to modify the code in order to interface SIMECA with the optically thick disk model described previously. We have implemented three changes:

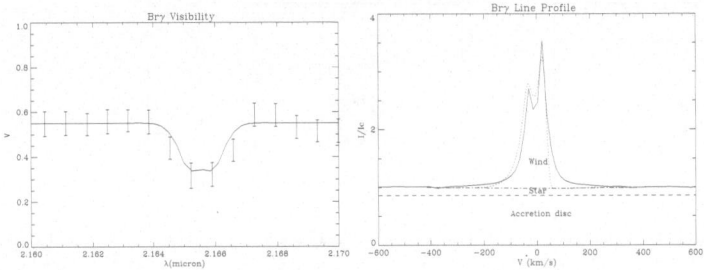

Fig. 4. *Left panel:* The visibility observed with AMBER (points with error bars) and the modeled outflowing wind model (full line). *Right panel:* double peaked Brγ profile observed by ISAAC (full line) and modeled with the outflowing wind model (dotted line). We have also plotted the cumulative contribution of the accretion disk (dashed line) and of the star (dash-dot).

1. The wind is no longer computed from the equator to the pole, but the computation occurs in a bipolar cone defined by a minimal angle allowing the disk to be present (see sketch in Fig. 2). We used an minimum angle of 4 degrees. The equatorial terminal velocity corresponds therefore to the terminal velocity at this minimal angle from the equatorial plane at the interface between the accretion disk and the stellar wind.
2. The disk hides the receding part of the wind. In Fig. 2, the part of the wind which is not visible from the observer is not taken into account in the outgoing flux.
3. Although the disk emission contributes less than 1% compared to the star flux in the visible (i.e. also in the Hα and Hβ lines) and can be neglected, at $2.1656\,\mu m$ the disk emission is 6.4 times larger than the stellar flux. This contribution decreases the normalized Brγ line intensity and also must be accounted for in the computation of the visibilities.

We find a successful simultaneous fit to the Hα, Hβ and Brγ line profiles compatible with the observations (see Fig. 4) and the outflowing wind model reproduces the AMBER measured drop in visibility across the Brγ line.

We are able to reproduce quite well the shape of the Hα and Hβ lines and the double peaked emission of the Brγ line. The peak asymmetry of the Brγ line is also reproduced thanks to the introduction in the SIMECA code of the opacity of the disk (point 2 of SIMECA modifications). Nevertheless the agreement is not perfect in the red wing of the profile probably due to our ad-hoc way of interfacing of the wind and the disk.

4 Discussion

The modeling presented in the previous section, although rather successful, brings new questions on the physics of the circumstellar environment of intermediate-mass young stars.

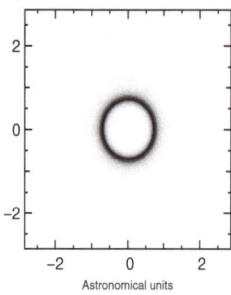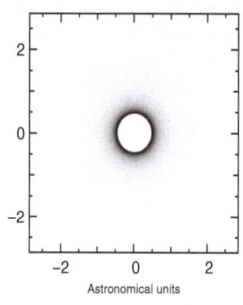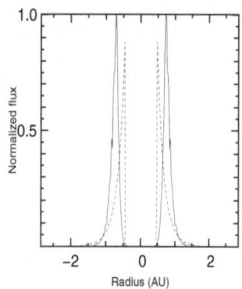

Fig. 5. Pole-on intensity maps of the wind Brγ emission (left panel) and of the *K*-band disk continuum emission (center panel). Right panel shows a radial cut of these intensity maps with the Brγ wind in solid line and the continuum disk in dashed line.

Figure 5 shows the pole-on intensity maps of the disk model in the continuum and of the wind in the Brγ line, as well as their respective intensity profile. This is a graphical explanation of the visibilities observed by AMBER: the wind angular extension in the Brγ line is larger than the disk apparent size and therefore the visibility is smaller within the line.

Can the result obtained with these observations constrain the nature of the wind? We recall that in protoplanetary disks, two main classes of disk wind models have been proposed depending of the geometry of the magnetic field lines: the disk wind [13, 14] and the X-wind [15]. We cannot conclude with the present state of data since we are unable to recover a precise kinematical map of the wind. More resolution with AMBER will help to answer the question, especially using the 10000 spectral resolution mode.

Also the modeling presented in this paper does a reasonably good job in reproducing nearly all the observational data and produces fiducial physical parameters for the circumstellar environment of MWC,297. However, we derive an inclination of $\sim 20\,$deg for the system, which is not consistent with a near edge-on orientation as proposed by [1]. The later is inferred from the photospheric lines that indicate a $350\,$km s^{-1} projected rotational velocity. An inclination of 20 deg would lead to a rotation above the break-up velocity.

In conclusion, we can claim that the models of disk and wind are compatible and are probably very close from the reality. A complete and self-consistent modeling of the environment is out of the scope of the paper but would allow to better constrain the relationship between the disk and the wind at least from the observational point of view.

5 Conclusion

We have presented first spatially resolved observations of the disk / wind interaction in the young stellar system MWC 297 with the VLT interferometer equipped with the instrument AMBER. We have observed that the continuum visibility in the K-band drops from 0.50 to 0.33 in the Brγ emission line of MWC 297. The spectrum obtained with AMBER is consistent with a double peaked spectrum observed with ISAAC on the VLT, where the peaks are roughly separated by about 60 km s^{-1}.

We have successfully modeled the circumstellar environment of MWC 297 using an optically thick geometrically thin disk and an outflowing stellar radial wind having a increasing outflowing velocity starting from the surface of the disk up to the pole. This combined model is able to reproduce many observational features like the shape of the SED over more than three decades of the wavelengths, the continuum visibilities obtained not only by AMBER but also by other infrared interferometers like IOTA and PTI, the spectral visibilities in the Brγ emission line as well as the Hα, Hβ and Brγ line profiles.

We are not yet able to constrain the exact nature of the wind and the type of connection with the disk, but we expect that future data obtained with AMBER at a higher spectral resolution will give new kinematical information on this interesting and intriguing region.

References

1. Drew, J.E., Busfield, G., Hoare, M.G., Murdoch, K.A., Nixon, C.A., Oudmaijer, R.D.: MWC 297, B1.5Ve: a zero-age main-sequence star in the Aquila Rift. MNRAS **286** (1997) 538
2. Petrov, R.G., et al.: Introducing the near infrared VLTI instrument AMBER to its users. Ap&SS **286** (2003) 57
3. Millour, F., Tatulli, E., Chelli, A., Duvert, G., Zins, G., Acke, B., Malbet, F.: Data reduction for the AMBER instrument. In: New Frontiers in Stellar Interferometry. Edited by Wesley A. Traub. Proceedings of the SPIE, Volume 5491, pp. 1222 (2004). (2004) 1222
4. Ohnaka, K., Beckmann, U., Berger, J.P., Brewer, M.K., Hofmann, K.H., Lacasse, M.G., Malanushenko, V., Millan-Gabet, R., Monnier, J.D., Pedretti, E., Schertl, D., Schloerb, F.P., Shenavrin, V.I., Traub, W.A., Weigelt, G., Yudin, B.F.: JHK'-band IOTA interferometry of the circumstellar environment of R CrB. A&A **408** (2003) 553
5. Pezzuto, S., Strafella, F., Lorenzetti, D.: On the Circumstellar Matter Distribution around Herbig Ae/Be Stars. ApJ **485** (1997) 290
6. Mannings, V.: Submillimetre Observations of Herbig Ae/be Systems. MNRAS **271** (1994) 587
7. Millan-Gabet, R., Schloerb, F.P., Traub, W.A.: Spatially Resolved Circumstellar Structure of Herbig AE/BE Stars in the Near-Infrared. ApJ **546** (2001) 358
8. Eisner, J.A., Lane, B.F., Hillenbrand, L.A., Akeson, R.L., Sargent, A.I.: Resolved Inner Disks around Herbig Ae/Be Stars. ApJ **613** (2004) 1049
9. Malbet, F., Bertout, C.: Detecting T Tauri disks with optical long-baseline interferometry. A&AS **113** (1995) 369

10. Malbet, F., Lachaume, R., Berger, J.P., Colavita, M.M., Di Folco, E., Eisner, J.A., Lane, B.F., Millan-Gabet, R., Ségransan, D., Traub, W.A.: New insights on the AU-scale circumstellar structure of FU Orionis. A&A **437** (2005) 627
11. Stee, P., de Araujo, F.X.: Line profiles and intensity maps from an axisymmetric radiative wind model for Be stars. A&A **292** (1994) 221
12. Stee, P., de Araujo, F.X., Vakili, F., Mourard, D., Arnold, L., Bonneau, D., Morand, F., Tallon-Bosc, I.: γ Cassiopeiae revisited by spectrally resolved interferometry. A&A **300** (1995) 219
13. Blandford, R.D., Payne, D.G.: Hydromagnetic flows from accretion discs and the production of radio jets. MNRAS **199** (1982) 883
14. Casse, F., Ferreira, J.: Magnetized accretion-ejection structures. IV. Magnetically-driven jets from resistive, viscous, Keplerian discs. A&A **353** (2000) 1115
15. Shu, F., Najita, J., Ostriker, E., Wilkin, F., Ruden, S., Lizano, S.: Magneto-centrifugally driven flows from young stars and disks. 1: A generalized model. ApJ **429** (1994) 781

Interferometry of M8E-IR with MIDI - Resolving the Dust Emission

M. Feldt[1], I. Pascucci[2], O. Chesneau[3], D. Apai[2], Th. Henning[1], Ch. Leinert[1], H. Linz[1], A. Men'shchikov[4], and B. Stecklum[5]

[1] Max Planck Institute for Astronomy, Königstuhl 17, D-69117 Heidelberg, Germany mfeldt@mpia.de
[2] Department of Astronomy/Steward Observatory, 933 N Cherry Ave., Rm. N204, Tucson AZ 85721-0065, U.S.A. pascucci@as.arizona.edu
[3] Observatoire de la Côte d'Azur, Boulevard de l'Observatoire B.P. 4229, F-06304 NICE Cedex 4, France Olivier.Chesneau@obs-azur.fr
[4] Department of Astronomy and Physics, Saint Mary's University, , Canada amenshch@ap.stmarys.ca
[5] Thüringer Landessternwarte Tautenburg, Sternwarte 5, D-07778 Tautenburg, Germany stecklum@tls-tautenburg.de

Summary. We report on interferometry of the high-mass young stellar object M8E-IR performed with MIDI at the ESO-VLTI. The observations were carried out using the UT1-UT3 and UT2-UT3 baselines. The visibilities in the 8...13 micron range derived from the spectrally dispersed fringes indicate the presence of circumstellar emission presumably originating from warm dust. We resolve the MIR emission of M8E-IR and derive an extension between 15 and 30mas which correspond to linear sizes of 30-50AU at the distance of the source. These are the scales at which circumstellar disks around massive stars are expected. We discuss the properties of the dust envelope based on the results of our preliminary model and in the context of supplementary NACO observations.

1 Introduction

M8E IR is a high-mass young stellar object, presumably of spectral type B0 ZAMS [5]. The distance to this highly obscured object is 1.8 kpc. Such young, high-mass stars provide an ideal laboratory to remedy the major riddle of massive star formation - whether it is achieved through accretion similarly as for the low-mass stars, or through "protostellar mergers" (e.g. [1]). We performed MIDI observations of M8E IR to directly detect emission from warm dust surrounding the star and measure if the distribution of this dust is spherically symmetric or elliptical. The latter case would of course hint to a flattened structure which, at the resolution scale of the interferometer of 21 mas or 40 AU, would very likely be an accretion disk.

Table 1. MIDI observations of M8E IR

Date	Mode	Baseline	Proj. Length	P.A.
Jun 2004	Grism-HiSens	UT1-UT3	98 m	42°
			93 m	45°
Aug 2004	Grism-HiSens	UT2-UT3	47 m	40°
Mar 2005	Grism-HiSens	UT3-UT4	48 m	95°

2 Observations and Data Reduction

Five observations of M8E IR were obtained in 2004 and 2005. The dates and parameters of these observations are summarized in Tab. 1. The observing dates were targeted to obtain visibilities at two baseline lengths and two position angles across the source. For visibility calibration, HD 169916 was observed before and after M8e IR. In June 2004, HD 168454 was additionally observed about two hours later.

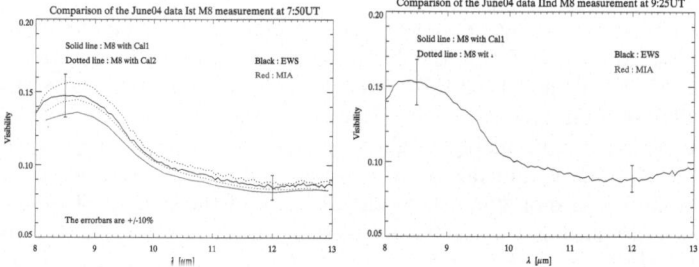

Fig. 1. Comparison of data reduction pipelines and calibrators.

For data reduction, the standard MIDI reduction packages MIA and EWS were used [3]. Fig. 1 shows a comparison of the two reduction techniques and the calibrators taken. The left graph in Fig. 1 shows the data taken in June 2004 in the first measurement at 7:50 UT calibrated with HD 169916 observed before (Cal1) and after (Cal2) M8E IR. The visibilities agree to within 10%. The right graph in Fig. 1 shows the same comparison for the second observation of June 2004, observed at 9:25 UT. This time, Cal2 was HD 168454, and it was observed in SciPhot mode, not HiSense. It is clearly visible that this reduces the resulting visibility by a factor of 0.85.

The data reduction using the two automated pipelines is still not fully under control and we report our results here with a note of caution.

3 Results

Fig. 2 shows the final calibrated visibilities of M8E IR. Two interesting comparisons can be made: First comparing visibilities measured on a ∼ 48 m

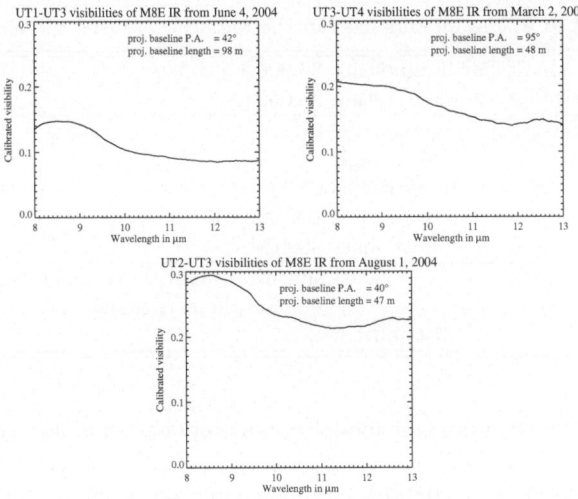

Fig. 2. Visibility of M8E IR.

baseline it appears that the fringe contrast is a factor of $\sim$ 1.5 higher at P.A. 40° than at 95°. Secondly comparing measurements made along P.A. $\sim$ 40°, the visibility at 98 m baseline is lower by a factor of 2 than on the 47 m baseline. This second fact is of course to be expected for a partially resolved object.

4 Modelling

For interpretation of the results, a radiative transfer model was calculated using the code described in [4]. The code models a "flared disk" as can be seen

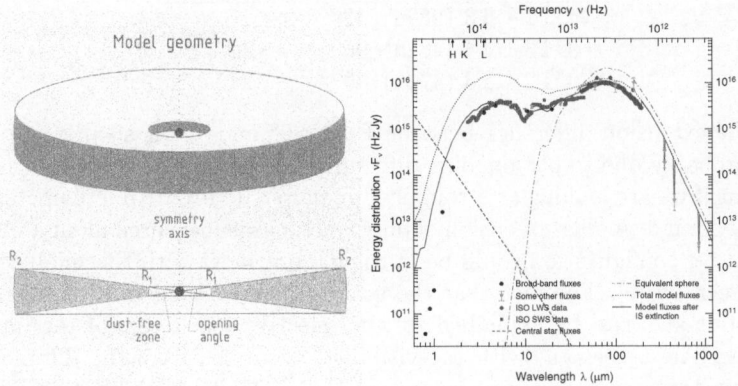

Fig. 3. Geometry of the radiative transfer model and the resulting SED.

Table 2. Parameters used in the radiative transfer model.

Stellar luminousity 30,000 $L_\odot$
Stellar eff. temperature 30,000 K
Flared disk opening angle 90°
Viewing angle 44°
Inner radius 25 AU
Outer radius 100,000 AU
Disk mass 3500 $M_\odot$
Dust silicate cores with ice mantles + amorphous carbon
Size distribution $\sim n^{-4.2}$

in Fig. 3. The most important model parameters are summarized in Tab. 2. Fig. 3 also shows the SED resulting from the radiative transfer model compared to measured broad band fluxes. The code not only produces an SED, but also intensity maps at chosen wavelengths, on which model visibilities can be calculated and the compared to the measured ones. This is shown in the next section.

5 Discussion

The interpretation of the measurement results is twofold. Calculating the FWHM of a Gaussian extended source at 10 μm using

$$V(u) = e^{-3.56FWHM^2}, \tag{1}$$

where $V(u)$ is the visibility, and $FWHM$ is measured in units of the interferometer resolution λ/B, we get the following source extensions:

Measurement	P.A.	FWHM [mas]
UT1-UT3 (98 m)	42°	17
UT3-UT4 (48 m)	95°	30
UT2-UT3 (47 m)	40°	27

The extension difference between P.A.s 40° and 42° is surprisingly large and probably due to our ongoing difficulties in the data reduction. It is thus unclear if we are looking at a roughly circular structure with a diameter of $\sim$ 30 mas, or if a smaller extension of only $\sim$ 17 mas is measured along P.A. 42°. The latter configuration could be in rough agreement with the small component seen by [6]. The fact that the scales we se are generally smaller than Simon et al.'s can be explained by the selective sensitivity of the interferometer. The agreement with the visibility calculated from the RT model is also less than perfect, as can be seen in Fig. 4. Model visibilities are plotted for the major and minor axes of the resulting ellipse, the measurements are

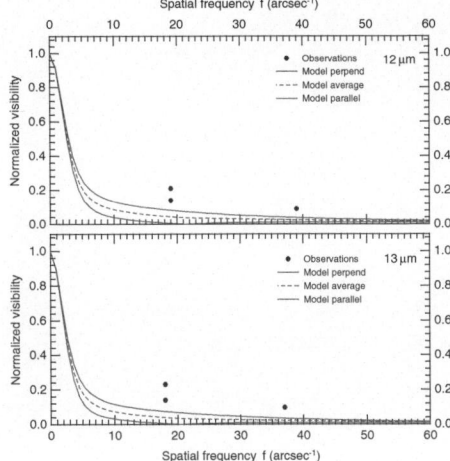

Fig. 4. Comparison of observed and model visibilities at $12\,\mu$m and $13\,\mu$m.

plotted as dots. Modelled visibilities are generally too low by a factor of $\sim$2, while the difference in visibilities between the two viewing angles is about correct.

6 Conclusion

Both the model and the data reduction need further refinement. However, if it is confirmed that we are looking at a clearly elliptical structure with axis dimensions of about 30 AU and 60 AU, it is definitely clear that we are looking at warm dust in a true accretion disk around this massive YSO. This would make a major differnece from the discoveries reported so far which usually detect flattened structures on scales of tens of thousands of AU (e.g. [2]).

References

1. J. Bally, H. Zinnecker, AJ **129**, 2281 (2005)
2. R. Chini, V. Hoffmeister, S. Kimeswenger, et al., Nature **429**, 155 (2004)
3. R. Köhler, http://www.strw.leidenuniv.nl/ nevec/MIDI/ ,2005
4. A.B. Men'shchikov, Th. Henning, A&A **318**, 879 (1997)
5. M. Simon, L. Cassar, M. Felli, J. Fischer, M., Massi, D. Sanders, ApJ **278**, 170 (1984)
6. M. Simon, D.M. Peterson, A.J. Longmore, J.W. Storey, A.T. Tokunaga, ApJ **298**, 328 (1985)

Observing T Tauri Stars in the Mid-Infrared with MIDI

Th. Ratzka and Ch. Leinert

Max-Planck-Institute for Astronomy, Heidelberg, Germany ratzka@mpia.de

Summary. Observations of T Tauri stars are part of the Guaranteed Time Observations (GTO) of the MID-infrared Interferometric instrument (MIDI) operated by ESO at the VLTI. Here we give a brief overview of the preliminary results obtained for some of the brighter T Tauri stars that are in the reach of MIDI in the present state of the VLTI without external fringe tracking.

1 The Origin of the Silicate Features

All observations that are presented here have been performed in the so-called HIGH SENS mode, i.e. the total flux of the individual sources required to normalise the visibilities were determined after finishing the interferometric measurements [1]. Since a prism with $\lambda/\Delta\lambda \approx 30$ was used to obtain dispersed visibilities, the photometric measurements provide low-resolution mid-infrared spectra. In Fig. 1 two instructive examples are shown. RY Tau (left) exhibits a pronounced silicate emission typical for warm dust reemitting the stellar irradiation, e.g. a thin upper layer of a disk seen under a low inclination angle. In contrast the spectrum of the northern component of the binary Haro 6-10 shows a deep silicate absorption indicating a cooler layer of absorbing dust in the line-of-sight.

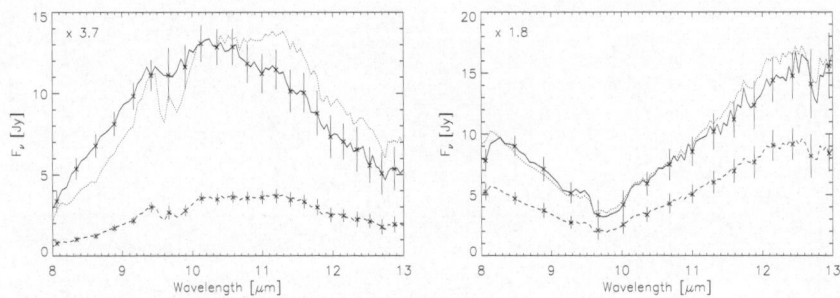

Fig. 1. The mid-infrared spectrum (solid line), the correlated flux (dashed), and the correlated flux multiplied with the scaling factor given in the upper left (dotted) for RY Tau (left) and Haro 6-10 N (right).

When multiplying these two spectra with the dispersed visibilities one gets the correlated fluxes that basically represents the flux emitted by the region that cannot be resolved by the interferometer. After scaling the correlated flux with a factor, the shape of the spectrum originating in the inner region can be compared with that of the spectrum of the total source. In the case of RY Tau the projected baseline had a length of 36 m. Therefore, the correlated flux originates in a region closer than 5 AU to the central source. Interestingly, the emission peak is shifted here to longer wavelengths. This may be caused by a higher abundance of larger or even crystalline particles in the inner region. The correlated flux of Haro 6-10 determined with a projected baseline of 29 m shows no such significant differences when compared with the integrated flux of the source.

2 The Geometry of a Close Binary

Z CMa is a close binary with a separation of only 0.1" at a position angle of 120° [2]. The south-eastern component is a FUor-like T Tauri star. It shows an absorption spectrum interpreted as that of an optically thick accretion disk. The north-western component is most likely a Herbig Be star surrounded by an asymmetric dust envelope. Since its visible and near-infrared light is escaping from the envelope probably only by scattering off the walls of a jet-blown cavity the T Tauri component is brighter in the visible wavelength range while in the near-infrared the intermediate-mass star is dominating.

In Fig. 2 two calibrated visibilities are plotted after substituting the wavelength by the spatial frequency u along the position angle of the projected baseline B, i.e. $u = B/\lambda$. To derive the binary parameters of Z CMa a nonlinear least squares fit of the simple function

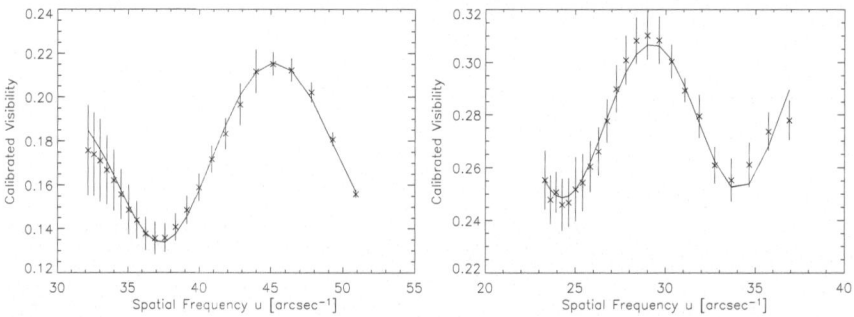

Fig. 2. The calibrated visibility, the calibration errors, and the best fit for the Z CMa data taken with a projected baseline length of 85.8 m along a position angle of 78.1° (left) and 62.2 m along a position angle of 110.4° (right).

$$f = (a_0 + a_1 u) \cdot \frac{\sqrt{1 + a_2^2 + 2a_2 \cos{(2\pi a_3 u)}}}{1 + a_2} \tag{1}$$

has been performed (see [3]). While in the image space the brightness distribution of the binary is convolved with the brightness distribution of the extended structures surrounding these stars, e.g. disks or envelopes, this contribution is according to the convolution theorem a multiplicative term in the Fourier space. The linear part of equation (1) is just a simple approach to compensate the visibility loss due to these extended structures. The nonlinear part of equation (1) fits the binary parameters. The flux ratio of the both components is given by a_2, while the separation of the components is given by a_3. This fit is valid as long as the fitted quantities are not wavelength-dependent. In the narrow wavelength range covered by MIDI this condition should be fulfilled for the separation. However, the flux ratio seems to suffer from that simple assumption. Therefore, the error derived from the two fits is large: $a_2 = 0.155 \pm 0.075$.

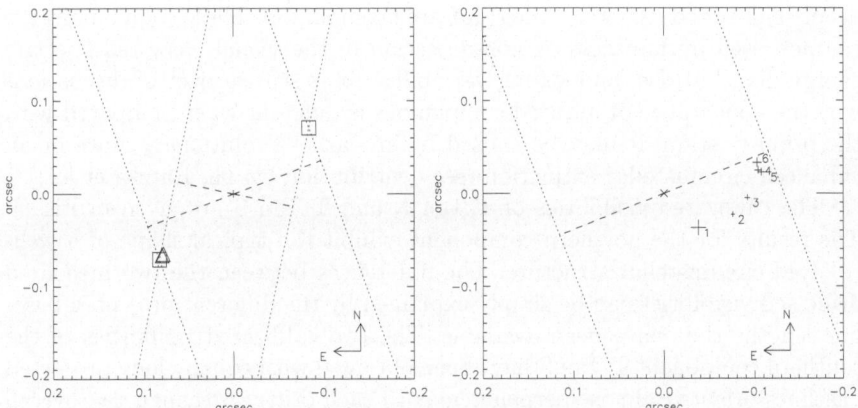

Fig. 3. *Left*: Reconstruction of the position of the T Tau component of Z CMa (squares) based on the two measured projected separations. The Herbig Be star is centred in the origin of the coordinate system. Overplotted are the positions derived with the adaptive optics system NAOS/CONICA in the J-, and Ks-band (triangles). *Right*: The measurements of the positions of T Tau Sb with respect to T Tau Sa. The labelled near-infrared positions are adopted from [4] and references therein. The crosses correspond to the errorbars. Overplotted is the projected separation derived from our mid-infrared interferometric measurement.

In Fig. 3 the position of the companion relative to the primary is reconstructed. The projected separations along the position angles of the projected baselines are indicated by the dashed lines. The primary is located in the origin of the coordinate system. The possible positions of the companion (squares) are determined by the intersections of the lines of constant projected separations (dotted lines) that are perpendicular to the projected separation vectors. However, with the visibility alone we cannot decide which of the two reconstructed positions is the right one. But is is still possible even with a two-telescope interferometer to extract the for this purpose required phase information (see contribution by R. Tubbs).

A resolved observation of this companion with a separation of 109 mas in the mid-infrared would require a single-dish telescope with a diameter of approximately 20 m when working diffraction-limited.

3 The Triple System T Tauri

T Tau has grown more and more complex as observations improve. The optical visible northern component T Tau N is the prototype of a classical T Tauri star. A close binary with a separation of only 0.1" is located 0.7" south of it. While T Tau Sb appears to be a 'normal' active low-mass pre-main-sequence star with a moderate extinction, T Tau Sa is the prototypical deeply embedded ($A_V \approx 35$ mag) infrared companion. These companions are characterised by faintness or non-detection in the visual, very red spectral energy distributions, and strong variability. It is still subject of discussions why the appearance of infrared companions is different when compared with the primary stars. It may be caused by an earlier evolutionary state or an enhanced circumstellar extinction (see contribution by Th. Ratzka et al.).

The calibrated visibilities of T Tau N and T Tau S are given in Fig. 4. The results for the northern component exhibit the typical shape of a well-resolved circumstellar structure. The differences between the two measured dispersed visibilities can be simply explained by the different projected baseline lengths that have been available. The two calibrated visibilities of the southern component T Tau S have been obtained with equally long projected baselines oriented almost perpendicular to each other. Although the overall shape of the visibilities is quite similar, one of them is multiplied with a binary signal originating in the known close pair Sa-Sb. Therefore, equation (1) has been fitted again to the data to derive the binary parameters (Fig. 4). The projected separation along the position angle 111.4° is visualised in Fig. 3. The true position angle of the companion should be close to this value, because no significant binary signal is present in the other calibrated visibility of T Tau S. Overplotted are the already in [4] published near-infrared positions. The orbits shown in that paper with semi-major axis of $\sim 0.10''$ and $\sim 0.12''$ and periods of $\sim 20\,\mathrm{yr}$, $\sim 40\,\mathrm{yr}$, respectively include the positions 4, 5, and 6 as westernmost part. Since the here derived projected separation is a lower limit to the true separation, our data favour an even wider orbit. Unfortunately, we cannot determine without a phase, whether the flux ratio of 0.456 ± 0.021 is measured with respect to T Tau Sa or T Tau Sb.

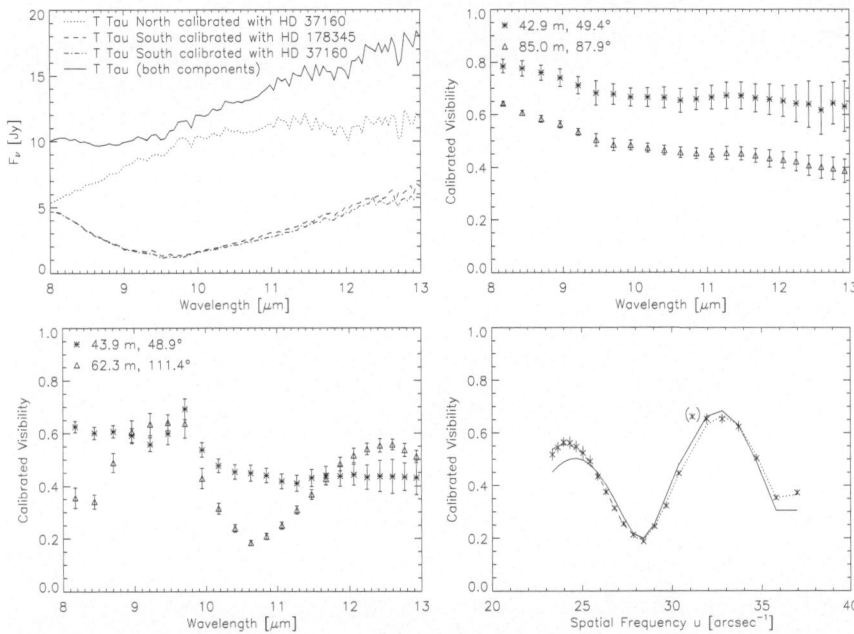

Fig. 4. *Upper Left*: The spectra for the whole T Tau system (solid) and both its northern (dotted) and southern component (dashed). *Upper Right*: The calibrated visibilities for T Tau N. *Lower Left*: The calibrated visibilities for T Tau S. *Lower Right*: The model fit of (parts of) the binary signal given in the lower left panel.

4 The Next Steps

The work on T Tauri stars is still in progress. In a next step the visibilities have to be explained not only by geometrical analysis, but also by radiative transfer models reproducing the spectral energy distributions. The hard constraints set by the visibilities will serve as a test for the quality of the models (see contribution by Th. Ratzka et al.). Furthermore, detailed studies of grain evolution in the inner circumstellar disk similar to those of Herbig Ae/Be stars [1] appear feasible also for low-mass T Tauri stars.

References

1. Ch. Leinert, R. van Boekel, L.B.F.M. Waters, et al: A&A, **423**, 573 (2004)
2. C.D. Koresko, V.W. Beckwith: AJ, **102**, 6, 2073 (1991)
3. Th. Ratzka: High Spatial Resolution Observations of Young Stellar Binaries. PhD Thesis, Ruperto-Carola University, Heidelberg, Germany (2005)
4. T.L. Beck, G.H. Schaefer, M. Simon: ApJ, **614**, 235 (2004)

Keck Interferometer Observations of the Young Spectroscopic Binary Haro 1-14c

Gail H. Schaefer[1,2], Michal Simon[2], and L. Prato[3]

[1] Space Telescope Science Institute, 3700 San Martin Drive, Baltimore, MD 21218 gschaefer@stsci.edu
[2] Dept. of Physics and Astronomy, SUNY Stony Brook, Stony Brook, NY 11794-3800
[3] Lowell Observatory, 1400 West Mars Hill Rd., Flagstaff, AZ 86001

Summary. Using the Keck Interferometer, we obtained the first measurement of the angular separation of Haro 1-14c, $\sim$ 15 mas. Modeling shows that continued interferometric measurements, combined with the spectroscopically determined orbital parameters, can yield a dynamical measurement of the component masses with sufficient precision to enable meaningful tests of theoretical calculations of pre−main-sequence evolution.

1 Introduction

Haro 1-14c (HBC 644) is a pre−main-sequence (PMS) binary in the Ophiuchus star forming region (SFR). The system was first identified as a single-lined binary, through spectroscopic observations in visible light, with a period of 591 days [1]. High-resolution infrared spectroscopy detected the secondary and yielded a mass ratio of 0.31 and an average flux ratio in the H-band of $\sim$ 0.4 [2]. Based on the K3 spectral type of the primary [3], which has an estimated mass of $\sim$ 1.2 $M_\odot$ [4], the mass ratio predicts a mass of $\sim$ 0.4 $M_\odot$ for the companion. Thus, the secondary falls within the mass region where there are few reliable measurements of dynamical masses for young stars and theoretical tracks of PMS evolution are particularly discrepant [5, 6].

Table 1 lists the orbital elements for Haro 1-14c, based on the radial velocity measurements [1, 2]. To determine the individual masses, the orbital inclination is needed. Spatially resolved observations of an orbit provide the required geometry and angular scale. At the average distance of the Ophiuchus SFR ($\sim$ 140 pc), the projected semi-major axis of Haro 1-14c, 1.5 AU, is resolvable by the current generation of long baseline optical/IR interferometers. In this talk, we present our first measurement of the angular separation of Haro 1-14c with the Keck Interferometer [7, 8]. With continued observations, the combined spectroscopic and astrometric measurements will yield the masses of the components as well as the distance to the system.

Table 1. Spectroscopic Binary Orbital Elements

P (days)	591.3 ± 0.3	$a_1 sini$ (Gm)	54.2 ± 0.8
$T_\circ$ (MJD)	45375.4 ± 3.0	$a_2 sini$ (Gm)	174.8 ± 7.8
e	0.617 ± 0.008	q	0.310 ± 0.014
ω	$232.90 \pm 0.55°$		

2 Observations and Data Analysis

We observed Haro 1-14c with the Keck Interferometer in the K-band on 2004 June 1. The interferometer combines the light of the two 10-meter telescopes at the W. M. Keck Observatory and has a baseline of 85 m, oriented 38° east of north. We interspersed observations of the target between measurements of unresolved calibrator stars. Figure 1 shows the calibrated fringe visibilities obtained for Haro 1-14c.

The squared visibility, V^2, of a binary is given by

$$V_{bin}^2 = \frac{1 + r^2 + 2r \cos[2\pi(u\Delta\alpha + v\Delta\delta)]}{(1+r)^2} \tag{1}$$

where $(\Delta\alpha, \Delta\delta)$ gives the separation between the two unresolved stars, and r is the flux ratio (i.e. [9]). The (u, v) coordinates define the projected baseline on the sky at the position of the source.

To measure the separation of Haro 1-14c at the time of the observations, we performed a χ^2 minimization by stepping through ranges of the flux ratio and the two-dimensional binary separation $(r, \Delta\alpha, \Delta\delta)$. At each step through the grid, we computed modeled V^2 values at the same hour angles and (u, v) coordinates as the observations. The χ^2 between the measured and modeled visibilities was calculated and the minimum in the χ^2 surface located. The

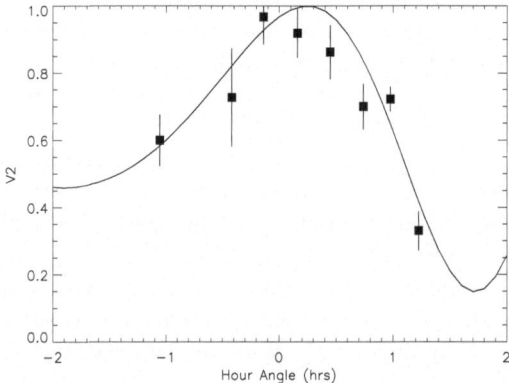

Fig. 1. Calibrated fringe visibilities for Haro 1-14c on 2004 June 1. The overplotted curve is the best-fit binary model, derived through χ^2 minimization (see text).

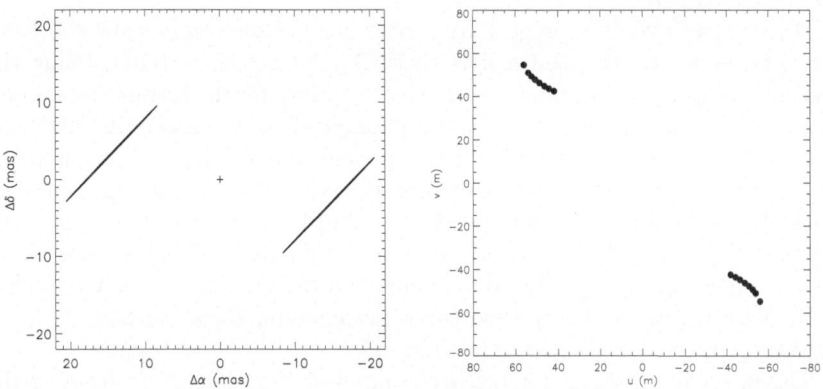

Fig. 2. *Left:* χ^2 surface for the binary separation. The contours show the 1 σ uncertainty level, corresponding to $\Delta\chi^2 = 1$. *Right:* (u, v) plane coverage for Haro 1-14c as observed on 2004 June 1 at the Keck Observatory.

uncertainties in $\Delta\alpha$, $\Delta\delta$, and r were determined by examining the range of parameters that produced a variation of 1 from the minimum χ^2.

The best-fit model for the visibility measurements is overplotted in Figure 1; it corresponds to a separation of $\Delta\alpha = 15 \pm 8$ mas, $\Delta\delta = 3 \pm 8$ mas, and a flux ratio of $r = 0.44 \pm 0.12$. These values yield a projected separation of 15^{+8}_{-3} mas at a position angle of $78°\ ^{+24}_{-41}$. A strong correlation between $\Delta\alpha$ and $\Delta\delta$ can be seen in a two-dimensional cross-cut through the χ^2 surface shown in Figure 2. The long and narrow error ellipses arise from the limited (u, v) coverage during time of observations. At the location of the Keck Observatory, the low altitude in the sky of Haro 1-14c (DEC $-24°$) produces little rotation in the uv-plane over the course of 2.5 hours. Effectively, the cross-cut in Figure 2 shows the extremely good resolution along the direction of the projected interferometer baseline. The χ^2 surface is symmetric about the origin because of the $180°$ ambiguity in the position angle inherent in the V^2 measurements.

3 Discussion and Future Prospects

The measured separation of ~ 15 mas is consistent with the published spectroscopic orbital parameters [2]. The precision of the eventual astrometric orbit derived from continued V^2 measurements with the Keck Interferometer will be limited primarily by the (u, v) coverage at the declination of Ophiuchus. Additional rotation in the uv-plane can be gained by increasing the length of time that Haro 1-14c is observed during future sessions. By adding simulated measurements to the observed data set from 2004, we found that extending the observing time from 2.5 hours to 3.5 hours reduced the uncertainties in the separation by a factor of ~ 2.

To illustrate what to expect from continued observations with the Keck Interferometer, we assumed a hypothetical orbit for Haro 1-14c. Using the spectroscopic orbital parameters, we selected values for the inclination i, semimajor axis a (in angular units), and the longitude of the ascending node, Ω, that give realistic estimates for the component masses, M_1 and M_2, and the distance to the system, while also reproducing the visibilities measured in 2004. We found that values of $i = 62°$, $a = 0.014''$, and $\Omega = 73°$ provide estimates of $M_1 = 1.4$ $M_\odot$, $M_2 = 0.44$ $M_\odot$, and a distance of 124 pc, consistent with the spectral types of the components and the distance to the Ophiuchus SFR. Together with the spectroscopic parameters, these estimated values yield the apparent orbit shown in Figure 3.

Based on this assumed orbit, we computed the relative position of the binary at monthly intervals during April through July in 2005-2007. For each date, we modeled the expected V^2 at 20 min intervals over 3.5 hours centered on transit and added Gaussian noise. Holding the known spectroscopic parameters fixed, we derived the remaining astrometric orbital elements by directly fitting an orbit to the simulated V^2 through χ^2 minimization. Fitting the orbit directly to the visibilities allows for easier handling of the tightly correlated uncertainties in $\Delta\alpha$ and $\Delta\delta$. Figure 4 shows the simulated V^2 measurements and the model orbit. The results of the simulation suggest that the orbital inclination can be measured with sufficient precision to determine the masses of Haro 1-14c to better than 10%. Therefore, with continued mapping of the astrometric orbit using ground-based interferometry and additional spectroscopic measurements to improve the precision of the mass ratio, observations of Haro 1-14c obtained over the next 2-3 years will contribute dynamical masses to use in conducting meaningful tests of PMS evolutionary tracks.

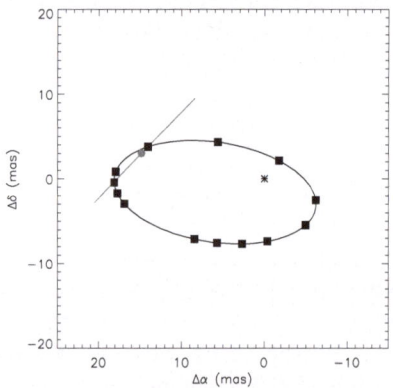

Fig. 3. Hypothetical orbit for Haro 1-14c. The gray circle represents the measurement with the Keck Interferometer. The squares represent modeled positions over the next three years.

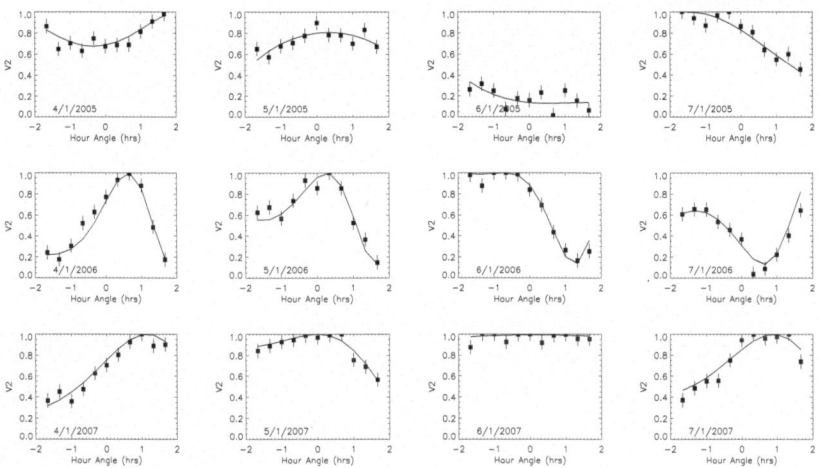

Fig. 4. Simulated V^2 measurements for the dates indicated in Figure 3. The solid line represents the best-fit orbit determined through χ^2 minimization.

Acknowledgements. We thank the Keck Interferometer staff for their efforts in obtaining these observations. We are grateful to Rafael Millan-Gabet and Rachel Akeson for their thorough support of our program. This work was supported in part by NSF grants AST 02-05427 (M.S. and G.S.) and 0444017 (L.P.) and JPL contract 1269936 (G.S.). The data were obtained at the Keck Observatory from time allocated to NASA through a partnership with Caltech and University of California. The Observatory was made possible by the generous financial support of the W.M. Keck Foundation. We wish to recognize the Hawaiian community for the opportunity to conduct these observations from the summit of Mauna Kea. This research has made use of software produced by the Michelson Science Center.

References

1. Reipurth, B. et al. 2002, AJ, 124, 2813
2. Simon, M. & Prato, L. 2004, ApJ, 613, L69
3. Herbig, G. H. & Bell, K. R. 1988, Lick Obs. Bull., 1111
4. Baraffe, I., Chabrier, G., Allard, F., & Hauschildt, P. H. 1998, A&A, 337, 403
5. Hillenbrand, L. A. & White, R. J. 2004, ApJ, 604, 741
6. Simon, M. 2001 in IAU Symp. 200, The Formation of Binary Stars, ed. H. Zinnecker & R. D. Mathieu (San Francisco: ASP), 454
7. Colavita, M. M., Wizinowich, L., & Akeson, R. L. 2004, Proc. SPIE, 5491, 454
8. Wizinowich, P. L. et al. 2004, Proc. SPIE, 5491, 1678
9. Boden, A. F. 1999, in Principles of Long Baseline Interferometry, Course Notes from the 1999 Michelson Summer School, ed. P. R. Lawson (Pasadena, CA: Michelson Science Center), 9

Preliminary Physical Orbit of the HD 98800 B System

Andy Boden[1], Anneila Sargent[2], Rachel Akeson[1] and John Carpenter[2]

[1] Michelson Science Center, California Institute of Technology
bode@ipac.caltech.edu
[2] Division of Physics, Math, and Astronomy, California Institute of Technology

Summary. As a part of a larger program to measure physical properties of PMS binary systems, we have observed the PMS quadruple system HD 98800 (aka TW Hya 4A) with the Keck Interferometer (KI), and resolved the two separate spectroscopic binary components of the system. In particular we have observed the HD 98800 North (B) binary component on multiple epochs, and integrating our data with the double-lined spectroscopic orbit from Torres et al [15] has allowed us to estimate a physical orbit for the B system. This orbit in turn yields relatively accurate (8%) mass estimates of the component stars, and an independent system distance and component luminosity estimates. Comparisons between the estimated physical parameters for the low-mass HD 98800 B components with mass-luminosity models of PMS stars, and implications for the age of HD 98800 are discussed.

1 Introduction

Among the areas where our understanding of stellar structure is most uncertain is in pre-main sequence (PMS) stars, particularly for low-mass systems [9, 5]. Providing empirical mass and luminosity constraints on models of PMS stars is critical for improving their accuracy, and the reliability of these models in turn is critical to our understanding of individual PMS systems in particular, and the process of star formation in general. So experimental determinations of masses and luminosities for PMS stars are of fundamental importance in constraining our understanding of star formation and early stellar evolution.

HD 98800 (HIP 55505, TWA 4A) is a well-studied quadruple star system in the TW Hya association. The system was first detected as a visual binary by [7]; at current epoch the separation is roughly 0.8" on approximately a N-S line [10]. Torres et al [15] established that both visual components are themselves spectroscopic binaries, finding a 262-day single-lined orbit for the primary/South/A component, and a 315-day double-lined orbit for the secondary/North/B component. The system exhibits a strong mid-infrared (IR) excess longward of 7 μm [16, 18], lithium signatures, but no signs of active accretion [12, 17]. The mid-IR excess, Li, HD 98800's putative membership in the TW Hya association, and the Hipparcos distance estimate of 46.7 ± 6.2 pc (establishing component luminosities) lead to the consensus that the system is PMS. Soderblom et al [13] has termed the system "post-T Tauri", and estimates a system age range of 5 – 20 Myr with a most likely value of 10 Myr based primarily on Li abundance. Multi-band IR imaging studies established that the mid-IR excess is associated with the double-lined B subsystem, most likely in the form of a circumbinary disk [8, 10].

Here we report on observations of the double-lined HD 98800 B binary subsystem made with the Keck Interferometer (KI, [3]) and the Hubble Space Telescope Fine Guidance Sensor (FGS). These observations resolve the B subsystem and allow us to estimate the visual and physical orbits (in combination with radial velocity measurement from [15]), and determine the component dynamical masses and luminosities.

2 Observations and Orbit Estimation

The interferometric observable used in this work measurements is the fringe contrast or *visibility* (squared) of an observed brightness distribution on the sky. KI (described in detail in [3]) was used to make the interferometric measurements presented here. Analysis of such data on a binary system is discussed in detail in previous work (e.g. [1, 6]).

HD 98800 B was observed in conjunction with calibration objects by KI in K-band ($\lambda \sim 2.2\mu$m) on five nights between 18 April 2003 and 22 April 2005, a dataset spanning roughly two years and 2.3 orbital periods. Our total KI calibrated data set includes 34 calibrated visibility scans on HD 98800 B. We note that as the Keck Telescopes separately resolve the HD 98800 A-B system (P2001), and the KI beam combiner is fed by single-mode fiber, no light from HD 98800 A falls on the fringe camera when the device is measuring HD 98800 B. Consequently no special provisions are necessary in processing KI observations of HD 98800 B.

The HD 98800 system was also observed by the Hubble Space Telescope Fine Guidance Sensors (FGS) in its "FGS-TRANS" mode and F583W filter on 20 epochs between 1996 and 2002. Of course, the FGS are also interferometers, but unlike the KI observations the FGS-TRANS data have been processed into estimated Ba-Bb separations by methods described in [4]. HD 98800 represents a challenging target for FGS observation: the B subsystem separation is near the resolution limit of the FGS, and (unlike the KI observations) the visual A component flux must be accounted for in data reduction. Because of these difficulties, of the 20 epochs only 11 of the measurements were deemed viable for triple-star analysis, and are used in the orbital analysis.

We estimate the visual and physical orbit of HD 98800 B by integrating the astrometric datasets described above with the double-lined radial velocity (RV) data on B presented in Torres et al Table 1 [15]. Figure 1 depicts our relative visual and spectroscopic orbit model of the HD 98800 B subsystem. The upper panel depicts the relative visual orbit model, with the primary (Ba) component rendered at the origin, and the secondary (Bb) component rendered at periastron. We have indicated the phase coverage of our KI V^2 data on the relative orbit with points (they are *not* separation vectors); the phase coverage of the V^2 data is sparse relative to other similar analyses [1]. Because of this sparse phase coverage of the V^2 data, our initial visual orbit determination we constrained orbital parameters measured by RV (i.e. e, ω, P) to their spectroscopic values, and found an orbit model that phased-up acceptably with the RV solution. Based on that constrained initial estimate, we then fully integrated the V^2, FGS, and RV data. The size of the HD 98800 B components are estimated and rendered to scale. The lower panel depicts the integrated double-lined spectroscopic orbit model and radial velocity data from [15]. Details of the orbit analysis are given in [2].

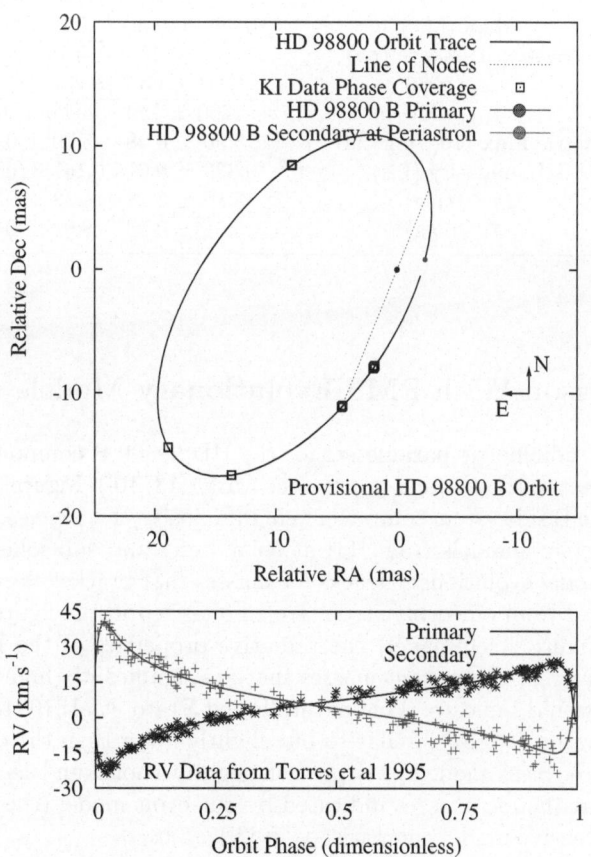

Fig. 1. Orbit Model for the HD 98800 B Subsystem

3 Physical Parameters of the HD 98800 B Subsystem

Component physical parameters derived from the model orbit and Spectral Energy Distribution (SED) modeling are summarized in Table 1. Chief among these results are dynamical mass estimates for the primary (Ba) and secondary (Bb) components, and independent distance estmate and resulting component absolute magnitude and luminosity estimates. Further details of the physical parameter estimation are given in [2].

Table 1. Selected Physical Parameters for HD 98800 B Components

Physical Parameter	Ba Component	Bb Component
a $(10^{-1}$ AU$)$	4.47 ± 0.13	5.36 ± 0.13
Mass $(M_\odot)$	0.699 ± 0.064	0.582 ± 0.051
System Distance (pc)	42.2 ± 4.7	
π_{orb} (mas)	23.7 ± 2.6	
T_{eff} (K)	4200 ± 150	4000 ± 150
Bolometric Flux $(10^{-9}$erg cm^{-2} s$^{-1})$	5.96 ± 0.28	3.00 ± 0.15
Luminosity $(L_\odot)$	0.330 ± 0.075	0.167 ± 0.038
M_K (mag)	3.80 ± 0.25	4.38 ± 0.25
M_V (mag)	6.91 ± 0.26	8.02 ± 0.27

4 Comparison With PMS Evolutionary Models

Our inferred radiometric parameters for the HD 98800 B components are in good agreement to those found in previous work ([13, 10]). Figure 2 shows the position of the B subsystem components in luminosity/T_{eff} space, along with PMS evolutionary models from [11] at solar (left) and sub-solar ([M/H]=-0.3, right). Model evolutionary tracks for masses that bracket the component masses inferred from our orbit model are emphasized in the figure. For solar metalicity (Figure 2 left panel) the radiative properties of the Ba and Bb components predict component masses that are significantly higher than our orbit model would indicate. This naturally led Prato et al [10] to infer a B orbital inclination ($\sim 58°$) similar to but slightly lower than the orbit model presented here. Siess models [11] predict slightly cooler and less luminous components at the low masses indicated by our orbit model (the reader can find more extensive model comparisons in [2]).

The apparent model/observation discrepancy is reduced if we consider the possibility that elemental abundances for HD 98800 components are subsolar. Soderblom et al [13] attempted an abundance estimate for HD 98800, and argue for solar abundance with an uncertainty of ~ 0.2 dex. Figure 2 right

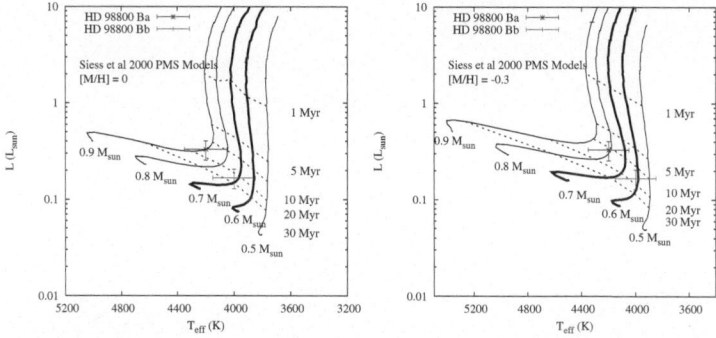

Fig. 2. Comparison of HD 98800 B Components with PMS Evolutionary Tracks from Siess et al 2000

panel shows the same comparison of component properties with Siess models at [M/H] = -0.3 [11]. The match between the lower abundance models at our inferred component masses and radiometric properties is improved, with both components matching the relevant mass tracks within the temperature error bars for both the lower abundance models. From superimposed isochrones in Figure 2 we infer the HD 98800 age is in the 8 – 20 MYr range, consistent with previous findings [13].

References

1. Boden, A., Creech-Eakman, M., and Queloz, D. 2000, ApJ 536, 880.
2. Boden et al. 2005, ApJ 635, 442.
3. Colavita, M. et al 2003, ApJ 592, L83.
4. Franz, O., et al. 1998, AJ 116, 1432.
5. Hillenbrand, L. & White, R. 2004, ApJ 604, 741.
6. Hummel, C. et al. 2001, AJ 121, 1623.
7. Innes, R. 1909, Transvaal Obs. Circ 1, 1.
8. Koerner, D. et al 2000, ApJ 533, L37.
9. Palla, F. & Stahler, S. 2001, ApJ 553, 299.
10. Prato, L. et al 2001, ApJ 549, 590 (P2001).
11. Siess L., Dufour E., & Forestini M. 2000, A&A 358, 593.
12. Soderblom, D. et al 1996, ApJ 460, 984.
13. Soderblom, D. et al 1998, ApJ 498, 385 (S98).
14. Tokovinin, A. 1999, Ast. Lett. 25, 669.
15. Torres, G. et al. 1995, ApJ 452, 870 (T95).
16. Walker, H. & Walstencroft, R. 1988, PASP 100, 1509.
17. Webb, R. et al 1999, ApJ 512, L63.
18. Zuckerman, B. & Becklin, E. 1993, ApJ 406, L25.

Part IV

Science: Stars — Galactic centre, AGNs, astrometry, exo-planets and future targets

Resolving the Dusty Tori in AGN with the VLT Interferometer

Klaus Meisenheimer

Max-Planck-Institut für Astronomie,
Königstuhl 17, D–69117 Heidelberg, Germany
meise@mpia.de

Summary. The MID-infrared Interferometer (MIDI) at the VLT Interferometer (VLTI) is the ideal instrument to resolve the emission of the putative dust tori in nearby AGN. Indeed, the first successful MIDI observations of the brightest and nearest Seyfert II galaxies – NGC 1068 and the Circinus galaxy – already provided us with an unexpected wealth of information about the dust distribution and properties in the inner few parsec of AGN. MIDI observations of the closest radio galaxy – Centaurus A – reveal a unresolved $10\,\mu m$ source ($< 0.2\,pc$) which presumably represents the base of the radio jet. As soon as the VLTI will be fully operational, about a dozen extragalactic sources will be accessible to MIDI observations. This will allow us to tackle two of the most important issues of AGN physics: (1) are torus-shaped dust structures and our various viewing angles into them indeed responsible for the apparent differences in Seyfert I and II galaxies, and (2) how does the gas reservoir in these tori feed the accretion disks around the black holes ?

There are several lines of indirect evidence, that in many Active Galactic Nuclei (AGN) the central engine – a super-massive black hole fed by an surrounding accretion disk – is embedded into a axi-symmetric, geometrically thick structure of gas and dust: the so-called torus. This dusty torus is held responsible for the observed dichotomy between Seyfert galaxies of type I and II, respectively: in type I, an unobscured view along the torus axis allows us to look directly into the central engine, while in type II objects the torus is seen edge-on and thus blocks our view towards the very center.

1 Dusty Tori in AGN: what do we expect ?

There are many different models for the geometry and density distribution of those tori in the literature. They mostly try to explain the overall SEDs of Type I and type II AGN as well as their relative abundance by radiative transfer in disk- or torus-like dust distributions. However, as long as the dust emission of the dusty structure remains unresolved, it is impossible to decide between these models. In order to give a feeling how these dusty tori might look at various mid-infrared wavelength and spatial resolutions, I will refer to a recent model by my student Marc Schartmann and collaborators

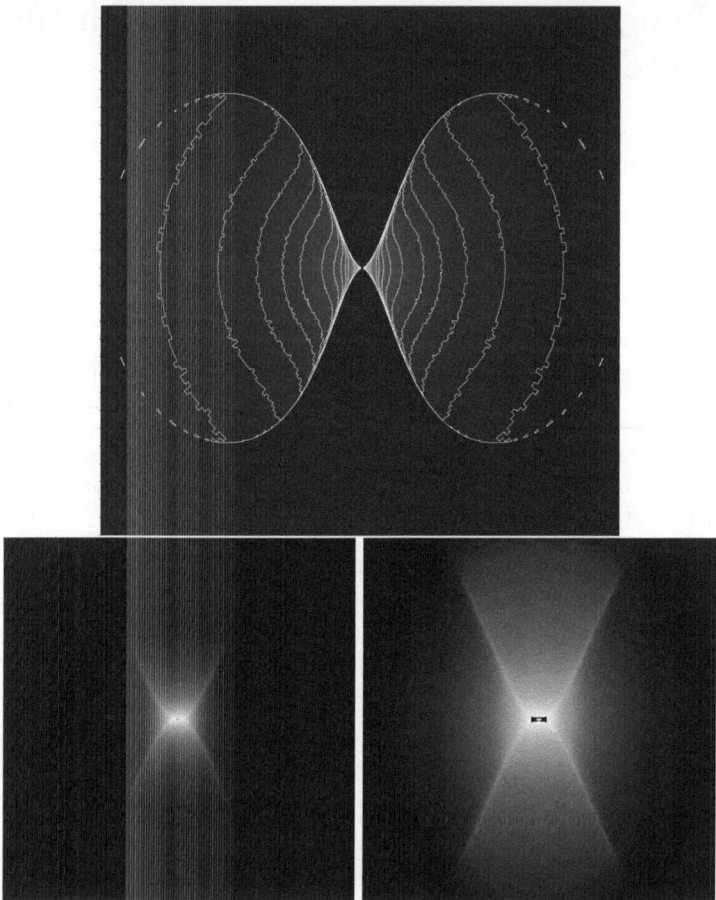

Fig. 1. Temperature distribution and mid-infrared emission of the torus model [6]. The upper panel shows the dust temperature which ranges from > 1000 K in the innermost part to 50 K at the outer edge. The lower panel shows the expected emission at 5 μm (left) and 13 μm (right) for a torus viewed under 60° from its symmetry axis.

[6]. Its basic geometry and temperature distribution is displayed in Fig. 1. Three-dimensional radiative transfer calculations predict SEDs and images at various wavelengths. In Fig. 2, the predicted image at 10μm is used to demonstrate what one might resolve with various telescopes: a diffraction limited 8 m telescope at $\lambda = 4.5\mu$m (*e.g.* NACO at the VLT) will hardly resolve the torus even in the closest Seyfert galaxies. Obviously, for observing dusty tori at mid-infrared wavelengths around 10 μm one needs a 100 m class telescope. In effect, this is what MIDI at the VLT Interferometer provides, although for full image quality, in practice the aperture plane needs to be filled

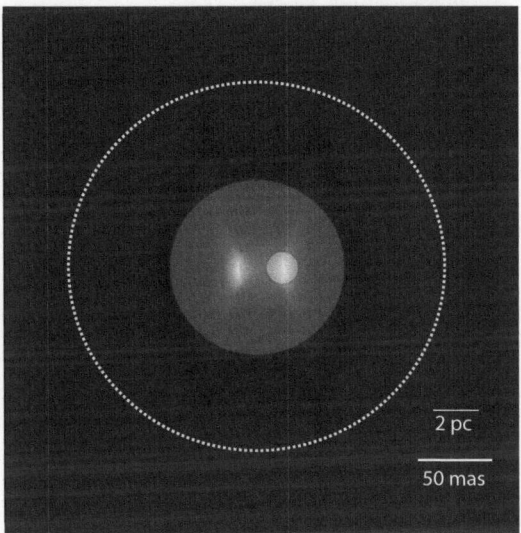

Fig. 2. Expected mid-infrared emission of the torus at a distance of 14 Mpc (like NGC 1068, NGC 4151) compared to the angular resolution provided by different telescopes. On top of the central part of the torus model (lower right panel in Fig. 1), the beam width of an diffraction limited 8 m telescope operating at $10\,\mu$m (dotted circle) and $4.5\,\mu$m (grey shaded circle) is shown. The light shaded circle gives the beam width of a 100 m telescope at $10\,\mu$m. Note that only a 100 m size telescope will be able to resolve sub-structures of the torus.

by a huge number of two-baseline observations with various combinations of the UTs.

2 Scientific questions to be answered by interferometry

The immediate aim of the VLTI observations of nearby AGN at mid-infrared wavelengths is to answer the following questions (ordered in ascending observational difficulty):

- Is there a compact AGN heated dust structure in the cores of AGN ?
- What is its size ?
- What is its shape ?
- What is its orientation with respect to the source axis (as defined by radio jets and other outflow phenomena) ?
- Are the dust structures similar in Seyfert I and Seyfert II galaxies ?
- Is the dusty structure clumpy ?
- What is its temperature distribution ?

The MIDI consortium has planned to spend 65 hours of its guaranteed time at UTs in order to find first answers to these questions (see section 4).

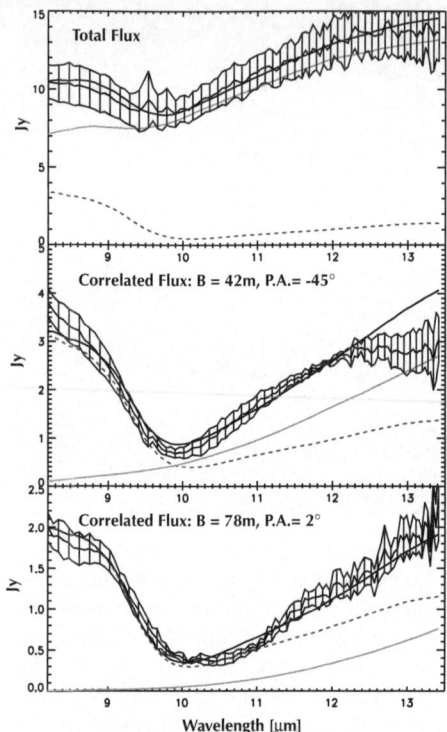

Fig. 3. Results of the MIDI observations of NGC 1068. The panels show the total flux in a 0.''6 aperture, and the correlated flux at baselines B = 42 m and B = 78 m, respectively (from top to bottom). In each panel the observed data (and their rms variation between different observations) are shown as hatched range, a black line shows the model fit which is composed of contributions of a hot component (dashed) and a warm component (grey/green).

3 First case study: the torus in NGC 1068

First MIDI observations of the proto-typical Seyfert II galaxy, NGC 1068, have been obtained during VLTI *Science Demonstration Time* (SDT).[1] Two independent baselines have been observed: a longer baseline between UT1 and UT3 (78 m) in June 2003 and a shorter baseline UT2–UT3 (42 m) in November 2003. The central peak of NGC 1068 was observed through a 0.''6 slit and the two interferometric output beams from the MIDI beam-combiner were dispersed by a prism. Fig. 3 shows spectra of the total flux and the

---

[1] The Science Demonstration Time was introduced to demonstrate the capabilities of VLTI in different areas of astronomy. It is coordinated by Francesco Paresce. The AGN observations are coordinated by Huub Röttgering. The observations of NGC 1068 have been carried out as joint effort of the VLTI and MIDI teams.

correlated flux obtained for the two baselines. They are dominated by a strong silicate (SiO) absorption feature between 9 and 11.5 μm which becomes much stronger in the interferometric spectra (corresponding to resolutions of about 25 and 13 mas, respectively). Jaffe, Meisenheimer *et al.* (2004) [3] model the results with two Gaussian brightness distributions of different blackbody spectrum, the symmetry axis of which is assumed to be aligned with the radio jet and which suffer a different amount of silicate absorption (see model fits in Fig. 3). They find a well resolved warm dust component of T $= 320 \pm 20$ K and size $d \times h = 49 \times 30$ mas^2 (i.e. 3.4×2.1 parsec at the distance of NGC 1068). Embedded into it is a hot component of T> 800 K, which is marginally resolved ($h = 0.7 \pm 0.2$ pc) in North-South direction but poorly constrained ($d \lesssim 1$ pc) along East-West (see Fig. 4). We interpret the 300 K component as the thermal emission of a geometrically thick dust distribution ("torus").

The hot component could naturally be explained by the emission of dust near sublimation temperature which is expected at the inner walls of an axial funnel which allows the radio jet and outflowing ionized gas to emerge along the source axis. The high silicate optical depth $\tau_{\rm SiO} = 2.1$ towards the hot component corresponds to $A_V \simeq 50$ and is a firm lower limit to the dust extinction towards the central engine. It is instructive to compare the model inferred from the MIDI observations with recent VLBA maps in the radio continuum (5 GHz) by Gallimore, Baum & O'Dell [1]: Fig. 5 shows that the radio continuum (most likely due to free-free emission of ionized gas) forms a disk-like structure with some indication of flaring towards its outer rim. This disk has the size of the hot component, suggesting that we indeed

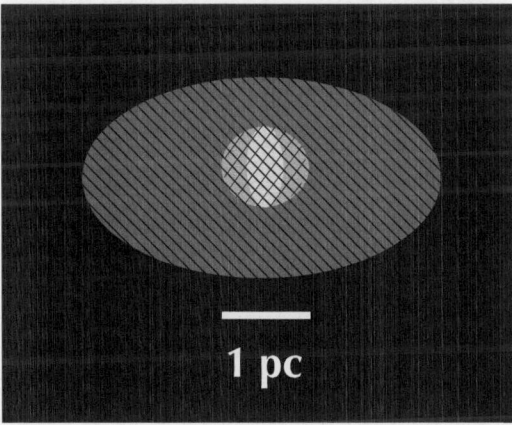

Fig. 4. The observational model of NGC 1068. The warm component ($T = 320$ K) is shown in orange, the hot component ($T > 800$ K) in yellow. The uncertainty in the East-West extension of the latter is indicated by two levels of yellow. The depth of the silicate absorption is represented by hatching.

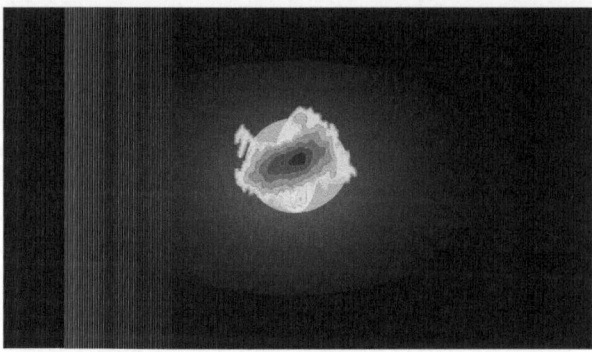

Fig. 5. The observational model of NGC 1068 compared with the radio continuum map from [1]. The brightness of the radio continuum is represented by grey shading.

witness the evaporation process at the inner wall of the funnel which feeds the central engine. Wittkowski et al. (2004) [9] have observed NGC 1068 with VINCI along one VLTI baseline at $\lambda = 2.1\,\mu$m. Together with earlier Speckle observations (Weigelt et al. 2005 [8]), they argue for a very compact ($\lesssim 3$ mas) K-band source which contributes about 15% to the central K-band component of 30 mas diameter and slightly elongated along $P.A. \simeq -20°$. Unless there exist dust-free holes in the torus, the large extinction towards the hot center makes it impossible for 2 μm radiation to escape directly. Therefore, I prefer the interpretation that in the K-band one sees (scattered) light from the upper funnel opening in the torus. Relative astrometry to a few mas accuracy will be required to pin down the relative location of the 2 μm and 9 μm emission.

4 The MIDI guaranteed time program on AGN

4.1 Overview

For the planning of the MIDI guaranteed time program we collected all southern AGN ($\delta \leq +20°$) with a total (large-aperture) N-band flux $S_N \gtrsim 1$ Jy. Preparatory observations with TIMMI2 and data from the literature were then used to select those 17 nearby AGN, the sub-arcsec core fluxes of which exceed $S_N(core) \simeq 400$ mJy at 10 μm (see http://www.eso.org/observing /proposals/gto/index.html for the complete list). Further observations have been carried out with NAOS-CONICA (NACO) resulting in diffraction limited images in the near-IR between the H and the M'-band (4.5μm) to derive highly resolved extinction maps and to identify the amount of thermal radiation from an unresolved core (< 100 mas).

Three of them (NGC 1068, NGC 3783, Mrk 463) were released to support the public SDT program. During the last year the MIDI team was able to

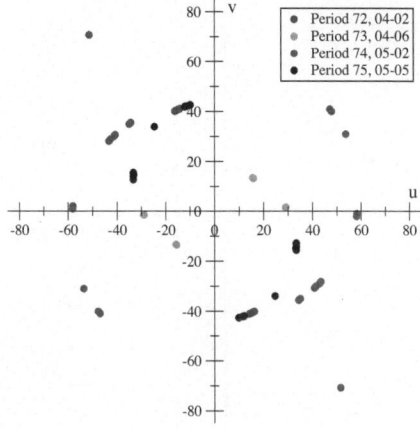

Fig. 6. Current UV coverage of Circinus with MIDI observations given in units of m for the projected baseline.

carry out successful interferometric measurements for the two brightest of the remaining AGN, the Circinus galaxy and Centaurus A.

4.2 Second case study: the Circinus galaxy

The Circinus galaxy is the closest AGN, which is classified as Seyfert II (at 4 Mpc distance). Archival HST observations show a very broad ionization cone but the very core remains hidden behind the very patchy dust structure in the disk of the galaxy. Thus, only our recent NACO observations (Prieto et al. 2004 [5]) allow a first glimpse of the core at $\lambda > 2\mu$m. Since the core seems marginally resolved already on the L-band observations, we choose the shortest baselines of the VLTI for the MIDI observations. Due to its southern declination ($\delta = -65°3$, Circinus allows us to reach a very nice coverage of the uv−plane (see Fig. 6).

Fig. 7 shows the results obtained with nearly orthogonal baselines UT2–UT3 and UT3–UT4 in four runs between February 2004 and May 2005. The left column displays total and correlated flux as shown for NGC 1068. The first thing to notice is the very low level of correlated flux $\langle S_{\mathrm{corr}} \rangle \simeq 400$ mJy which was sufficient to track the interferometric fringes with MIDI. Second, there is no indication that the silicate absorption feature is deeper in the correlated flux (*i.e.* at higher spatial resolution). This is more obvious in the right panels where we show the *visibility* $V := S_{\mathrm{corr}}/S_{\mathrm{tot}}$. In most cases, it seems almost flat between 8.3 and 13 μm, indicating that (i) the SiO absorption obvious in the total flux is a global feature (due to foreground material), and (ii) any evidence for a hotter inner component (as seen in NGC 1068) is weak if not absent. Although we are still far from understanding the

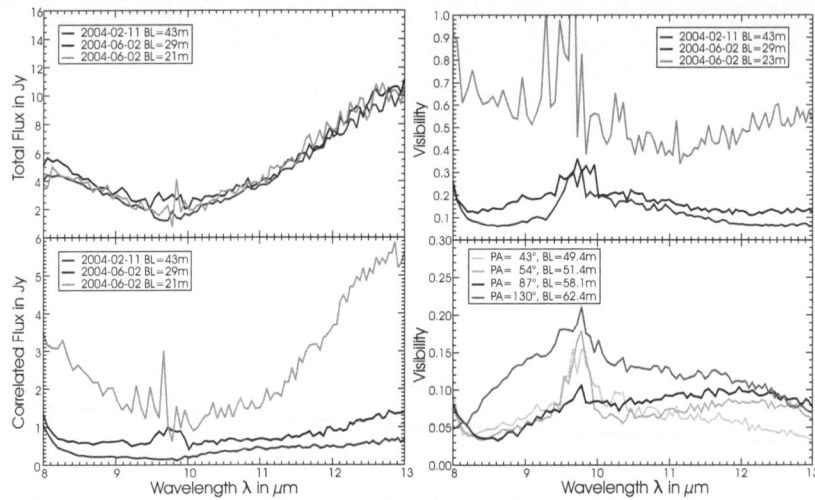

Fig. 7. Results of the first MIDI observations of the Circinus galaxy. *Left:* Total flux and correlated flux for three observations in 2004. *Right:* Visibility measured in these observations (top panel) plus further visibilty measurments obtained in 2005 (lower panel). Note that the region $9.5 < \lambda < 10\,\mu\mathrm{m}$ is strongly affected by atmospheric ozone absorption and should be discarded.

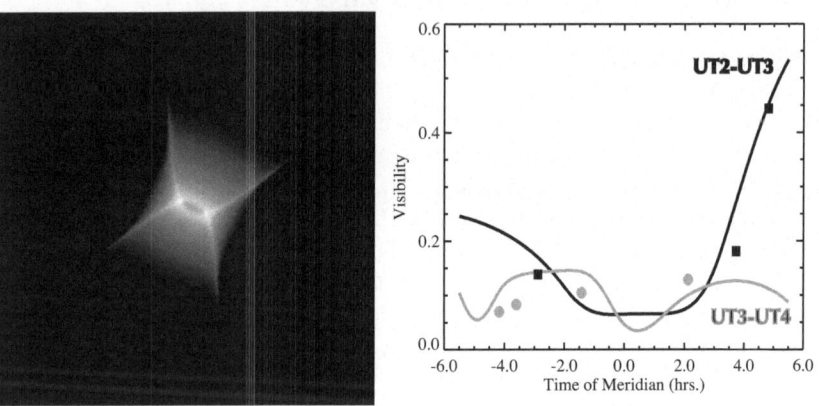

Fig. 8. Comparison between a torus model for Circinus with the observed visibilities at $\lambda = 11.5 \pm 0.5\,\mu\mathrm{m}$. *Left panel:* Image of the torus model [6] scaled to distance and luminosity of the Circinus core (image size: 18 pc). Inclination and orientation of the source axis are determined from the ionization cone. *Right panel:* Predicted visibility at $11\,mu\mathrm{m}$ as function of time for VLTI baselines UT2–UT3 and UT3–UT4. The observed visibilities are shown as ■ (UT2–UT3) and ● (UT3–UT4), respectively

wealth of information contained in Fig. 7, a first hint about the dust structure
in the core of Circinus can be obtained by comparing the visibility behavior
predicted by our torus model [6], scaled to the distance and luminosity of
Circinus with the observed visibility $V(11\mu m)$ in Fig. 8: although quantitative
discrepancies exist (which could be accounted for by fine-tuning the model),
the rapid increase of the visibility along the baseline UT2–UT3 at > 3 hours
after the Meridian, which reflects the small extent of the model emission
along $P.A. \simeq -20°$, is well reproduced. This indicates, that the dust torus of
Circinus is geometrically much thinner than that in NGC 1068 – in perfect
agreement with the fact that the ionization cone of the Circinus galaxy is
much less collimated than that of NGC 1068.

4.3 A special case: the radio galaxy Centaurus A

The faintest source which MIDI could measure so far is the famous radio
galaxy Centaurus A $(S_N(core) = 1.2\,\text{Jy})$. Two projected baselines of about
60 m were observed with UT2–UT4 in February. As demonstrated in Fig. 9,
the correlated flux can well be measured and leads to consistent results. Due
to the low contrast between the total flux in a 0″.6 aperture and the extremely
bright background, the total flux measurement is much more uncertain. Nevertheless, it is clear that the visibility of Centaurus A is on the level of 80%
at $\lambda < 9\mu m$, with some indication of a decrease towards the longest wavelengths. We interpret the measurements in terms of an unresolved source
$(d < 0.2\text{ pc})$ which might be surrounded by a faint and well resolved component of cool dust (T $< 300\,\text{K}$). If we refer to two wavelengths at either side
of the silicate feature, the observed slope of the spectrum is $S_\nu \sim \nu^{-0.65}$.

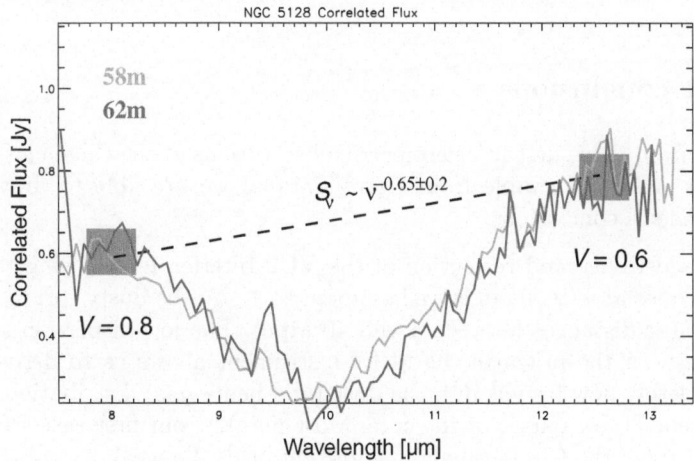

Fig. 9. MIDI observations of Centaurus A: Correlated flux measured with two
orientations of UT2–UT4.

The limited wavelength coverage of the MIDI observations alone would not allow to decide between a thermal or non-thermal origin of the unresolved radiation. However, when considering the overall spectrum from the radio to the near-IR at sub-arcsec resolution (Fig. 10), it is likely that the unresolved $8\ldots13\,\mu m$ core of Centaurus A is a synchrotron source, presumably the base of the radio jet seen on VLBI maps.

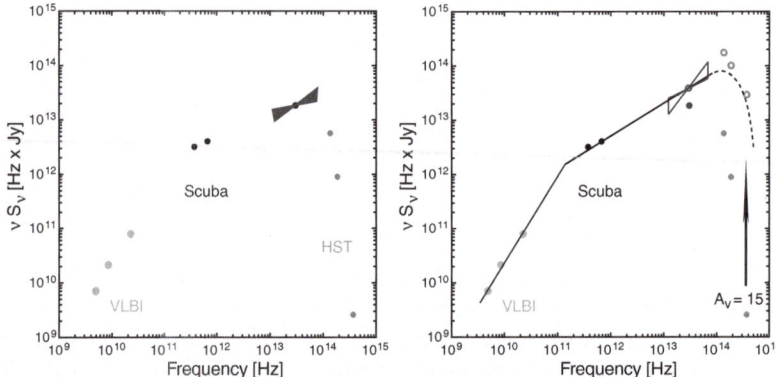

Fig. 10. The overall spectrum of the core of Centaurus A. *Left panel:* Core measurement with MIDI between 8.5 and $12.5\,\mu m$ (slope and its error) and at other wavelengths observed by VLBI ($\simeq 3$ mas, [7]) and the HST (100 mas, [4]). The Scuba points [2] refer to a $10''$ aperture. *Right panel:* intrinsic spectrum (corrected for $A_V = 15$) and its interpretation as a synchrotron spectrum with high-frequency cut-off, which is self-absorbed below $\nu \simeq 3 \times 10^{11}$ Hz.

5 First conclusions

Already from these first interferometric observations of the three closest and brightest AGN observable from Cerro Paranal we are able to draw some rather general conclusions:

– The sensitivity and resolution of the VLT Interferometer using the unit telescopes at $\lambda \simeq 10\,\mu m$ is well adapted to study the dusty tori in nearby AGN (at distances between 4 and 40 Mpc). The low-dispersion spectral mode with the prism in the MIDI instrument allows us to derive some completely new insight into the dust (and hence gas) distribution within the central few parsec of the core. Most notably, our first case studies of NGC 1068, the Circinus galaxy and Centaurus A reveal:

– Compact AGN-heated dust structures of radius $r_{300K} \leq 2pc \simeq 4r_{sub}$ do indeed exist in Seyfert II galaxies.

- Some of them are geometrically thick (torus-like: $h/r > 0.5$) as predicted by the unified schemes and also expected from the tight collimation of ionization cones and outflows. However, in some cases (*e.g.* Circinus) the dust distribution seems to form a thinner structure (disk-like: $h/r < 0.3$) which could naturally account for a large observed opening angle of the ionization cone.
- In one case one finds evidence for a very hot compact component $r \simeq r_{sub}$ which can be understood as emission from the hot, inner walls of the funnel which should open along the axis of the torus.
- In the radio galaxy Centaurus A, there are no hints of a massive circumnuclear dust distribution. That implies that even in cases where a high dust column obscures our view into the core region on the hundred pc scale, there might exist pretty "naked" (gas and dust free) AGN cores.

In summary, the wide variety of circum-nuclear dust structure seen already in the three closest AGN, puts a big question mark behind over-simplified "unified schemes" which assume that all AGN are intrinsically the same.

6 Future interferometric insight into the cores of AGN

After the successful MIDI observations of AGN have proven that interferometry at near- and mid-infrared wavelengths are not only feasible but allow completely new insight into AGN physics, I would like to outline the next steps ahead by presenting four scientific questions which need to be addressed by future studies:

Question 1: Do Seyfert I galaxies indeed contain the same kind of tori as Seyfert II galaxies ? In order to answer this question it is not only necessary to observe a sample of Seyfert I galaxies but also to reach the dynamic range to resolve the extended torus emission from that of the unresolved core.

Question 2: What is the internal structure of these tori ? Can we find evidence of their clumpiness – expected from physical models ? What is their temperature distribution ?

Question 3: How does the torus feed the accretion disk ? This will require a resolution of about a lightyear (0.3 pc) and accurate relative astrometry over a wide range of wavelengths between the radio (VLBI) and the mid- and near-IR.

Question 4: What is the structure of the broad line region (BLR) ? Can we see BLRs in Seyfert II galaxies directly ? The BLRs of Seyfert galaxies have sizes of several dozen lightdays (0.03 pc). This corresponds to 0.1 mas for a typical nearby Seyfert I galaxy at $z = 0.01$, that is 100 times the resolution achieved with the VLTI at $10\,\mu$m and still 20 times its resolution at $2\,\mu$m.

I am confident that the current and second generation instruments at the VLTI are well capable to answer the first two questions, almost a decade before any filled-aperture 100 m telescope (like OWL) will become operational.

Questions 3 and 4 need resolutions which are well beyond the capacity of any telescope of only 100 m diameter. Here only larger arrays can help. Thus I conclude, that at least for AGN physics, the way to go is definitely not the way towards ELTs but rather towards telescope arrays which are ten to a hundred times larger than the VLTI – that is an array of about a dozen 10 m class telescopes spread over several kilometers. Whether these telescopes will be linked by fibers (A. Quirrenbach, these proceedings) or by heterodyne techniques (R. Schieder, these proceedings) will be shown by the future. Whatever happens, I have no doubt that the future of AGN physics will shift even more into interferometry (which dominates already the radio and millimetre regime), and that our first steps in infrared interferometry currently undertaken at the VLTI will prove pivotal over the next years.

Acknowledgements All MIDI results have been obtained in close collaboration with Walter Jaffe (Leiden) and my student, Konrad Tristram. Nothing of this would have been possible without the VLTI team, of whom I would like to mention A. Glindemann, A. Richichi, S. Morel, M. Schöller, M. Wittkowski, and the MIDI team: Ch. Leinert, R. Waters, U. Graser, G. Perrin, O. Chesneau and R. Köhler. Most of the insight into the physics and radiation of AGN tori, I owe to my student Marc Schartmann and his co-supervisors: M. Camenzind, S. Wolf and Th. Henning.

References

1. J. Gallimore, S. A. Baum and C.P. O'Dea: ApJ **613**, 794 (2004)
2. T.G. Hawarden, G. Sandel, H.E. Matthews et al.: MNRAS **260**, 844 (1993)
3. W. Jaffe, K. Meisenheimer, H. Röttgering et al.: Nature **429**, 47 (2004)
4. A. Marconi, E.J. Schreier, A. Koekemoer at al.: ApJ **528**, 276 (2000)
5. A. Prieto, K. Meisenheimer, O. Marco et al.: ApJ **614**, 135 (2004)
6. M. Schartmann, K. Meisenheimer, M. Camenzind, S. Wolf and Th. Henning: A&A **437**, 861 (2005)
7. S.J. Tingay, D.L. Jauncey, J.E. Reynolds et al.: AJ **115**, 960 (1998)
8. G. Weigelt, M. Wittkowski, Y.Y. Balega et al.: A&A **425**, 77 (2004)
9. M. Wittkowski, P. Kervella, R. Arsenault at al.: A&A **418**, L39 (2004)

A New Analysis of MIDI Observations of the Nucleus of NGC 1068

Anne Poncelet[1,2], Guy Perrin[2] and Hélène Sol[1]

[1] LUTH, Observatoire de Paris, 92195 Meudon Cedex
[2] LESIA, Observatoire de Paris, 92195 Meudon Cedex
firstname.lastname@obspm.fr

Summary. We report the results of the new analysis made on MIDI data of the active galactic nucleus (AGN) of NGC 1068. The visibility measurements and the MIDI spectrum are well reproduced by a simple model of radiative transfer between two spherical components whose angular sizes are $\sim$ 35 and 83 mas, and temperatures are $\sim$ 366K and 226K respectively. This new approach allows to obtain the evolution of the optical depth of the extended component as a function of wavelength in the N-band, highlighting the probable presence of silicates in its composition. This confirms that MIDI has actually observed the distribution of dust around the core of the AGN.

1 Introduction

According to the standard unified scheme of AGNs, NGC 1068 harbors a Seyfert type II nucleus, meaning that the radiation coming from the central engine is obscured by a dusty torus surrounding it. The first aim of interferometric studies of AGNs is to test this paradigm since interferometry allows to observe closer to the core of the nucleus and to resolve the source of the thermal IR emission. First interferometric and speckle K-band observations of NGC 1068 are reported by Wittkowski et al. [12] and Weigelt et al. [11]. The observed structures seem to be north-west elongated and have sizes of $\sim$18×40 mas.

NGC 1068 was observed with MIDI during four nights of Science Demonstration Time (SDT) in June and November 2003. Visibility measurements are quite small accross the N-band (see Fig. 1), meaning that the nucleus of NGC 1068 is well resolved by MIDI. Jaffe et al. [6] modeled the nucleus with two elliptical gaussian disks and derived sizes of $\sim$ 30×49 mas for a warm component ($T\sim$350K) and 10×12 mas for an inner hot component ($T>$800K).

Here, we present a new analysis of MIDI data independently reduced with the software for data reduction developed at the Paris Observatory, and the modeling carried out on them.

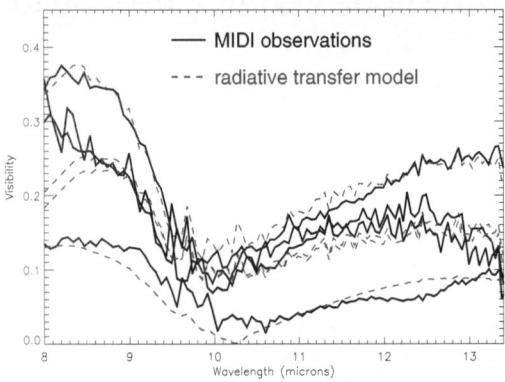

Fig. 1. Visibility measurements of 2003 and fits by the radiative transfer model.

2 Different steps in data modeling

MIDI observations of NGC 1068 available up to now correspond to only four different configurations of the projected baseline (PB), which sets limitations on the number of free parameters to be retrieved from the modeling.

In order to get the typical size of the nucleus and its evolution accross the full N-band, we first considered a single uniformly radiating disk model. The diameter of the source taken wavelength dependent is here the only free parameter. Applying this model to data corresponding to different configurations of the PB, differences appeared in the values of diameters then derived. Therefore, to try to put in evidence any elongation of the source, we computed a model of a single elongated disk. This led to a degeneracy of parameters showing that data do not contain information on shape and orientation. Therefore we restricted ourselves to spherical symmetry and by taking into account the evolution of the one uniform disk diameter between 30 and 60 mas accross the N-band, we then considered a two uniformly radiating disks model. This model improved the fits of visibilities and as a result, the inner disk has a smooth evolution around 30 mas accross the band, while the extended disk shows two size ranges at low and high wavelengths ($\sim$55 and 160 mas).

The next logical step in the modeling was to consider radiative transfer between the two spherical components. The inner compact component is simply described by a black body radiation at the temperature T_{in}. It is surrounded by a thermally radiating layer at temperature T_{layer} which have an optical depth responsible for absorption. Thus, the emerging radiation from the source is given by:

$$I(\lambda, \theta) = B(\lambda, T_{in})e^{-\tau(\lambda)/cos\theta} + B(\lambda, T_{layer})[1 - e^{-\tau(\lambda)/cos\theta}]$$

if $\oslash_{in} > \oslash_{layer} sin\theta$ (θ being the angle between the line of sight and the radius vector on the layer).

$$I(\lambda, \theta) = B(\lambda, T_{\text{layer}})[1 - e^{-2\tau(\lambda)/cos\theta}]$$

otherwise (where $B(\lambda,T)$ is the Planck function). Here, the five parameters considered are the diameter and the temperature of each component, and the optical depth which is the only parameter taken wavelength dependent.

3 Results

The χ^2 minimization rapidly converges to values of 35 mas for the diameter of the inner compact component and 83 mas for the extended layer. The temperatures are, on the contrary, not constrained by visibility measurements and this leads to several minima on the χ^2 hyper-surface. We therefore had to use an additional constrain given by the MIDI spectrum of NGC 1068 over the N-band in order to choose the optimal couple of temperatures. In this way, it provided: $T_{\text{in}} \sim 366$ K and $T_{\text{layer}} \sim 226$ K. With these parameters, the model accounts well for visibility measurements in the full N-band (except at 8 microns; see Fig. 1). Examples of fits of visibility points at fixed wavelengths are presented on Fig. 2. These figures highlight the lack of visibility points at low spatial frequency, needed to put stronger constrains on the modeling. The fit of the spectrum is presented on Fig. 3.

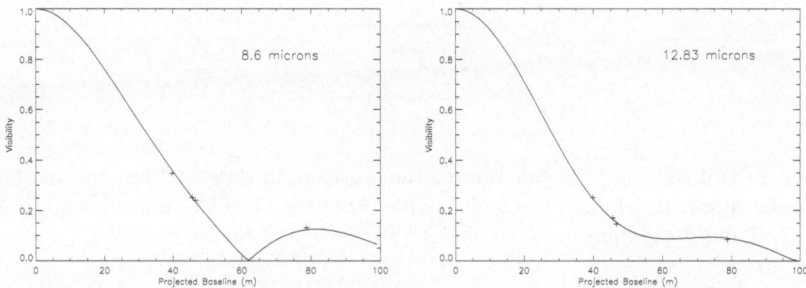

Fig. 2. Example of fits by the radiative tranfer model of the MIDI visibility points as function of the PB at given wavelengths.

Nevertheless, the originality of the study resides in its capability to provide the evolution of the optical depth as function of wavelength in the N-band, without taking any a priori assumptions (about its variation law or the composition of the extended layer for instance). Looking at the shape of the optical depth as a function of wavelength (see Fig. 4), it appears that the strong slope around 9μm and the bump around 10μm can be the signature of the presence of amorphous silicates in the layer. This shows that what we are observing is actually the distribution of dust around the core of the nucleus of NGC 1068. Thus, in order to reproduce this shape we used the code DUSTY [5], which describes the complete radiative transfer in a

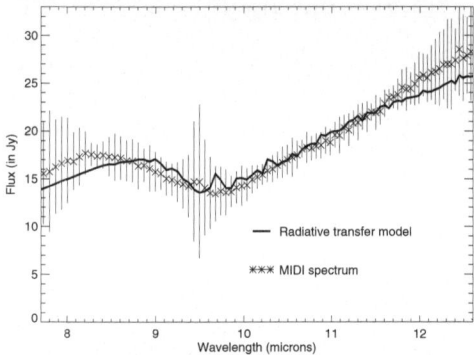

Fig. 3. Fit of the MIDI spectrum with the radiative transfer model.

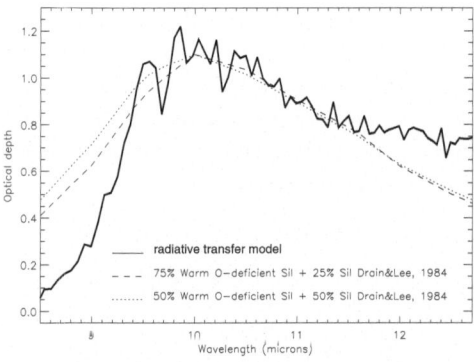

Fig. 4. Optical depth as function of the wavelength derived from the radiative transfer model (thick line); preliminary fits using the DUSTY code are superposed (dashed and dotted lines).

spherical layer made of dust. Preliminary fits obtained with different simple compositions of amorphous silicates from Draine & Lee [2] and Ossenkopf [9] are presented on Fig. 4. Discrepancies at the edges of the band are probably due to the fact that, contrary to DUSTY, we do not consider the scattering by dust grains.

4 Discussion

These results are in good agreement with those derived from the approach made by Jaffe et al. [6], except for temperatures. Nevertheless, this is due to the differences in the assumptions we respectively took (concerning the geometry and especially the optical depth).

However, according to the sublimation radius of dust, which are ~ 5 and 9 mas for graphite grains and silicates respectively (Bavairnis et al. [1]), silicates are actually able to enter in the composition of the dusty layer.

Next, following the approach of Krolik [7] and assuming thermal emissions by spherical black bodies to derive the typical angular size of each component inside an AGN, we find temperatures of the order of 260 and 160 K for typical sizes of 35 and 83 mas. Moreover, according to measurements of the spectrum carried out at the Subaru telescopes on an aperture of 290×180 mas and with a 0.1" resolution, Tomono et al. [10] derived temperature of ~ 234 K and an optical depth of ~ 0.92 at 9.7 μm. All these values are completely consistent with our results.

Another point of interest is the location of H_2O masers with regard to elements around the core of the AGN. Indeed, according to masers emission models (Maloney et al. [8]) and observations in the nucleus of NGC 1068 (Greenhill et al. [3], [4]), H_2O masers seem to trace the inner edge of the distribution of dust. This is actually consistent with our modeling. More precisely, masers seem to be confined in an area located between the sublimation radius of graphite grains and the inner edge of the dusty layer we have modeled.

As a conclusion, this simple radiative transfer model accounts well for the fits of the MIDI visibilities and spectrum. The derived simple description of the nucleus of NGC 1068 seen in the N-band is consistent with other observations. The main result resides in the evolution of the optical depth of the layer accross the full N-band which allows to conclude that MIDI actually observed the dust surrounding the core of NGC 1068. Besides, further observations of NGC 1068 with MIDI and AMBER will be mandatory to get the elongation and a better physical description of the dust. VISIR and NACO observations will provide low frequency points, to make the link with larger scales dust distribution.

References

1. Bavairnis, R. 1987, ApJ, 320, 537
2. Draine, B.T. & Lee, H.M. 1984, ApJ, 285, 89
3. Greenhill, L.J., Gwinn, C.R., Antonucci, et al. 1996, ApJ, 472, L21
4. Greenhill, L.J., Gwinn, C.R. 1997, Ap&SS, 248, 261G
5. Ivezìc, Z., Nenkova, M. & Elitzur, M., 1999, User Manual for DUSTY, University of Kentucky Internal Report
6. Jaffe, W., Meisenheimer, K., Röttgering, H.J.A., et al. 2004, Nature, 429, 47
7. Krolik, J., H., 1999, Active Galactic Nuclei, Princeton Series in Astrophysics
8. Maloney, P.R. 2002, PASA, 19, 401
9. Ossenkopf, V., Henning, Th. & Mathis, J.S. 1992, A&A, 261, 567
10. Tomono, D., Doi, Y., Usuda, T. et al. 2004, A&A, 417, L1
11. Weigelt, G., Wittkowski, M., Balega, Y.Y., et al. 2004, A&A, 425, 77
12. Wittkowski, M., Kervella, P., Arsenault, R., et al. 2004, A&A, 418, L39

IRS 3 - The Brightest Compact MIR Source in the Galactic Center

A. Eckart[1], J.-U. Pott[1,2], A. Glindemann[2], T. Viehmann[1], R. Schödel[1], C. Straubmeier[1], C. Leinert[3], M. Feldt[3], R. Genzel[4], and M. Robberto[5]

[1] I. Physikalisches Institut; University of Cologne Zülpicher Straße 77; 50937 Köln (Germany) eckart@ph1.uni-koeln.de
[2] ESO (Germany)
[3] MPIA (Germany)
[4] MPE (Germany)
[5] STScI (USA)

Summary. In a recent VLTI experiment we partially resolved the dust enshrouded star IRS 3 in the central light year of our galaxy. This observation is the first step in investigating both IRS 3 in particular and the stellar population of the Galactic Center in general with the VLTI at highest angular resolution. Here some of the scientific issues that can be addressed by a complete MIDI dataset on IRS 3 in the mid infrared are outlined. A bright compact source like IRS 3 is also technically essential for future VLTI phase-reference experiments at 10 μm in order to investigate other nearby sources, e.g. the Sgr A* black hole. For this purpose it is important to know the strength and compactness of IRS 3 on the longest baselines.

Due to its proximity ($\sim$ 8 kpc), the center of our galaxy (GC) offers a unique variety of experiments and observations, which grows together with the technical progress and the commissioning of new instruments. At the present point, the angular resolution of the VLTI already allows us to resolve the dust envelopes of some stars in the GC. The nature of star formation and evolution close to a super-massive black hole is of broad astrophysical interest. A structural analysis on the scale of tens to hundreds of AUs opens the way for a detailed study of stellar properties, as well as of the interaction between a star and the GC environment.

The unusually large number of massive, young stars in the stellar cluster at the GC (e.g. Genzel et al., 2003; Eckart et al., 2004; Moultaka et al., 2004) are indicative of an active star formation history despite the tidal forces exerted by the gravitational potential of the central SBH. The presence of numerous stars in short-lived phases of their development, such as dust-producing Wolf-Rayet (WR) stars, indicates that the most recent star formation episode took place not more than a few million years ago. IRS 3 with its 1-2 arcsec extended mid infrared (MIR) excess is one of the most prominent of these sources (Viehmann et al., 2005; Moultaka et al., 2004). Observations of IRS 3 are an excellent starting point for infrared, interferometric measurements in the central stellar cluster. Pott et al. (2004) review the technical aspects of VLTI-GC observations, which are ideally suited to study the capabilities of the new instruments close to the system limits under normal observing conditions.

Rieke, Telesco,& Harper (1978) and Becklin et al. (1978) argued that IRS 3 is a dust-enshrouded supergiant with a compact circumstellar dust shell. Gezari et al. (1985) found that IRS 3 is the most compact and (together with IRS 7) hottest MIR source (T~400 K) in the central cluster, with total integrated flux densities of about 30 Jy at 8 μm to 12 μm.

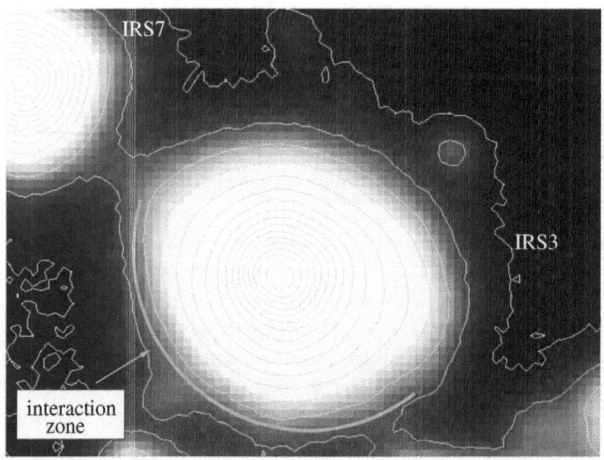

Fig. 1. ISAAC M-band image of IRS7 and IRS3. The interaction zone with the wind from the IRS16 cluster to the southeast is also indicated. The image is about 5"×3.7" in size.

Given its high luminosity of ~ 5 · $10^4 L_\odot$ IRS 3 may in fact be a star at the very tip of the Asymptotic Giant Branch (AGB). These most luminous dust-enshrouded AGB stars will stay at a high luminosity during their entire mass loss phase. Their mass-loss rates, derived from observations, span a range from about 10^{-7} to $10^{-3} M_\odot$ yr^{-1} (van Loon et al. 1999) with wind velocities of the order of 10-20 km/s (e.g.Bergeat & Chevallier 2005) . In general the more luminous and cooler stars are found to reach higher mass-loss rates. This is in agreement with model calculations (Schröder, Wachter & Winters 2003). For a synthetic sample of more than 5000 brighter tip-AGB stars a collective mass-loss rate of $5.0\times10^{-4} M_\odot$ yr^{-1} was found. Of these, 20 are carbon-rich super-giants with a large IR excess and a mass-loss rate well in excess of $10^{-6} M_\odot$ yr^{-1}, including 10 dust-enshrouded, extreme tip-AGB stars seen in their short-lived (~30 000 yrs) super-wind phase with a mass loss of $>10^{-5} M_\odot$ yr^{-1}. They produce about 50% of the collective mass-loss of the whole sample.

However, the nature of IRS 3 is not identified unambiguously yet. A recent identification of a carbon-rich WR star of type WC5/6 as a near infrared counterpart of IRS 3, based on the detection of a 2.11 μm He I/C III line (Horrobin et al. 2004), is probably applicable to a K∼15 faint star ∼120 mas east of the bright source. However, given the fact that most other dust enshrouded sources in the central stellar cluster have been associated with hot and luminous young stars, an identification of IRS 3 with a massive WR star in its dust forming phase cannot be fully excluded either. Extensive mass loss associated with bright continuum emission takes place in the WC stage. Products of helium burning are dredged up to the surface, enhancing the carbon and further depleting the hydrogen abundance.

As shown by Viehmann et al. (2005), the dust shell of IRS 3 is interacting with the GC ISM. They find the photo-center of IRS 3 in the ISAAC M-band image shifted by ∼160 mas to the NW with respect to the L-band image. About 1″ to the southeast of IRS 3 high-pass filtered L- and M-band NAOS/CONICA images show a sharp interaction zone of the outer part of the dust shell with the wind arising from the IRS 16 cluster of hot, massive Helium stars.

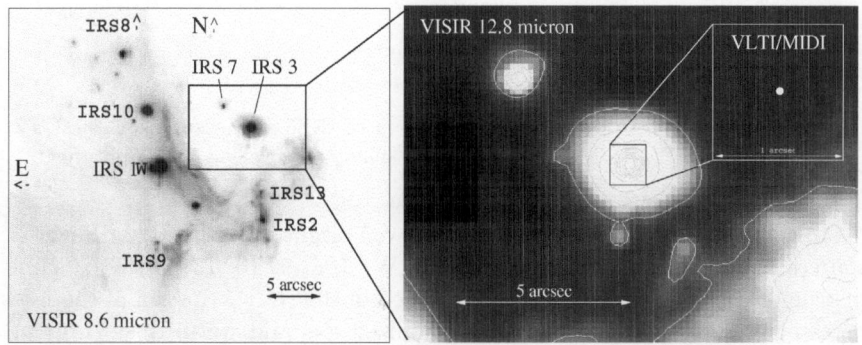

Fig. 2. VISIR MIR images with the observed targets indicated and an inset demonstrating the scale on which the current UT2-UT3 VLTI/MIDI data detected a compact source with a visibility of about 25%. IRS 8 is located north of the shown image.

We designed a VLTI experiment with MIDI (N-band, 8-12 μm) to investigate the dust shell of IRS 3 (see also Pott et al. 2005 ESO Messenger). The lower spectroscopic resolution used (R=30) offers dispersed visibility data over the entire N-band, as well as a spectrum of the uncorrelated flux density. The first VLTI detection of a star in the GC was achieved in June 2004: We partially resolved IRS 3 with VLTI using MIDI on the 47 m UT2-UT3 baseline (see Figs. 2 and 3).

It was found that ∼25% of the flux density of IRS 3 are concentrated in a compact (i.e. unresolved) component with a size of ≤40 mas (i.e. ≤300 AU). This agrees with the interpretation that IRS 3 is a luminous compact object in an intensive dust forming phase. In general, the visibility amplitude was found to increase with wavelength by ~ 0.05. Although the uncertainty of a single visibility value (5-10%) seems to be too large to unambiguously identify such a trend, the error on the slope (i.e. wavelength dependent variation) of a visibility dataset over the entire N-band is of the order of 1% only. This trend indicates that the compact portion of the IRS 3 dust shell is extended and only partially resolved on the UT2-UT3 baseline. We also find indications for a narrower width of the 9.3 μm silicate line towards the center, indicating the presence of fresh unprocessed small grains closer to the central star in IRS 3 (van Boekel et al., 2003).

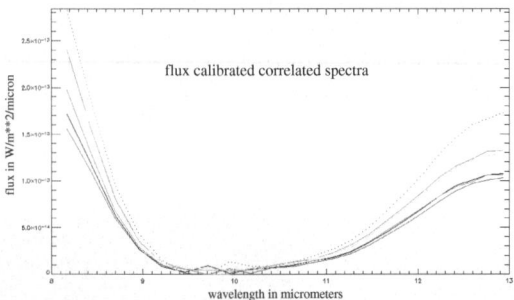

Fig. 3. Flux calibrated and correlated spectra of IRS3 as measured on the UT2-UT3 baseline with VLTI/MIDI. The 9.3 μm silicate absorption line is prominent.

In addition, the remaining six of the seven brightest (N-band) MIR excess sources in the GC were observed (IRS 1W, 2, 8, 9, 10, 13) with the same instrument setup. Most of them are located in the Northern Arm of the ISM or associated with the mini-spiral of ionized gas and warm dust (Fig. 2). They appear to be hot stars with strong, fast winds that create bow shocks as they plough through the gas and dust of the mini-spiral (Tanner et al., 2002). The MIR emission associated with theses sources arises most probably from these bow shocks that can be resolved at 2 μm (Tanner et al. 2005). The most important observational aspects for some of these sources are covered in Pott et al. in this edition. All of these sources have not yet been detected interferometrically and therefore, IRS 3 can be regarded not only as the hottest but as well the most compact bright source of MIR emission within the central stellar cluster.

We will conduct further VLTI/MIDI observations of IRS 3 in ESO Period 75. Main goal will be to increase the uv-plane coverage (Fig. 4). We will use the UT3-UT4 (62 m) and UT1-UT4 (130 m) baselines which are complementary to UT2-UT3 in terms of length and orientation. The final dataset will cover an angular resolution of about 13 mas at the longest baseline up to 100 mas. These observations will be ideally suited to provide information about the radial structure and symmetry of the correlated flux, about the inner edge of the dust shell, as well as about a possible binary character of

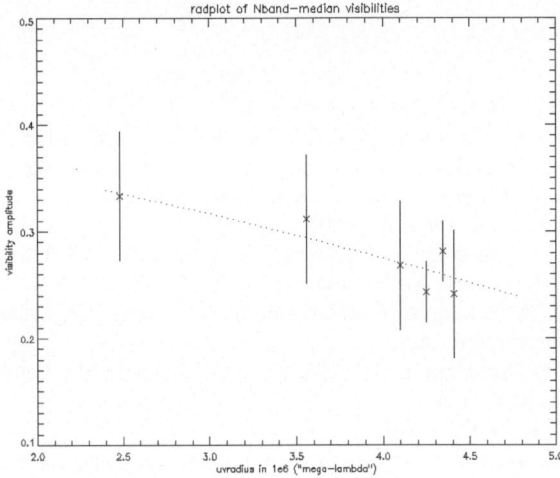

Fig. 4. Median 8-12 μm visibility of IRS 3 as a function of projected baseline length. Currently the uncertainty of the visibilities is above all affected by the instrument calibration. Therefore the errors are given by the standard deviation of the instrument calibration over the entire night.

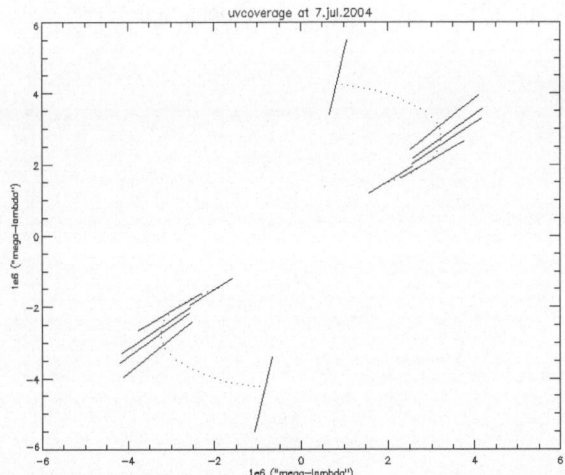

Fig. 5. The uv-coverage of the observations in P73 is shown. The dotted line indicates the change of projected baseline length due to earth rotation. Whereas the earth rotation curve is calculated at a central wavelength of 10.34 μm the solid lines are showing the uv-coverage of each dispersed, calibrated visibility dataset.

IRS 3. Collisions of winds in binary systems may support dust formation through density increase and rapid cooling of the material.

References

1. Bergeat, J.; Chevallier, L., 2005, A&A 429, 235;
2. Becklin, E.E., et al. 1978, ApJ 219, 121;
3. van Boekel, R., Waters, L. B. F. M., Dominik, C., et al. 2003, A&A, 400, L21;
4. Eckart, A., Moultaka, J., Viehmann, T., et al. 2004 ApJ 602, 760;
5. Genzel, R., Schödel, R., Ott, T., et al. 2003, ApJ 594, 812;
6. Gezari, D.Y. et al. 1985, ApJ 299, 1007;
7. Horrobin, M., Eisenhauer, F., Tecza, M., et al. 2004, AN 325, 88;
8. Moultaka, J., Eckart, A., Viehmann, T., et al. 2004, A&A 425, 529;
9. Pott, J.-U., A. Eckart, A. Glindemann, T. Viehmann, Ch. Leinert, et al. 2005, ESO Messenger, 119, 43;
10. Pott, J.-U., Glindemann, A., Eckart, A., Schoeller, M., Leinert, Ch., et al. 2004, SPIE 5491, 126P
11. Pott, J.-U. et al. this edition.
12. Rieke, G.H. Telesco, C.M., Harper, D.A., ApJ 220, 556, 1978;
13. Schröder, K.-P.; Wachter, A.; Winters, J. M., 2003, A&A 398, 229;
14. Tanner, A. et al. 2002, ApJ 575, 860;
15. Tanner, A. et al. 2005, ApJ, 624, 742;
16. Schödel, R.; Ott, T.; Genzel, R.; et al. 2002, Nature 419, 694;
17. van Loon, J.Th.; Groenewegen, M. A. T.; de Koter, A.; et al. 1999, A&A 351, 559;
18. Viehmann, T., Eckart, A. Schödel, et al. 2005, 433, 117;

Scientific Prospects for VLTI in the Galactic Centre: Getting to the Schwarzschild Radius

T. Paumard[1], G. Perrin[2], A. Eckart[3], R. Genzel[1,4], P. Léna[2], R. Schödel[3], F. Eisenhauer[1], T. Müller[5], and S. Gillessen[1]

[1] Max-Planck Institut für extraterrestrische Physik (MPE), Garching, DEU
[2] Observatoire de Paris – site de Meudon, FRA
[3] 1.Physikalisches Institut der Universität Köln, DEU
[4] Department of Physics, University of California, Berkeley, USA
[5] Institute of Astronomy and Astrophysics, Dept. of Theoretical Astrophysics, University of Tübingen, DEU

Summary. The centre of our Galaxy is by far the closest galactic nucleus to the Earth. Not surprisingly, Sgr A*, the super-massive black hole candidate lying at its gravitational centre, is the black hole (BH) candidate which offers the best spatial resolution in terms of Schwarzschild radius R_S, thanks to the good compromise between its mass ($\simeq 3.6 \times 10^6$ $M_\odot$) and distance ($\simeq 8$ kpc). For a 3.6×10^6 $M_\odot$ BH, $R_S = 14$ $R_\odot$ ($\simeq 8$ µas at 8 kpc). This is still two orders of magnitudes smaller than the spatial resolution offered by the VLTI, but it is the right scale for studies involving only astrometry with PRIMA. We will show how this property can be used to tremendously improve the knowledge of the central mass distribution, constrain the properties of the BH, and even directly probe the space-time around this object.

1 Introduction

At the very centre of the Galaxy lies the super-massive BH candidate Sgr A* [3, 6], surrounded by a parsec-scale star cluster consisting of an old spheroidal stellar population intermixed with another population of early type stars of unknown origin, possibly confined within two disks [5, 7, 12], and containing several dozen Wolf-Rayet stars as well as a group (GCIRS16) of half a dozen luminous blue supergiants in a transitional phase; and by a complex of ionised gas clouds (Sgr A West or the "Minispiral"), itself contained within the Central Cavity of a Cirumnuclear Disk (CND). The central arc-second contains the core (or "cusp") of the star cluster, consisting mostly of OB main-sequence stars, the S-stars [2]. The proper motion and radial velocity measurements of these stars fit very nicely on Keplerian orbits, allowing for a precise determination of the distance to the Galactic Centre (GC), $R_0 = 7.62 \pm 0.32$ kpc, and of the mass included within the periastron of these stars (16 light hours or 1400 R_S for S2, the best determined star), $3.61 \pm 0.32 \times 10^6$ $M_\odot$. At the current point, there is no explanation for such a concentration of mass less exotic than a super-massive black hole. However, this mass is not yet constrained observationally to be contained within its Schwarzschild radius R_S. Furthermore, [9] have shown that the data currently

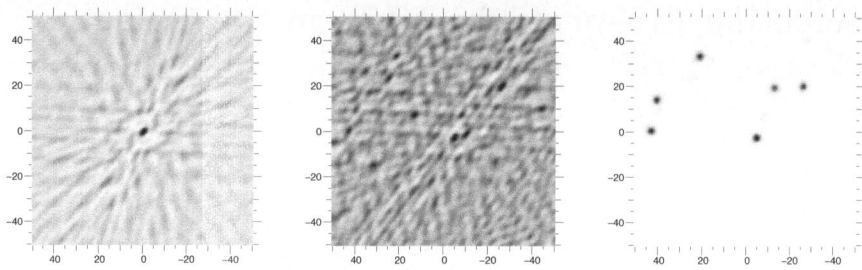

Fig. 1. One-night simulated observation of a model cluster containing six stars with a dynamical range of 1 magnitude. Left: synthesised beam. Middle: raw synthesised image. Right: reconvolved CLEANed image (axes in mas)

available are still consistent with the presence of an extended mass component accounting for up to $\simeq 10\%$ of the total mass, in addition to a central point mass. The core of the star cluster is likely to contain stellar residuals, accumulated during the lifetime of the Galaxy due to mass-segregation, so that the mass-to-luminosity ratio of the cusp may be significantly higher than 1. The central super-massive dark object coincides very precisely with Sgr A*, a point-like source in the radio range, also detected in X-rays and in the near infrared (NIR). The X-ray and NIR source is variable, showing so called flares a few times a day, outbursts of energy that last about one hour.

In the following, we will use simulations of VLTI observations to show how this facility can be used to determine the orbits of cusp-stars even closer to the BH than the S-stars, and possibly of the flares themselves. These experiments will provide very interesting information on the astrophysical environment of Sgr A*, such as the mass-to-light ratio of the cusp, but their outcome are expected to be much deeper: they will allow for the first time testing – and hopefully validating – the General Relativity (GR) in a completely unexplored regime, which is of primary importance for cosmology and fundamental physics. However, these observations are quite demanding and will require a specifically optimised instrument. A design of a general purpose faint object imager that would suite our observational requirements is presented in the instrumental part of this workshop (Eisenhauer et al.).

2 Probing stellar dynamics in the relativistic regime

Measuring the proper motion of stars even closer to Sgr A* than the famous S-stars would allow to prove in an indisputable manner that most of the mass sensed by the S-stars (at a few thousand R_S) is indeed contained within the periastron of these closer stars, or a few hundred R_S. Furthermore, comparing the mass estimates at these two radii will give a measurement of the mass of the cusp, and hence of its mass-to-light ratio. The orbits of the stars we

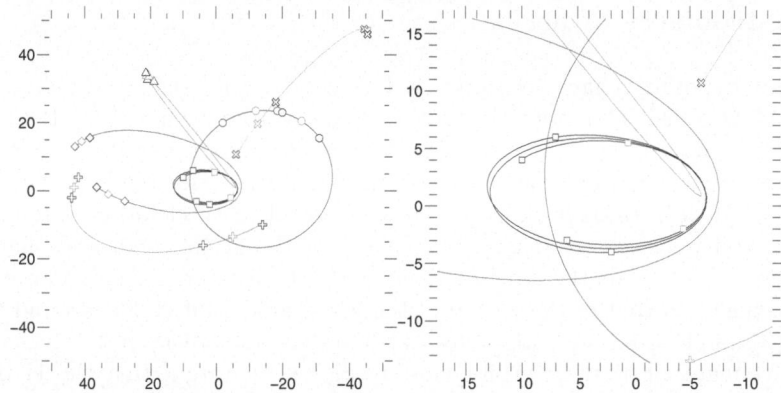

Fig. 2. Orbits of six stars in the relativistic potential well of Sgr A*, over 15 months (axes in mas). The (initial) orbital parameters are those of S-stars, 10 times downscaled. The symbols represent the crude astrometry retrieved from the simulated observations (brightest pixel on a mas grid): the precession is already significant

will find in the diffraction limit of the VLT around Sgr A* will quickly show periastron shifts due to two effects: a retrograde shift due to the extended mass component, i.e. the mass of the cusp; and a relativistic prograde shift (Fig. 2), well known for the case of Mercury, which historically provided one of the first compelling tests of the GR. [13] have discussed these effects in the case of the GC. For a semi-axis of the order of 1 mpc (25 mas), the relativistic effect is easily detected after only one year.

According to the star counts and population study of [5], a few ($\simeq 4-10$) stars brighter than $m_K = 19.5$ are expected within the central 100 mas. In order to assess whether VLTI observations of such stars would allow determining their orbits, we have simulated these observations, using a model cluster of six stars with a dynamical range of 1 magnitude, located at the GC and evolving in the relativistic potential well of Sgr A*, and real VLTI baselines for the 4 UTs on selected dates. We have assumed that the instrument will be able to provide reasonably accurate measurements (normal errors of 1% on the squared visibilities and 2° on the phases have been added), with 5 spectral elements in the K band, each data set requiring in total 3 hours (including acquisition and calibration). We have found that the number of free parameters (18) in the problem (6 stars) prevented model-fitting techniques from being useful. We have therefore been led to use image synthesis techniques (Fig. 1). Although only the two brightest stars are clearly seen on the synthesised image, the standard deconvolution method "CLEAN" allows recovering the 6 stars. Very crude astrometry (brightest pixel) on the deconvolved images for May, June and July 2005 and 2006 is sufficient to observe the relativistic periastron shift for some of the stars (Fig. 2).

3 Astrometry of flares

Several hypotheses have been proposed concerning the origin of the flares of
Sgr A*: star-disk interactions [11] rather far away from the BH ($10^2 - 10^4 \, R_S$);
sudden heating of hot electrons (e.g. by magnetic reconnection) in a permanent jet [8]; and synchrotron emission from electrons accelerated in the central
part ($\lesssim 10 \, R_S$), through processes such as turbulent acceleration, reconnec-
tions, and weak shocks (e.g. [14]). On top of the bright, hour-scale flares,
smaller variations on the order of 17 minutes are observed [4]. This timescale
corresponds to that of the period of the last stable orbit (LSO) around the
BH (27 min for a $3.6 \times 10^6 \, M_\odot$ Schwarzschild BH), and are natural if the flares
involve material on these orbits, close to Sgr A*: [4] argue that the 17 min
period may correspond to that of the LSO of a Kerr BH of spin parameter
$a = 0.52$ given its well constrained mass. However, [11] mention that their
star-disk interaction model can qualitatively interpret these short-timescale
variations as the emission of the transiently heated stellar photospheres during the disk passages.

The same instrument as that proposed above for measuring proper motion
of cusp stars will also be able to measure the astrometry of the flares of
Sgr A*, as an offset to a known star about 2″ away, with an accuracy of
order 10 μas or 1 R_S in only less than 1 minute integration time. This will
lead to unambiguous determination of the location of the flares, and therefore
of their nature. The material responsible for the flares is likely to move at
least at a few percent of light speed (15 μas min^{-1}), and therefore to yield
significant proper motion within the duration of a flare. Of the three above-
mentioned flare models, the third is that with the most scientific potential, as
the material involved would be orbiting the BH at only a few R_S. In principle,
in that case, it would be possible to trace the orbit of the flaring material,
and therefore to probe the space-time in the very strong gravitational field
of a super-massive black hole.

We have used relativistic ray tracing [10] to investigate the apparent trajectory of a bright patch of material orbiting a Schwarzschild BH on its LSO,
which is affected by several relativistic effects: strong lensing and multiple
images, Doppler effect, gravitational redshift and beaming effect (Fig. 3). We
have then used a Monte-Carlo approach to study the output of observations
of this orbit at a 10 μas ($\simeq 1 \, R_S$) accuracy. We find that each single flare
allows deriving the orbital parameters, hence comparing the plane of the accretion disk to that of the Galaxy. Furthermore, if each flare really traces the
same orbit, it is possible to co-add individual observations in order to reduce
the error bars, and to find the signature of the strong relativistic effects on the
integrated curve. These results are in good agreement with those obtained in
parallel by [1].

Finally, we find that a photon-efficient, phase-referenced, multiway combiner will enable the VLTI not only to address the physical conditions pre-

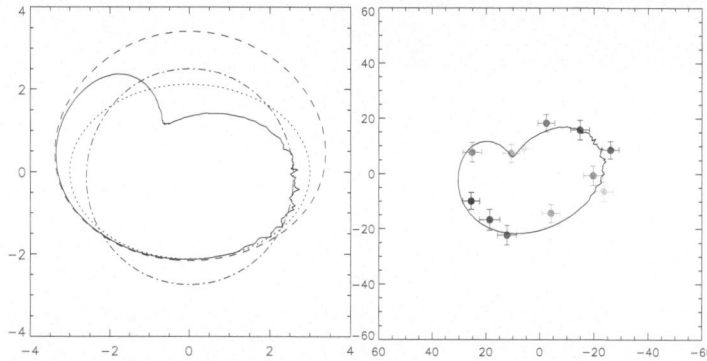

Fig. 3. Appearance of the LSO of a Schwarzschild BH. Left: (axes in R_S) real, unlensed orbit at 45° inclination (*dots*); primary (*dashes*) and secondary (*dash-dots*) images of the orbit; apparent trajectory of the centroid, including all RG effects (*solid line*) for a red spectrum $\nu L_\nu \propto \nu^{-3}$,[2]. Right: co-adding ten individual ovservations of flares makes the error bars shrink and allows finding the signature of the GR effects (axes in µas)

vailing in the Galactic Centre at a few R_S of Sgr A*, but also to perform extremely important tests of the GR in a completely unexplored regime.

References

1. Broderick, A. & Loeb, A. 2005, MNRAS, 363, 353
2. Eisenhauer, F., Genzel, R., Alexander, T., et al. 2005, ApJ, 628, 246
3. Genzel, R., Eckart, A., Ott, T., & Eisenhauer, F. 1997, MNRAS, 291, 219
4. Genzel, R., Schödel, R., Ott, T., et al. 2003a, Nature, 425, 934
5. Genzel, R., Schödel, R., Ott, T., et al. 2003b, ApJ, 594, 812
6. Ghez, A. M., Klein, B. L., Morris, M., & Becklin, E. E. 1998, ApJ, 509, 678
7. Horrobin, M., Eisenhauer, F., Tecza, M., et al. 2004, Astron. Nachr., 325, 88
8. Markoff, S., Falcke, H., Yuan, F., & Biermann, P. L. 2001, A&A, 379, L13
9. Mouawad, N., Eckart, A., Pfalzner, S., et al. 2005, Astron. Nachr., 326, 83
10. Müller, T. 2005, PhD thesis, University of Tübingen, Germany
11. Nayakshin, S., Cuadra, J., & Sunyaev, R. 2004, A&A, 413, 173
12. Paumard, T., Genzel, R., Martins, F., et al. 2005, ApJ, 643, 1011
13. Rubilar, G. F. & Eckart, A. 2001, A&A, 374, 95
14. Yuan, F., Quataert, E., & Narayan, R. 2004, ApJ, 606, 894

Beyond the VLTI

Andreas Quirrenbach

Leiden Observatory, P.O. Box 9513, NL-2300 RA Leiden, The Netherlands

Summary. A large next-generation interferometer could provide images with $\sim 10\,\mu$as resolution, which would revolutionize many fields of astrophysics. A straw-man concept for such a facility is presented, and the most important technological challenges for its realization are discussed.

1 Introdution: The VLTI and Beyond

The present volume is a tribute to the VLTI, which has already made important contributions to stellar and extragalactic astrophysics, and which is expected to remain the pre-eminent interferometric facility for the foreseeable future. Nevertheless, it is not too early to start thinking about potential successors to the VLTI. If a next-generation facility is to become reality in the middle of the next decade, the technology development has to start immediately, and possible organizational and financial arrangements will have to be addressed soon. Planning for the long-term future could also have implications for the VLTI itself, because the scientific, technical, and operational foundations of a next-generation array would have to be laid with the VLTI.

The VLTI has created many opportunities to address important astrophysical problems at hitherto inaccessible angular resolution and sensitivity; the impressive early results from the first generation of its instruments have just started to show the enormous potential that can be realized by the full exploitation of the present capabilities, by future more advanced instruments, and by completing the infrastructure. Nevertheless, a few limitations of the VLTI are inherent in the Paranal site and in the facility constructed on the mountain. Most importantly, it will be (almost) impossible to extend the baselines beyond the current 200 m limit, and the number of telescopes that can be combined simultaneously is limited to eight. In contrast, many high-priority scientific topics require baselines in the range 1 km ... 10 km (e.g., Surdej et al. 2004), and arrays with ~ 20 telescopes are needed for true synthesis imaging. In addition, high sensitivity with good sky coverage is an important attribute of any general-purpose instrument. No existing or currently planned facility comes close to providing these capabilities, but one can reasonably assume that a true successor to the VLTI should be designed to these specifications.

Table 1. Summary of strawman ELSA parameters and characteristics.

Parameter	Value	Comment
Number of telescopes	27	Needed for snapshot imaging
Array geometry	Y-shape	One of many possibilities
Telescope phasing	Autonomous	Adaptive optics / laser guide stars
Array co-phasing	External	Dual-star operation
Sky coverage	$\gtrsim 10\%$	At R band, near Galactic pole
Telescope diameter	10 m	Needed to get good sky coverage
Efficiency	25%	To limit telescope size
Wavelength range	$0.5 \ldots 20\,\mu m$	Could be reduced to $0.5 \ldots 2.2\,\mu m$
Beam transport	Fiber bundles	To limit facility cost
Cost	$\lesssim 400\,M€$	Design-to-cost target figure

2 ELSA: A Strawman Concept for an Extremely Large Synthesis Array

A strawman concept for an Extremely Large Synthesis Array (ELSA) going far beyond the capabilities of the VLTI could consist of 27 ten-meter telescopes and baselines of up to 10 km (Quirrenbach 2004, see also Tab. 1). It appears that a facility with ELSA's capabilities could be built today with existing technologies, but the cost would probably be prohibitively high. A technology roadmap for ELSA must therefore provide solutions that are not only *technically feasible*, but also *affordable*. In this context it will be interesting to explore to which extent cost-reduction approaches that are being investigated for the design and construction of large monolithic telescopes – such as ESO's OWL concept – can also be applied to ELSA. It should be emphasized that the parameters and features of ELSA as presented here are not based on any detailed trade-off or design study, but rather meant as an initial step to stimulate further discussion.

3 Science with a Next-Generation Array

With baselines up to $B = 10$ km and operating down to $\lambda = 0.5\,\mu m$, ELSA would deliver images with $10\,\mu as$ resolution, two orders of magnitude better than any other telescope contemplated at the moment. Combined with a sensitivity (for compact objects) that equals or surpasses present-day large monolithic telescopes, this spectacular angular resolution enables a wealth of completely new observing programs in many different areas of astrophysics:

– Imaging of Jupiter-size objects at a distance of ~ 10 pc with 16 resolution elements across the disk;
– High-quality images of stellar surfaces (90 resolution elements across the disk of Solar-type stars at 10 pc);

- Binary stars (Roche lobe overflow, mass transfer, accretion);
- Pre-main-sequence disks (temperature and density laws, vertical structure, flaring, magnetic fields, jets and outflows, gaps created by planets);
- Three-dimensional motions of stars in globular clusters;
- General-relativistic precession of orbits of stars near the Galactic Center;
- Baade-Wesselink distances of pulsating stars, novae, and supernovae;
- Extragalactic stellar populations in crowded regions;
- Detailed images of broad-line regions in active galaxies and geometric distances of quasars;
- Shapes and Doppler factors of the afterglows of gamma-ray bursts.

Precise narrow-angle astrometry would be another interesting application of ELSA. The atmospheric limit of the astrometric error for very long baselines scales with $\theta L_0^{1/3}/B$, where θ is the angle between the target and an astrometric reference star on the sky, L_0 the outer scale of atmospheric turbulence, and B the baseline length (Shao & Colavita 1992). L_0 is generally believed to be of order 100 m (e.g., Quirrenbach 2002), which means that ELSA could achieve an accuracy considerably better than 1 μas, which would allow the detection of terrestrial planets through the reflex motion of their parent stars, and to measure their masses dynamically.

4 Key Technologies for ELSA

To achieve good sky coverage, it must be possible to co-phase ELSA on rather faint stars ($R \approx 20$), so that array elements with ~ 10 m diameter are needed. Allocating half of the project cost to the telescopes would mean that each telescope should not cost more than about 7.5 M€, including enclosures and the adaptive optics system. It will therefore be necessary to capitalize on advances such as those needed by OWL and ALMA: cheap mass production of primary mirror segments, standardized elements for the mechanical structure, and minimization of non-recurrent design and engineering effort through replication of identical elements. The interferometer elements will need only a small field-of-view, which should be achievable with a spherical primary mirror; this minimizes the cost of the mirror segments. 27-fold reproduction would reduce the price of the AO systems. Applying the "standard" $D^{2.7}$ scaling law of telescope cost with diameter (e.g., Stepp et al. 2002) to the OWL concept (100 m for 1,000 M€) suggests a price of only 2 M€ for a 10 m telescope built with the same design and construction principles. A target of 7.5 M€ for the ELSA elements thus appears to be not totally unreasonable.

Because of diffraction effects and the desired field size, the minimum diameter of the optical elements in a classical "bulk optics" system increases with the length of the beam train. For a propagation length of 10 km, field of 2″, and operation in the near-infrared, a beam diameter of 50 cm is needed.

The optics and vacuum system are thus clearly a cost driver for large interferometric arrays. Another problem is the poor transmission due to the many
optical elements between the telescope and the beam combiner. These difficulties could be solved by relaying the light with single-mode fibers from the
telescopes to a central facility; this technique is being tested by the 'Ohana
project on Mauna Kea (Perrin et al. 2002). A fiber delay compensator could
be constructed from fiber segments with lengths of 1 m, 2 m, 4 m, . . . ; selecting
an appropriate chain of these segments can provide any desired delay in steps
of 1 m. The remaining delay could be taken out in a fiber which is stretched
mechanically to the desired length. The largest challenge for this concept is
the development of fibers with extremely low dispersion, or of appropriate
dispersion compensation schemes. In addition, nearly lossless switches would
be needed for selecting the sets of discrete fibers. If these requirements should
turn out to be too demanding, one could consider as a fall-back a hybrid solution with constant balanced fiber lengths in the interferometer arms, and
bulk-optics delay compensation (Glindemann et al. 2002).

The beams from the individual array elements can be combined in many
different ways; each beam combination technique has its own advantages and
drawbacks (see e.g. Quirrenbach 2001). In its largest configuration, ELSA will
be a very dilute array ($B/D \approx 1,000$), similar to long-baseline radio interferometers. For such arrays, image-plane beam combination is very inefficient;
one should thus combine the light in the pupil plane. The field-of-view is in
this case limited to $\sim R$ spatial resolution elements across, where $R \equiv \lambda/\Delta\lambda$
denotes the spectral resolution of each wavelength channel. In the more com-
pact configurations of ELSA, it would be desirable to image a larger field.
This requires Fizeau beam combination in the image plane, i.e., the exit pupil
of the interferometer must be a scaled replica of the input pupil.

5 Site Selection

Finding a good site for an Extremely Large Synthesis Array is a difficult
task. It is obvious that one needs a reasonably flat plateau of considerable
size, and such plateaus tend to have poorer seeing than the best mountain
tops. One candidate site, for which data from a systematic site evaluation
campaign are available, is Llano de Chajnantor, the location of the ALMA
millimeter array at an altitude of 5000 m in the Chilean Andes. Typical values
in the range $1'' \ldots 1''5$ have been reported for the seeing at the Chajnantor
plateau itself, with substantially better seeing at a location 100 m above the
plateau (Giovanelli et al. 2001). This indicates that a rather large fraction of
the turbulence occurs in the boundary layer just above the plateau, probably
due to katabatic winds off the surrounding mountain slopes. The boundary-
layer seeing is easier to correct with adaptive optics than high-altitude seeing,
and it does not significantly contribute to anisoplanatism. This suggests that
Llano de Chajnantor should offer acceptable seeing conditions for ELSA.

It has also been suggested that Antarctica offers attractive sites for interferometry, in particular Dome C (Marks et al. 1998, Lloyd et al. 2002, Lawrence et al. 2004). During the winter, very little high-altitude turbulence is present above the Antarctic plateau; the isoplanatic angle is thus much larger than at mid-latitude sites. Since the diameter of the telescopes in ELSA is largely driven by the requirement to co-phase the array, there is a substantial advantage in going to a site with superb high-altitude seeing. One could thus trade off the cost of constructing and operating ELSA in the harsh Antarctic environment against the substantial savings in telescope size. It is very likely that the Antarctic option would only be attractive for a fiber-coupled interferometer, because otherwise the infrastructure cost would be dominated by the beam transport and delay line tubes.

References

1. Giovanelli, R., Darling, J., Sarazin, M., Yu, J., Harvey, P., et al. (2001). *The optical/infrared astronomical quality of high Atacama sites. I. Preliminary results of optical seeing.* PASP **113**, 789-802

2. Glindemann, A., Bauvir, B., van Boekel, R., et al. (2002). *Growing up – the completion of the VLTI.* In *Scientific drivers for ESO future VLT/VLTI instrumentation.* Eds. Bergeron, J., & Monnet, G., p. 279-288

3. Lawrence, J.S., Ashley, M.C.B., Tokovinin, A., & Travouillon, T. (2004). *Exceptional astronomical seeing conditions above Dome C in Antarctica.* Nature **431**, 278-281

4. Lloyd, J.P., Oppenheimer, B.R., & Graham, J.R. (2002). *The potential of differential astrometric interferometry from the high Antarctic plateau.* PASA **19**, 318-322

5. Marks, R.D., Vernin, J., Azouit, M., Manigault, J.F., & Clevelin, C. (1998). *Measurement of optical seeing on the high Antarctic plateau.* A&AS **134**, 161-172

6. Perrin, G., Lai, O. Woillez, J., Guerin, J., Reynaud, F., et al. (2002). *'OHANA phase II: a prototype demonstrator of fiber linked interferometry between very large telescopes.* In *Interferometry for optical astronomy II.* Ed. Traub, W.A., SPIE Vol. 4838, p. 1290-1295

7. Quirrenbach, A. (2001). *Optical interferometry.* ARAA **39**, 353-401

8. Quirrenbach, A. (2002). *Site testing and site monitoring for extremely large telescopes.* In *Astronomical site evaluation in the visible and radio range.* Eds. Vernin, J., Benkhaldoun, Z., & Muñoz-Tuñón, C., ASP Conference Series Vol. 266, p. 516-522

9. Quirrenbach, A. (2004). *Design considerations for an Extremely Large Synthesis Array.* In *New frontiers in stellar interferometry.* Ed. Traub, W.A., SPIE Vol. 5491, p. 1563-1573

10. Shao, M., & Colavita, M.M. (1992). *Potential of long-baseline interferometry for narrow-angle astrometry.* A&A **262**, 353-358

11. Stepp, L., Daggert, L., & Gillett, P. (2002). *Estimating the costs of extremely large telescopes.* In *Future giant telescopes.* Eds. Angel, J.R.P., & Gilmozzi, R., SPIE Vol. 4840, p. 309-321

12. Surdej, J., Caro, D., & Detal, A., Eds. (2004). *Science cases for next generation optical/infrared interferometric facilities. Proceedings of the 37th Liège International Astrophysical Colloquium.* Liège University

The Power of Optical and Infrared Interferometry - From Dreams to Reality

Thomas Henning

Max Planck Institute for Astronomy,
Königstuhl 17, D-69117 Heidelberg, Germany
henning@mpia.de

It is a great honour and a special pleasure to give the concluding remarks for this very exciting meeting. Optical and infrared interferometry has seen a tremendous development from the early suggestions of Fizeau in 1867 to measure stellar diameters and the construction of a separate-element stellar interferometer by A.A. Michelson and F.G. Pease to long-baseline interferometry with 8m class telescopes. Technical progress has been especially rapid during the last decade due to improvements in optical and infrared detectors, precision optics, photonics, and mechanical control systems which enables one to measure and track phase variations produced by the Earth's atmosphere. Visibility measurements of faint objects have become possible and closure-phase imaging with multi-telescope arrays has been successfully demonstrated. During this process we all learned that control systems and optical hardware have to be well integrated and cannot be treated as separate components.

Interferometry was an integral part of the Paranal Observatory from the early planning of this facility. This concept is now bearing fruits with interesting scientific results coming from the first instruments VINCI, MIDI, and AMBER. In fact, AMBER is still under commissioning and will start to become available to routine observations by the community in fall 2005. More than 170 workshop participants have demonstrated the huge interest of the international astronomical community in these results.

The last year has seen the foundation of the European Interferometry Initiative, fostering the networking in the field of optical and infrared interferometry. The Fizeau Exchange Visitors Program very nicely supports this networking idea. We are presently developing visions for interferometry in Europe which include optical and infrared interferometry with very long baselines, larger arrays with higher imaging efficiencies and nulling experiments at the Antarctic Plateau. Not all of these experiments will be performed within ESO. Large projects with interferometric capabilities and European participation such as the Large Binocular Telescope (LBT) are in the final construction phase. The LBT with its maximum baseline of 23.8 m will provide excellent uv coverage and imaging capabilities. Good imaging capabilities are essential for objects with complex structure where a priori structural information is not available. The Keck nuller is presently used to search for exozodiacal dust clouds and the LBT nuller will hopefully join soon. Interferometry is a rapidly evolving field. In such a situation ESO has to demonstrate that it can keep its leadership for interferometry in Europe. With the ATs coming soon and the unique PRIMA facility ESO is in a good position, but must develop a vision for the future.

During the last couple of days we have seen interferometry results, covering a broad range of scientific topics, from asteroids to AGNs, with VLTI results in the spotlight. Interferometry now goes far beyond measuring stellar diameters, limb darkening, and binaries, although we have seen fascinating new results on extremely fast rotators, accurate mass determination for very low-mass stars and members of compact stellar groups such as HD 98800, the calibration of the period-luminosity relation for Cepheid stars, and the precise distance to the Pleiades. Interferometric measurements in the infrared, together with millimeter interferometry, are now driving our understanding of the physical and chemical structure of protoplanetary disks. We are now able to constrain the inner disk edge, the nature of disk winds, and the radial dust composition. For the first time, we have found direct observational evidence for radial mixing in protoplanetary disks with MIDI observations of dust mineralogy and grain size. Interferometry has had a similar impact on our understanding of mass loss from AGB stars and LBVs, as has been demonstrated for the key objects IRC+10216 and η Carinae during the meeting. We are now able to characterize the infrared objects in the central star cluster of our Galaxy and to resolve the AGN dust tori with the first MIDI observations of such sources. The ground-breaking MIDI observations of NGC 1068 and Circinus A are a particularly impressive result coming from the VLTI and MIDI teams.

Information encoded in observations with high spatial resolution allow to construct physical source models, using advanced numerical tools such as multi-dimensional radiative transfer and MHD calculations. With radiative transfer theory meets interferometric observations; with spectroscopic and imaging capabilities interferometry also provides direct constraints for theoretical models. We have all learned that infrared interferometry is not an isolated tool, but has to be an integral part of astronomical observations. This has been nicely demonstrated by the η Carinae example where HST, AO data and interferometric measurements haven been combined to reach a comprehensive understanding of the source structure. Although imaging a source provides more direct information than just visibilities, with MIDI we have also realized that visibilities measured at various wavelengths carry a lot of very useful information. Coupling interferometry with spectroscopy is the way to go for many applications.

The discussion of future projects and the next generation of VLTI instruments showed the potential of the field for the investigation of stellar surfaces, protoplanetary disks, extrasolar planets, the Galactic Centre and space time around its black hole, and the central regions of AGNs. A massive astrometric exoplanet survey with the dual-feed facility PRIMA will complement radial-velocity efforts both on planetary mass and orbital distributions. In space, we expect ESA's Darwin mission and NASA's Terrestrial Planet Finder(s) to become reality and to close the far-infrared wavelength gap for observations with high spatial resolution. On the ground, extension of interferometry with large telescopes to visible wavelengths and to larger baselines will be a challenge for the coming years.

This gives me the opportunity to thank all the participants and speakers for coming and actively contributing to the success of this meeting. I would like to thank Bavaria for providing good beer and sunshine, and the Max Planck Institute for Extraterrestrial Physics for the meeting facilities. I would like to extend my thanks to both the Scientific and Local Organizing Committees for organizing a productive and stimulating meeting. This meeting has demonstrated the scientific potential of optical and infrared interferometry with exciting new results.

Part V

Instrumentation: Concepts for future interferometric fringe tracking

Multiple Beam Fringe Tracking at VLTI

M. Gai, D. Bonino, L. Corcione, D. Gardiol, M. G. Lattanzi, D. Loreggia,
G. Massone[1] and S. Menardi[2]

[1] Istituto Nazionale di Astrofisica – Osservatorio Astronomico di Torino, V.
Osservatorio 20, 10025 Pino T.se (TO), Italy gai@to.astro.it
[2] European Southern Observatory, , K.Schwarzschild-Str.2, 85748 Garching b.
München, Germany smenardi@eso.org

Summary. The combination of more than two beams is appealing from the scientific standpoint due to the increased sampling of the spatial frequency plane. Each telescope beam is affected by an independent piston noise, which must be measured and corrected in order to achieve stable observing conditions on the science targets. We discuss fringe sensor implementation concepts, taking into account aspects evidenced from the current experience at VLTI.

1 Introduction

The potential of higher spatial and spectral resolution, and of high efficiency in the u-v plane coverage, have been discussed in other contributions to this ([1]) and other dedicated workshops ([2]). Significant performance improvements are achievable at VLTI with the proposed second generation instruments (e.g. [3] and [4]), whereas future large arrays may benefit of new concepts for combination ([5]), beam transport ([6]) and available technology ([7]).

The case considered is that of combination of four equivalent telescopes, i.e. either Unit (UT) or Auxiliary (AT) Telescopes, as this is possible with simple upgrade of the current VLTI infrastructure. Mixed AT and UT combination is conceptually feasible, but requires careful evaluation of the trade-off between achievable performance and technical constraints (i.e. costs). The interferometric signal is affected by residual wavefront error (WFE) and spectral variation of transmission and phase, for each beam and independent polarisation component. At least one independent piston term on each telescope must be measured and corrected by the Fringe Tracking (FT) loop. The FT problem is linear with the number of individual apertures of the interferometer.

In Section 2 we review the effects of residual atmospheric turbulence expected on VLTI beams, after AO correction. In Section 3 we remind some of the basic technical functions required by an interferometric instrument and we discuss possible combination concepts for the simple case of four beams, which might correspond to a reasonable upgrade of the VLTI facility. Finally, in Section 4 we draw our conclusions.

2 Atmospheric limitations

Interferometry with individual aperture size larger than the atmospheric coherence length r_0 requires an adaptive optics (AO) system, and the performance depends on AO efficiency.

The input WFE variation results in a variable fraction of the nominal photon flux actually coupled to the fibre optics (or pinholes) used by many instruments as spatial filter; a reasonable estimate is provided by the Strehl ratio, scaling with wavelength λ.

An evaluation of the impact of residual WFE on the UT was performed, based on a realistic data set from ESO (courtesy F. Delplancke). The input WFE data set is composed of 3200 instances, with time resolution 2.4 ms, and spatial resolution of 64 points over the pupil diameter. The RMS WFE on the pupil has average value 250 nm and standard deviation 32 nm.

Also, the photometric visibility associated to the independent intensity fluctuations, using half of the WFE data set for each input beam, is derived with the expression $V_{ph} = 2 \cdot \sqrt{I_1 \cdot I_2}/(I_1 + I_2)$, where I_1 and I_2 are the individual beam intensities. The large WFE cases induce both low flux and low visibility, with double impact on interferometric performance.

The histograms of Strehl ratio (left) and photometric visibility (right) are shown in Fig. 1; the average Strehl ratio is 60%, with RMS 7.5%, and the mean visibility is 99.7%, with RMS 0.7%. The distribution is not normal, and the fraction of events below the 3σ threshold is respectively 1.5% and 2%.

In H band, due to the shorter wavelength, the degradation is larger. The same WFE data set results in a Strehl ratio with average 39% and RMS 8.5%, and visibility with mean 98.4% and RMS 3.4%. The beam intensity drops below 10% several times per second!

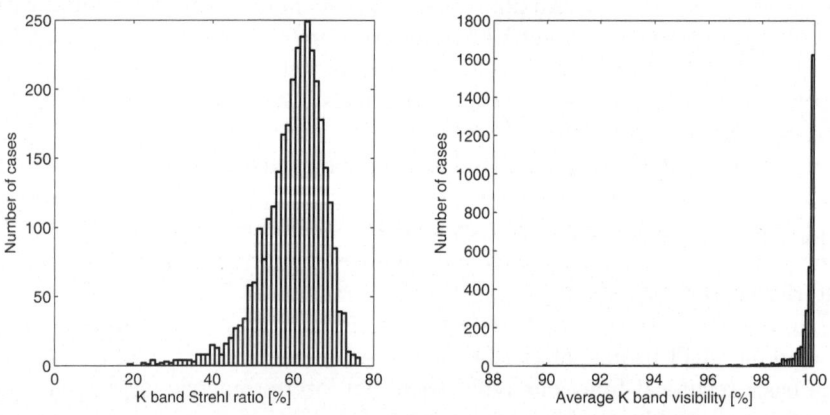

Fig. 1. Histogram of K band fibre coupled flux and visibility.

Also, the actual spectral distribution of the coupled flux is variable, inducing variations to the measured effective wavelength. This increases the measurement noise, due to the discrepancy with respect to a static signal model, and may contribute to the systematic error. The contribution to OPD noise for the PRIMA FSU has been estimated in about 19 nm RMS.

Future improvements in AO technology will provide evident benefits to interferometric instruments on this respect: higher effective throughput and stable beam intensity.

Apart flux fluctuation, AO and spatial filtering produce flat local wavefront sections; the relative phase between apertures fluctuates due to atmospheric turbulence on the time scale of the coherence period t_0, ranging from few milliseconds in the visible range to few ten ms in the near infrared (NIR) and few hundred ms in the thermal infrared (MIR). The OPD between two telescope beams is affected by this phase fluctuation, also defined as atmospheric piston. Typical RMS values at Paranal, in average conditions, are $\sim 20\mu m$ (design value for both FINITO and PRIMA FSU).

Also, two stars are affected by a degradation of the correlation between their piston disturbance, i.e. a differential OPD (DOPD) noise, increasing with their angular separation θ and again decreasing at longer wavelength.

The telescope beams are transferred in air, inducing significant longitudinal dispersion, which must be taken into account in the overall OPD and group delay (GD) measurement. In general, longitudinal dispersion has to be accounted for and compensated in the fringe sensor design.

3 Beam combination

The concepts of the current generation of VLTI fringe sensors (FINITO and PRIMA FSU) have been described in previous papers, including a detailed signal model ([8]). They are applicable in a straightforward way to the four beam combiner concept; design optimisations are in any case required.

The basic functions are similar: beam alignment, combination and detection. The telescope beams must be injected into the instrument with fine adjustment of the lateral and angular position, through a suitable alignment sub-system. Then the beams are to be combined with each other, and the interferometric outputs must be fed to the detector. An important auxiliary function shared by most instrument is spatial filtering, implemented either by fibre optics or pinholes, removing part of the residual WFE and related coherence degradation at the expense of intensity fluctuations, as mentioned above. Fibre optics is also convenient for beam transportation within the instrument, in particular between the beam combiner and the detector. Integrated optics components made available by recent technical developments (e.g. beam splitters and combiners) are also quite easily connected by fibres.

Transversal atmospheric dispersion compensation (TADC) is useful to maximise the flux injected into the instrument over the selected spectral range. The longitudinal (LADC) component affects interferometric visibility, due to the OPD spread over the spectral bandwidth, increasing with the zenithal distance. Both can be compensated by appropriate devices at the instrument input; FINITO implements no compensation, whereas the PRIMA FSU has LADC capability. A new fringe sensor may implement both TADC and LADC for optimal sensitivity.

Spectral dispersion is useful to reduce the visibility degradation induced by significant variation over the bandwidth of the atmosphere and component characteristics (phase and transmission). Also, it provides a convenient method for GD measurement, by evaluation of the phase relationship over the spectral range. The PRIMA FSU has limited spectral resolution, implemented by a central "white light" band and two narrow side bands, whereas FINITO does not have dispersion and uses spatial modulation for identification of the central fringe corresponding to zero GD. Spectral dispersion appears to be a more convenient and robust technique, reducing the sensitivity to high frequency noise associated to time modulation of the internal OPD, and may be optimised for the requirements of future instruments.

The two polarisation components of the input beams are independent, and the corresponding interferometric combinations are in general out of phase, depending on instrumental contributions, and the cumulative interferogram may be affected by visibility degradation. To prevent this, it is necessary either to ensure symmetrical instrumental polarisation, which is not trivial, or to separate the components. In FINITO, one polarisation is used for photometry, whereas in the PRIMA FSU the two components are shifted by $\pi/2$ to generate the ABCD interferometric outputs, separated after combination.

Interface to metrology is crucial for accurate astrometric measurements, in order to identify and correct the instrumental phase contributions. Also, space allocation in the crowded VLTI combination lab is a significant issue for new instruments.

The combination of four beams (1, 2, 3, 4) could be made over all six possible baselines, but adopting a pair-wise scheme only four combinations (1-2, 2-3, 3-4, 4-1) are required. This saves about 30% of the components at the level of beam splitting and combination, and the higher flux fraction provides a potential improvement in sensitivity of about 0.44 magnitudes (or 22% on noise). The conceptual schematic of a possible symmetric combination of four beams, and a possible layout, are shown in Fig. 2. It takes advantage of the two polarisation components of each input beam for each combination, with optimal usage of the available photons. An appropriate trade-off on the spectral dispersion (i.e. GD measurement accuracy) vs. sensitivity can be made on the detection system, also depending on the detector performance. The scheme is compact and highly modular, using several copies of the same basic components. Also, the concept can be easily scaled to e.g. six beams, with simple geometry modifications.

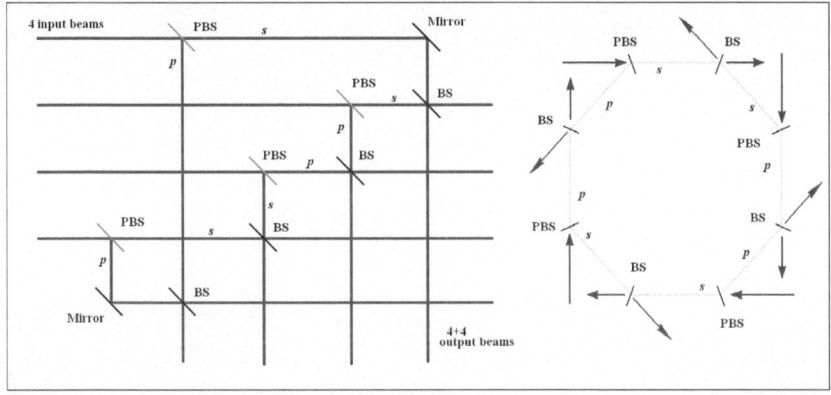

Fig. 2. Conceptual schematic and layout of a symmetric four beam combination.

4 Conclusions

Four beam combination is a case in which conventional bulk optics may still be practical, with the advantage of using proven technology. Technological improvement on a number of aspects may provide significant benefits to interferometric instrumentation, either at VLTI or in possible future facilities.

Acknowledgements This work has been performed within the OPTICON programme. OPTICON is funded by the European Commission's Sixth Framework Programme under contract number RII3-CT-2004-001566. We acknowledge contributions to the OATo group activity from the Italian National Institute for Astrophysics (INAF, ref. 0330909) and from the Ministero dell'Istruzione, dell'Università e della Ricerca (ref. 2003027003).

References

1. A. Quirrenbach: *these proceedings*
2. J. Surdej, D. Caro, A. Detal (Eds.): *Science Cases for Next Generation Optical/IR Interferometric Facilities* , Proc. 37th Liège Intern. Astroph. Colloquium
3. B. Lopez, S. Wolf, M. Dugué: *these proceedings*
4. F. Malbet, J-P. Berger, P. Kern, P. Garcia: *these proceedings*
5. E. Ribak: *these proceedings*
6. G. Perrin: *these proceedings*
7. A. Tünnermann: *these proceedings*
8. M. Gai et al.: Proc. SPIE 5491-61, 2004, 528.

Multiple-beam Fringe Tracking for the VLTI

F. Cassaing, F. Baron, I. Mocoeur, L. M. Mugnier, G. Rousset, and
B. Sorrente[1]

ONERA / Optics Department, BP72 92322 Châtillon, France
`frederic.cassaing@onera.fr`

Summary. The performance of interferometric instruments is strongly linked to
the quality of piston stabilization. Next generation VLTI instruments plan to use
4 to 8 beams simultaneously. In the current VLTI implementation, the maximum
number of beams that can be phased using FINITO and PRIMA/FSU simultaneously is 5. Therefore, a new fringe sensor is required for the VLTI.

When cophasing a large number of beams, other approachs than the classical
complex pairwise combination should be considered. Focal-plane wavefront sensing
allows the cophasing of a large number of beams with a very simple opto-mechanical
device and can also measure high-order modes, such as tip/tilt/defocus/... This
solution has been selected in other projects such as DARWIN or Earth observation.

Experimental results, obtained with a laboratory breadboard tested on the
BRISE bench at ONERA, confirm the validity of this approach for nanometric
measurements and make such a simple fringe sensor an attractive component for
the 2nd generation VLTI.

1 Multiple-beam fringe tracking at VLTI

Most projects for VLTI 2nd generation instruments plan to use at least
4 beams (4 ATs dedicated to interferometry, or 4 UTs for maximum sensitivity), or even 8 beams (4 UTs+4 ATs). To reach their maximum performance, these instruments require active correction of the Optical Path
Difference (OPD) between the beams and of the WaveFront Error (WFE) of
each beam.

Correction devices are available at VLTI: all telescopes include a tip/tilt
actuator, UTs are equiped with a deformable mirror (DM), and a dedicated
location exists in the ATs. Six delay lines are available and provision has
been made for 8. The situation is different for the sensors. To the best of our
knowledge:

- The OPD between 2 pairs of beams can be measured with PRIMA/FSU
  (in K) and between 3 beams simultaneously with FINITO (in H).
- The WFE can be measured on the UTs by MACAO (in the visible). But
  the path in the tunnel (turbulence and static WFE) is not seen. There is
  no wave-front sensor for the ATs.
- Tip/tilt in the lab can be measured (for ≤ 4 beams) with IRIS (IR).

The performance of 2nd generation instruments would thus be increased by a new sensor in the VLTI lab that would allow the simultaneous measurement of the OPDs between 4 to 8 beams and their tip/tilt. The WFE can even be measured, at least for the lowest modes, to allow pre-compensation by MACAO or *a posteriori* calibration or real-time correction with DMs in the ATs. This single sensor would work in a single IR band to save photons for the scientific instruments.

Such a multi-beam multi-purpose sensor has already been investigated at ONERA for two other projects with similar requirements. The first one is DARWIN [1], where the real-time cophasing (piston/tip/tilt with nanometric accuracy) of the 6 sub-apertures is one of the main challenges. Measurement of the WFE (from defocus to spherical aberration) during operation is also required for calibration. A laboratory breadboard DWARF (DarWin AstRonomical Fringe sensor) has been defined and validated [2].

The second project deals with Earth imaging from the GEO orbit [3]. For a good on-ground resolution, a large deployable telescope is required, with a sufficiently-filled aperture. A critical issue is then to measure differential piston/tip/tilt, for typically 10 sub-apertures, on very extended scenes.

2 Solutions for multiple-beam fringe sensing

Two solutions are most classically used for fringe sensing. The first one is coaxial (or amplitude) combination in a pupil plane, with a beam splitter. This pairwise combination is widely used in stellar interferometry, often with temporal modulation [4], or with spatial modulation [5] as proposed by [6]. There are several drawbacks for this solution: the complexity quickly increases with the number of beams, only unresolved (or slightly resolved) sources can be used, and only the differential WFE can be measured. The complexity can be reduced by integrated optics, but in this case tip/tilt or WFE can not be measured.

The second solution is multiaxial (or wavefront) combination in a focal plane. In this case, an arbitray number (typically smaller than 20) of parallel beams are arranged, with a non-redundant configuration [7], in front of a focusing device and a detector. The complexity is then transfered on data processing, to solve for the inverse problem: given the observed image in the focal plane, what is the phase distribution in the pupil plane that gave birth to it ? Several approaches can be used:

– Phase retrieval: when the object is known (unresolved reference star), only the focal plane is used. Although phase retrieval is known for centro-symmetric pupils (case of monolithic telescopes) to lead to a sign ambiguity on the even part of the phase, a non-redundant multiple-beam configuration allows to uniquely solve for piston/tip/tilt [8].

– Phase diversity: 2 images are taken, in the focal plane and in an extra-focal plane (or with any other known aberration). In this case, the two data sets allow to fully retrieve the two macro-unknowns (object and phase): phase ambiguities are resolved and the object can also be estimated [9].

Different algorithms have been validated at ONERA: an iterative MAP (Maximum a Posteriori) resolution, and an analytical approximate algorithm for real-time piston/tip/tilt measurement by phase retrieval.

The pupil-plane and focal-plane approaches were compared in the framework of DARWIN and Earth observation. Performance are comparable, but a focal-plane sensor was selected for both applications, because of the simple hardware, the large number of beams, the ability two measure the absolute value of high-order modes, or the operation on complex objects.

3 Experimental results on the BRISE bench

A multiple-aperture testbed has been built to validate the two cophasing algorithms: Phase Retrieval for unresolved sources, and Phase Diversity for extended scenes. The main components of BRISE (Banc Reconfigurable d'Imagerie sur Scènes Etendues, described in [10]) are:

– An unresolved (monomode fiber + He-Ne laser) and an extended (photographic plate + white cell) object.
– An aperture mask with three sub-abertures and three planar mirrors on piezo-electric calibrated piston/tip/tilt plateforms.
– The cophasing sensor, which simultaneously records the focal and extrafocal images of each object in a single frame of a CCD camera.
– A control software and an efficient isolation against environmental disturbances (air turbulence, vibrations, thermal drifts).

Figure 1 shows experimental results for linearity. A 30-point piston slope from -500 nm to +500 nm is applied on a given sub-aperture. The estimated piston with phase retrieval (left, with three measurements for each piston value) or phase diversity (right) is linear between $\pm\lambda/2$ and wrapped as expected. A saturation effect appears for phase diversity, that remains to be fully understood; however, this is not an issue in closed loop. Because diffraction is chromatic, whereas our numerical model is monochromatic, the spectral bandwidth is an important parameter to optimize. Measurements were made at $\lambda = 650$ nm with $\Delta\lambda=40$ nm for phase diversity and $\Delta\lambda=$ 10, 40 and 80 nm for phase retrieval. This shows that a rather large spectral range ($\Delta\lambda= \lambda/8$) can be used.

Figure 2/left shows experimental results for repetability, in the case of phase retrieval with $\lambda=650$ nm. Values agree with the simulation and clearly show the photon-noise $1/\sqrt{N}$ law. For low fluxes, the sensor is dominated by the detector noise. But the 0.75 nm repetability specified for DWARF is reached in the correct magnitude range.

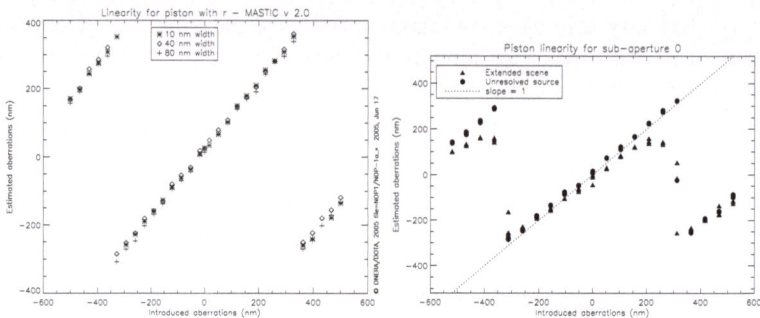

Fig. 1. Piston Linearity for phase retrieval and different bandwidths (left) and comparison between phase diversity and phase retrieval (right).

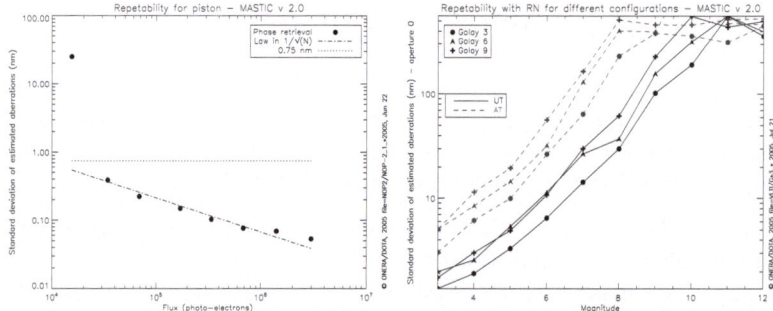

Fig. 2. Piston repeatability with phase retrieval estimated for DARWIN (left) and VLTI (right).

Figure 2/right shows very preliminary simulation results in the case of VLTI. The number of sub-apertures was varied between 3 and 9 (based on Golay configurations for simplicity). The piston measurement error slightly increases with the number of apertures. For a $\lambda/10$ accuracy in K (200 nm), the limiting magnitude is about 9-10 (resp 7-8) for the UTs (resp ATs).

4 Conclusion

This communication has shown that a simple setup (telescope + focal-plane detector fed by all the beams with a good aperture configuration) is an attractive solution to measure piston and a few WFE modes on a large (3 to a few tens) number of sub-apertures with an arbitrary (even fully resolved) object. The first experimental tests performed on BRISE have validated nanometric piston/tip/tilt repeatability with 3 sub-apertures. Additional tests with more modes/apertures are under way.

Such a compact device is an attractive solution for a multi-purpose 2nd generation sensor at VLTI, that would perform all real-time measurements in the lab (piston/tip/tilt/high order modes) with an arbitrary number of beams (3 to 8) in a single IR band. It could also, with a (densified) homothetic pupil-mapping, work with very extended objects.

References

1. http://sci.esa.int/home/darwin/index.cfm.
2. F. Cassaing, F. Baron, E. Schmidt, S. Hofer, L. M. Mugnier, M. Barillot, G. Rousset, T. Stuffler and Y. Salvadé, *DARWIN Fringe Sensor (DWARF): Concept Study*, In *Towards Other Earths*, vol. SP-539, pp. 389–392. ESA (2003), Conference date: April 2003.
3. L. Mugnier, F. Cassaing, B. Sorrente, F. Baron, M.-T. Velluet, V. Michau and G. Rousset, *Multiple-Aperture Optical Telescopes: some key issues for Earth observation from a GEO orbit*, In *5th International Conference On Space Optics*, vol. SP-554, pp. 181–187, Toulouse, France, CNES/ESA, ESA (2004).
4. M. Shao, M. M. Colavita, B. E. Hines, D. H. Staelin, D. J. Hutter, K. J. Johnston, D. Mozurkewich, R. S. Simon, J. L. Hershey, J. A. Hughes and G. H. Kaplan, *The Mark III Stellar Interferometer*, Astron. Astrophys., 193 (March 1988), pp. 357–371.
5. M. Gay et al., *Multiple beam fringe tracking at VLTI*, In *The Power of Optical/IR Interferometry: Recent Scientific Results and 2nd Generation VLTI Instrumentation*, edited by A. Chelli and F. Delplancke., ESO Astrophysics Symposia. Springer Verlag (2005).
6. F. Cassaing, B. Fleury, C. Coudrain, P.-Y. Madec, E. Di Folco, A. Glindemann and S. Lévêque, *An optimized fringe tracker for the VLTI/PRIMA instrument*, In *Interferometry in optical astronomy*, edited by P. J. Léna and A. Quirrenbach, vol. 4006, pp. 152–163, Bellingham, Washington, Proc. Soc. Photo-Opt. Instrum. Eng., SPIE (2000).
7. M. J. E. Golay, *Point Arrays Having Compact, Nonredundant Autocorrelations*, J. Opt. Soc. Am., 61 (1971), pp. 272–273.
8. F. Baron, F. Cassaing and L. Mugnier, *Alignement des pupilles d'un télescope multi-pupilles*, In *19ième Colloque sur le Traitement du Signal et des Images*, edited by J.-M. Chassery and C. Jutten. GRETSI (September 2003).
9. F. Baron, F. Cassaing, A. Blanc and D. Laubier, *Cophasing a wide field multiple-aperture array by phase-diversity: influence of aperture redundancy and dilution*, In *Interferometry in Space*, edited by M. Shao, vol. 4852, Hawaii, USA, Proc. Soc. Photo-Opt. Instrum. Eng., SPIE (2002).
10. B. Sorrente, F. Cassaing, F. Baron, C. Coudrain, B. Fleury, F. Mendez, V. Michau, L. Mugnier, G. Rousset, L. Rousset-Rouvière and M.-T. Velluet, *Multiple-Aperture Optical Telescopes: cophasing sensor testbed*, In *5th International Conference On Space Optics*, vol. SP-554, pp. 479–484, Toulouse, France, CNES/ESA, ESA (2004).

Part VI

Instrumentation: 2nd Generation
Instrumentation for the VLTI – Proposals

APerture Synthesis in the MID-Infrared with the VLTI

B. Lopez[1], S. Wolf[2], M. Dugué[1], U. Graser[2], Ph. Mathias[1], P. Antonelli[1],
J.-C. Augereau[8], J. Behrend[3], N. Berruyer[1], Y. Bresson[1], O. Chesneau[1],
C. Connot[3], K. Demyk[13], E. DiFolco[4], A. Dutrey[6], S. Flament[1],
Ph. Gitton[4], A. Glazenborg[9], A. Glindemann[4], M. Heininger[2],
Th. Henning[2], K.-H. Hofmann[3], Y. Hugues[1], W. Jaffe[10], S. Jankov[7],
S. Kraus[3], S. Lagarde[1], Ch. Leinert[2], H. Linz[2], K. Meisenheimer[2],
L. Mosoni[2], J.-L. Menut[1], U. Neumann[3], A. Niedzielski[12], F. Przygodda[2],
F. Puech[4], T. Ratzka[2], R. Rohloff[2], A. Roussel[1], D. Schertl[3],
F.-X. Schmider[7], B. Stecklum[11], E. Thiébaut[5], F. Vakili[7], K. Wagner[2], and
G. Weigelt[3]

1. Observatoire de la Côte d'Azur
2. Max Planck Institute for Astronomy of Heidelberg
3. Max Planck Institute for Radio-Astronomy of Bonn
4. European Southern Observatory
5. Observatoire de Lyon
6. Observatoire de Bordeaux
7. Lab. Astro. Univ. de Nice
8. Obs. de Grenoble
9. ASTRON Netherlands
10. Univ. of Leiden
11. Thüringer Landessternwarte Tautenburg
12. Univ. of Torun
13. Université de Lille.

Summary. Our objective is to develop of the mid-infrared imaging for the VLTI.
Several areas of astrophysics will benefit of this new capability. APreS-MIDI comprises a beam combiner which interfaces with the current MIDI instrument. It thus constitutes an extension to the two-beam interferometric instrument MIDI by increasing the number of recombined beams up to four. This extension provides better uv-coverage (6 visibility points measured in one set) and moreover will allow measurement of 4 closure phase relations thus providing for the first time aperture synthesis images in the mid-infrared spectral regime.

The mid-infrared spectral domain is very relevant for the study of the environments of various astrophysical sources. The science cases are broad. There are illustrated in the first part of this article by the perspective of imaging the circumstellar discs of young stellar objects. The APreS-MIDI instrument is described in a second part of this article.

1 Science cases

The image reconstruction in the 8-13μm will allow to address qualitatively new questions in numerous scientific areas, such as in the cases of :

1. The Circumstellar Environment of Young Low and Intermediate Mass Stars,
2. The Multiplicity of Young Stars,
3. The Physical Conditions of Massive Star Formation,
4. The Dust and Winds from Evolved Stars,
5. The Hot Stars Environments,
6. The Inner Regions of Active Galactic Nuclei.

At this point, however, we decided to present only parts of the science case studies concerning circumstellar environment of young low and intermediate mass stars. The complete science case study for APReS-MIDI is published in a separate document (Wolf et al. 2005).

Circumstellar environment of young low and intermediate mass stars:
The material in the circumstellar disks comprises the building blocks for future planetary systems (e.g., Lissauer 1993, Beckwith et al. 2000). These proto-planetary disks have now been imaged from the optical/near-infrared to the millimeter wavelength range around low-mass YSOs (T Tauri stars, e.g. Dutrey et al. 1994, 1998; Burrows et al. 1996), intermediate-mass YSOs (Herbig Ae/Be stars, e.g. Mannings & Sargent 1997, 2000), and around a possible massive star (Fontani et al. 2004; see also Chini et al. 2004, Shepherd et al. 2004, Schreyer et al. 2002).

Circumstellar disks are key subject of observations by all available ways. Unfortunately, the innermost region of disks with a radius of a few AU, where planet formation is expected to take place, can only be marginally investigated so far by the study of the spectral energy distribution (SED), high-resolution images obtained with the Hubble Space Telescope or by using adaptive optics (e.g., differential polarimetric measurements with NAOS/CONICA – e.g., Apai et al. 2004). This will remain the case until ALMA will be in operation with its longest baselines (2012), except if one can provide high angular resolution images in the mid-infrared wavelength range. Figure 1 is a montage explaining which part of a disk is sampled depending on the instrument/telescope in use. This figure clearly reveals the importance of a mid-infrared VLTI imager, like APReS-MIDI with respect to existing and future instruments.

Given the typical distance of nearby star-forming regions of $\sim$140-200 pc, the dominant observable quantity originating from the inner disk region ($r <$ 10 − 20AU) is the emission of mid-infrared continuum radiation by warm dust. One of the first exciting results achieved with MIDI was to show the difference in the dust grain evolution in the "inner" disk (represented by the correlated flux) vs. the outer disk (net flux), using the low-resolution spectroscopy observing mode (van Boekel et al. 2004). However, given the

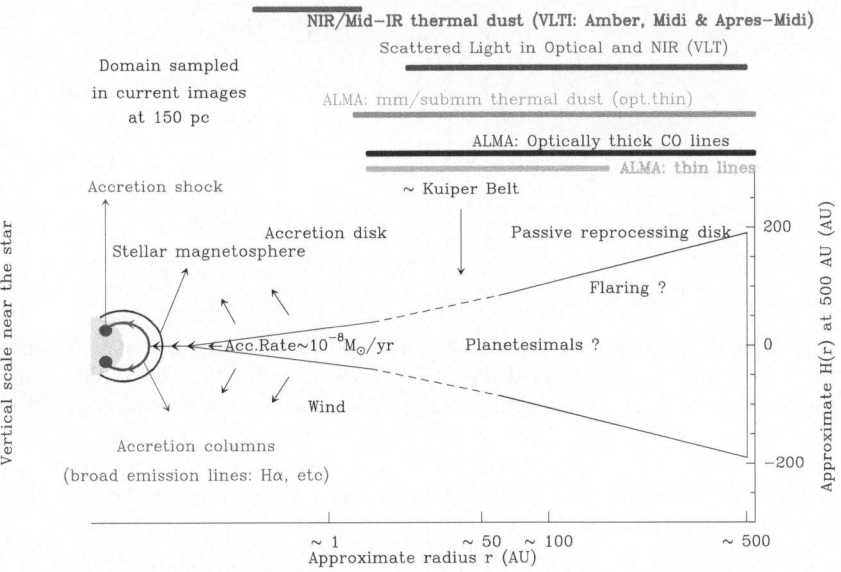

Fig. 1. A proto-planetary disk surrounding a T Tauri star at a distance of 150 pc. The region of the disk sampled by the current and future telescopes are shown. By comparison of our Solar System, the approximate location of the Kuiper Belt is given. This montage emphasizes the complementarity of APReS-MIDI with future interferometers such as the Atacama Large Millimeter Array (ALMA).

intrinsic limitations of the analysis of MIDI observations (interpretation of visibilities), we have only weak constraints on the structure and size of the emitting region. The situation becomes even more difficult if MIDI visibilities are used to constrain more than the geometrical parameters, such as the inner emissivity profile of circumstellar disks. Only high angular resolution images at mid-infrared wavelengths can disentangle the various disk models.

Having the ability to obtain (reconstructed) images, instead of visibilities only, will dramatically improve this situation. In contrast to the ambiguous interpretations of visibilities as obtained with MIDI, an imaging capability is aimed at answering decisive questions about the inner regions of young circumstellar disks, such as:

[1] What is the surface brightness profile in circumstellar disks around T Tauri / HAe/Be stars?
While MIDI allows to derive the "mean" disk size and the approximate inclination of the disk, imaging will allow to consider that circumstellar disks which are not seen exactly face-on do show a brightness profile which cannot

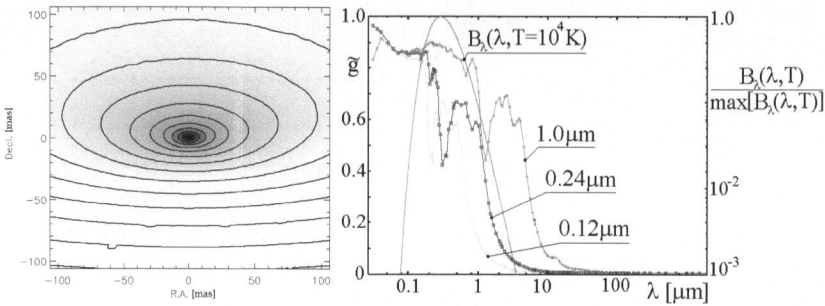

Fig. 2. *Left:* 10µm intensity map of the inner 30AU×30AU region of a circumstellar T Tauri disk in an assumed distance of 140 pc; inclination angle: 60°. Due to the flared surface of the disk, the overlayed iso-intensity contour lines are clearly not centered on the star. *Right:* Stellar photospheric SED of a Herbig Ae star vs. the wavelength-dependence of the scattering parameter *g* for different grain sizes, resulting in a dependence of the efficiency to transport radiation in the upper layers of circumstellar disks.

be described by iso-brightness contours, centered on the location of the central star, as assumed in the previous data analysis (see Fig. 2 for illustration).

An APReS-MIDI like facility will permit the derivation of the radial temperature profile of the hot dust on the disk surface and the inner disk rim. This profile will provide information about the radial and vertical structure of the disk, and thus also about the importance of viscous and accretion heating (in addition to the stellar heating) and thus on the interior density structure in the planet-forming region of the disk.

[2] Does the brightness profile show evidence for dust grain growth and sedimentation on the mid-plane?
The most reliable conclusions about grain growth are based on the millimeter slope in the SED of circumstellar disks (Beckwith et al. 1990) and more recently on images of dust disks provided by millimeter interferometry (e.g. Butterfly Star in Taurus - Wolf et al. 2003; CQ Tau - Testi et al. 2003; Natta et al. 2004). These images clearly reveal grain growth up to particles of cm size. What is clearly missing, however, are *a)* observational constraints on the region, where dust grain growth is presumably fastest (first qualitative results have been obtained with MIDI, based on a detailed analysis of the profile of the 10µm silicate feature; van Boekel et al. 2004), and *b)* a detailed knowledge of the vertical distribution of the dust particles. Fig. 2[right] shows that the scattering parameter of dust grains significantly changes from sub-micron to micron-size (or larger) grains in the wavelength range of stellar emission. Together with the dust density structure, the forward vs. backward-scattering behavior of dust grains determines the temperature structure of

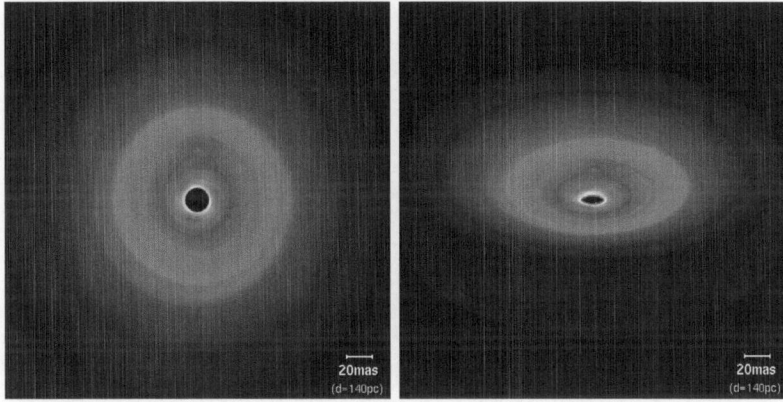

Fig. 3. Simulated 10μm images of the inner region (radius 20 AU) of a circumstellar T Tauri disk, with an embedded Jupiter-mass planet at a distance of 5.2 AU from the central star (Wolf & Klahr 2005). The left/right image shows the disk under an inclination of 0°/60°. For both inclinations, the hot region around the planet above the center of the disk, indicated as bright areas in these reemission images, is clearly visible. Assuming a distance of 140 pc, the corresponding 20 mas scale is indicated in the lower right edge of both images.

the disk which in turn controls the vertical structure of the disk which can be obtained with APReS-MIDI.

[3] Is the inner disk structure modified by early stages of planet formation?
The inner region of circumstellar disks is expected to show large-scale (sub-AU - AU sized) density fluctuations / inhomogenities. The most prominent examples are predicted long-lived anti-cyclonic vortices in which an increased density of dust grains may undergo an accelerated growth process - the first step towards planet formation (Klahr & Bodenheimer 2003). Locally increased densities and the resulting locally increased disk scale height have direct impact on the heating of the disk by the central star and are expected to show up as local brightness variation (due to increased absorption / shadowing effects) in the mid-infrared images (for illustration, see Fig. 3).

As simulations of dust settling show, the disk flaring and thus the ability to absorb stellar radiation even at large distances from the star depends on the grain size distribution remaining in the upper disk layers (e.g., Dullemond & Dominik 2004). Conclusions about the importance of this effect may be derived by comparing the average intensity profile for a large sample of sources.

[4] Is there indirect or even direct evidence for the presence of planets?
Once (proto-)planets have been formed, they may significantly alter the surface density profile of the disk and thus cause signatures in the disk that are

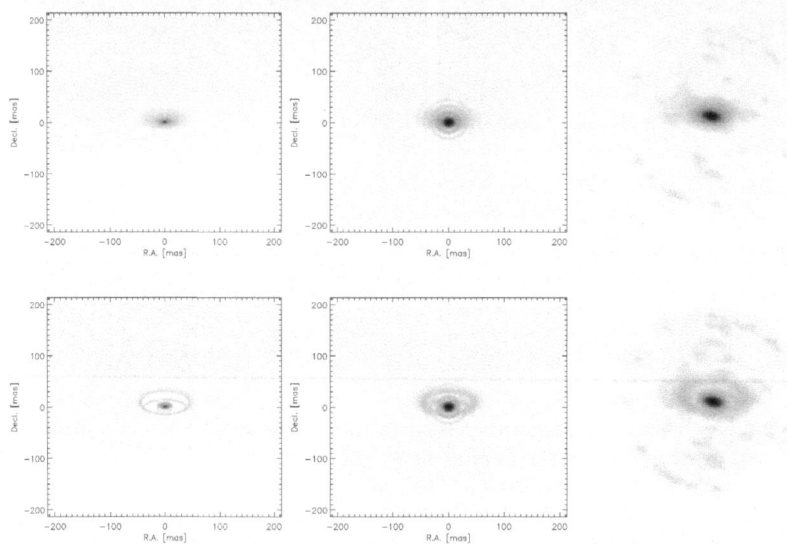

Fig. 4. Simulated 10μm images of a circumstellar disk around a T Tauri star without (top) and with (bottom) a 4 AU gap in a distance of 4 AU from the central star. Disk inclination: $60°$ (for comparison: $0° =$ face-on); assumed distance: 140 pc. *Left:* Original image; *Middle:* Reconstructed image, assuming the maximum possible uv-coverage to be obtained with the VLTI and maximum possible baselines; *Right:* Reconstructed image, based on the constraints given by the APReS-MIDI module.

much easier to find than the planets themselves. The appearance and type of these signatures depend on the mass and orbit of the planet, but even more on the evolutionary stage of the circumstellar disk. While the spatial structure of optically thick, young circumstellar disks around Herbig Ae/Be and T Tauri stars is dominated by gas dynamics, the much lower optical depth and gas-to-dust mass ratios in debris disks (Zuckermann, Forveille, & Kastner 1995; Dent et al. 1995; Artymowicz 1997; Liseau & Artymowicz 1998; Greaves, Coulson, & Holland 2000) make the Poynting-Robertson effect and stellar wind drag, in addition to gravitation, responsible for the resulting disk density distribution.

Hydrodynamic simulations of gaseous, viscous protoplanetary disks with an embedded protoplanet show that the planet may open and maintain a significant gap (e.g., Bryden et al. 1999; Kley 1999, 2000; D'Angelo et al. 2003). The feasibility to detect such a feature is illustrated in Fig. 4. The minimum mass for a planet to open a gap (Paardekooper & Mellema 2004) was found to be $0.05\,M_{\mathrm{Jupiter}}$.

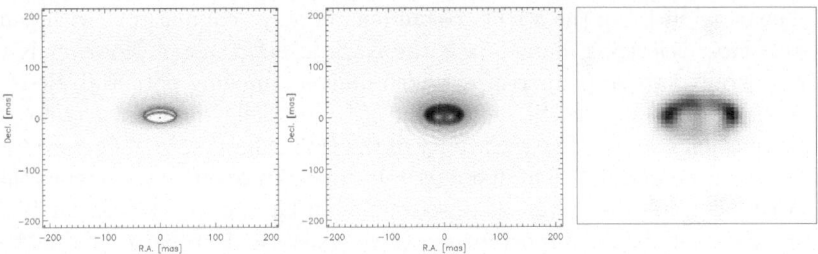

Fig. 5. Simulated 10μm image of the inner region of a T Tauri circumstellar disk with a cleared inner region, seen under an inclination of 60° (assumed distance: 140 pc). *Left:* Original image; *Middle:* Reconstructed image, assuming the maximum possible uv-coverage to be obtained with the VLTI and maximum possible baselines; *Right:* Reconstructed image (not scaled), based on the constraints given by the APreS-MIDI module.

[5] What is the status of disk clearing within the inner few AU?
According to the temperature and luminosity of the central star, the sublimation radius for dust grains is in the order of 0.1 - 1.0 AU (T Tauri - Herbig Ae/Be stars). In contrast to these values, a significantly larger inner dust disk radius of ~4 AU has been measured in the 10 Myr old protoplanetary disk around TW Hydrae (Calvet et al. 2002). Other examples are the object Coku Tau/4 with an evacuated inner zone of radius ~10 AU (D'Alessio et al. 2005, Quillen et al. 2004) and GM Aur with a significant decrease of the dust reemission inside ~4 AU around the central star (Rice et al. 2003). The confirmation of these indirectly (via SED modelling) determined gaps, as well as the test of other disks for the existence / non-existence of similar gaps will provide valuable constraints on the evolution of the planet-forming region and thus on the process of planet formation itself (see Fig. 5 for an illustration of the feasibility to detect a large inner Gap with APreS-MIDI).

2 Description of the APreS-MIDI instrument

2.1 International context of the mid-infrared interferometry

The VLTI offers a unique facility in the world: important telescope sizes (UTs), long baselines (up to 200 meters), and modularity of the configuration (ATs). APreS-MIDI is a four beam combiner interfaced to the current MIDI instrument. APreS-MIDI stands for Aperture Synthesis in the MID-Infrared. The development of APreS-MIDI will directly provide a significant upgrade of the scientific benefits/returns of the VLTI, as well as an important transition toward the next generation instruments.

The capabilities of the VLTI are unique: its only "competitor" with comparable large telescope diameters is the Keck interferometer. However Keck possesses only two large apertures, which limit its imaging potential. Besides these inteferometers, the ISI instrument of UC Berkeley allows a 3 telescope recombination scheme. The sensitivity of the ISI3 is limited by its heterodyne mode operation and its 1.6 m diameter telescopes. In practice this means that the VLTI can observe hundreds of sources, compared to a few tens for ISI.

In this context, an important expertise exists in Europe in the field of mid-infrared interferometry. The success of the MIDI instrument (Leinert et al. 2004), developed by Germany, the Netherlands and France, has opened the development of mid-infrared high angular resolution astrophysics since June 2003. About fifteen years ago, the first direct detection interferometry at 10 microns was demonstrated at the SOIRDETE interferometer at the Plateau de Calern. Our community has presently the possibility to develope the mid-infrared imaging instrument of the VLTI.

2.2 Principle of the recombiner

APreS-MIDI (see Fig. 6) is a module inserted in front of MIDI. The main optical element is a segmented mirror with 4 reflecting faces. Images of the star produced by the 4 different telescopes are focussed on the different faces. The role of this mirror is to send the 4 beams along the same axis into the MIDI entrance window. The small tilt angle between the pupils aiming at producing the fringe pattern is directly linked to the angle which remains between the beams after the reflection onto the segmented mirror.

2.3 Evaluated performances

The evaluated performances are the following :

- Recombination mode: densified images and pupil plane with tilt angle, 3 or 4 telescope beams recombination allowed (4 ATs or 4 UTs).
- Angular resolution: 10 mas at 10 microns
- Spectral mode: using 2-3 filters, R=30.
- Field of view : corresponding to Airy disc sizes. At 10 microns : 0.25 arcsec with UTs; 1 arcsec with ATs.
- An external fringe tracking system is required for fringes stability. Theoretical following values for the sensitivity are assumed for 3 T mode (3 visibility and one closure phase), Visibility = 1, and for a signal to noise of 50 :
  - Basic mode (10 seconds) : 2.0 Jy with UTs, 40. Jy with ATs
  - Fringe track sequence of 15 minutes : 0.2 Jy with UTs, 4. Jy with ATs
- Expected imaging performance: one aperture synthesis image from 10 to 70 data sets.

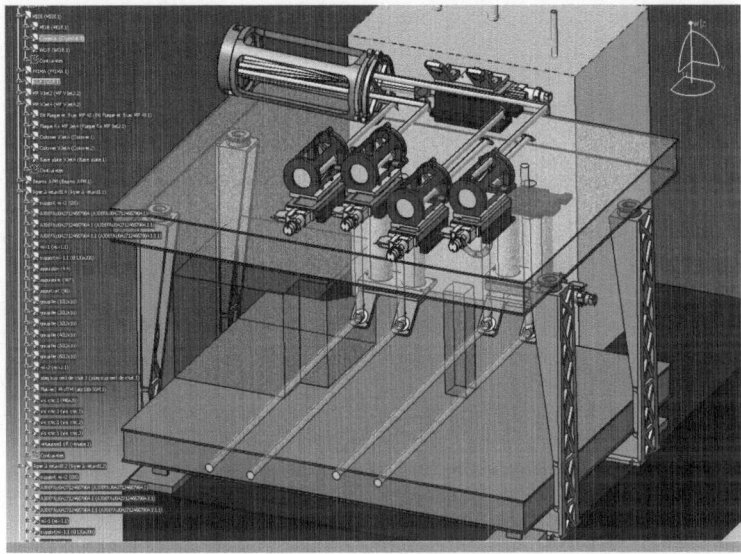

Fig. 6. A 3D Catia view of the APreS-MIDI module.

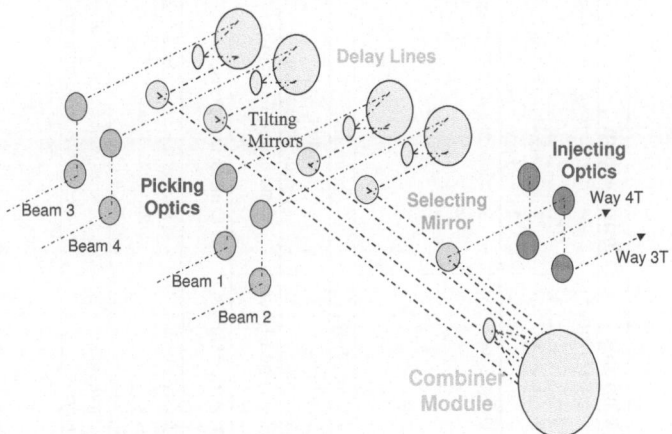

Fig. 7. Optical elements of APreS-MIDI.

2.4 Interfacing APreS-MIDI

The main part of the instrument APreS-MIDI will be located on a bench
above the optical table of MIDI. Some additional elements will be also placed
in the MIDI cryostat. The following figure (Fig. 7) shows a conceptual schema
of the instrument and the table (Tab. 1) gives the name, the composition and
the function of the different modules.

Tab. 1. Details and functions of the optical elements.

Modules	Composition	Function
Feeding Optics	- 4 flat mirrors or dichroics (ESO)	- *Deflect the VLTI beams toward the MIDI table*
Picking Optics	- 2 motorized periscopes - 2 fixed periscopes	- Bypass the MIDI warm optics - Inject the VLTI beams in APreS-MIDI warm optics
Delay Lines	- 4 independent motorized cat's eyes	- Place the pupil at the pupil mask position in the MIDI cryostat - Optimise the OPD - During fringe acquisition, engender an OPD superior to the coherence length in order to adjust the OPD pair by pair of beam
Tilting Mirrors	- 4 flat mirrors	- Produce a tilt angle between the beams
Combiner Module	- 1 collimator - 1 pyramidal mirror	- Produce a tilt angle between the beams - Place the beams on the same averaged axis
Selecting Mirror	- 1 motorized flat mirror	- Select the observation mode (observation with 3 or 4 telescopes)
Injecting Optics	- 2 motorized periscope	- Inject the beams come from the APreS-MIDI warm optics into the MIDI cryostat
Image masks	- *Image masks placed in the pinholes slider of the MIDI cryostat*	- *Reduce the background* - *Determine the fringes sampling*
Photometric Optics	- *Lenses placed in the dispersion slider of the MIDI cryostat*	- *Produce an image plane for a part of the light in order to monitor the flux ratio between the beams*

2.5 Spectral performances

The spectroscopic capability of APreS-MIDI is limited. APreS-MIDI will observe in the relatively broad N spectral band which extends from about 8 to 13 microns of wavelength. We propose two methods for obtaining spectral information with APreS-MIDI. The first method uses the MIDI filter wheel. APreS-MIDI may observe efficiently 2-3 spectral channels for one set of visibility/filter observations without adding too much overhead to the observing sequence. The second method presently foreseen consists in the use of the MIDI Grism for dispersing the pupil image onto the detector. Only

5 spectral channels could be observed simultaneously by this method. The feasibility study shows that only APreS-MIDI in its 3 beams mode will be feasible when using this spectral method. It shows in addition that the pupil camera in front of the detector has to be changed for optimization.

After the implementation of APreS-MIDI, the original MIDI will remain a very efficient spectro-interferometer combiner, in this sense is very complementary to APreS-MIDI, which is designed primarily for imaging. For this reason the APreS-MIDI engineering study concluded that it is important to have an efficient switch from APreS-MIDI to MIDI.

References

1. Apai, D., Pascucci, I., Brandner, W., Henning, Th., et al. 2004, A&A, 415, 671
2. Artymowicz, Annu. Rev. Earth Planet. Sci. 1997, 25, 175
3. Bate, M.R. 2000, MNRAS, 314, 33
4. Bate, M.R., Bonnell, I.A. 1997, MNRAS, 285, 33
5. Bate, M.R., Lubow, S.H., Ogilvie, G.I., Miller, K.A. 2003, MNRAS, 341, 213
6. Beckwith, S. V. W., Henning, Th., Nakagawa, Y. 2000, In Proceedings of Protostars and Planets IV, 533
7. Beckwith S.V.W., Sargent A.I., Chini R.S. and Guesten R. 1990, AJ, 99, 924
8. Bryden, G., Chen, X., Lin, D.N.C., Nelson, R.P., Papaloizou, J.C.B. 1999, ApJ, 514, 344
9. Burrows et al., 1996, ApJ., 473, 437
10. Calvet, N., D'Alessio, P., Hartmann, L., Wilner, D., Walsh, A., Sitko, M. 2002, ApJ, 568, 1008
11. Carciofi, A.C., Bjorkman, J.E., Magalhaes, A.M. 2004, ApJ, 604, 238
12. Chini, R., Hoffmeister, V., Kimeswenger, S., et al. 2004, Nature, 429, 155
13. D'Alessio, P., Hartmann, L., Calvet, N., et al. 2005, ApJ 621, 461
14. D'Angelo, G.D., Kley, W., Henning, Th. 2003, ApJ, 586, 540
15. Dent, W.R.F., Greaves, J.S., Mannings, V., Coulson, I.M., Walther, D.M. 1995, MNRAS, 277, L25
16. Dutrey, A., Guilloteau, S., Simon, M., 1994, A&A, 286, 149
17. Dutrey, A., Guilloteau, S., Duvert, G., Prato, L., Simon, M., Schuster, K., and Ménard, F., 1998, A&A 338, L63
18. Fontani, F., Cesaroni, R., Testi, L., Walmsley, C.M., Molinari, S., Neri, R., Shepherd, D., Brand, J., Palla, F., Zhang, Q. 2004, A&A, 414, 299
19. Greaves, J.S., Coulson, I.M., Holland, W.S. 2000, MNRAS, 312, L1
20. Klahr, H., Bodenheimer, P. 2004, ApJ, 582, 869
21. Kley, W. 1999, MNRAS, 303, 696
22. Kley, W. 2000, MNRAS, 313, L47
23. Liseau, R., Artymowicz, P. 1998, A&A, 334, 935
24. Lissauer, J. J. 1993, ARA&A, 31, 129
25. Lubow, S.H., Seibert, M., Artymowicz, P. 1999, ApJ, 526, 1001
26. Mannings, V. & Sargent, A.I., 1997, ApJ, 490, 792
27. Mannings, V. & Sargent, A.I., 2000, ApJ, 529, 391
28. Natta, A., Testi, L., Neri, R., Shepherd, D.S., Wilner, D.J. 2004, A&A, 416, 179

29. Quillen, A.C., Blackman, E.G., Frank, A., Varnie, P. 2004, ApJ, 612, L137
30. Paardekooper, S.-J., Mellema, G., 2004, A&A 425, 9
31. Rice, W.K.M., Wood, K., Armitage, P.J., Whitney, B.A., Bjorkman, J.E. 2003, MNRAS, 342, 79
32. Schreyer, K., Henning, Th., van der Tak, F.F.S., Boonman, A.M.S., van Dishoeck, E.F. 2002, A&A, 394, 561
33. Shepherd, D.S., Borders, T., Claussen, M., Shirley, Y., Kurtz, S. 2004, ApJ, 614, 211
34. Testi, L., Natta, A., Shepherd, D.S., Wilner, D.J., A&A, 403, 323
35. van Boekel, R., Min, M., Leinert, Ch., Waters, L.B.F.M., Richichi, A., et al. 2004, Nature, 432, 479
36. Zuckermann, B., Forveille, T., Kastner, J.H. 1995, Nature, 373, 494
37. M. Dugué and the APreS-MIDI team: APreS-MIDI, a 4 beam recombiner. The Power of Optical/IR Interferometry: Recent Scientific Results and 2nd Generation VLTI Instrumentation, this volume.
38. J.-L. Menut et al.: Model experiment for APreS-MIDI. The Power of Optical/IR Interferometry: Recent Scientific Results and 2nd Generation VLTI Instrumentation, this volume.
39. Ch. Leinert et al., 2004, SPIE 5491, 19.

VITRUV - Imaging Close Environments of Stars and Galaxies with the VLTI at Milli-Arcsec Resolution

Fabien Malbet[1], Jean-Philippe Berger[1], Paulo Garcia[2], Pierre Kern[1], Karine Perraut[1], Myriam Benisty[1], Laurent Jocou[1], Emilie Herwats[1,3], Jean-Baptiste Lebouquin[1], Pierre Labeye[4], Etienne Le Coarer[1], Olivier Preis[1], Eric Tatulli[1], and Eric Thiébaut[5]

[1] Laboratoire d'Astrophysique de Grenoble, BP 53, F-38041 Grenoble cedex 9, France Fabien.Malbet@obs.ujf-grenoble.fr
[2] Centro de Astrofísica da Universidade do Porto, Rua das Estrelas, 4150-762 Porto, Portugal
[3] Université de Liège, Liège, Belgique
[4] CEA-LETI, Grenoble, France
[5] Centre de Recherche en Astrophysique de Lyon, Lyon, France

Summary. The VITRUV project has the objective to deliver milli-arcsecond spectro-images of the environment of compact sources like young stars, active galaxies and evolved stars to the community. This instrument of the VLTI second generation based on the integrated optics technology is able to combine from 4 to 8 beams from the VLT telescopes. Working primarily in the near infrared, it will provide intermediate to high spectral resolutions and eventually polarization analysis. This paper summarizes the result from the concept study led within the Joint Research Activity *advanced instruments* of the OPTICON program.

1 Introduction

The VLT interferometric facility is unique in the world, since it offers giant 8m telescopes, 2m auxiliary telescopes and the necessary infrastructure to combine them. With four 8m unit telescopes (UTs) equipped with adaptive optics systems, four 1.8m auxiliary relocable telescopes (ATs) equipped with tip-tilt correction, a maximum separation of 130m for UTs and 200m for ATs, 6 available delay lines, slots foreseen for 2 more ones, a dual feed capability (PRIMA) and a complete system control, the VLTI is the best site to propose the first optical interferometer to deliver routinely aperture synthesis images like the millimeter wave interferometers are already doing for more than 10 years. The quality of the images will be as good as the ones delivered by the IRAM Plateau de Bure Interferometer with six 15m antennas and a maximum baseline of 500m at 1-3mm. The VLTI will be for a long time the only facility with 10m class telescopes able to provide images with 1mas angular resolution in optical wavelength.

We propose a second generation instrument for the VLTI, called VITRUV, aimed at taking the best profit of the imaging capability of the array, especially within the PRIMA framework. The science objectives of VITRUV are focused on the kinematics and morphology of compact astrophysical objects at optical wavelengths like the environment of AGN, star forming regions, stellar surfaces and circumstellar environments. The instrument will deliver aperture synthesis images with spectral resolution as the final data product to the astronomer.

The specifications can be summarized as:

- beam combiners for 4T and 8T operation,
- a temporal resolution of the order of 1 day,
- 2 or 3 spectral resolutions from 100 to 30000,
- image dynamics from 100 to 1000,
- a field of view up to1 arcsec
- initial wavelength coverage from 1 to 2.5 microns that could be extended from 0.5 to 5 microns.

The technology that is contemplated at this stage is integrated optics because it offers simplicity, stability especially for phases, operational liability, and high performances. This technology has been already successfully validated on the 3 telescope IOTA interferometer where the system routinely delivers visibilities and closure phases for 3 baselines [1, 2] and on the VLTI to replace the fiber coupler of VINCI [3].

2 Science objectives

The science cases definition methology was to concentrate in a few fields where VITRUV can make a substantial contribution, without being fully exhaustive. We list here the four science cases for the VITRUV concept where significant advance can be achieved:

1. Studying the formation of stars and planets by direct spectro-imaging of their inner disk regions from the orbits of Mercury to Neptune. The goal is to study the structure of the inner disks of young stars (see Fig. 1) as well as the launching of jets and winds (see Fig. 2).
2. Imaging the magnetic and convective hallmarks of stellar surfaces (see e.g. review by Mozurkevitch, 1996, [9]).
3. Connecting the geometries of the close environments of evolved stars with their progenitors (see Fig. 3) and imaging the environmenet of microquasars (see Fig. 4).
4. Probing the close environment of active galactic nuclei (see Figs. 5 and 6) and supermassive black holes (see Fig. 7).

Further details are given in the *VITRUV Science Cases*.

Nota Bene: For reasons of space availibility, it has not been possible to publish the entire Science Case in this volume. The reader can retrieve the prepared preprint on `astro-ph/0507580`.

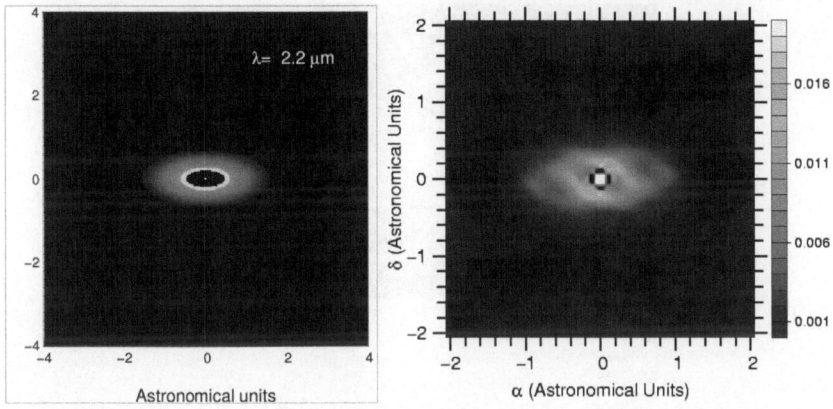

Fig. 1. Simulation of a protoplanetary disk around a Herbig Ae/Be star (Tatulli et al. 2004, [4]). Left: puffed-up inner wall model (star removed). Right: VLTI aperture-synthesis reconstructed image (including the central star). The scale of 1 mas corresponds to 1.5 AU and the ring is about 15 times fainter than the central star.

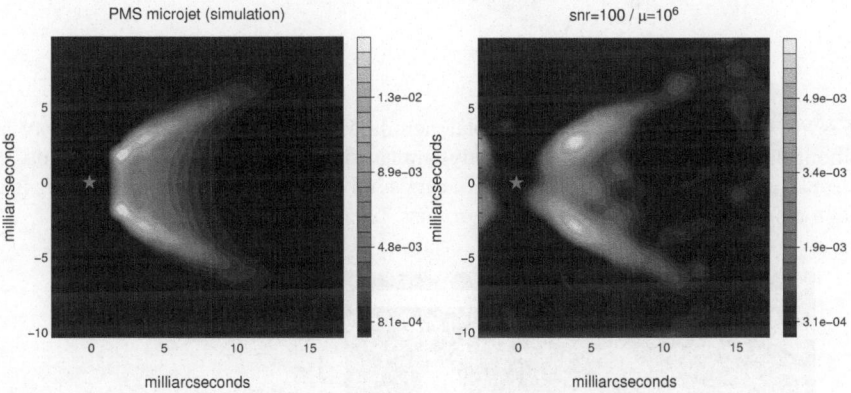

Fig. 2. Paβ emission from the inner region of a MHD disk wind in a T Tauri star (left) and VLTI aperture-synthesis reconstructed image by Thiébaut et al. (2003, [5]). The star contribution, which is on average has a surface brightness 100 times the jet has been removed.

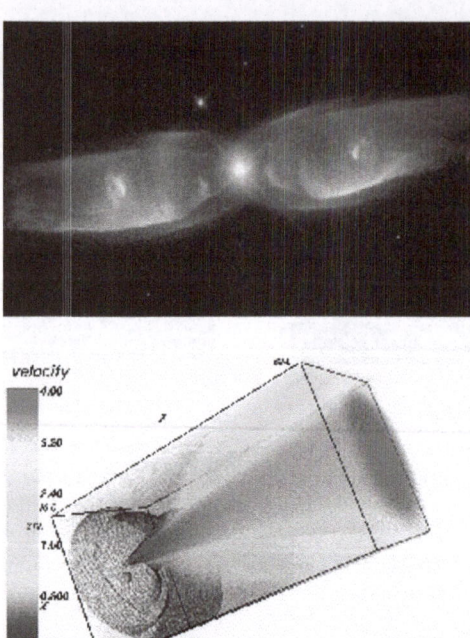

Fig. 3. Top: HST image of the bipolar nebula M 2-9. The long side of the f.o.v is about 30000 AU. Bottom: the hydrodynamical simulation by Garcia-Arredondo & Frank (2004) using an accreting interacting symbiotic-like binary with P=20 yr [6]. The model box includes only the innermost 320×160×160 AU.

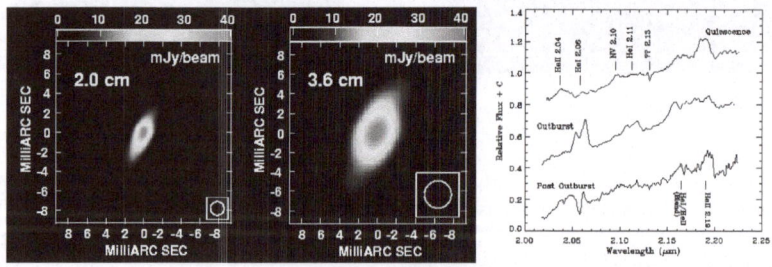

Fig. 4. Left: VLBA images of the microquasar GRS 1915+105 at 2.0 and 3.6 cm on April 2, 2003 showing the compact jet. The convolving beams are 1.4 and 2.8 mas, respectively. 1 mas corresponds to 12 AU at 12 kpc distance. (Fuchs et al. 2003) Right: K-band spectra of Cyg X-3, each displayed with an arbitrary flux offset for clarity. At the top a quiescent spectrum, in the middle a spectrum taken during a very extreme radio/X-ray flaring of the system, at the bottom a "post-outburst." spectrum (Hanson et al. 2000).

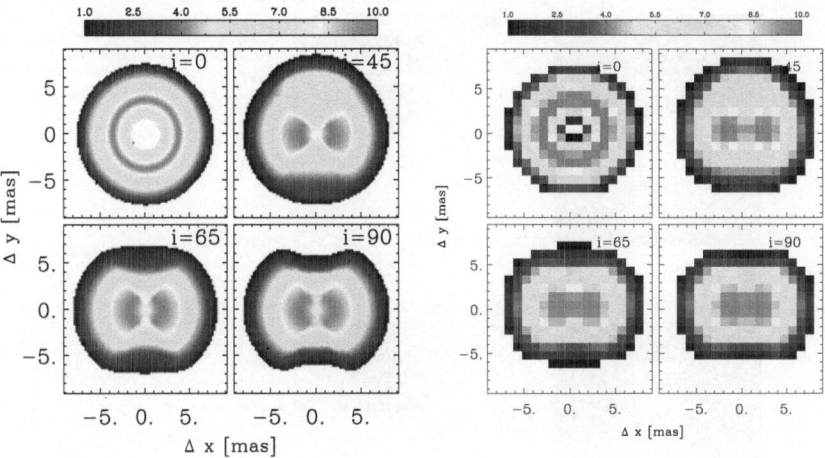

Fig. 5. Left: Model images of the NGC 1068 torus (Granato et al. 2004, [7]) seen at different inclinations in the K band. i is the inclination of the torus pole axis with respect to the line of sight (with $i = 0$ the torus is seen face on). Images are in units of 10^{-10} erg s^{-1} cm^{-2} Å^{-1} except for the $i = 0$ image which is in units of 10^{-10} erg s^{-1} cm^{-2} Å^{-1}. Right: same images as before but convolved with a PSF with 2 mas FWHM and re-binned to 1 mas pixels.

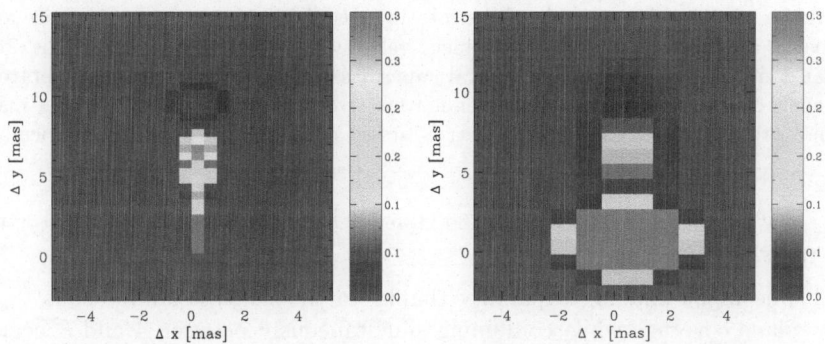

Fig. 6. Left: Model image of the jet of 3C273 in the K band. Image is in units of mJy. This image assumed a conical jet with a simple power law distribution for the particles. The density and magnetic field distributions along the jet axis are tuned to reproduced real VLA radio observations (Mantovani et al. 1999, [8]). Right: same model image as before but with a 2 mass spatial resolution and pixels of 1 mas.

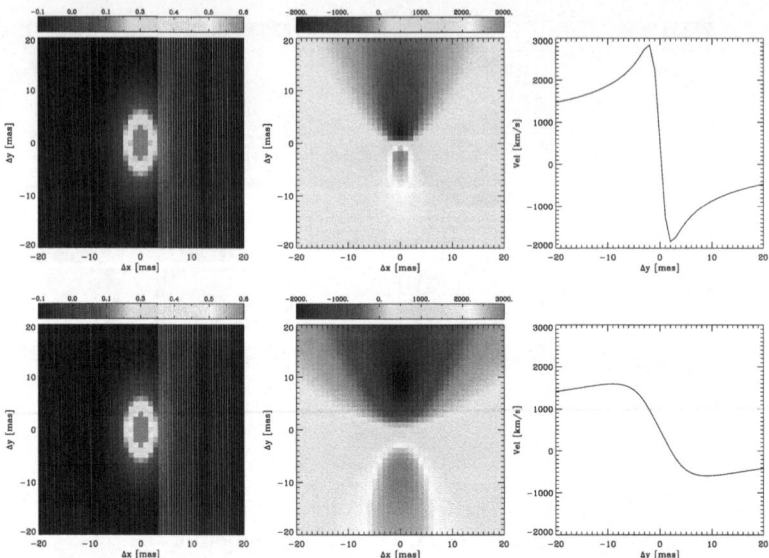

Fig. 7. Left panels: Line images (e.g. Brγ) of the rotating gas disk of Centaurus A which is inclined by 60 deg w.r.t. the line of sight and has the major axis along the y axis. The line images have pixel sizes of 1 mas and spatial resolution of 2 mas. Middle panels: velocity fields of the rotating gas disk with the same pixel size and spatial resolution as before. Left panels: velocity curves along the major axis. Top panels refers to the case of a central supermassive BH of 10^8 $M_\odot$, while bottom panels consider the case of an extended massive dark objects with 10^8 $M_\odot$ mass and core radius of 5 mas. The two cases are clearly distinguishable kinematically.

The top level technical requirements has been deduced from these science requirements:

- One night imaging capability thanks to the largest simultaneous (u,v) plane coverage while combining simultaneously between 4 and 8 beams (can be downgraded to 6)
- Spectral resolution from 100 up to 30,000
- Sensitivity and spectral resolution requires phase stabilization (fringe tracker) and eventually dual beam referencing (PRIMA).

VITRUV has been designed to be the instrument for phase reference imaging (PRIMA) with the telescopes available on the site. In addition it can perform phase closure imaging in the case where there is no adequate reference star in the field. Therefore it is important to consider VITRUV in the PRIMA development scheme although this service is not mandatory for relatively bright targets (up to $K = 11 - 13$ with the UTs).

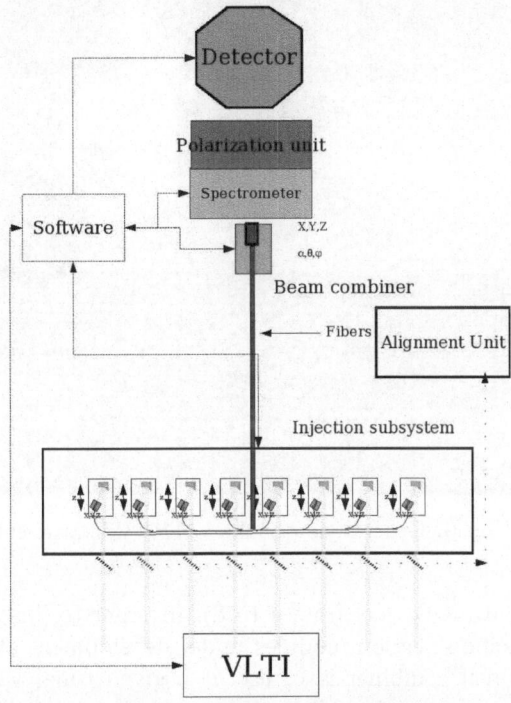

Fig. 8. Vitruv concept

3 VITRUV concept

VITRUV is a non-direct imaging instrument. It measures electromagnetic field complex coherence at different spectral resolutions, different polarization states for a maximum of 28 baselines (8 telescopes). Visibilities, phases and closure phases are the raw observables. It is a single-mode instrument with field of view capability of a few Airy disks. The data product is the reconstructed spectral image cube at all wavelengths.

VITRUV is designed to simplify operation as much as possible. A self-aligned integrated optics (IO) beam combiner is at the heart of the instrument concept. At this stage of the study we plan to have four beam combiners to cover the J, H, K bands:

– optimized 4-way beam combiner for J/H band
– 5 to 8 way beam combiner for J/H band
– optimized 4-way beam combiner for K band
– 5 to 8 way beam combiner for K band,

and two additional ones to extend to R/I and L bands. Each 8-way beam combiner has a maximum of 8 inputs and can be used with whatever combination

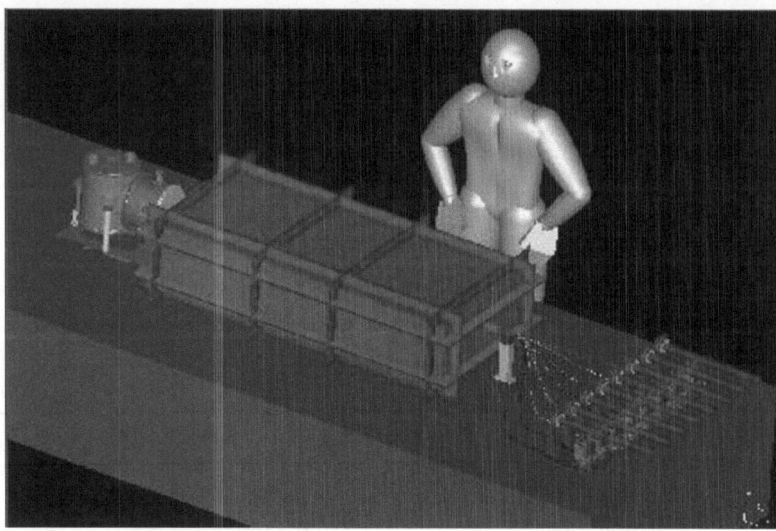

Fig. 9. General view of the VITRUV instrument

of telescopes is available (i.e. from 4 to 8). In order to limit the size of the instrument, specific injection modules under development at LAOG will allow selecting which combiner is in use at a given time. VITRUV concept has a spectral resolution capability with an optional polarization state capability which will be traded-off between astrophysical goals and instrument complexity.

The VITRUV instrument (see Fig. 9) can be described by 6 subsystems as follows.

3.1 Injection subsystem

This system has two functions: (1) to inject VLTI beams into VITRUV beam combiner fibers, and (2) to select the chosen combiner thanks to a motorized translation. Focus will be made thanks to off-axis parabolae in order to discard chromaticity effects. At the focus of each parabola a miniaturized fiber positioner [10] holds N fibers, N being the number of beam combiners (4 in the JH/K only version). Fiber numerical apertures will be chosen in order to optimize an average coupling covering the VITRUV band pass. A 1D translation allows us to select between the N fibers. A miniaturized fiber positioner allows on-sky flux optimization. Each injection module is located on a translation stage designed to equalize optical paths for internal fringe acquisition purposes.

3.2 Integrated optics beam combiners

The optical beam combining function will be achieved thanks to integrated optics (IO) technologies that allow embedding single mode optical circuits in glass or silica chips. As of today we plan to use two different types of beam combiners, one for 4-way beam combination and one for 5 to 8-way beam combination. The beam combination principle (coaxial or multiaxial) is different between the two in order to maximize the signal-to-noise ratio [11].

For the JH/K only version of VITRUV the 4 beam combiners are stacked together (or integrated in the same chip, not defined as of today) on a remotely controlled motorized 6-axes positioner which allows us to position the beam combiner output in front of the detector. Only one beam combiner is operational at the same time, selected by the injection module (i.e. on wavelength band) and, if required, by the combiner positioner.

The total number of degrees of freedom for VITRUV is $8 + 56 = 64$ dof.

3.3 Spectrometer

Preliminary design of the spectrometer will be mainly constrained by science case requirements and technical issues. It will include motorized axes in order to select spectral resolution mode. IO combiners add a supplement interesting property in the sense that the chip can act as an entrance slit for a spectrometer and does not require optical anamorphous transform.

3.4 Detector

The detector choice is still at its premises. The JHK detector will be a nitrogen cooled array with low readout noise but fast enough to be able to read quadrants and sample spectrally dispersed fringes within the atmosphere perturbations eventually compensated by the fringe tracker.

Performant detectors can also now be found for the R/I part of the spectrum and L band. The proposed visible extension requires an additional development for a low noise fast read out detector or photon-counting detector, with performances close to the performances of an AO visible wavefront sensor (see OPTICON development in JRA2 based on an EEV detector 288x288).

3.5 Software

Software development is required to ensure control of all VITRUV motorized elements, camera readout and data acquisition, interface with VLTI software, data reduction, image reconstruction. We do not expect that the VITRUV instrument control software to be very different from the AMBER software except for the number of input beams. We plan to use the maximum of the heritage from the AMBER software.

Like for AMBER the data reduction software (DRS) will be part of the package, but in addition we are working in collaboration with the Jean-Marie Mariotti Center (JMMC, see http://mariotti.fr) to provide image reconstruction software to provide reconstructed images to the users.

3.6 Polarization control

As of today two VITRUV instrumental modes that will deal with polarization issues are contemplated: *polarization split at the output* recording two linear polarizations and relax constrains due to the use of birefringent waveguides. With proper calibration this mode should allow to provide information on the degree of linear polarization of the source; *a full polarization analyzing module* (should the science case demonstrate its importance) where linear and circular differential polarization states will be measured.

4 VITRUV within the VLTI infrastructure

VITRUV is an instrument that requires a full and operational VLTI infrastructure.

UT operation requires adaptive optics capability which is already available with the MACAO systems. The availability of AO systems on AT would significantly improve the capabilities of the instrument, mainly to reach shorter wavelengths ideally down to the R band.

The considered concept allows long exposure acquisition requested for the high spectral resolution mode. Therefore the sensitivity and the spectral coverage of the instrument will be highly improved with fringe tracking capabilities[6]. In addition with the PRIMA facility if adapted to all telescope subsystems, the imaging capability can be pushed towards fainter sources with a bright reference nearby.

5 Expected performances

The performances of the VITRUV instrument depend on the sensitivity of the fringe tracker. We assume here that this fringe tracker can work up to K=11. Since the fringe tracker is a low pass filter, we assume that the piston correction is almost perfect over a few seconds.

We have computed SNR curves with a perfect external PRIMA fringe tracker. The limiting magnitude of this PRIMA fringe tracker should be the

---

[6] This study can be developed in parallel if this fringe tracking facility exists, but if not then it should be added to the project.

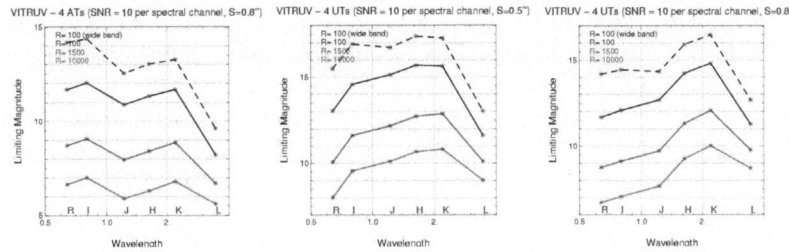

Fig. 10. Limiting magnitude through the VITRUV wavelength coverage. Left panel with 4 ATs, middle panel with 4 UTs and an average seeing of 0.8", and right panel with 4 UTs and an exceptional seeing of 0.5".

one computed by AMBER in the low resolution mode, i.e. $K = 11 - 13$ depending on average seeing conditions. We assumed in our calculation that the AO guide star is bright and is $V = 5$. The conditions remain about the same up to $V = 13$. Since the fringes are stabilized, we used a 100s elementary exposure time. We have not yet fully investigated the dual feed option in simulations (losses due anisoplanetism).

For imaging the requested visibility accuracy does not need to be very stringent. In mm radio interferometry, maps are produced with 10% visibility errors. More important is the (u, v) coverage. A typical visibility accuracy below 1% and phase accuracy of the order of 1 degree is sufficient and already achieved [12, 3, 1, 2].

The field of view of VITRUV is fundamentally limited by the FOV accessible by the injection fibers, i.e. in K 250 mas with the ATs and 60 mas with the UTs. Most of the science which is contemplated focuses on compact objects within this limit. However like for radio interferometry, we know that we can extend the FOV by performing mosaicing (Tatulli, 2004).

We focus the project on objects with a bright central reference, but no so bright that we need to cancel it. Basically a dynamic range between the faintest and the brightest features in the image between 100 and 1000 are achievable. We have started to build a software-based end-to-end simulator of VITRUV which allows us to investigate this issue with more accuracy.

6 Project management

At this stage, it is difficult to predict an accurate evaluation of the resources required for the project.

Cost. A first rough estimation for an instrument operable in the 1-2.5μm range leads to a cost of 1.2 MEuros. The additional cost to extend the wavelength range (0.6 - 4 μm) is 1.1 MEuros. This extension includes an additional camera for the visible and some technological development for the integrated optics components.

Manpower. The 1-2.5 μm instrument requires a manpower support of 56 FTE for its design and construction. The required additional manpower is 22 FTE for a 0.6-4 μm extension.

Schedule. The required time to achieve the full manufacturing of the 1-2.5 μm instrument is 3.5 years. Six additional months will be required to achieve the extended version, taking into account that main additional developments for the proposed extension will be carried out in parallel during the whole project.

Collaborations. For the moment, a formal consortium has not yet been established. We are waiting for the end of the selection process to start working on this issue. We are very open to various forms of collaborations.

References

1. Monnier, J.D., Traub, W.A., Schloerb, F.P., Millan-Gabet, R., Berger, J.P., Pedretti, E., Carleton, N.P., Kraus, S., Lacasse, M.G., Brewer, M., Ragland, S., Ahearn, A., Coldwell, C., Haguenauer, P., Kern, P., Labeye, P., Lagny, L., Malbet, F., Malin, D., Maymounkov, P., Morel, S., Papaliolios, C., Perraut, K., Pearlman, M., Porro, I.L., Schanen, I., Souccar, K., Torres, G., Wallace, G.: First Results with the IOTA3 Imaging Interferometer: The Spectroscopic Binaries λ Virginis and WR 140. ApJ **602** (2004) L57–L60
2. Kraus, S., Schloerb, F.P., Traub, W.A., Carleton, N.P., Lacasse, M., Pearlman, M., Monnier, J.D., Millan-Gabet, R., Berger, J.P., Haguenauer, P., Perraut, K., Kern, P., Malbet, F., Labeye, P.: Infrared Imaging of Capella with the IOTA Closure Phase Interferometer. AJ **130** (2005) 246–255
3. LeBouquin, J.B., Rousselet-Perraut, K., Kern, P., Malbet, F., Haguenauer, P., Kervella, P., Schanen, I., Berger, J.P., Delboulbé, A., Arezki, B., Schöller, M.: First observations with an H-band integrated optics beam combiner at the VLTI. A&A **424** (2004) 719–726
4. Tatulli, E., Thiebaut, E.M., Malbet, F., Duvert, G.: Imaging young stellar objects with AMBER on the VLTI. In: New Frontiers in Stellar Interferometry, Proceedings of SPIE Volume 5491. Edited by Wesley A. Traub. Bellingham, WA: The International Society for Optical Engineering, 2004., p.117. (2004) 117–+
5. Thiébaut, E., Garcia, P.J.V., Foy, R.: Imaging with Amber/VLTI: the case of microjets. Ap&SS **286** (2003) 171–176
6. García-Arredondo, F., Frank, A.: Collimated Outflow Formation via Binary Stars: Three-Dimensional Simulations of Asymptotic Giant Branch Wind and Disk Wind Interactions. ApJ **600** (2004) 992–1003
7. Granato, G.L., De Zotti, G., Silva, L., Bressan, A., Danese, L.: A Physical Model for the Coevolution of QSOs and Their Spheroidal Hosts. ApJ **600** (2004) 580–594
8. Mantovani, F., Junor, W., Valerio, C., McHardy, I.: Results of VLBI monitoring of 3C273 at 22 GHz and 43 GHz. New Astronomy Review **43** (1999) 737–740
9. Mozurkewich, D.: Interferometric imaging of stellar surfaces (review). In: IAU Symp. 176: Stellar Surface Structure. (1996) 131–+

10. Preis, O., Pichon, L., Delboulbe, A., Kern, P.Y., Magnard, Y., Ventura, N.:
    Three-dimensional micropositioning device for optical fiber guided by a piezo-
    electric tube. In: New Frontiers in Stellar Interferometry, Proceedings of SPIE
    Volume 5491. Edited by Wesley A. Traub. Bellingham, WA: The International
    Society for Optical Engineering, 2004., p.1379. (2004) 1379-+

11. Lebouquin, J.B., Berger, J.P., Labeye, P., Tatulli, E., Malbet, F., Rousselet-
    Perraut, K., Kern, P.: Comparison of integrated optics concepts for a near-
    infrared multi-telescope beam combiner. In: New Frontiers in Stellar Interfer-
    ometry, Proceedings of SPIE Volume 5491. Edited by Wesley A. Traub. Belling-
    ham, WA: The International Society for Optical Engineering, 2004., p.1362.
    (2004) 1362-+

12. Berger, J.P., Haguenauer, P., Kern, P., Perraut, K., Malbet, F., Schanen, I.,
    Severi, M., Millan-Gabet, R., Traub, W.: Integrated optics for astronomical
    interferometry. IV. First measurements of stars. A&A **376** (2001) L31–L34

VIDA: A Direct Spectro-Imager for the VLTI

Olivier Lardière[1] and Jean Schneider[2]

[1] Collège de France, OHP, F-04870 St-Michel-l'observatoire, France
   ladiere@obs-hp.fr
[2] LUTH, Observatoire de Paris-Meudon, F-92000 Meudon, France
   schneider@obspm.fr

Summary. In order to exploit the unique full forthcoming VLTI infrastructure (4 to 8 telescopes with AO and cophasing system), we propose VIDA, a near-IR direct imaging instrument using an all-to-one beam combiner and single-mode fibers. The pupil densification technique is used to concentrate all the flux in the field accessible by the VLTI array in one observation. This optimal use of photons provides more luminous and contrasted snapshot images than the Fizeau mode. Thanks to its snapshot capabilities ant its better sensitivity, this innovative instrument should open new investigation fields for the VLTI, as faint extragalactic sources studying and hot jupiters coronagraphic imaging.

1 Introduction

The next generation instrumentation for long-baseline stellar interferometry relies, in general, on the availability of a cophased array involving a larger number of apertures (cf. other proposals in these proceedings) in order to map and study more complex or fainter sources.

In such conditions, direct imaging seems to be the natural next step for stellar interferometry. Indeed, the transition to come, from coherenced to cophased interferometry, is very analogous to the transition we have already lived through, from speckle interferometry to adaptive optics imaging with single-dish telescopes (Fig. 1).

Thanks to this comparative vision, we understand that an all-to-one beam combination for long-exposure direct imaging seems to be the most elegant and efficient solution for observing with cophased arrays. Persisting in using techniques initially developed to recover a part of the phase lost in the turbulence, such as closure phase, is a conservative attitude inducing unjustified complications.

2 General description of the instrument

High angular resolution snapshot imaging seems now possible with an all-to-one beam combiner and a cophasing system installed on current interferometers. The current VLTI is able to combine light from 3 telescopes

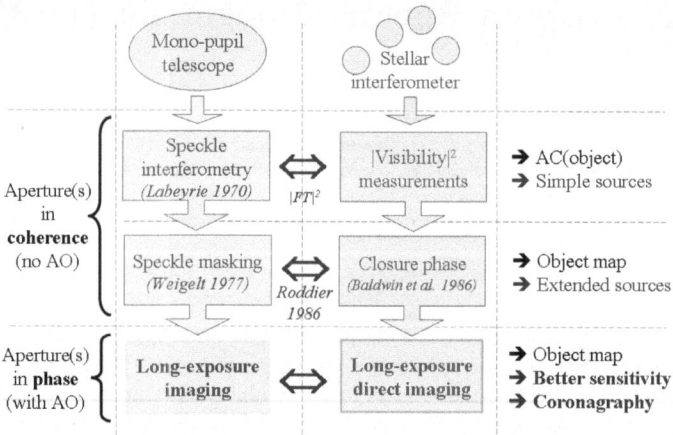

Fig. 1. Evolution and analogies of high angular resolution techniques used on single telescopes and in stellar interferometry: Speckle interferometry [2] and speckle masking [3] have been used on single telescope to recover a part of the information lost in the turbulence. Now, long-exposure imaging with adaptive optic systems is preferred to have a better sensitivity and to allow coronagraphy. Stellar interferometers know the same evolution: visibility modulus measurements and closure phase are equivalent to the speckle interferometry and the speckle masking respectively [4]. Then, direct imaging on cophased arrays is clearly the next step for stellar interferometry.

coherently, but the combination of 4 to 6 cophased beams is foreseen in subsequent phases. The full VLTI array will consist of 8 telescopes (4 UTs and 4 ATs) with adaptive optics or tip-tilt correctors (MACAO and STRAP), 6 delay lines, a dual-feed facility and a phase-referenced metrology system (PRIMA).

In order to exploit this unique forthcoming VLTI infrastructure, we propose here a new instrument for direct sprectro-imaging in the near-IR wavelengths (possibly in the visible) referred to as VIDA for "Vlti Imaging with a Densified Array".

Whereas the Fizeau imaging mode (homothetic mapping) gives very low dynamic-range images when the aperture is highly diluted, the "densified pupil" or "hypertelescope" imaging mode [8] concentrates most light into the central interference peak, providing very luminous images (Fig. 2) allowing direct imaging of very faint sources, and also stellar coronagraphy. This promising imaging mode has been validated in laboratory and on the sky [9], [10].

Far from inducing field loss, as it is often reproached, an appropriate pupil densification actually makes an optimal use of collected photons, making the direct imaging field (DIF) equal to the field accessible to the array ($\sim N \times N$

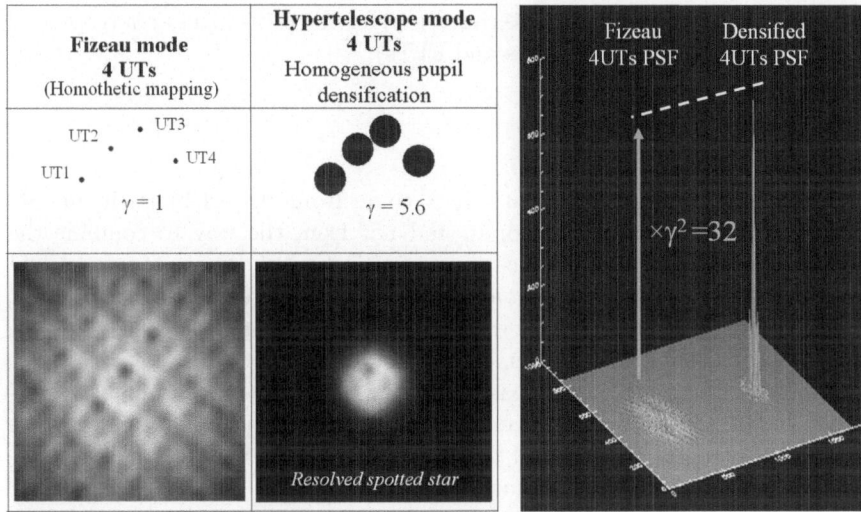

Fig. 2. Left: Exit pupil and direct imaging of a resolved spotted star with the VLTI (simulation by F. Patru). Right: Densified 4UT PSF compared to the Fizeau 4UT PSF; the central interference peak is magnified by a factor 32.

pixels, with N the number of apertures). This effect is well visible on the Fig. 2 showing the Fizeau and densified images of an extended source.

Moreover, the flux of a non-resolved source is spread over only few detector pixels, whatever the aperture number, reducing drastically the contribution of the read-out noise. For this reason, VIDA will be significantly more sensitive than the Fizeau mode, and *a fortiori* than the classical pair-wise observation mode requiring several detectors or a large amount of pixels to sample properly all the frequencies (AMBER case). With 4 UTs, VIDA offers a luminosity gain of 32 ($\sim$ 4 *mag.*) compared to a Fizeau imaging mode, and 6 magnitudes compared to AMBER (2UTs).

Thanks to its high sensitivity (limiting magnitude is 20 in K, cf. sec. 6) and its coronagraphic capabilities, VIDA will open new investigation fields for the VLTI by studying extragalactic sources and hot jupiters. Moreover, VIDA should also open the doors of the stellar interferometry to a larger community by providing direct exploitable images to astronomers. In this way, VIDA is a return to a more natural and conventional telescope concept, resembling a masked Extremely Large Telescope (ELT). Finally, the main objectives of this proposal is to :

– exploit the unique full forthcoming VLTI infrastructure (4 to 6 and maybe 8 telescopes),
– push back the frontiers of stellar interferometry with direct imaging of faint extragalactic sources and also planetary systems by coronagraphy,

– prepare the science and the technologies for the post-VLTI area (cophased large arrays, hypertelescopes and ELTs).

3 Science Cases

The fundamental limitations of VIDA come from the VLTI itself, mainly the number of available telescopes, and not from the way to combine the beams. Then, the general science cases of VIDA will be those of other VLTI instruments, classically compact and bright sources, as envelopes of hot stars, inner regions of protoplanetary disks around TTauri or Herbig Ae/Be stars, close binary stars, bright AGNs, etc.

However, thanks to its snapshot imaging capability and its higher sensitivity, VIDA will open new scientific fields for interferometry concerning the extrasolar planets (Fig. 3) and the faint extragalactic sources. Among the most original science cases taking the advantages of VIDA, we can cite:

– Day-night flux and spectra variations of hot jupiters (atmospheric winds, thermalization...).
– Hot jutiters around red giant stars (such planets cannot be detected by coronagraphy because the parent star is too big, but they could be seen by standard imaging with VIDA thanks to the high angular resolution).
– Detection of binary planets, binary brown dwarfs and planet satellites by direct imaging or differential astrometry with the VIDA internal dual-feed mode (VIDA could see the wobble due to a 10 Earth-mass moon on a Titan-like orbit around a Saturn-like planet at 5 pc).
– Dynamical derivation of planet masses (planet masses are not well known because they are derived from spectra and models, a possible solution is to find a low mass companion on high resolution images).
– Search for additional images of known gravitationally lensed QSOs, where an even number of images is observed, while the theory predicts an odd number (e.g. a 5th image is predicted on HS 0810+2554).
– Detection of quasar optical jets (e.g. 3C273).

All these objectives completes perfectly those of current or planned VLT/VLTI instruments. In summary, VIDA will be an extension of AMBER toward the extragalactic faint sources, and also an extension of VLT-Planet-Finder toward the very high angular resolution.

4 High-level requirements

4.1 Interface with infrastructure

The VIDA instrument will be located in the VLTI focal laboratory, and will be "plugged" behind the beam compressor by single-mode optical fibers.

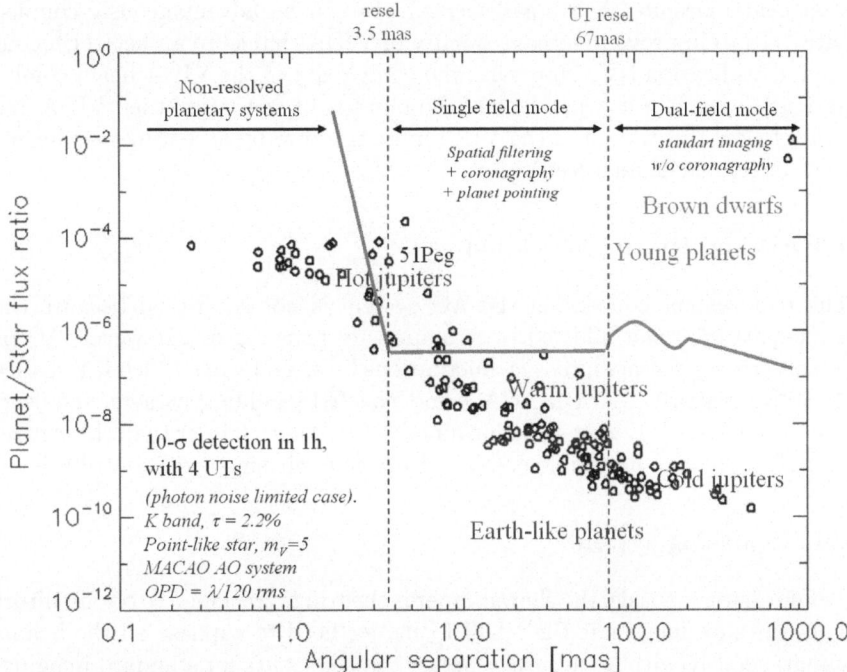

Fig. 3. Brightness ratio vs. angular separation diagram of known extrasolar planets. The detection profile of VIDA (3σ in 10 hours) limited by the star photon noise is plotted for the densified pupil mode (4UTs, MACAO, perfect coronagraph, $\lambda/120\ rms$ cophasing errors).

As other interferometric instruments, VIDA will require a stabilization of telescope beams and fields and correction of the atmospheric dispersion.

VIDA can be interfaced with the PRIMA infrastructure to take the advantage of the dual-feed and the metrology system. Thus, VIDA will be able to perform phase referencing on two separate fields using more than 2 telescopes. This will allow direct imaging on very faint sources and differential astrometry (Fig. 7).

VIDA requires optical fibers and a fringe tracker for long-exposure like VITRUV, a spectro-imager by aperture synthesis proposed by Malbet et al. (these proceedings) for the VLTI. Then VIDA could be advantageously coupled with VITRUV to take the advantage of the fiber injection modules, and possibly the spectrometer and the IR camera.

4.2 Number and type of telescopes

The VIDA concept uses an all-to-one beam combination scheme which can manage simultaneously any number of apertures of any size (ATs and/or UTs, dual field). Despite their smaller size, ATs can be advantageously coupled with UTs to improve the image quality of VIDA thanks to an heterogeneous pupil densification [12]. Moreover, the complexity of the VIDA beam combination scheme is not dependent of the number of apertures, then VIDA can easily follow the growth of the VLTI infrastructure by accepting new beams (new apertures or dual-feed).

4.3 AO and tip-tilt correction

The requirement concerning the AO system is not so critical because the fringe pattern is not affected by sub-aperture phase errors. However, AO or tip-tilt errors generate intensity fluctuations between beams which will reduce the PSF contrast. With MACAO and the IRIS guiding camera, intensity fluctuations reach 12%, generating an halo 1000 times fainter than the central peak. This halo is negligible compared to the cophasing error contribution.

4.4 Cophasing system

As for other proposals, the showstopper is the fringe tracking. To obtain direct long-exposure images at the VLTI focus, we need to cophase all the beams simultaneously with an accuracy better than $\lambda/4$ rms for standard imaging. However the cophasing requirement is more severe for high-contrast imaging and coronagraphy. A $\lambda/120$ rms accuracy is needed to meet the science goals of VIDA. An internal accurate FSU will be needed for coronagraphy, as a 2nd stage of the PRIMA FSUs.

5 Concept analysis

VIDA is a direct imaging instrument for the near-IR (extension to visible and to L band are possible) using an all-to-one beam combination scheme and single-mode fibers. One fiber set is planned for J/H bands, and another set for K band. To increase the sensitivity and allow coronagraphy, the output pupil is densified according the hypertelescope or the IRAN [13] concepts (both are equal when fibers are used for the beam combination). Figure 4 shows the general optical layout of VIDA, which consists in six subsystems:

- injection modules,
- fiber beam combiner,
- internal focal plane FSU,
- coronagraph,
- differential static delay lines for off-axis pointing,
- focal instrumentation.

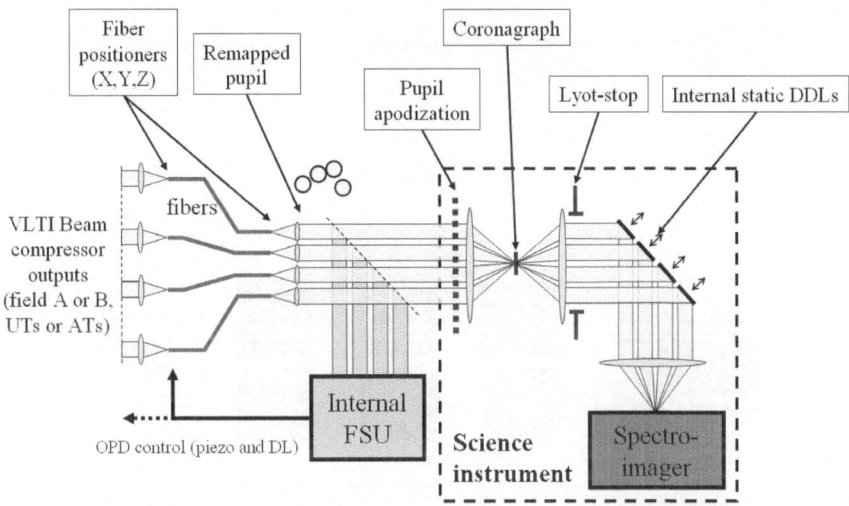

Fig. 4. General optical layout of the VIDA instrument.

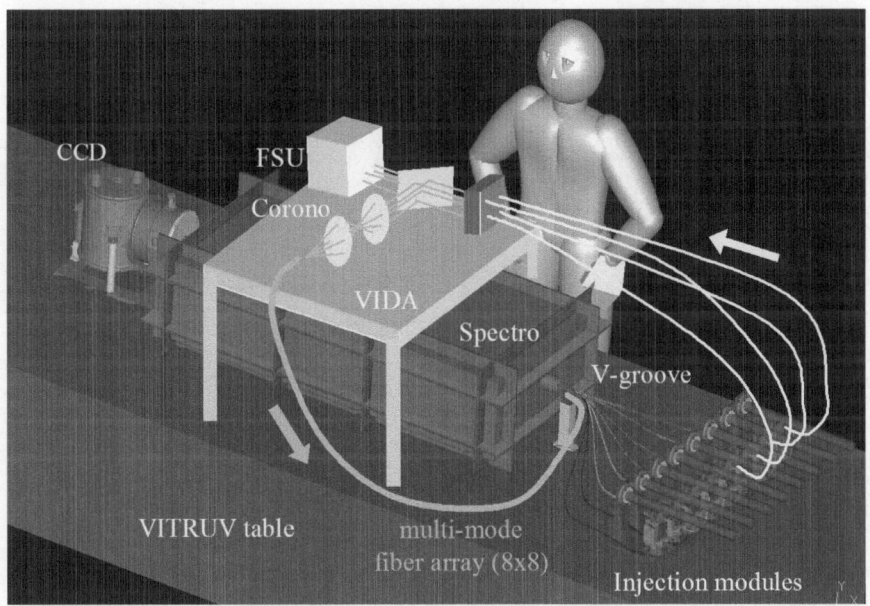

Fig. 5. Possible implementation of VIDA on the VITRUV instrument (drawing adapted from Malbet et al.).

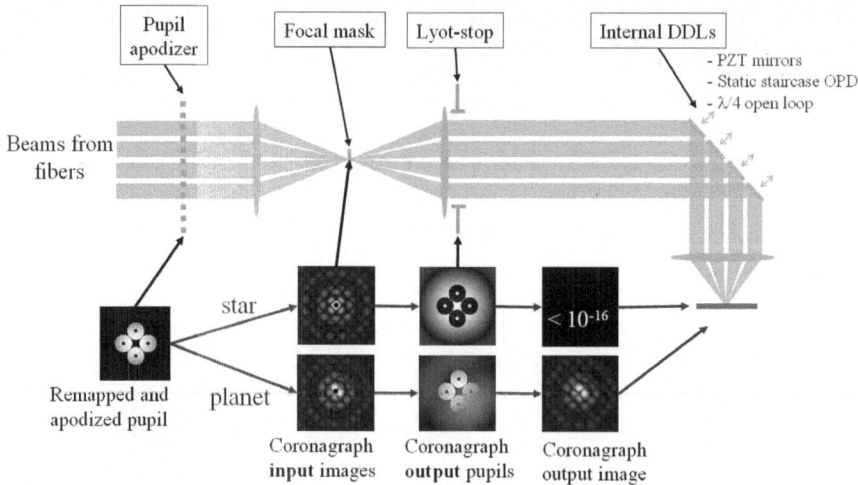

Fig. 6. Coronagraphy on the VLTI with a Roddier phase mask and a pupil reconfiguration and apodization. The star extinction is total and the planet transmission is 64% (separation= 1 *resel* (4 *mas* in K), intensity scale in power 0.3). A known planet outside the field can be recentered by introducing a static staircase OPD between the beams with the internal small DDLs.

As the high-level requirements of VIDA are close to those of VITRUV (sec. 4), we can propose VIDA as a "direct imaging module" of VITRUV in order to reduce the cost of VIDA, as well as the risk taken by ESO. Figure 5 shows a possible implementation of VIDA on the VITRUV bench. As the VITRUV injection module can select a chosen combiner by motorized translation, we can add a new position for feeding the VIDA combiner. In the same way, the VIDA output image plane can be transmitted to the entrance slit of the VITRUV spectrometer by a small array of multi-mode fibers.

A coronagraph is planned for analyzing known hot jupiters. If there are no wavefront corrugations and if the star is not resolved, a total starlight extinction is possible with a Roddier phase mask (Fig. 6). A Lyot coronagraph is an alternative solution for resolved stars.

Stellar coronagraphy is the most demanding application concerning the cophasing: $\lambda/120\ rms$ is needed. To met this requirement, an internal FSU is planned and has to be installed as close as possible to the coronagraph in order to reduce the non-common path. Focal plane FSUs proposed for the VLTI, such as the "phase-retrieval" technique developed by ONERA [14] and the new "dispersed speckles" technique developed by the LISE group [15], are very promising and interesting for VIDA.

As the FSU should be fed by the same fibers than the science channel, a beam-splitter (for the K band) or a dichroic (sensing in J, science in H) is needed to share the flux between the two channels. This beam-splitter can

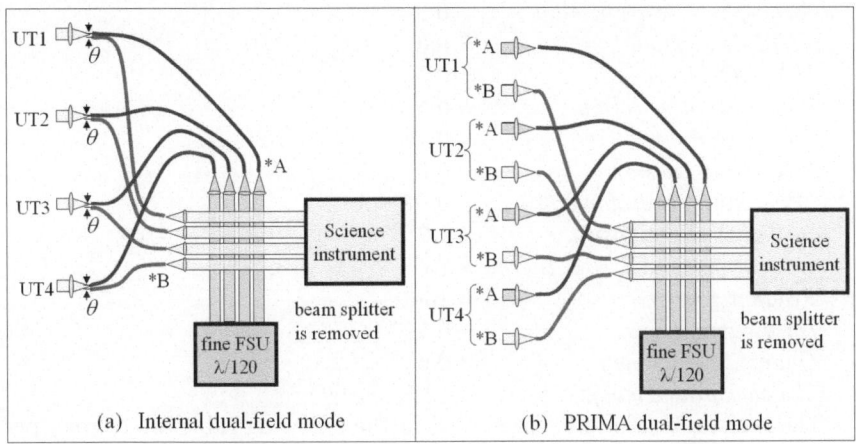

(a) Internal dual-field mode   (b) PRIMA dual-field mode

Fig. 7. Internal and PRIMA dual-field modes of VIDA. (a): the internal dual-field mode can manage 2 fields separated from λ/d up to the field transmitted by the VLTI coudé train (2" for UTs, 8.9" for ATs). The faint source (*B) is recentered by internal static DDLs. (b): The PRIMA dual-field mode can explore larger angular separations up to the isoplanetic angle (10" in K); the faint source (*B) is then recentered by the PRIMA DDLs.

be removed when VIDA is used in dual-field modes (internal or PRIMA) as shown on figure 7.

6 Performances

6.1 Limiting magnitude

In practice, performances will be mainly limited by the FSU sensitivity. The limiting magnitude of the "dispersed speckle" FSU is about V=15 for standard imaging in K ($\lambda/10$ rms), and about V=9 for stellar coronagraphy in K ($\lambda/120$ rms).

To compute the intrinsic limiting magnitude in the K band of VIDA (with a perfect FSU or in the dual-feed mode), we have considered the following parameters :

Atmosphere transmission	$= 0.8$
VLTI throughput	$= 0.3$
MACAO Strehl	$= 0.5$
Polarization selection	$= 0.5$
Fiber injection	$= 0.75$
Fiber transmission	$= 0.9$
"Pistonic" Strehl for $\lambda/10$	$= 0.67$
$\Rightarrow$ **Total throughput**	**=2.7%**
Central pixel encircled energy	$= 73\%$
Readout noise	$= 16e^-$
Dark current	$= 0.1e^-/s$
Quantum efficiency	$= 0.6$
Sky background mag.	$= 12.5/arcsec^2$

The limiting magnitudes presented in the following table are defined per one resel (resolution element, i.e. 4 mas in K), for SNR=10 in a 10s exposure time and with a maximum pupil densification ($\gamma = 5.6$ for UTs, and $\gamma = 26$ for ATs in a UT-like array) :

	Imaging	Spectro-imaging		
		R=100	R=1500	R=30000
4 UTs ($\gamma = 5.6$)	20.5	17.6	14.6	11.4
4 ATs ($\gamma = 26$)	17.5	14.2	11.3	8.1

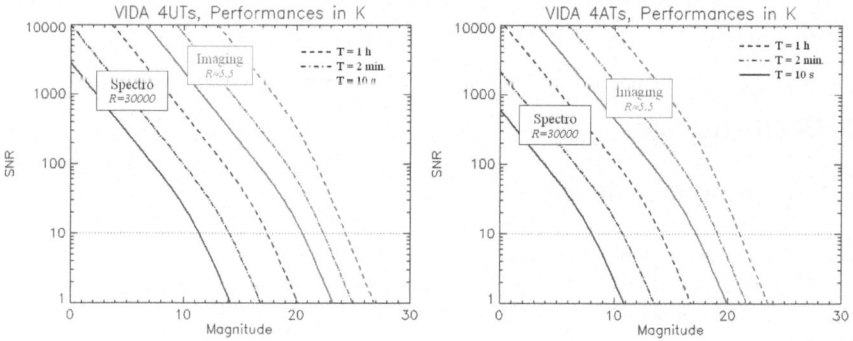

Fig. 8. SNR reached in imaging and spectroscopic mode after 10s/2min/1h of integration with 4 UTs (left) and with 4 ATs (right).

Figure 8 plots the SNR as a function of the magnitude. VIDA is so sensitive because 73% of the flux of a point-like source falls on one pixel. In other words, VIDA is equivalent to a 130m (and more) telescope for the resolution, and to a 16m telescope with a 2% throughput for the collecting surface.

6.2 Field of view

As we told in section 2, the field is not limited by VIDA itself, but by the few number of telescopes involved in the VLTI. All other VLTI instruments are confronted also to this limitation. With 4 UTs, the accessible field diameter after one observation is about 4 resels, whatever the combination scheme. The goal of the pupil densification is just to concentrate all the flux in the field accessible by the VLTI array, in order to maximize the sensitivity without losing information. The DIF of VIDA is then 15mas in K band with 4UTs. To map extended sources larger than the DIF, super-synthesis is possible using a stack of direct images taken at different hour angles [16].

6.3 Dynamic range

The contrast of direct images provided by VIDA will be mainly limited by the cophasing accuracy. The intensity ratio between the central peak of the PSF and the background halo is $N/(1 - S)$, with N the number of apertures and S the "pistonic" Strehl ratio. This formula gives a contrast of 10 for $\lambda/10\ rms$, 200 for $\lambda/40\ rms$ and 1500 for $\lambda/120\ rms$.

7 R&D plan and schedule

A test-bench, referred to as SIRIUS is under development at OCA to validate in laboratory a beam combiner and densifier using single-mode fibers [17]. The goals of this prototype is to:

- compare performances of direct imaging vs. closure phase on different sources,
- validate the fiber beam combiner,
- check AO, tip-tilt and piston specifications,
- test the internal FSU in close loop,
- test the coronagraph and speckle noise reduction techniques,
- test software tools on real images.

For the internal FSU, first results from ONERA test-bench (BRISE) developed for the DARWIN mission are expected soon, as well as from the "dispersed speckle" technique. Lessons learned from VLT-PF study should help for the coronagraph development.

If VIDA is selected for a feasibility study before the end of 2005, the first direct image obtained with the VLTI is planned for 2010, and for 2012 with coronagraph and internal FSU. This schedule assumes that VIDA is a module of another instrument, as VITRUV.

8 Conclusion

To conclude, we would like to remind that if the VLTI is cophased, then direct imaging becomes the more natural and efficient way to observe, and does not induce any information loss. In such conditions, the VLTI will be equivalent to a 130m telescope with a 16m full aperture, and will be very sensitive (K=20.5 in only 10s).

Moreover, VIDA will offer unique capabilities, as snapshot imaging, coronagraphy and differential astrometry (with the PRIMA dual-feed). VIDA will open new scientific fields for interferometry concerning the extrasolar planets and the extragalactic sources.

Lastly, VIDA should make the interferometry more popular and should be a first step toward larger cophased arrays, such as hypertelescopes and ELTs.

Acknowledgments The authors are grateful to the whole working group involved in the VIDA concept study: F. Patru, V. Borkowski, D. Mourard, F. Martinache, A. Labeyrie, P. Antonelli, L. Arnold, D. Bonneau, Y. Bresson, M. Carbillet, F. Cassaing, O. Chesneau, J-M. Clausse, J. Dejonghe, L. Delage, S. Lagarde, B. Lopez , L. Mugnier, F. Reynaud, G. Rousset, A. Schutz, Ph. Stee and A. Spang.

References

1 Baldwin, J. E., Haniff, C. A., Mackay, C. D., & Warner, P. J. 1986, *Nature*, **320**, 595
2. Labeyrie, A. 1970, *A&A*, **6**, 85
3. Weigelt, G. P. 1977, *Optics Communications*, **21**, 55
4. Roddier, F. 1986, *Optics Communications*, **60**, 145
5. Lardiere, O., Mourard, D., Patru, F., & Carbillet, M. 2004, *Proc. SPIE*, **5491**, 415
6. Lardière, O. 2004, *EAS Publications Series*, **12**, 299
7. Lardiere, O., Labeyrie, A., Mourard, D., Riaud, P., Arnold, L., Dejonghe, J., & Gillet, S. 2003, *Proc. SPIE*, **4838**, 1018
8. Labeyrie A., 1996, *A&A Suppl. Ser.*, **118**, 517.
9. Pedretti, E., Labeyrie, A., Arnold, L., Thureau, N., Lardiere, O., Boccaletti, A., & Riaud, P. 2000, *A&A Suppl. Ser.*, **147**, 285
10. Gillet, S., et al. 2003, *A&A*, **400**, 393
11. Malbet, F. et al. 2005, *these proceedings.*
12. Labeyrie, A. 2004, *Proc. SPIE*, **5382**, 205
13. Vakili, F., et al. 2004, *A&A*, **421**, 147
14. Cassaing, F. et al. 2005, *these proceedings.*
15. Borkowski, V. et al. 2005, *these proceedings.*
16. Carbillet, M., et al. 2002, *A&A*, **387**, 744
17. Patru, F. et al. 2005, *these proceedings.*

UVES-I: Interferometric High-Resolution Spectroscopy

Andreas Quirrenbach[1], Simon Albrecht[1], Ramon Vink[1], Oskar von der Lühe[2], Josef Hron[3], and Günter Wiedemann[4]

[1] Sterrewacht Leiden, P.O. Box 9513, NL-2300 RA, Leiden, The Netherlands
quirrenb@strw.leidenuniv.nl
[2] Kiepenheuer-Institut für Sonnenphysik, Schöneckstr. 6, D-79104 Freiburg,
Deutschland
[3] Institut für Astronomie der Universität Wien Türkenschanzstraße 17, A-1180
Wien, Austria
[4] Hamburger Sternwarte, Gojenbergsweg 112, D-21029 Hamburg

Summary. A combination of high spatial and spectral resolution in optical astronomy would enable completely new observational approaches to many open problems in stellar and circumstellar astrophysics. In this paper we will show that by combining the VLTI and the UVES spectrograph one could get access to crucial information about atmospheric structure and rotation of stars on a fast timescale and for a fraction of the cost of a "normal" instrument for the VLTI, which would have to be built from scratch.

1 Introduction

The UVES-I concept combines high spatial and spectral resolution by providing a fiber link from the VLTI to the UVES spectrograph. The general philosophy of the project is to take advantage of previous investments in the VLTI infrastructure and in the UVES instrument, to leave existing hardware untouched, and to implement the links between existing major systems in the simplest way possible. UVES-I can thus create new capabilities for the VLTI at one tenth of the cost of a complete new instrument.

UVES-I will combine the light from two VLTI Auxiliary Telescopes. It will work in conjunction with a PRIMA Fringe Sensing Unit, which stabilizes the optical path difference between the two interferometer arms to a fraction of a wavelength in the visible. The UVES-I hardware consists of three components:

- A beam combiner on a $45\,\mathrm{cm} \times 85\,\mathrm{cm}$ breadboard that accepts two input beams and feeds four outputs carrying the fringe signals into fiber feeds;
- Four fibers that connect the VLT Interferometer lab to UVES;
- A fiber head that feeds the light from the four fibers into UVES (similar to the existing head with the fibers from FLAMES).

The UVES instrument will be operated in one of its standards modes, and produce data in its standard format. The fact that the four fibers carry interferometric fringe signals rather than light from four separate stars is essentially "transparent" to UVES. This reduces the amount of instrument control and data reduction software drastically. The impact on the single-telescope use of UVES will be minimal, as UVES-I can be used during dark time while UT2 is used with a different instrument.

Taking advantage of the unique infrastructure and instrumentation of the Paranal Observatory, UVES-I can thus bring the following unparalleled capabilities to the VLTI in a time- and cost-efficient manner:

–   High spectral resolution ($R = 60,000$) interferometry, sufficient to resolve absorption lines even in many late-type stars;
–   Access to the rich spectral information in the visible wavelength range, including TiO absorption bands and the important Hα line;
–   A factor of 1.5 to 2 improved angular resolution compared to existing instruments, through the use of shorter wavelengths.

Interferometry at high spectral resolution is a new tool for stellar astrophysics, which provides hitherto inaccessible information on stellar rotation properties, atmospheric structure, and surface features. Although restricted to observations of relatively bright stars (down to $V = 7 \ldots 9$, depending on stellar color), UVES-I can have a profound impact on a large number of open questions in stellar astrophysics.

2 Scientific Case

2.1 Late-Type Giant Stars

Measuring the variation of the stellar diameter with wavelength, or even better wavelength-dependent limb darkening profiles, provides a sensitive probe for the structure of strongly extended atmospheres of cool giant stars. Such data can be directly compared with predictions of theoretical models, and provide qualitatively new tests of state-of-the-art three-dimensional stellar model atmospheres (Quirrenbach & Aufdenberg 2004). These models make predictions for the emergent spectrum at every point of the stellar disk. To compare model predictions with data from traditional spectroscopy, they have to be integrated over the full disk first. In contrast, interferometric spectroscopy gives access to the center-to-limb variation of the emergent spectrum, and is thus naturally suited for comparisons with model atmospheres.

A first step in this direction has been made with the Mark III Interferometer, by measuring the diameters of a sample of cool giant stars in two filters centered at 712 nm and 754 nm (Quirrenbach et al. 1993, 2001). These two filters probe a deep TiO absorption band and the relatively uncontaminated continuum, respectively (see Fig. 1). Many stars are found to be substantially larger at 712 nm than at 754 nm. It is easy to understand the principle behind this effect: we effectively measure the diameter of the $\tau = 1$ surface

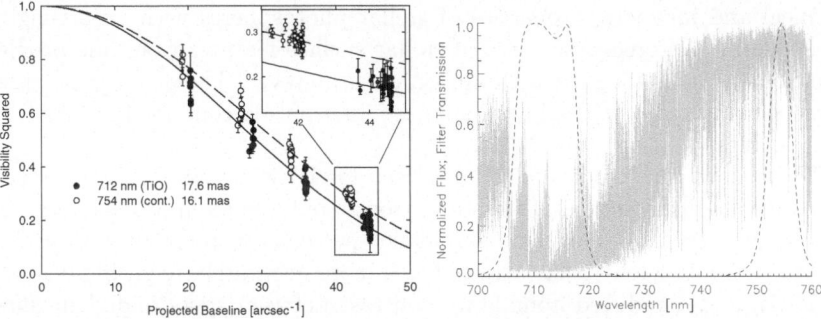

Fig. 1. Mark III observations of the K giant β Peg. The diameter at 712 nm is about 10% larger than the diameter in a 754 nm filter. The right panel shows the two filter transmission curves superimposed on a synthetic spectrum.

of the star, and the height of that surface varies with opacity and therefore with wavelength. The atmospheres of cool giants are so distended that this height variation is observable with an interferometer. It is clear, however, from that the Mark III observations average over a wide range of TiO absorption strengths because of the large spectral widths of the filters. Interferometric spectroscopy with UVES-I will provide much more detailed information on the diameter and limb darkening profiles as a function of TiO absorption depth, and thus substantially better constraints on the theoretical models.

Variations of the apparent diameter and limb darkening profile with wavelength are even more pronounced in Mira stars than in "normal" giants. Here UVES-I will enable detailed investigations of the pulsation and wind acceleration mechanisms. Again, high spectral resolution is required to sample a large range of depths in the stellar atmosphere.

2.2 Cepheids

Limb darkening curves in a spectral line provide a direct measurement of the projection factors of Cepheid pulsations, which relate the true velocity of the pulsation to the observed radial velocity curve. Uncertainties in the "p factor", which presently has to be computed from theoretical models, are a serious limiting factor in current estimates of Cepheid distances with the Baade-Wesselink method (Sabbey et al. 1995, Marengo et al. 2002). Interferometric spectroscopy can thus eliminate one of the important contributions to the error budget for distances to Cepheids and other variable stars.

2.3 Generalized Doppler Imaging

Classical Doppler Imaging (DI) has been developed to a very powerful tool (e.g., Rice 2002, Kochukhov et al. 2004). This technique allows mapping the chemical and magnetic properties of stellar photospheres with surprisingly small details. The reconstruction of stellar surface features from line profile variations alone is plagued with ambiguities, however. These can to a large extent be resolved by the additional phase information contained in interferometric data (Jankov et al. 2001).

Up to now, line profiles have been used for DI which are based on average atmospheric structures; this can obviously only be an approximation, in particular in regions of extreme abundance peculiarity. Tools are now available to compute such stellar atmospheres more accurately (e.g., Shulyak et al. 2004), and a reduced abundance contrast between spots and their surrounding is expected. UVES-I data will allow a direct check of the models, because abundance analyses can be performed for individual surface regions of prominent chemically peculiar stars. The same approach is also applicable for other stars with inhomogeneous surface properties, like active cool giant stars, as interferometry allows to study individual surface regions. The fact that interferometry can isolate the active regions will in such observations compensate for the lower signal-to-noise compared to single-telescope spectra, which always average over the whole stellar surface.

2.4 Rotational Axes

Stellar rotation induces a phase difference between the red wings and the blue wings of stellar absorption lines. Measuring the position angle of the phase gradient allows determining the orientation of the stellar axis on the sky (Petrov 1989, Chelli & Petrov 1995). More detailed modeling of the interferometric signal can also provide the inclination of the stellar rotation axis (Domiciano de Souza et al. 2004). UVES-I will thus open a way to determine the orientation of stellar rotational axes in space.

To know the orientation of the stellar axis in space is of particular interest in double or multiple star systems. One can thus check whether the rotation axes of binaries are aligned with each other, and with the orbital rotation axes of the systems. Here the orientation of the rotational axes contains information about the origin and evolution of the system. One can also search for (partial) alignment of rotation axes in star forming regions and stellar clusters.

2.5 Extrasolar Planets

The orientation of the stellar rotation axis will be of special interest for stars which harbor planets, since the mutual inclination between the orbital plane of the companion and the rotation axis of the star can provide insights into the formation process of the planet. If the orbital evolution of planetary systems is dominated by few-body scattering processes, one might expect to find orbits that are not aligned with the stellar angular momentum (Lin &

Ida 1997, Papaloizou & Terquem 2001). In the near future, astrometric orbits will become available from ground-based and space-based astrometry (VLTI-PRIMA, SIM, GAIA). Combining this information with UVES-I observations will provide the relative inclination between the orbital and equatorial plane for a large number of planets in a variety of orbits.

2.6 Differential Rotation

Along with the oscillation spectrum, differential rotation is a powerful diagnostic of the interior structure of a star. Unfortunately, observations of differential rotation are difficult with classical spectroscopy, and degeneracies exist between inclination, limb darkening, and differential rotation (e.g., Gray 1977). These degeneracies can be resolved by the additional information from interferometric spectroscopy (Dominiciano de Souza et al. 2004). High spectral resolution is of the essence for this application, giving UVES-I a large edge over AMBER.

2.7 Circumstellar Matter

Velocity-resolved interferometric observations in emission lines can be used to determine the structure and velocity field of disks around pre-main-sequence objects and around Be stars (e.g., Quirrenbach et al. 1997, Quirrenbach 1997). It is possible to determine the disk opening angle and the rotation law in the disk, to measure the location of the inner edge of the disk, and to obtain detailed information on asymmetries caused by spiral waves. Interferometric observations of winds and outflows from pre-main-sequence stars and from evolved objects can be used to determine their extent and overall geometry, and to probe sub-structure such as clumps and shells. These observations will not need the full spectral resolution offered by UVES-I, but access to the Hα line is of critical importance.

3 Instrument and Infrastructure

3.1 Telescopes, Adaptive Optics

UVES-I will work in the wavelength range between 0.6 μm and 1.0 μm. Therefore it will work in conjunction with the ATs and not the UTs as the MACAO adaptive optics systems of the UTs do not deliver well-corrected wavefronts in the visible.

Even in good seeing conditions, the ATs do not deliver a coherent wavefront, either. In principle there are three possibilities: (a) No AO or tip-tilt correction is available. This means that the aperture has to be stopped down to about 25 cm (corresponding to r_0 at 800 nm, where r_0 is the Fried parameter at 800 nm). (b) Tip-tilt correction is performed with the existing tip-tilt unit for a 75 cm subaperture (corresponding to $3\,r_0$ at 800 nm). This could be achieved by placing a common pupil stop in front of the tip-tilt sensor, fringe tracker, and UVES-I. This is assumed as the default in the SNR calculations below. A pupil stop with variable size would be highly desirable to adapt to changing seeing conditions, and to optimize for different observing wavelengths. (c) If MACAO-like high-order AO systems were implemented on the ATs, the full 1.8 m aperture could be used. This would give a sensitivity gain of 1.9 magnitudes compared to the numbers given in Sect. 4.3, but here we do not consider this option further because of the implied cost.

3.2 Fringe Tracking

UVES-I will rely on the PRIMA FSU for fringe tracking. With the pupil stop in place the fringe tracking sensitivity is reduced by 1.9 magnitudes with respect to the nominal AT case. The expected fringe tracking noise is 100 nm for a star with $m_K = 8$, and 50 nm at $m_K = 6$ (see Delplancke 2004, Fig. 5-12). This corresponds to visibility losses of 14% and 2.4% at 800 nm, which is easily tolerable.

All UVES-I targets will be sufficiently bright for on-source fringe tracking, i.e., only one of the PRIMA channels is needed. PRIMA will be commanded to keep the fringes stable at a nominal wavelength within the UVES-I wavelength range. This wavelength may be chosen either close to the center of the UVES-I range (0.8 μm), or at the wavelength of a specific spectral line (e.g., Hα), if only a small wavelength range around that line is of interest.

3.3 Dispersion Compensation

Since the VLTI delay lines are filled with air, there will be an imbalance between the pathlength in air between the two interferometer arms, which gives rise to a variation of the optical pathlength difference with wavelength due to the dispersive nature of air. For a short observation with UVES-I this leads to a slow variation of the phase with wavelength, which can easily be calibrated out. During the course of a several-minute observation, however, the amount of air imbalance changes; this leads to a loss of fringe contrast at wavelengths away from the nominal fringe tracking wavelength. (If the FSU keeps the fringe phase constant at a specific wavelength λ_0, dispersion will cause the phase to rotate at any other wavelength.) During a 15-minute integration, the OPD can change by several meters, which leads to a differential fringe rotation of several μm between the blue and red edges of a 0.6 μm to 1.0 μm wavelength band.

The dispersion problem could be circumvented by restricting the exposure time and/or the baseline geometry, but the first of these options would reduce the limiting magnitude and observational efficiency, whereas the second would limit the coverage of the uv plane and complicate scheduling. It is

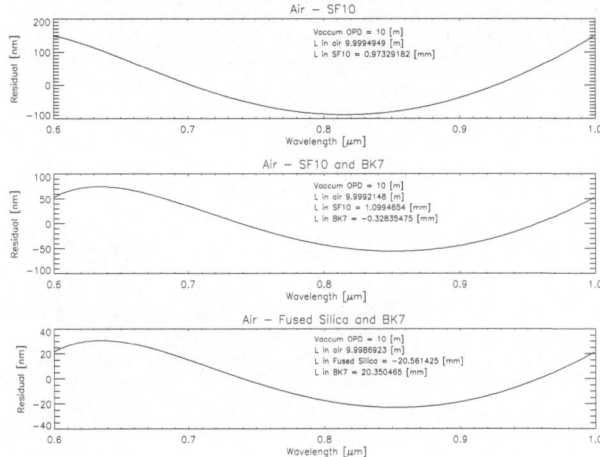

Fig. 2. Residual dispersion after accumulation of a 10 m change of the total delay, using an atmospheric dispersion compensator made of SF10, SF10/BK7 and BK7/Fused Silica. The combination of SF10 and BK7 has been chosen for UVES-I.

therefore foreseen to equip UVES-I with an atmospheric dispersion compensator. Viable solutions for a single-material dispersion compensator exist; SF10 would have to be chosen for the material (see Fig. 2, upper panel). For extreme parameters the residual uncompensated optical pathlength would reach 150 nm, however; we therefore prefer an arrangement where two different materials are used in the two interferometer arms. A combination of fused silica and BK7 gives the best theoretical performance, but the required glass thickness is relatively high (see Fig. 2, lower panel). We therefore favor a SF10/BK7 combination, which provides compensation to ≤ 70 nm even for extreme delay rates with much thinner glasses (see Fig. 2, middle panel).

3.4 Beam Combiner

The primary observable in an interferometer is the complex visibility, i.e., the amplitude and phase of the coherence function of the radiation received by the two telescopes (e.g. Quirrenbach 2001). To obtain the full information on the visibility phase and amplitude, one can measure the four fringe quadratures, i.e., one has to count the photons at phases 0, $\pi/2$, π, and $3\pi/2$. These four bin counts are commonly called A, B, C, and D; the square of the visibility V^2 and the fringe phase ϕ can then be estimated using

$$V^2 = 4 \cdot \frac{(A-C)^2 + (B-D)^2}{(A+B+C+D)^2} \quad , \quad \phi = \arctan\left(\frac{A-C}{B-D}\right) \quad . \quad (1)$$

For any given wavelength λ, the full interferometric information is thus contained in the four intensities A(λ), B(λ), C (λ), and D(λ) carried in the

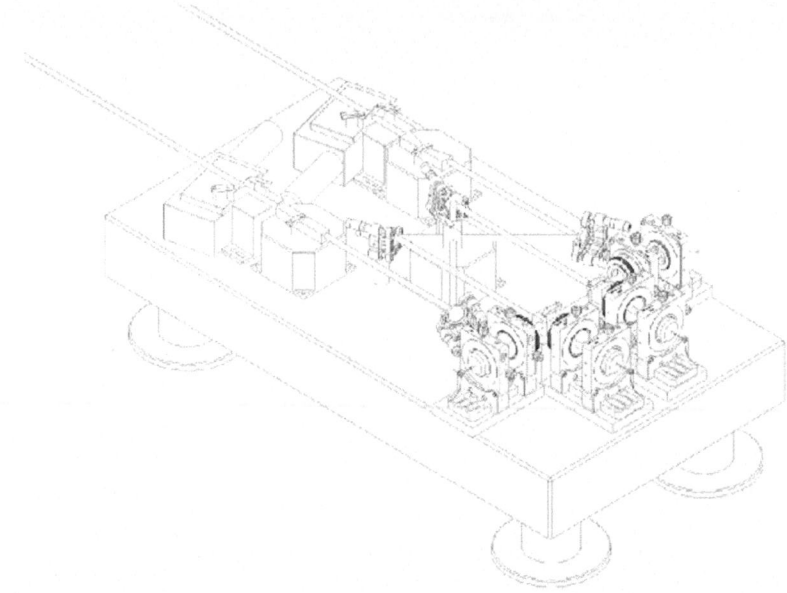

Fig. 3. UVES-I beam combination table. The two stellar light beams enter from the top left, and pass through the atmospheric dispersion compensator and achromatic phase shifter before being combined at the central beam combiner. The two polarization states are separated, and the resulting four beams coupled into optical fibers; the four fiber holders are visible at the bottom right.

output beams of the beam combiner. Applying the fringe estimators on a wavelength-by-wavelength basis, one can thus derive the amplitude and phase of the visibility as a function of λ. The phase will usually be corrupted by turbulence in the Earth's atmosphere, but differential phases $\phi(\lambda_1) - \phi(\lambda_2)$ provide very valuable observables, such as phase differences between the red and blue wings of spectral lines. In this context it is worth noting that phenomena on scales much smaller than the "resolution limit" λ/B of the interferometer with baseline B are accessible in this way, because differential phases can easily be measured with a precision of a few degrees or even a fraction of a degree, depending on the signal-to-noise ratio.

In UVES-I the beams pass first through the atmospheric dispersion corrector. Then one beam is sent through a K-prism, which introduces a $\pi/2$ phase shift of one polarization state with respect to the other. A glass compensator is used in the other beam. Two flat mirrors direct the beams to the main beam combiner. The combined beams which have a phase shift of π relative to each other are sent to polarizing beam splitters by a second pair of flat folding mirrors (see Fig. 3).

The requirements on the precision of the phase shift are rather loose; one can easily tolerate up to $10°$ errors as long as they are stable. Similarly, there are no stringent requirements on the splitting ratios and polarization purity of the beam splitter cubes. It is therefore possible to use standard commercial components. The preliminary design of the UVES-I beam combiner uses predominantly commercial off-the-shelf components arranged on a 45 cm × 85 cm breadboard (see Fig. 3). This compact design ensures stability and minimizes the space taken up in the VLT Interferometer lab.

Due to its high spectral resolution, the coherence length of UVES-I is very large (≥ 3 cm). The drift of the optical path difference during one exposure must not be larger than ~ 50 nm, however; therefore the UVES-I breadboard should be mounted as rigidly as possible to the PRIMA FSU. This eliminates the necessity of a metrology system. A convenient arrangement of the UVES-I beam combiner is next to the PRIMA FSUs. Two dichroic beam splitters in front of a PRIMA fringe sensing unit can be used to divert the light shortward of 1μm to UVES-I.

3.5 Fiber

The four fringe quadratures A, B, C, D are injected into optical fibers with $70\,\mu$m core diameter (compared to $120\,\mu$m in the existing FLAMES-UVES link). This will provide a spectral resolution of $\sim 60{,}000$, while still making alignment of the fiber couplers easy. The total fiber length required to connect the VLTI lab to the Nasmyth platform of UVES is about 150 m. Note that after the combiner one can freely manipulate the beams without having to pay attention to differences in optical pathlength, because the phase relations between the beams are defined in the beam combiner. There is no need to use single-mode fibers; multi-mode fibers will do just as well.

3.6 UVES

The fiber interface to the UVES spectrograph will be similar to the existing link from FLAMES. The fibers from the interferometric beam combiner will be glued to a head identical to the FLAMES head. The UVES-I head will be mounted close the FLAMES head; a pair of periscope mirrors will direct the light to the UVES slit. The second of these mirrors can be folded away to clear the path towards the FLAMES head. Through this arrangement it is possible to leave the existing FLAMES feed untouched.

4 Performance

Typical exposure times will range from about 1 min to 15 min. This is sufficiently long to ensure that the detector read noise is not dominant (see Sect. 4.3), and sufficiently short to allow compensation of the atmospheric dispersion accumulating during the exposure with a relatively simple device (see Sect. 3.3.).

4.1 Overall Transmission and Efficiency

The current configuration of the VLTI ATs contains a dichroic beam splitter sending the visible light to the tip-tilt system, and passing the infrared to the delay lines and instrument. This beam splitter reduces the VLTI transmission in the visible considerably (Koehler & Gitton, Figure 3.2.10-2). For UVES-I this dichroic should be replaced by a 50-50 or better 10-90 beam splitter. The sensitivity estimates given below are based on this change, and assume a total VLTI transmission of 20% (Koehler & Gitton, Figure 3.2.10-3).

The UVES-I beam combiner has 12 optical surfaces, which will be coated with anti-reflection coatings optimized for the wavelength range from $0.6\,\mu$m to $1.0\,\mu$m. With the conservative assumption of 95% efficiency per surface, the resulting overall throughput of the beam combination table is 55%.

The transmission of the UVES-I fiber link depends on the fiber coupling efficiency as well as the bulk transmission losses in the fiber. The former should be similar to that of FLAMES, but the fiber length will be larger – about 150 m compared to 40 m for the FLAMES-UVES link. CeramOptec fibers, which also have been used for the FLAMES link, have a transmission $\geq 80\,\%$ over the whole wavelength range considered. By using them we can safely assume that the UVES-I fiber link will have $\sim$90% of the throughput of the FLAMES link, measured from input coupler to output coupler.

4.2 Wavefront Quality

The system visibility of an interferometer, i.e., the fringe contrast measured on a point source, depends on the wavefront quality of the two beams and on the degree to which they overlap correctly at the beam combiner[1]. Factors influencing the system visibility include alignment, aberrations in the instrument optics, tracking (tip-tilt) errors, and higher-order wavefront errors due to atmospheric seeing. We expect that a system visibility of order 0.5 can be reached.

4.3 Signal-to-Noise Ratio

The total difference between FLAMES and UVES-I in sensitivity is 8.0 magnitudes (pupil diameter reduced by $9 \cdot 10^{-3}$, two telescopes instead of one, VLTI transmission 20% versus 80%, UVES-I beam combiner transmission of 0.55, four fringe quadratures, and system visibility of 0.5). As most targets need only be resolved partly, the visibility will only be marginally reduced relative to unresolved targets. Applying this scaling, the following sample SNR estimates per spectral element (0.004 nm) near 800 nm are obtained:

---

[1] We do not consider the possibility of increasing the system visibility with a spatial filter here, because the introduction of such a filter would impose much more stringent requirements on the alignment.

Table 1. Sample targets for UVES-I with spectral type, V magnitude, and S/N

Sp. type	V mag.	Exp. time [s]	SNR	Sample Key Target
M2	2	60	104	β Peg (M giant, diameter(λ), V=2.4)
K2	2	60	61	β Gem (K giant, oscillations, V=1.2)
G2	2	60	55	α Cen A,B (seismology, V=0.0)
A2	2	60	38	ζ Tau (Be, Hα disk, V= 3.0)
M2	5	300	58	o Cet (Mira, variable, V$\geq$3.0)
K2	5	300	33	ε Eri (dust disk, planet, V=3.7)
G2	5	300	30	L Car (Cepheid, V=3.4)
A2	5	300	21	β Pic (dust disk, V=3.8)
M2	7	900	39	VY CMa (extreme mass loss, V=7.9)
K2	7	900	22	HR 1099 (RS CVn, V=5.9)
G2	7	900	20	HD 209458 (transiting planet, V=7.7)
A2	7	900	14	HR 3831 (roAp, V=6.2)
M2	9	900	14	YY Gem (low-mass ecl. binary, V=9.1)

It should be pointed out that not all observing programs will require the full spectral resolution provided by UVES-I. In those cases it will be possible to obtain a higher SNR per desired spectral element by binning the data.

Due to the differential nature of most measurements to be made with UVES-I it should be possible to reach the theoretical photon noise limit just as in traditional high-resolution high-SNR stellar spectroscopy. For example for a measurement of the phase change across a spectral line the adjacent continuum provides an excellent calibration; co-adding exposures taken over a few hours will give an SNR of up to 1,000 per spectral element on bright stars; this enables measuring phase changes with wavelength as small as a milliradian. This means that UVES-I will be sensitive to structures that are ~ 100 times smaller than the resolution limit of the interferometer – it will not be possible to image such small structures, but models of such small structures can be constructed and tested with UVES-I.

5 Conclusion

The combination of long-baseline interferometry with high-resolution spectroscopy in one instrument gives access to hitherto unobservable properties of stellar surfaces and stellar rotation. The infrastructure available at ESO's VLT facility – the VLTI with fringe-tracking capabilities, Auxiliary Telescopes and the high-resolution spectrograph UVES designed for single-telescope use – provides an opportunity to implement interferometric high-resolution spectroscopy with few additional components: only a simple beam combiner and a fiber link would be needed. Since none of the items that make most instrument projects time-consuming and expensive (detector, dewar, electronics, spectrograph optics, ...) are required, UVES-I could be implemented quickly and at low cost.

Acknowledgement We would like to thank the ESO staff in Garching, in particular Luca Pasquini and Gerardo Avila, for their support and many helpful suggestions.

References

1. Chelli A, Petrov RG. 1995. A&AS 109:401-415
2. Delplancke F. 2004. VLT-SPE-ESO-15700-3051
3. Domiciano de Souza A, Zorec J, Jankov S, Vakili F, Abe L, Janot-Pacheco E. 2004. A&A 418:781-794
4. Gray DF. 1977. ApJ 211:198-206
5. Jankov S, Vakili F, Domiciano de Souza A, Janot-Pacheco E. 2001. A&A 377:721-734
6. Kochukhov O, Drake NA, Piskunov N, de la Reza R. 2004. A&A 424:935-950
7. Koehler B, Gitton B. 2002. VLT-ICD-ESO-15000-1826
8. Lin DNC, Ida S. 1997. ApJ 477:781-791
9. Marengo M, Sasselov D, Karovska M, Papaliolios C, Armstrong JT. 2002. ApJ 567:1131-1139
10. Papaloizou JCB, Terquem C. 2001. MNRAS 325:221-230
11. Petrov RG. 1989. In *Diffraction-limited imaging with very large telescopes,* ed. DM Alloin, JM Mariotti, NATO ASI Vol. 274, pp. 249-271
12. Quirrenbach A. 1997. In *Science with the VLT Interferometer,* ed. F Paresce, pp. 163-170. Berlin/Heidelberg: Springer-Verlag
13. Quirrenbach A. 2001. ARAA 39:353-401
14. Quirrenbach A, Aufdenberg J. 2004. In *Modelling of stellar atmospheres,* ed. N Piskunov, WW Weiss, DF Gray, p. E68. IAU Symp. 210
15. Quirrenbach A, Bjorkman KS, Bjorkman JE, Hummel CA, Buscher DF, Armstrong JT, Mozurkewich D, Elias NM, Babler BL. 1997. ApJ 479:477-496
16. Quirrenbach A, Mozurkewich D, Armstrong JT, Buscher DF, Hummel, CA. 1993. ApJ 406:215-219
17. Quirrenbach A, Mozurkewich D, Armstrong T, Buscher D, Hummel, C. 2001. In *Galaxies and their constituents at the highest angular resolutions,* ed. RT Schilizzi, SN Vogel, F Paresce, MS Elvis, pp. 304-305. IAU Symp. 205
18. Rice JB. 2002. AN 323:220-235
19. Sabbey CN, Sasselov DD, Fieldus MS, Lester JB, Venn KA, Butler RP. 1995. ApJ 446:250-260
20. Shulyak D, Tsymbal V, Ryabchikova T, Stütz C, Weiss WW. 2004. A&A 428:993-1000

VEGA: A Visible Spectrograph and Polarimeter for the VLTI

D. Mourard[1], P. Antonelli[1], A. Blazit[1], D. Bonneau[1], Y. Bresson[1],
J. M. Clausse[1], A. Domiciano[2], M. Dugué[1], R. Foy[3], P. Harmanec[4],
M. Heininger[2], K.-H. Hofmann[2], S. Jankov[5], P. Koubsky[4], S. Lagarde[1],
J. B. Lebouquin[6], P. Mathias[1], A. Meilland[1], N. Nardetto[1], R. Petrov[5],
K. Rousselet-Perraut[6], D. Schertl[2], Ph. Stee[1], I. Tallon-Bosc[3], M. Tallon[3],
E. Thiébaut[3], F. Vakili[5], and G. Weigelt[2]

[1] Observatoire de la Côte d'Azur, Dépt. GEMINI, 06130 Grasse, France
[2] Max-Planck-Institute für Radioastronomie, Auf den Hügel 69, D-53121 Bonn, Deutschland
[3] CRAL UMR 5574 CNRS/ENS Lyon/UCBL, Observatoire de Lyon, 69561 Saint-Genis-Laval cedex, France
[4] Astronomical Institute, Academy of Sciences of the Czech Republic, 251 65 Ondřejov, Czech Republic
[5] Université de Nice-Sophia Antipolis, LUAN, Parc Valrose, F-06108 Nice Cedex 02, France
[6] LAOG, 414, rue de la Piscine, Domaine Universitaire 38400 Saint-Martin d'Hères, France

Summary. The ESO/VLTI has now clearly a position of world leader in the domain of ground-based optical interferometry. With four 8.2 m telescopes and two (four) 1.8 m telescopes, the Paranal Observatory is without any doubt the best optical interferometric facility in the world. Since many years, it has attracted the major part of the European interferometric community and with the opening of MIDI and AMBER, the astronomers have now access to 'general user' interferometric instruments in the thermal and near infrared. This paper describes a project for a second generation focal instrument of the VLTI, named *VEGA* for *Visible spEctroGraph and polArimeter*. The goal is to give access to the visible wavelength region, with spectroscopic and polarimetric capabilities, taking advantage of the coherent field of view of the VLTI. It is a unique scientific field for the VLTI. For example, a 200m interferometer operating in the visible will be able to resolve structures of the order of 0.5 mas or 0.1 AU at the distance of the Ophiuchus cloud.

1 Introduction

Interferometry has been intensively performed at long wavelengths, starting with the radio interferometers about 50 years ago since it was easier to guide radio wavelengths in cables while keeping the phase information or using a local oscillator and a correlator to recombine "a posteriori" the beams over intercontinental distances. In the optical domain, a lot of work has been done at IR and near-IR wavelengths since it was technically easier, or we must say, less difficult to recombine directly the optical beams, since the coherence length is larger and the turbulence slower. Although, the visible domain of the electromagnetic spectrum was not explored at the same level as near or mid infrared, some very nice and important results have been however obtained with the GI2T interferometer in south of France [11] and also with the NPOI array in Flagstaff, USA [1] or the SUSI interferometer in Australia [5]. We will present in this paper the science cases of a new but already existing and tested instrument: the REGAIN focal instrument which was designed and built for the Grand Interféromètre à 2 Télescopes (GI2T) in southern France. This instrument, called **VEGA** *(Visible spEctroGraph and polArimeter)* in its VLTI adaptation, will open new astrophysical fields possible in the visible domain. It will provide a spectral resolution up to 30000 at 0.6 μm and a spatial resolution of less than 1mas for up to four telescopes in the $X - \lambda$ dispersed-fringe mode (i.e., spatial information in one direction and wavelength information in the perpendicular direction). A polarimetric device (SPIN) [15] measuring simultaneously the polarization in two directions either circular or linear is also implemented in this instrument. A multiple band-passes mode (called COURTES mode) is dedicated to wide field and high sensitivity observations at four simultaneous wavelengths for three telescopes. It will open the extragalactic domain up to magnitude 13 in the visible. Since **VEGA** was used on the sky on 1.5 m telescopes it is also very well suited for the 1.8 m VISA array on the VLTI and will only need minor adaptations for the injection of the VLTI beams. This paper will focus first on some of the most promising science cases only possible with this visible instrument and second on the technical aspect of this project.

2 Science drivers for a visible VLTI instrument

2.1 Extragalactic programs

Active Galactic Nuclei

Studies of the morphology and kinematics of AGN is a key objective of visible VLTI observations. Classical interferometry will allow studies of structures in the inner narrow line region (NLR) (e.g., possibly torus or jet structures), whereas differential interferometry (e.g., measurements of the wavelength dependence of the visibility function with high spectral resolution) will provide results on the size, structure, and kinematics of the broad line region (BLR). The important advantages of interferometry in the visible are the higher spatial resolution, the higher emission line signal and the photon-counting detectors. For the BLR studies by differential interferometry, a preliminary estimation of the SNR shows that:

$$SNR \propto \sqrt{flux}.\frac{Baseline}{\lambda}.\frac{I_{line}}{I_{continuum}} \tag{1}$$

It indicates that the ATs in Hα would do about eight times better than the UTs in Brγ. These SNR values are given for single mode operation. The Strehl ratio in the visible with the ATs is assumed to be half the value in the K band with the UTs. This allows much more precise photocenter displacements $\epsilon(\lambda, t)$ measures and much stronger constraints on the BLR kinematics and therefore on the central mass or fainter limiting magnitude for a given accuracy. A visible instrument working with the current telescopes at Paranal (telescopes+AO) must be operated in the multimode regime. Under these conditions, UTs will be only 4 times more efficient than the ATs and the limiting magnitude, defined as the limit where one gets 1 photon per speckle and per single short-exposure, does not depend on the telescope but only on the seeing conditions. Therefore, at this limit, differential measurements will permit high signal to noise ratio by evaluating a large number of single exposures containing a large number of individual speckles. Calculations have been made showing that under good seeing conditions (0.5"), with a wide reference channel of 50 nm and a science channel of 0.1 nm, a 1% accuracy could be achieved on a V=11.2 source in 20mn. Under excellent seeing conditions (0.3"), 45 mn are necessary on a V=12.3 source. If one considers a wide reference channel of 50nm and a science channel of 1 nm, a V=15 source could be measured with a 5% accuracy in 1.2 h under good seeing conditions (0.5"). This time goes down to 26 mn under excellent seeing conditions (0.3").

Cosmological distance scales

Interferometry can improve the accuracy of the zero point of the Period-Luminosity relation of Cepheids, which is a fundamental parameter for the distance scale in the Universe. Such measurements are based on simultaneous measurements of radial velocity curves and angular diameter variations [7]. It has been shown [10] that the gain in resolution permitted by the visible substantially increases the precision of the diameter measurement and therefore can give access to a few tens of Cepheids, instead of just a few stars in the near-infrared. In addition to the Period-Luminosity relation, many other important Cepheid projects will be performed, such as the study of surface brightness and velocity structure variations through the pulsation cycle and the subsequent improvement of the models of pulsating atmospheres. For the brightest Cepheids, the resolution with 200 m baseline in the visible is large enough to obtain detailed information on the atmospheric structure. Measurements of the temporal variation of the limb-darkening profiles allow studies of the stellar photosphere. Adding the spectral dimension permits to track the pulsation in the different layers of the atmosphere and study in greater detail its mechanism [4]. Furthermore, studying in a line profile the variation of the spectrum $s(\lambda, t)$ together with that of the photocenter displacement $\epsilon(\lambda, t)$ measured by differential interferometry allows the extension of the Cepheid program to unresolved objects. These two last applications require spectral resolutions between 20 000 and 40 000, which is much easier to achieve in the visible [14]. Coupled with a high-sensitivity instrument, this technique could allow reaching direct determination of the angular diameter of a small sample of LMC Cepheids [12].

2.2 Stellar activity

Spots and Doppler imaging

Interferometry can contribute to the study of stellar activity by detecting and mapping star spots. A limited number of stars have features large enough to allow interferometric imaging. The gain in resolution allowed by the visible will substantially increase the number of targets. For instance, one of the main result of helioseismology is the location of the tachocline which is thought to be the engine of the solar magnetic field that gives rise to spots. In the Sun, spots are relatively small (1% of the solar radius), and, usually being formed in the equatorial region, tend to move to largest latitude. It seems that no solar spots have been observed close to the poles. Using stellar rotation, it is possible to reconstruct the stellar surface through inversions of high-resolution spectroscopic observations such as Doppler Imaging technics, based on the line profile variation as a function of time in spectrally resolved lines. However, many examples of such sruface reconstruction show that spots are located mostly close to the pole, which contrasts to what is observed on the Sun. Is this result due to a selection effect, or to a different mechanism? In any case, classical Doppler imaging leads to ambiguity in the reconstruction process, that can only be solved by direct imaging as can be provided by *VEGA*.

Asteroseismology

Asteroseismology studies the temporal variation of the photometric flux or of the spectrum $s(\lambda, t)$ to identify the resonant modes of stars. Since these modes propagate through the entire star (and the low order modes actually go through or close to the center), asteroseismology allows the exploration of the third dimension of the star. As in the case of Doppler Imaging, the combination of the spectral information of asteroseismology with differential interferometry can solve much improved results. As for the study of stellar spots the study of the spectrum alone yields a lot of problems, mainly to measure very small velocity variations, to identify the modes, to access higher orders and solve windowing effect. For almost all pulsating stars, the angular diameter is below 1 mas, thus strongly reducing the contribution of resolved interferometry. However, in the case of non-resolved stars, it has been shown that adding the photocenter displacement $\epsilon(\lambda, t)$ to $s(\lambda, t)$ solves many problems. The demand for instrumental stability is relaxed, since for example it is possible to calibrate $\epsilon(\lambda, t)$ using $s(\lambda, t)$ as a reference which has been recorded at the same time by the same pixels in the same instrument. This also allows a lower dependence on a perfect modelling of the non-oscillating spectrum. The element of spatial resolution introduced by $\epsilon(\lambda, t)$ permits to access higher modes. The fact that many modes affect very differently $s(\lambda, t)$ and $\epsilon(\lambda, t)$ strongly helps identifying them. This last feature might also somehow ease the temporal coverage and window problems, but this remains to be studied. As for stellar spots, Doppler Imaging can be applied to the reconstruction of the pulsation modes. It has been recently demonstrated [6] by fully simulating the image reconstruction process that the combination of $s(\lambda, t)$ and $\epsilon(\lambda, t)$ is far more discriminating than the use of spectroscopic data alone. The technique could be applied to solar type stars with a dedicated instrument optimized for this type of observations with the ATs: it would be some kind of Echelle spectrograph allowing to combine information from all lines in the visible spectrum. Such an instrument could only exist in a somehow later future but *VEGA* would yield access to pulsating B stars and possibly δ Scuti stars. As for stellar spots, the need for photosphere lines, high spectral resolution and a spatial resolution allowing to approach resolving the stellar diameter point toward a visible instrument. For the observation of stellar spots and for asteroseismology there are significantly more interesting spectral features below 0.6 μm than between 0.6 and 0.8 μm. This is a reason to lower the wavelength accessible to a visible VLTI instrument as much as possible.

2.3 Other stellar physics programs

A large number of stellar physics programs are very well adapted to the visible domain. First the gain in angular resolution could be used for increasing our knowledge of fundamental parameters: diameters, effective temperatures, rotation, masses in the case of binary systems. Dedicated programs are foreseen for the circumstellar studies of Mira stars and active hot stars. With *VEGA*, it will be also possible to study in detail the physical relation between the environment and the internal structure of young and evolved stars.

3 Technical description of the proposition

3.1 General description

By offering access to the visible band, a high spectral resolution and an extended field of view with a high sensitivity, *VEGA* is fully complementary to the existing or planned near-infrared instruments in many aspects. Both infrared and visible measurements are mandatory for a correct and complete radiative transfer modelling. The spatial resolution will be between 2 and 4 times better, which could be decisive for many programs. On one hand, many of the astrophysical signals investigated by interferometry have stronger signatures at visible wavelengths. On the other hand, new physical processes involved in stars will be investigated, opening wider the field of astrophysical domains covered by the VLTI. A high spectral resolution is easily reachable in the visible domain, which is mandatory for progress in stellar activity studies. A multi-way combiner (or multi-axial) is highly desirable to extend the scientific programs of the first generation and to allow a coherent field of view of a few Airy disks.

Our previous experience [11] has lead to the definition of an instrument combining two modes of operation for a better suitability to the science cases.

– A **high spectral resolution** operation with a dispersed fringe mode (called $X - \lambda$)
  – Simultaneous combination of 4 telescopes (see figure 1), giving access simultaneously, for one measurement, to 6 squared visibilities, 6 differential phases and 3 closure phases.
  – Spectral resolution of 1500, 5000 and 30000 with a spectral range from 0.5 to 0.9 μm (if possible from 0.4 to 1.0 μm).
  – Polarimetric capabilities allowing to simultaneously record the interferograms in two polarizations, either linear or circular.
– A **high sensitivity operation** with a multiple band-passes mode (called COURTES)
  – Simultaneous combination of 3 telescopes.
  – Spectral resolution of 100 to 2500 and spectral range: from 0.5 to 0.9 μm (if possible from 0.4 to 1.0 μm).
  – 4 simultaneous field of view of 1.5"x1"(at four different wavelengths).

3.2 Interfaces with the VLTI

In the $X - \lambda$ mode, the instrument is able to recombine up to 4 telescopes simultaneously and to produce dispersed fringes at different spatial and wave frequencies (Fig.1). The beam combination is done in an image plane, in a multi axial and multimode configuration.

In the COURTES mode, 3 telescopes are recombined and form 4 simultaneous large field-of-view images (1.5"x1") at different wavelengths. As indicated in the Interface Control Document of the VLTI, a change of the M9 mirror of the Coudé trains will be necessary to send part of the visible light into the interferometric laboratory. As part of the visible spectrum is used for the AT/STRAP sensors, it will be necessary to study and build new dichroic coating for these mirrors.

As the instrument is designed to sample the whole seeing pattern, an improvement of the adaptive optic devices (STRAP) is not necessary. Our instrument is based on the Michelson Multimode Measurement, developed successfully on the GI2T [2, 3]. The multimode approach leads by definition to a random atmospheric phase for each coherent cell of the atmosphere. Thus, cophasing has no sense and it is not necessary to use a 'super' external fringe tracking device. The requirement for **VEGA** is a coherence tracker, typically at the

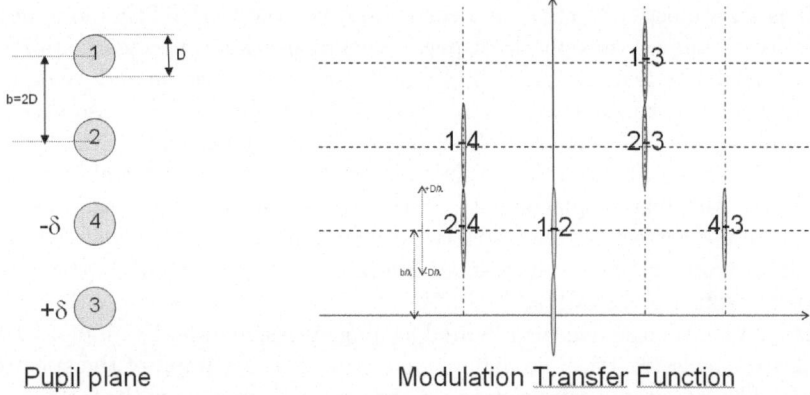

Pupil plane   Modulation Transfer Function

Fig. 1. The special 4 beams $X - \lambda$ mode of *VEGA*. The 4 pupils are arranged in a linear, compact and redundant configuration. Optical path difference offsets are set on beams 3 and 4 so that in the two dimensional Modulation Transfer Function the high frequency energies are separated both in the spatial or in the spectral conjugate axis.

level of 1/500 of the coherence length. The high spectral resolution of the instrument leads to a coherence length between 0.9 and 18 mm. The requirement for the OPD stabilization is defined as the coherence length divided by the number of spectral channels [8]. In our case, it corresponds in the worst case to a stabilization of about 2 μm, which is much more than the foreseen performance of the VLTI fringe tracker. The real time processing of the data will be able to give a measurement of the optical path differences, at the level of a fraction of the coherence length. If necessary this real time processing could be used to achieve the coherencing of the VLTI.

VEGA is equipped with a polarimetric device allowing the simultaneous recording of both polarizations, either in linear or circular configuration [15]. Thus a good knowledge of the 'instrumental' polarization of the VLTI infrastructure is necessary if one want to perform spectro polarimetric interferometric measurements on celestial sources. The current information given in the ICD document are encouraging but as specific measurements on the VLTI are foreseen, it will be very important to extend these measurements to the visible domain, especially with respect to the coating properties. The 180° of field rotation for telescopes located on one side of the tunnel with respect to the telescopes located on the other side is not a problem for the recombining scheme of *VEGA*.

Due to the large field of view of the instrument, calculated offset corrections between the spectral bandpass of the image tracker systems and the visible band of the spectrograph are enough in all cases for the pointing. For the $X - \lambda$ mode, even in the lowest spectral resolution, it has been shown that a longitudinal atmospheric dispersion compensation is not necessary [16] as

well as an atmospheric refraction correction. For the COURTES mode however, longitudinal atmospheric dispersion compensation is necessary [16]. We
have already studied, designed, built and tested such a system and we do not
see any difficulties for that. In the COURTES mode, we have also studied
the need for the atmospheric refraction correction by calculating the largest
zenithal distance as a function of the wavelength and of the spectral band (for
a 1/10 of Airy disk displacement). This leads to the conclusion that we can
accept to not correct the atmospheric refraction for the current configuration
of the instrument with only a small limitations of the sky coverage for the
largest spectral bandwidths.

Finally, we have designed the required feeding optics, in order to adapt 4 VLTI
beams to the geometrical configuration of the entrance pupil of the spectrograph. These feeding optics combine a configuration module with static delay
lines as well as beam compressors and pupil reimaging, which is mandatory
for eliminating any Fresnel diffraction effects [9] in the spectrograph, as already done on GI2T/REGAIN by the successful use of the same Variable
Curvature Mirror on the LAROCA delay line as on the VLTI. **VEGA** could
be easily installed in the VLTI focal laboratory at the current location of the
VINCI experiment.

3.3 Subsystems breakdown

– Calibration and Alignment Unit: We use a laser diode, a white light source
  and a spectral lamp. These sources are coupled to the spectrograph by
  the use of a beam splitter and an optical fiber. A flat mirror allows us to
  switch on the sources in the instrument.

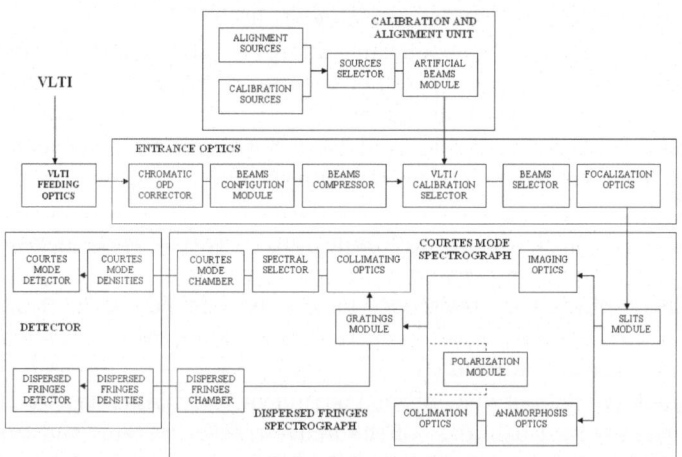

Fig. 2. Functional analysis and product breakdown of **VEGA**

- Entrance optics: This part of the instrument has different functions. First we designed a Chromatic OPD corrector, then a set of flat mirrors to arrange the geometry of the different beams. A beam compressor allows one to reduce the diameter of the individual beams to the 5mm diameter of the spectrograph and to re-image the pupil plane at the correct distance. Shutters will allow to select up to 4 beams for feeding the instrument. The entrance optics subsystem ends by the parabolic mirror forming the interferometric focus on the entrance slit of the spectrograph.
- Spectrograph: Two modes of operation have been designed with respect to the science program. These two modes use the same gratings module but the collimation and the chamber are different in the dispersed fringes and in the Courtes mode. Note that in the $X - \lambda$ mode, a polarimetric device is also used optionally.
- Detectors: Each mode of the spectrograph feed the light to a dedicated intensified CCD detector working in the photon counting regime. New developments have been made during the last years and we are now reaching very good quantum efficiency, fast operation and very good spatial resolution [Blazit et al., in preparation].

4 Performances

4.1 Assumptions

For the limiting magnitude and the signal-to-noise ratio (SNR) calculations [13], we have evaluated the global transmission to be 1.2%, with the following repartition: VLTI=0.1, *VEGA*=0.4, Detector=0.3. For the VLTI we have used a conservative M9 dichroic (50/50). The instrumental visibility is assumed to be 0.8. We consider an exposure time of 20 ms and a total integration time of 1800s in order to avoid a too large smearing of spatial frequencies. The maximum coherent spectral bandwidth is taken as 50 nm. The median seeing is 0.8" (best 20% of time = 0.5") and the wavelength is set at 650 nm.

4.2 Results

With the above assumptions, the following results (figs. 3, 4, 5 and 6) are obtained.

4.3 Summary of performances

VEGA will offer a dispersed-fringe mode able to combine up to four telescopes. In the highest spectral resolution, the spectral band is $\Delta\lambda$=6.7 nm and the width of an individual spectral channel is $\delta\lambda$=0.02 nm. In this configuration, we reach in 30 mn a limiting magnitude (SNR=10) of 9 for the AT

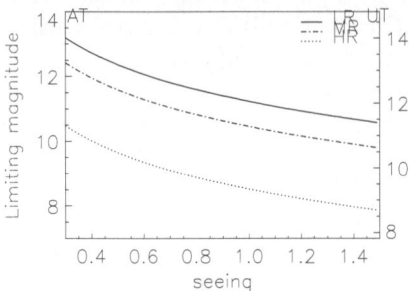

Fig. 3. Limiting magnitude (AT or UT) of the dispersed fringe mode as a function of the seeing. We assume a SNR of 10 in the differential visibility measurement with a reference channel of 50 (resp. 40 and 6.7) nm and a science channel of 0.4 (resp. 0.13 and 0.02) nm in the low (resp. medium and high) resolution mode.

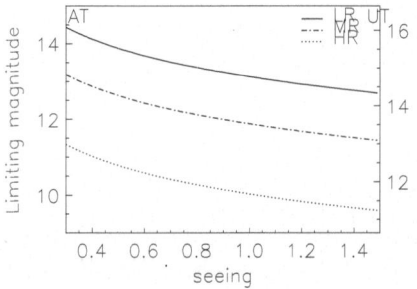

Fig. 4. Limiting magnitude (AT or UT) of the Courtes mode as a function of the seeing. We assume a SNR of 10 in the differential visibility measurement with a reference channel of 36 (resp. 12 and 2) nm and a science channel of 1.0 (resp. 0.3 and 0.06) nm in the low (resp. medium and high) resolution mode.

and the median seeing conditions and 10 under the best seeing conditions. This configuration allows us also to simultaneously measure two polarizations, either linear or circular.

In the multiple bandpass mode, 3 telescopes are recombined simultaneously and we form 4 images of 1.5"x1" at different wavelengths. The lowest spectral resolution allows to select a wide reference spectral channel of $36nm$ and a narrow scientific channel of $1nm$. This high sensitivity mode allows to reach in 30mn a limiting magnitude (SNR=10) of 13.5 with the ATs under median seeing conditions and of 15.5 with the UTs under best seeing conditions.

We wish also to point out that these calculations have been made with the same algorithm as the one used for the GI2T estimations. This algorithm has been validated on the sky of the Calern Observatory by actual observations.

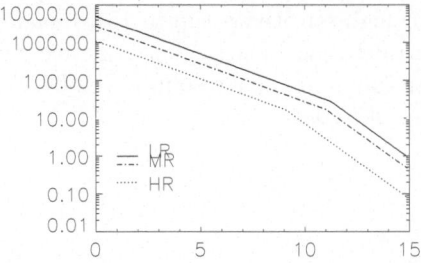

Fig. 5. SNR of the dispersed fringe mode as a function of magnitude. We assume a differential visibility measurement with a reference channel of 50 (resp. 40 and 6.7) nm and a science channel of 0.4 (resp. 0.13 and 0.02) nm in the low (resp. medium and high) resolution mode. Median seeing conditions are assumed.

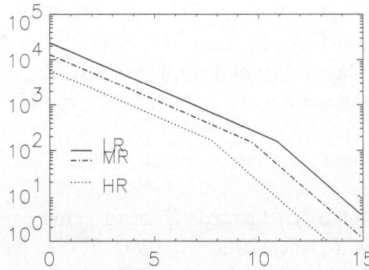

Fig. 6. Signal to noise ratio computation of the Courtes mode as a function of magnitude. We assume a differential visibility measurement with a reference channel of 36 (resp. 12 and 2) nm and a science channel of 1.0 (resp. 0.3 and 0.06) nm in the low (resp. medium and high) resolution mode. Median seeing conditions are assumed.

The only important hypothesis done is that the optical path difference is stabilized for duration of 30mn at the level of 1/500 of the coherence length, which corresponds in the worst case to 2μm.

5 Conclusions

VEGA is a unique opportunity to implement rapidly visible scientific programs on the VLTI. This project does not need any additional heavy infrastructure development, such a fringe tracker or adaptive optics devices. By using the multi-speckle approach and the coherencing operation of the

VLTI, differential interferometry allows us to reach either high spectral resolution (R=30000) or high sensitivity (up to 15). Imaging of stellar surfaces and probing the central Broad Line Region of active galactic nuclei clearly need high spectral and high angular resolution. *VEGA* and the *VLTI* are dedicated to this scientific goal.

References

1. Armstrong, J.T., Mozurkewich, D., Rickard, L.J. et al, 1998, Astrophys. J., 496, 550
2. Bério P., Mourard D., Chesneau O. et al., 1999, JOSA-A Vol.16
3. Bério P., Mourard D., Pierron M. et al., 2001, JOSA-A Vol.18
4. Breitfellner M.G., Gillet, D., 1993, A&A, 277, 524
5. Davis J., Tango W. J., Booth A.J. et al., 1998, MNRAS, 303, 773
6. Jankov, S., Vakili, F., de Souza, A. D. and Janot Pacheco, E., 2001, A&A, 377, 721
7. Kervella P., Bersier D., Mourard D. et al., 2004, A&A 423, 327
8. Koechlin L., Lawson P.R., Mourard D. et al., 1996 Appl. Opt. 35, 3002
9. Mékarnia D. and Gay J., 1989 J. Opt. 20, 131
10. Mourard D., "Observations of Cepheids with the VLTI," ESO Workshop June 96
11. Mourard D., Thureau N., Abe L. et al., 2001, C.R.Acad.Sci. Paris, t2, S. IV, 35
12. Mourard D., Nardetto N., Lagarde S. et al., this conference
13. Petrov R., Roddier F. and Aime C., 1986, JOSA-A, Vol. 3
14. Petrov R.G., 1989, NATO ASI series C274, 249
15. Rousselet-Perraut K., Vakili F. and Mourard D., 1996, Optical Engineering, 35, 2943
16. Thureau N., 2001, J. Opt. A: Pure Appl. Opt. 3, 440

BOBCAT - A Photon-efficient Multi-way Combiner for the VLTI

David Buscher, Fabien Baron, Julien Coyne, Chris Haniff, John Young

Astrophysics Group, Cavendish Laboratory, University of Cambridge, UK

Summary. We describe a concept for a near-infrared (JHK) bulk-optics combiner designed for efficient model-independent imaging of faint sources. The combiner is designed to accommodate (through a reconfigurable switchyard) any number of input beams from three to six. The instrument will include its own group-delay fringe tracker in addition to the "science" beam combiner. This will mean that fringe tracking will be possible using baseline "bootstrapping" while science data can be being secured on much longer baselines where the fringe visibility may be very low. We expect that a photon-efficient optical design will allow faint ($K \sim 13$) science targets to be observed, including AGN. Our proposed instrument will allow the full imaging potential of the VLTI infrastructure to be realised.

1 Science case

The scientific productivity of VLTI would be greatly enhanced if model-independent images could be made routinely. For all science programmes the existing VLTI capability of measuring visibilities and closure phases on a small number of baselines/triangles suffices when one is sure a source can be correctly represented by one of a small number of competing models, each with only a few unknown parameters. However, for many astrophysical problems the situation is more complex — models may have many parameters, the number of competing models may be large, or no available model may fit the interferometric data. Under these circumstances it can become possible to draw completely erroneous conclusions if an incorrect model is used. The ability to reconstruct model-independet images is the key to allowing reliable scientific conclusions to be drawn.

The first-generation VLTI instruments (particularly AMBER) do in principle allow the reconstruction of model-independent images, but this would require multiple reconfigurations of the ATs and/or the VLTI beam relay optics. Hence it is often difficult for potential observers to justify the amount of telescope time needed for imaging. Even if time is awarded, only a small fraction of the visibility phase information is likely to be measured, leading to lower-quality images for the same (u, v)-plane coverage compared with an instrument that can interfere more beams simultaneously.

We propose here a beam combiner designed to make effective use of the beams from up to six telescopes simultaneously, and thereby make rapid imaging with the VLTI a reality. The instrument makes use of a proven bulk-optics combiner design together with a dedicated fringe tracker, and has the name "Bulk Optics Beam Combiner And Tracker" or BOBCAT.

Space does not permit the elaboration of the full science reference mission for an imaging interferometric instrument, which is detailed in our report for the EU JRA4 "WP 1.1 Advanced Instruments Initial Matrix Document", dated March 24th 2005. Our reference misson concentrates on four main fields of research, young stellar objects (YSOs), the study of stellar multiplicity, the late stages of stellar evolution (dust shells), and active galactic nuclei. In all cases, we concentrate on the science that can be offered by *imaging* of moderately complex objects as opposed to measurement of model parameters of simple (i.e. few-parameter) models. A notable feature of this mission is that it is very broad, in that it covers a large range of astrophysical research, and is therefore likely to appeal to a wide range of ESO astronomers.

2 Derived top-level requirements for imaging

Given a science case which rests on the imaging of complex sources, we can ask how this top-level science requirement flows down to top-level requirements on the implementation. The major requirements can be summarised as (a) adequate (u, v)-plane coverage (b) adequate phase information, and (c) a real-time fringe-tracking system which is appropriate for resolved sources. We discuss each of these requirements briefly below.

The relationship between (u, v)-plane coverage and image quality is a topic which has been well covered in studies of radio synthesis imaging, and these are equally applicable to the optical/IR domain. "Rules of thumb" have been developed which capture the overall requirements for a given imaging scenario. The most important of these is that if one is intending to make an image with a given number N of "filled pixels" (this is a rough measure of how many resolution-element-sized regions in the image are emitting significant flux) then one needs to have visibility amplitude and phase measurements for least N *independent* points in the (u, v) plane. "Independent" points are those which are at least λ/θ apart in the Fourier plane, where λ is the mean wavelength of observation and θ is a measure of the overall angular extent of the object.

Thus in order to make a 10×10 pixel image with an angular resolution of θ_0, one would need to make approximately 100 visibility measurements, spaced at intervals of about $\lambda/(10\theta_0)$ and spread over a region of radius λ/θ_0 in the (u, v) plane. With only 3 telescopes, we sample only 3 points in the (u, v) plane at any one time. Earth rotation synthesis will allow more data points to be collected over time, but the number of *independent* data points will not be large: typically the maximum image complexity that could potentially be reconstructed even after eight hours of rotation synthesis with an ideally-spaced array of 3 telescopes would be about 3×3 resolution elements.

Thus in order to make images of even modest complexity, either repeated relocation of the telescopes or combining the beams from more telescopes is required. The latter is likely to be a much more efficient use of telescope time: increasing the number of telescopes combined from 3 to 4 increases the number of (u, v) points sampled by a factor of 2, and going to six beams increases the (u, v) sampling by another factor of 2.5, i.e. a 6-telescope array would be 5 times faster than a 3-telescope array in covering the (u, v) plane.

It is commonly recognised that in order to reconstruct images from interferometers, measurements of visibility phase information are required: measurements of visibility amplitudes alone are not usually sufficient to reliably make images. Experience from radio astronomy over the last 50 years has shown that the two main techniques for recovering phase information in the presence of phase perturbations (as are present in all ground-based optical/IR interferometric measurements) are closure phase and phase-referencing. Closure phases require the simultaneous measurement of fringes on three or more baselines, whereas phase referencing relies on simultaneous measurements of phases on the science target and a nearby unresolved reference source. In the optical/IR regime, measurement of closure phases to accuracies of fractions of a degree requires quite simple hardware, whereas phase referencing at the same level, e.g. with the PRIMA instrument, requires a number of complex, precise and expensive subsystems in order to work. Typically therefore, phase referencing can be applied on only a few baselines at a time, thereby restricting the available (u, v) coverage. Nevertheless, phase referencing is thought by some astronomers to be preferable to using closure phases, because they have concerns related to the well-known result that (except in a limited range of circumstances) there is no unique way of deriving a set of object phases from a set of closure phases.

We can use simulations to get an idea of the relative importance of the various forms of phase information and (u, v) coverage to the quality of the images that can be reconstructed. Figure 1 shows the results from simulations of imaging with high signal-to-noise data (SNR>100:1 per visibility data point) as a function of two variables: the (u, v) coverage and the source of the phase information. Interferometric data were generated simulating 6-hour Earth-rotation synthesis observations of a test source (an extended eliptical star with a binary companion). Data were generated corresponding to using arrays of either 4 or 6 telescopes. For each array, either phase-referenced data (in the simulations, simply the object phases with ~0.01 radians of added noise) or closure phase data (with identical amounts of noise) were generated, giving a total of 4 datasets. Each of these datasets were used for image reconstruction and the results are shown in Figure 1.

It is readily apparent from the figure that the difference in resulting image quality between using 4 and using 6 telescopes is much greater than that between using the phases resulting from phase referencing and closure phases. This confirms that the closure phase captures a large fraction of the phase

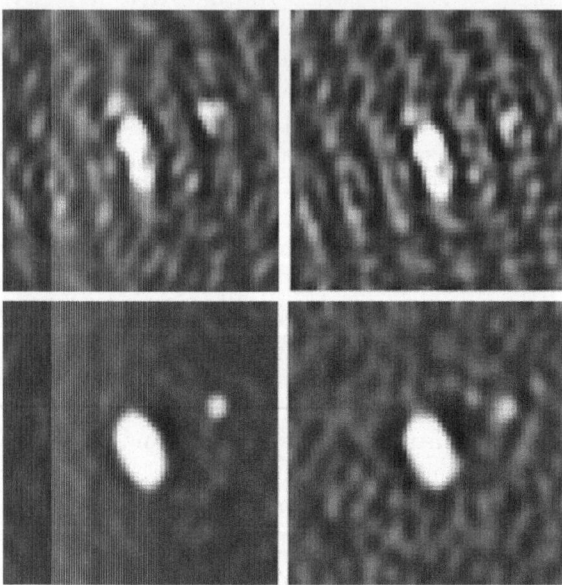

Fig. 1. Image reconstructions from simulated high-SNR data of an elliptical star with a companion which is approximately 3.4 magnitudes fainter than the primary. The upper images are reconstructed from simulated data using the beams from only 4 telescopes (i.e. 6 instantaneous baselines), while the lower images are reconstructed from an array of 6 telescopes (i.e. 15 instantaneous baselines). In each case an Earth-rotation synthesis of 6 hours duration was simulated. The leftmost images are reconstructed from uncorrupted phase data, simulating data from a phase-referenced system, while the rightmost images are reconstructed from closure-phase data. All images have the same greyscale levels. It can be seen that the difference between images reconstructed from 4-telescope data and those reconstructed from 6-telescope data is far greater than the difference between images constructed from phase-referenced data and from closure phase data.

information available, especially for larger numbers of telescopes, leaving the (u, v) coverage as the limiting factor in image quality. It is clear that given a choice between phase-referencing on a limited number of baselines and closure phases on a larger number of baselines, that the latter is to be preferred.

A less·obvious requirement implied by a science emphasis on imaging is that of fringe tracking. Implicit in an imaging as opposed to astrometric observation is the assumption that the science targets will likely be resolved many times over by the longest baseline in the array. A necessary consequence of resolving the target is that the fringe contrast is low on the longest baselines. Low fringe contrasts mean low signal-to-noise ratios for fringe tracking (the SNR scales as the square of the fringe visibility) which means that fringe tracking on the longest baselines is the weak link in the imaging process. The most general way around this problem is the use of baseline bootstrapping. This involves tracking fringes on a "chain" of short baselines (nearest-neighbour telescopes) and using this to infer the fringe motion on the longest baselines. Thus a general-purpose imaging interferometer must incorporate a bootstrapping fringe tracker. Furthermore, this fringe tracker should operate using group-delay tracking methods, since phase tracking methods are susceptible to short timescale fringe SNR "drop-outs" (due to, for example, AO system Strehl fluctuations). The likelihood of a dropout somewhere along the chain of telescopes involved in bootstrapping grows with the length of the chain, rendering group-delay tracking, which is much more resistant to such drop-outs, the most competitive technique.

To summarise therefore, the optimum imaging instrument allows the combination of the beams from a large number of telescopes, allows the measurement of closure phase information, and incorporates a bootstrapping group-delay fringe tracker. This is the concept presented here.

3 BOBCAT concept description

A high-level block diagram of the proposed instrument, assuming a full complement of 6 input beams, is shown in Figure 2. If an initial implementation using only 4 input beams were desired, the fast switchyard identified at the top of the figure would not need to be installed, but could be introduced later as a potential upgrade.

3.1 Waveband selection and switchyard

Up to six path-compensated beams will enter the instrument from the VLTI delay lines. The beams them immediately enter a fast switchyard incorporating dichroic mirrors, which has two functions:

– It reflects light in the chosen science band (one of J, H or K bands) from a subset of four of the entering beams into the four input ports of the science combiner.
– It serves to transmit either the H or K-band light from all of the input beams to the fringe tracking combiner.

The switchyard is designed to be reconfigurable in less than 10 seconds, and will allow at least one permutation of every combination of four beams from up to six input beams to be selected, i.e. all baselines and closure triangles will be accessible to the science combiner. By using three configurations of the switchyard, all 15 baselines available from 6 telescope beams (or 12 of 15 baselines with two configurations) will be measurable within one or two minutes.

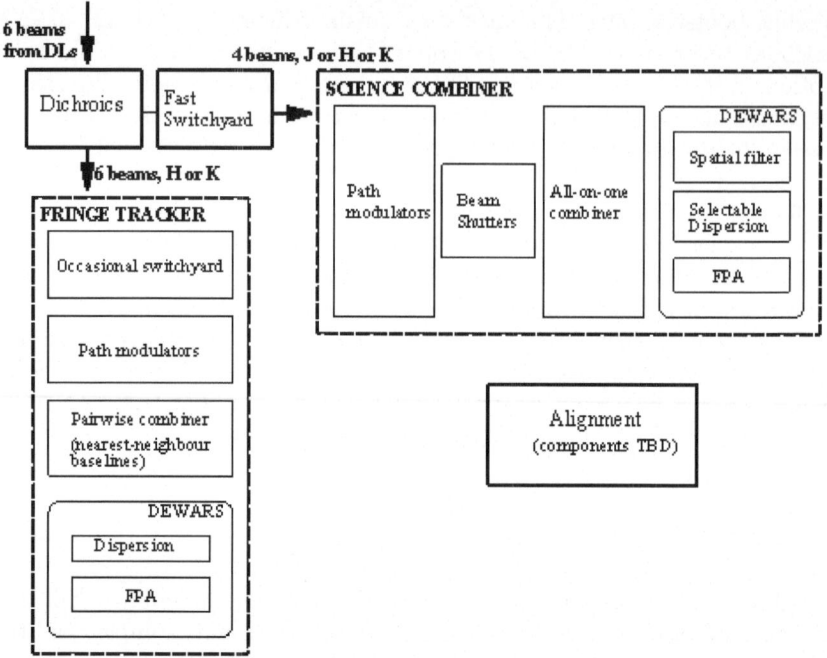

Fig. 2. Block diagram of the BOBCAT instrument concept assuming a maximum complement of six input beams. Light enters at the top, and is spectrally split between the science beam combiner at the right and the group-delay fringe tracking beam combiner to the bottom.

3.2 Internal group-delay fringe tracker

The fringe tracker will consist of a pairwise pupil-plane beam combiner that continuously measures fringes on the five (or equivalent for shorter chains) nearest-neighbour baselines in a bootstrapping "chain" of telescopes. This will most likely be implemented with the optical components mounted on discrete commercial or custom mounts. Low-noise hybrid FPA detectors will be used, such as those supplied by Rockwell or Raytheon, although only a small number (<100) of pixels will be used on any given device.

The fringe tracker will operate as a group-delay fringe tracker in either the H or K photometric band, with one of the two remaining bands from the set of J, H, and K being fed into the science combiner. The fringe tracker will be designed to maximize the raw signal-to-noise of the fringe measurement, as the data will not normally be used for science analysis. A small number (5) of spectral channels will be used. As the combiner will only serve to maintain the fringes within their coherence envelope, the OPD modulation of the input beams (provided internal to the instrument) will be relatively slow, and multiple short-exposure realisations of the fringes will be averaged to obtain each OPD estimate.

For a 6-beam instrument a total of 5 detector arrays will be required, each located in a separate dewar cooled to LN_2 temperatures and including a cold stop and a fixed dispersing element. This number may be revised downwards if further detailed studies suggest that more than two of the fringe-tracking outputs can be delivered to each focal-plane array or if fewer input beams are to be handled.

Within the group-delay tracker itself, there will be a separate additional switchyard to manage reconfigurations of the telescopes that feed the instrument. This will be adjusted when the VLTI array configuration is altered such that the short fringe-tracking baselines correspond to different pairings of the input beams as delivered to the instrument entrance port.

3.3 Science combiner

Our proposed implementation for the science beam combiner is a four-way pupil plane combiner. This will receive four input beams, and each of its four outputs will encode the full set of six fringe patterns appropriate to the four input beams. The choice of which input beams are fed to the science combiner will be determined by the instrument's "fast" switchyard which will allow rapid reconfiguration of the input beams.

Currently, we envisage using a high-stability optically-contacted beam combiner. A prototype combiner of this type has already been fabricated and tested in Cambridge (see Figure 3) and its behaviour and performance is well understood. Temporal modulation of the input beams so as to visualise the interference fringes will be provided by a separate set of PZT stacks actuating the mirrors feeding the combiner optics.

Each of the four combined beams will be fed to a cooled low-noise hybrid FPA via a spatial filter and a dispersing element. We expect the instrument will offer a choice of at least two, and possibly three, spectral resolution modes: low (R $\sim$ 30), medium (R $\sim$ 300) and high (R to be determined depending on science case and reference mission).

In this type of beam combiner, each of the four outputs provides interference signals on all six instantaneously-available baselines and so one detector and dewar interrogating a single output is sufficient to measure all the visibilities and closure phases. This type of design thus offers considerable flexibility in how the detectors and dewars are deployed; for example fewer than four detectors might be installed initially to reduce costs. Alternatively, detectors with fixed dispersion could be employed, with different spectral resolutions being provided at the different combiner outputs.

From a systems engineering approach, the choice of a four-way combiner rather than a six-way system has been based on a number of trade studies and serves to optimise the instantaneous fringe signal-to-noise ratio, to minimize risks associated with opto-mechanical stability, to minimize cross-talk between the different fringe signals, to simplify the optical design of the beam

Fig. 3. Photograph of a prototype optically-contacted near-infrared four-way beam combiner for COAST. Once fabricated, this combiner requires no internal alignment. The overall footprint of this device, which has four inputs and four outputs, is 20cm x 10cm. A similar device for the VLTI would be expanded in scale.

combiner and to take maximum advantage of tried-and-tested components and methodologies.

4 Performance

4.1 Imaging capabilities

The proposed instrument will permit imaging by phase closure. The choice of a 6-beam instrument allows the desired (u, v)-plane coverage to be obtained with the minimum observing time (given that only six delay lines are available). Science data is recorded on 6 baselines and 4 closure triangles (3 independent) simultaneously, and the fast switchyard allows all 15 baselines and 10 independent closure phases available from 6 telescopes to be selected in a few minutes. For the same (u, v)-plane coverage, combining more beams together secures a greater fraction of the phase information: for example, the 6-beam concept preserves 2/3 of the phase information, compared with 1/3 for a 3-beam instrument such as AMBER.

One key feature of our proposed instrument is the ability to take advantage of baseline and, in most cases, wavelength bootstrapping by using a separate optimized fringe-tracking beam combiner. This will crucially allow science measurements to be made on the very longest VLTI baselines by monitoring the atmospheric fluctuations on the much shorter nearest-neighbor baselines between closely spaced telescopes. Unless a bright point source reference is available within the isoplanatic patch, this will be the only way of measuring fringe amplitudes and phases for the resolved sources that will be the targets of many imaging studies.

4.2 Spectral coverage and resolution

The proposed instrument will be designed to record science data in any one
of the *J, H* or *K* near-infrared photometric bands. In parallel, either the *H* or
K photometric band will be used for fringe tracking. At each of the science
beam combiner outputs at least two different spectral dispersions will be
available: a low resolution mode with R~30, and an intermediate resolution
mode with R~300. If there is a suitably compelling science case, then a high
resolution mode will be considered as part of the conceptual design, although
in the absence of such a science case we have not yet chosen an appropriate
spectral resolution.

4.3 Limiting magnitude

The limiting magnitude for our proposed instrument will be determined by
the limiting sensitivity of the group-delay fringe tracking sub-system. If a
target is bright enough for the fringe tracker to operate, then arbitrarily long
integration times (perhaps spread over multiple nights, at the same sidereal
time) can be used to build up signal-to-noise in the science combiner. We
have computed the fringe tracker limiting magnitude for several modes of
operation, assuming the atmospheric parameters appropriate for "average"
and "excellent" seeing conditions circulated to all work-package teams (see
Table 1).

Table 1. Seeing parameters used for limiting magnitude and SNR calculations

Parameter (at $\lambda = 500$ nm)	Seeing FWHM (arcsec)	Fried Parameter r_0 (cm)	Coherence Time t_0 (ms)
"Average" seeing	0.8	12.6	6
"Best" seeing	0.5	20.2	15

We have assumed a basic integration of time of $2t_0$, and sky backgrounds
of 14.4 mag/arsec^2 and 13.0 mag/arsec^2 in the H and K bands respectively,
with the transmitted background from an Airy-disk sized area of sky being
split between two combiner outputs. Spatial and temporal wavefront decor-
relations due to the VLTI infrastructure have been included as described in
the VLTI Interface Control Document, VLT-ICD-ESO-15000-1826, Issue 3.0
(henceforth "ICD"). These have been augmented with additional throughput
losses, and spatial and temporal coherence loss factors for the instrument and
assumed exposure time, as listed in Table 2. The thermal background from
the VLTI optical train was calculated assuming an emissivity of $(1-\eta)$, where
η is the overall throughput. Finally, for ease of comparison, we have assumed
a point source target.

Table 2. Instrument parameters used in performance calculations.

	Fringe Tracking Combiner	Science Combiner
Instrument throughput (including detector QE)	40%	20%
Spatial wavefront error	$\lambda/20$ RMS @ 633nm	$\lambda/20$ RMS @ 633nm
Temporal phase jitter over coherent integration	$\lambda/20$ RMS @ 633nm	$\lambda/20$ RMS @ 633nm
Detector read noise	$3.0e^-$	$1.5e^-$
Visibility loss factor due to $2t_0$ integration	0.79	0.79

Detailed simulations of the group-delay tracking process were used to determinine the limiting photon flux required for group delay tracking in different observing conditions. In the process of deriving our limiting magnitudes quoted here, we have used a limiting flux values of twice those calculated from simulations in order to allow some margin of safety. For simplicity, values are given here for only two cases, a baseline between two UT's and a baseline between two AT's. In the UT case, we have assumed that the science target is also the reference star for the Adaptive Optics system. In order to calculate the Strehl delivered by the AO, we have assumed that the star has the colours of a G5 dwarf ($V\text{-}H = 1.3$ and $V\text{-}K = 1.5$). The ICD Strehl predictions as a function of guide star V magnitude for 0.65" seeing were used for both the "best" and "average" seeing cases.

Figures for the limiting sensitivity of the fringe-tracking subsystem for a number of the different observing modes are tabulated in Table 3 for average and best seeing conditions.

Table 3. Limiting sensitivities for the group-delay fringe tracking subsystem of the instrument.

	UT-UT (self-ref. AO)		AT-AT	
	H	**K**	**H**	**K**
"Average" seeing	13.5	14.4	12.7	12.7
"Best" seeing	13.6	14.5	13.9	13.5

It can be seen that the limiting sensitivities are adequate for the observation of many 10s of the brightest AGN, and hundreds of YSOs and other science targets.

4.4 Signal-to-noise ratio

In order to demonstrate the astronomical capability of the proposed instrument, we have also computed the signal-to-noise ratio for visibility amplitude estimation with the science beam combiner. With the type of combiner we are proposing, each of the four beam combiner outputs will deliver an independent estimate of the fringe signal. In the interest of clarity we have ignored the possibility of averaging the signals from multiple outputs (which could improve the SNR by a factor of between 2 and 4), and present below the signal-to-noise expected per single spectral channel and per single beam combiner output for various observing modes.

In deriving the values in Table 4, we have assumed an on-source incoherent integration time of 100 seconds, and have computed the signal-to-noise ratio expected for sources at three magnitudes (assuming Vega-type colours): one at the approximate sensitivity of the fringe tracker and also for 2 and 4 magnitudes brighter than this.

Table 4. Signal-to-noise ratios in 100s per spectral channel per beam combiner output for the science beam combiner, configured with a spectral resolution, R~30. Values are given for a baseline between two UTs, and for a baseline between two ATs.

	UT-UT (self ref. AO)						AT-AT					
	J		H		K		J		H		K	
	mag	SNR	mag	SNR	mag	SNR	mag	SNR	mag	SNR	mag	SNR
"Average"	13	0.03	13	1.5	13	10.5	13	0.01	13	0.13	13	0.54
seeing	11	1	11	18	11	53	11	0.5	11	3.3	11	11
	9	10	9	71	9	157	9	6	9	20	9	45
"Best"	13	0.05	13	2.4	13	13	13	0.3	13	1.3	13	3.2
seeing	11	1.6	11	21	11	55	11	4.3	11	14	11	22
	9	13	9	73	9	158	9	19	9	43	9	63

It can be seen from the table that, for the faintest sources under average seeing conditions, the instrument will be limited by the signal-to-noise in the science beam combiner, and not by failure of the fringe-tracking sub-system. It should be noted that several techniques may be used to improve the science signal-to-noise ratio for faint targets, which have not been included in the above calculations, including coherent or incoherent combination of signals from multiple beam combiner outputs, incoherent combination of signals from adjacent spectral channels, the use of measurements from the fringe tracker to phase up consecutive basic integrations and hence extend the effective coherence time in the science combiner, and the use of a bright AO reference star for the UTs (where available).

4.5 Dynamic range

The dynamic range of images reconstructed from the visibility amplitude and closure phase measurements will depend on the number of data points and their random and systematic errors. The dynamic range expected in a map derived from n visibility data can in general be described by the following approximate formula:

$$\text{DynamicRange} = \frac{\sqrt{n}}{\sqrt{\left(\frac{\delta A}{A}\right)^2 + \delta\phi^2}}$$

where δA and $\delta\phi$ refer to the visibility amplitude and phase errors respectively.

Assuming that a sequence of calibrated measurements of the 15 baselines (and associated closure phases) associated with a 6-element subset of the UTs and ATs can be secured in 30 minutes, and that observations take place between Hour Angles of ± 3 hours, a suitable value for n will be of order 180. If we further assume fractional amplitude errors and phase errors of, say, 5%, this gives a dynamic range of approximately 200:1, i.e. 6 magnitudes. Hence, we can expect features at least 5 magnitudes fainter than the brightest unresolved component in the image to be detected reliably in any interferometric map.

5 Conclusions

We have presented a beam combiner concept which has been designed to exploit the wide range of interferometric science that is available from using the VLTI for model-independent imaging. A balanced approach has been taken to providing *all* the instrumentation needed to perform useful imaging science observations, including a multi-way bootstrapping fringe tracker. Proven high-efficiency bulk optics technologies have been used throughout in order to allow the observation of faint targets. A key feature of the design is that it can make use of the four ATs and six delay lines of the VLTI to their maximum advantage, for bootstrapping fringe tracking and rapid (u,v) coverage. Thus, even if scientific arguments are put aside, there is a strong case for a multi-beam correlator for the VLTI purely on the basis of maximising the return on the existing investment in the infrastructure at Paranal.

Acknowledgements The authors would like to thank W. Warre and C. Coates for enhancing the depth of their understanding of the concept of complexity.

Near-Infrared Fiber Imager for the VLTI

Ralph Neuhäuser[1], Andreas Tünnermann[2], Marc Hempel[1], Bringfried
Stecklum[3], Jena-Peter Ruske[4], Eike Guenther[3], Artie Hatzes[3], Rolf Chini[5],
Roland Lemke[5], Günther Wuchterl[1], and Oskar von der Lühe[6]

[1] Astrophysikalisches Institut und Universitäts-Sternwarte, Schillergäßchen 2,
  D-07745 Jena, Germany rne@astro.uni-jena.de
[2] Institute for Applied Physics, Universität Jena, D-07745 Jena, Germany
[3] Thüringer Landessternwarte Tautenburg, D-07778 Tautenburg, Germany
[4] GC-Tec, D-07745 Jena, Germany
[5] Ruhr-Universität Bochum, Astronomisches Institut, Universitätsstr. 150,
  D-44780 Bochum, Germany
[6] Kiepenheuer-Institut für Sonnenphysik, Schöneckstr. 6, D-79104 Freiburg,
  Germany

Summary. For detection of massive close-in exo-planets around other stars, a
contrast of $\sim 0.3\%$ is sufficient and can be achieved with state-of-the-art single-
mode fibers as spatial filter. For lower-mass planets, much higher accuracy is necessary. Apart from the atmosphere, there are several sources of noise which degrade the phase stability of interferometers based on bulk optics by changes in
the optical path. For direct detection of low-mass exo-planets and for direct observation of convection on giant stars, our two main science goals, as well as for
secondary science goals such as lower-mass faint circumstellar disks around both
nearby low-mass stars as well as more distant high-mass stars and for the direct
detection of close spectroscopic binary (T Tauri) stars and to follow their orbits
for direct dynamical mass determination, we need more than two apertures to fill
the uv-plane, again very complicated with bulk-optics. Integrated optics allows to
solve these problems and to minimize the components. We have experiences in the
development and fabrication of integrated-optical circuits and modulators in various wavelength ranges, i.e. from visible to the telecommunication band, as well as
integrated-optical sensor systems like interferometers. Starting from two channel
interferometers for the J-band at $1.55\mu m$ based on Lithiumniobat, we will develop
multi-channel-interferometer systems for longer wavelengths. Such systems with
new materials and fibers for the near- and thermal infrared can then be used, e.g.,
for 2nd Gen VLTI instrumentation.

1 Introduction

We propose to build a Near-Infrared Fiber Imager (NIFI) based on integrated
optics for the Very Large Telescope (VLT) Interferometer (VLTI) at the
European Southern Observatory (ESO) on Cerro Paranal, Chile.

As detailed in the next section, we need to achieve *stable fringes for
long integration times* with visibilities going down to 10^{-4}, which is not yet
achieved with current-generation instrumentation. This is partly due to noise

inherent in bulk-optics. Hence, we suggest to build a new instrument based completely on integrated optics, i.e. fibers from the telescope feed to the interferometric light combination and onwards to the detectors.

We will first demonstrate this technique in the J- and/or H-band and will then adapt it to the K-band and possibly to the thermal IR later. We plan to combine one to three UTs plus ATs.

2 Science case(s)

We propose to build an instrument which is open to the whole astrophysical community and offers science cases in various fields of Galactic and Extra-galactic astronomy. Our own design driver for NIFI are long-term stability for very low visibilities to detect young hot low-mass extra-solar planets close-in to their host stars and convection on the surface of giant stars.

Our NIFI instrument will overcome stability problems known from bulk optics interferometers of the first generation. The topic of planet formation and direct observations of extrasolar planets is of large interest in modern astrophysics, in line with the origins theme, and also in connection with astro-biology, the ultimate question of life on other planets.

2.1 Extrasolar Planets and their Environments

The formation processes of stars and planets are one of the most impor-tant parts of astrophysical research. All extrasolar planets detected so far are Jupiter-like objects in very close orbits, discovered by the radial velocity technique, which gives a lower mass limit on the companions (hence, planet candidates). In addition, there are a very few planets (or candidates) de-tected directly, including GQ Lup b discovered by us (Neuhäuser et al. 2005) with Adaptive Optics Imaging at the ESO VLT; for such a planet detected directly, one cannot determine the mass directly by Keplers law, because the orbital period is too long (1000 yrs). Hence, neither for radial velocity planet candidates (indirect detection, close orbit) nor for direct detections (wide orbits) one can determine the mass directly, so that the planet nature remains somewhat uncertain. To overcome the limitations of resolution by conventional methods of planet detection like radial velocity surveys and di-rect imaging and to study close-in companions and earth-like planets, it is necessary to carry out interferometric observations. Our scientific goals are direct interferometric observations of extrasolar planets as well as the visual detection (resolving) of close binary stars of the T Tauri type - for direct mass determination, in order to test and calibrate theoretical evolutionary tracks and isochrones - partly developed by us. For the test and calibration of such theoretical models, convection need to be understood much better than so far; hence, we plan to observe convection directly on the surface of giants with NIFI. Furthermore, we will study massive young stars, analyse in detail the structure of circumstellar disks of nearby young late-type stars, where planets form.

Up to now more than 100 extrasolar planets have been detected indirectly, nearly all of them by radial velocity variations of their host stars. Since the inclination of the orbit remains unknown, these indirect detections only reveal the minimum masses and therefore such objects can only be considered as candidates. In the rare case of a transit the inclination, mass, radius, and the density of the transiting object can be derived and thus planets can be confirmed. Extrasolar planets can be divided into two categories: short-period planets with circular orbits and long period planets often on eccentric orbits. Among almost 150 extrasolar planets known to date, one sixth has periods with less than 10 days. The radial velocity technique is only capable of detecting close in planets since the gravitational pull on the host star is too weak for wider orbits. Another method to detect planets on orbits between 2 and 10 AU is via astrometry, i.e., the wobbling of the host star in both right ascension and declination. Furthermore, it is possible to detect extrasolar planets by direct imaging, which is very challenging due to the large brightness difference between star and planet. In the case of young planets this method is applicable because young planets still contract and thus are more easily detectable in the IR spectral range. This method has been applied successfully in the case of GQ Lup by our group (Neuhäuser et al. 2005): the 1 Myr young classical T Tauri star GQ Lup has a L2 companion, comparison with evolutionary models and model atmospheres suggest that the companion - GQ Lup b - has a mass between 1 to 3 Jupiter masses, implying that GQ Lup b is in fact a planet.

The direct detection of extrasolar planets is of major importance to significantly improve our understanding of the formation process of star and planet formation. Therefore it is important also to detect and analyse planets on close-in orbits. Furthermore, circumstellar disks have to be studied. Our group is currently studying the gas content and the dynamics of well-known circumstellar disks (Hempel & Schmitt 2003, Hempel et al. 2005) - important ingredients to estimate disk masses.

Unfortunately, conventional methods such as radial velocity searches for planets and spectroscopic surveys of circumstellar disks set a limit to the resolution of the observations. A completely different and very powerful method to characterize planetary systems is *interferometry*: a system consisting of two separate sources (i.e., star & planet) results in characteristic variations of the visibilitiy when observed at different baselines. Interferometry is capable of overcoming these limits and allows to study close-in extrasolar planets as well as to resolve the inner structure of circumstellar disks. Such observations are indispensable to develop a coherent picture of the process of planet formation. Other than the methods described before, interferometry offers much higher angular resolution and is capable of resolving objects with very large differences in brightness. For extrasolar planet studies, interferometers capable of measuring the fringe contrast with high stability are needed. The instruments currently installed at the interferometers do not offer the precision needed to detect low-mass extrasolar planets - a visibility contrast of 0.0001 is needed for this purpose. This can be achieved for long integration times only using integrated optics. Such devices are commonly used in communication technology; their main advantages over conventional bulk-optics are miniaturization of complex optical setups, high modulation- and transfer rates well in the Gbit/s range, and a much better insensitivity to disturbances leading to outstandingly stable fringes. All detection methods together can detect whole planetary systems, i.e. planets at different separations from the host stars, and only interferometry can determine the masses of the planets directly by measuring their light and detecting their position (orbit).

We are searching for planetary (and brown dwarf) companions around young stars by the direct imaging technique, namely around stars from roughly 1 to 100 million years age, all within 40 to 100 pc. Given the typical magnitude of such stars in the infrared and the typical magnitude of 5 Jupiter mass objects at the given age range (using Burrows et al. 1997), we can calculate the magnitude difference (or contrast) between star and planet, hence the visibility contrast to be achieved. For the planet (candidates) known so far, the detection of ϵ Eri b and ι Hor b may just be possible, with expected visibility contrast of 0.0001, but this is roughly the limit of the current instrumentation AMBER, which is actually not yet achieved. The most important part in the target selection is the young age of the primary star and, hence, the companion, too, so that its still contracting and, hence, emitting light itself. However, even if the companions cannot be detected by interferometry, given their minimum masses, we can certainly find upper mass limits from non-detections, which will be very useful in determining their true masses and their nature (below 13 Jupiter mass, then planet; or above 13 Jup mass, then brown dwarf).

2.2 K-Giant Surfaces

Due to their large surfaces, K giant single stars are perfectly suitable for VLTI measurements. Indeed, K giants are known to show surface structures: Choi et al. (1995) found CaII variability with periods between 60 – 160 days in 10 out of 12 K giant stars. This can be interpreted as rotational periods which are indirectly visible by means of CaII variability studies. Many K giants are also known to show variable radial velocities: Hatzes & Cochran (1993) found periods between 233 – 643 days in 3 K giant stars. Furthermore, Setiawan et al. (2004) found a long-term variability of the K giant HD 78647 correlated with variability in the line profiles.

To explain the observed variability of the K giant α Tau by means of stellar surface structures, roughly one third of the stellar surface would have to be covered with starspots resulting from magnetic activity. The angular diameter of 21 mas allows us to resolve such a starspot on α Tau. Planets as companions may also be responsible for long-term variability (Frink et al. 2002, Setiawan et al. 2003). The VLTI will be capable of testing this hypothesis, i.e., to detect planets via astrometric measurements. The expected astrometric signal is of the order of ~ 0.3 mas. Thus it is not detectable using one VLT UT alone, but it is feasible to resolve this with the VLTI. For the current instrumentation of first generation instruments at the VLTI only a very limited number of K giants is detectable. Our proposed instrument allows to detect far more such objects. Thus it will be possible to carry out statistical studies of K giant surfaces: up to now 135 K-type stars – almost all of them giants – are known to have an angular diameter above 5 mas making them potential targets for such a study. Our group (Setiawan & Hatzes) have carried out a survey of about 80 K giants studying variability and radial velocities. Thus we have a representative subsample of interesting objects.

3 Technology: Integrated Optics

The direct detection of extrasolar planets is based on sophisticated brightness measurements with respect to the enormous intensity ratio between central star and planet. With interferometric telescope systems like ESO-VLTI and Keck Interferometer it is possible to synthesise apertures, which have a substantial narrower PSF compared to single telescopes. For young warm exoplanets (1000 ... 1500 K) the intensity ratio between star and planet amounts to approximately $3 \cdot 10^{-5}$ to $3 \cdot 10^{-4}$ in the near infrared. They contract significantly and gain gravitational energy, which is emitted in the infrared leading to by some orders of magnitude higher brightness compared to old planets (i.e. Burrows et al. 1997, Wuchterl & Tscharnuter 2003). Consequently, an interferometer for investigation of extrasolar planets must allow to measure the contrast of an interference pattern with high accuracy, which implies a high phase stability. Simulations show, that the contrast error should be lower than 0.3 % for interferometric verification of planets similar to 51 Pegasi (de Foresto et al. 1997). In practise such precision can be reached using singlemode optical fibres as spatial filters (Perrin et al. 1997). But for the verification of lower-mass exoplanets a considerably higher precision is necessary.

There are some factors influencing the phase stability of bulk-optics interferometers besides atmospheric turbulences which lead to fluctuations and systematic changes of the optical path length, i.e., seismic activity and temperature gradients. At the Institute of Applied Physics of the University of Jena (IAP) we have carried out investigations to spatial filtering using singlemode fibres (Zeitner et al. 1999).

Since the investigation of the trajectory of exoplanets (and thus also its mass) is possible with relatively good accuracy using two apertures, for the direct imaging a combination of more apertures, that is a covering in the uv-plane is needed as good as possible. In the case of VLTI (four 8.2 m and some 1.8 m telescopes) maximum of 21 baselines are possible where interferometric signals can be measured. Optical designs for the combination of some apertures basing on bulk optics are extremely difficult and thus susceptible to systematic and random perturbations which influence the phase stability.

The astronomic precision in the range of microarcseconds, for which a reference object is needed (dual feed) will be increased by the improvement of the phase stability using the proposed integrated-optical concept. Integrated optics allows a considerable miniaturisation of complex optical systems. Especially the combination of integrated-optical circuits with optical fibres forms a completely atmospherically protected light path leading to substantial stability increase of the system.

In optical telecommunications integrated-optical components are used for fast light modulation, splitting and combining as well as multiplexing of data channels. They are fabricated in large quantities. Optical glasses, the ferroelectric crystal lithium niobate, compound semiconductors and silicon are used as substrate materials mostly. The waveguides are fabricated by micropatterning, ion exchange or metal indiffusion or by stacked deposition of semiconducting or oxidic layers with different composition followed by microfabrication.

The advantages of integrated-optical systems against bulk-optics are the small size, stability, very high data rates up to the 40 Gbit/s range, good reproducible interferometric characteristics and extinctions as well as the proof against outer influences to the optical path. Interferometric principles are used in modulators, switches, wavelength multiplexers and filters (Murphy 1999). For example an amplitude modulator is implemented as a two beam interferometer, where the optical path length can be controlled by means of electrooptics. For this purpose an electrode system is positioned in plane of a Mach-Zehnder-waveguide structure. The modulator transmission follows a quadratic cosine function if a voltage is applied.

Integrated-optical devices are available almost only for telecommunication purposes, where standard devices for wavelengths around the $1.3\,\mu m$ and $1.55\,\mu m$ wavelength are used. Other elements are not available except modulators for the 800 nm and 1060 nm range. In research institutes like the IAP waveguides and devices for a large number of wavelengths and basing on various substrates are investigated including also photonic crystal waveguides and fibres.

There exists a variety of possibilities to imply passive integrated-optical elements into interferometers for astronomic purposes. Especially two or more telescopes can be networked to interferometers with large base length. In Malbet 1999 interferometric schemes are presented in form of block diagrams. A part of these assemblies can be implemented as integrated or fibre optics advantageously. Successful experiments have been done using commercial passive structures in laboratory (Berger 1999). The up to now highest integration was realised in form of the combination of eight inputs using an integrated-optical interferometric structure by Berger 2000. A disadvantage is the relatively low spectral bandwidth.

For control of interference it is advantageous to combine more functions in the integrated interferometer. Especially the implementation of phase shifters allows an increase of measurement accuracy. These elements have not been investigated for astronomic purposes in the past and have been realised by mechanical means.

A problem is the efficient incoupling of stellar light into optical fibres. The technical possibilities and limits have been described by Shaklan and Roddier (1999) in theoretical way. However, experiments have been done mainly in laboratory using lasers and discharge lamps.

There are great experiences in the field of integrated-optical elements especially in glass and electrooptic materials like lithium niobate LiNbO3 and potassium titanyl phosphate (KTP). Diverse elements like interferometers for sensors, polarisers as well as electro-optic phase and amplitude modulators for wavelengths between 450 nm and 1550 nm have been realised (Ruske et al. 1995b, Ruske et al. 1999). Waveguides in KTP have an outstanding position in this field. Beside the electrooptic function they are able to guide light of a comparably large spectral bandwidth in singlemode (Rottschalk et al. 1997). This is the prerequisite for broadband electrooptical controllable interferometric devices for light modulation and colour mixing. The optical bandwidth amounts to be already 200 nm in the visible and much more in the infrared. These waveguides and modulators have been developed for colour image generation.

The (i) theoretical modelling and experimental verification of waveguides in single- and multimode technique with respect to operation ranges and guiding parameters for application in beam combination, beam splitting and modulation (Rottschalk et al. 1995, 1997), the (ii) concept of integrated-optical modulators for colour image generation with laser light in form of a power proofed and broadband solution (Ruske et al. 1995a, Ruske et al. 1999) and (iii) the fabrication of demonstrators for colour image generation systems (Ruske et al. 1998, Ruske et al. 1999) have been focal points in these projects. Interferometric modulators with contrast ratio of about 500:1 in the visible and 5000:1 in the infrared for modulation frequency of more than some 100 MHz and guided optical power up to 100 mW have been realized in reproducible way. These devices will find application in small displays and apparatuses for printing and photofinishing. Devices for applications in the near infrared (for example 830 nm, 1060 nm, 1550 nm) have been successful realized by means of industry cooperation.

Therefore it was possible for the first time to fabricate singlemode waveguides for the visible range and to show a precise amplitude modulation with high extinction using magnesium doped lithium niobate. These modulators can be used in laser based photofinishing systems, which are currently equipped with complex bulk optics.

The fabrication of waveguides and modulators for ultrashort pulses in the range of one picosecond and peak power of more than 120 Watts is possible, which is an international outstanding value. Such ultrashort pulse laser systems for use in micro material processing and biotechnology can be configured more efficient and simple (Ruske et al. 2003a,b).

Further increase of the guidable power in singlemode waveguides was successful by change of the waveguide geometriy. A special property of KTP allows an increase of the waveguide cross section thus causing a decrease of the optical power density in the waveguide. These waveguide requires a new modulation concept, which was demonstrated as asymmetric diffraction amplitude modulator. So it is possible to guide and modulate light of some watts in singlemode (Werner et al. 2003). Further investigations are due to material improvement for integrated-optical purposes, especially using upgrading of lithium niobate substrates by liquid phase epitaxial layer deposition. An important point are the activities concerning the use of integrated-optical devices in aerospace and astronomy. So flight proofed high-power gigahertz phase modulators have been realized for optical communication between geostationary and low earth orbit satellites. Its operation wavelength is 1064 nm, the transmitted power is in the range of 1 Watt that is extremely high for integrated devices. These devices have been realised for the first time (Ruske et al. 2003a).

To open the mid-infrared the use of photonic crystals and photonic crystal fibres is advantageous. Its function bases on the combination of allowed and forbidden zones for light transmission due to Bragg reflection on microstructures in planar as well as fibre geometry. Beside the increase of the IR transmission range it is possible to control the dispersion and to enlarge the guidable power in optical fibres. In planar structures it is possible to connect angled waveguides without bends thus enlarging the package density of integrated-optical circuits. The waveguides are prepared by microfabrication technique. They consist of a periodic sequence of high and low refracting structures in sub wavelength geometry, mostly as hexagonal sequence of holes in the substrate. By suitable arrangement of allowed and forbidden zones it is possible to force the light onto a predetermined path. Some work concentrates on photonic crystal waveguides with high index contrast mostly on the basis of semiconductors, for example GaAs against Al2O3 or air (Chow et al. 2001, Olivier et al. 2001). Other groups work with low index contrast in high refracting materials, for example GaAs against Ga1-xAlxAs (Bogaerts et al. 2001, Lalanne 2002). However the coupling to optical fibres is difficult due to the high index of refraction of the semiconductors. The losses are relatively high and amount up to 30 dB/mm.

Therefore, we favour photonic crystal waveguides with low index contrast (Tünnermann et al. 2004). Here a layer system is used which consists of a Nb2O5 layer on thermal oxidised silica. A cladding layer is formed by sputtered silica. So a symmetric layer waveguide is obtained, which guides the light in vertical direction. The lateral confinement is done by a photonic crystal structure made of 370 nm diameter holes with a depth of 1.1 μm hexagonal arranged with a period of 595 nm. It has been realised by a combination of electron beam lithography, reactive ion beam etching (RIE) and inductive coupled plasma etching (ICP). So singlemode waveguiding with an attenuation of 1.7 dB/mm was obtained. This relatively high value has to be associated with the some tenth millimetres short structures necessary for waveguide bending since devices are realised by small photonic structures combined with straight conventional low loss rib waveguides. These photonic structures are a future oriented possibility to increase the package density of integrated-optical devices. However, their fabrication is very difficult and can only be realised with the expensive electron beam lithography. The structures are two dimensional now, but stackable in future. With photonic structures the middle infrared can be opened up for integrated optics.

4 Summary

Infrared interferometry with integrated optics is an innovative method in extrasolar planet studies and necessary to obtain new insight of close-in companions. This is indispensable to solve long-standing questions concerning planet-formation, i.e., mass-distributions, metallicities, orbits, and atmospheres of extrasolar planets and brown dwarfs.

Direct detection and interferometric observations of extrasolar planets are important goals of european astrophysics, as stated by ESO, ESA, and EU panel long-term strategies. Around 2015, the DARWIN mission will be launched by the European Space Agency. The challenging main goal of this mission is to search for Earth-like planets and possible signs of life on extrasolar planets with a spaceborn interferometer. To do this in an efficient way, possible targets have to be detected. In the course of our proposed project we will be able to determine parameters of extrasolar planets and planetary systems with high accuracy and thus be able to provide a list of candidates for planets harbouring extrasolar life.

References

1. Berger J.-P., K. Rousselet-Perraut, P. Kern, F. Malbet, I. Schanen-Duport, F. Reynaud, P. Haguenauer, P. Benech, 1999, Astron. Astrophys. Supp. Ser. 139, 173
2. Berger J.-P., P. Benech, I. Schanen, G. Maury, F. Malbet, F. Reynaud, 2000, Proceedings SPIE, Vol. 4006, 986
3. Bogaerts W., Bienstman P., Taillaert D., Baets R., DeZutter D., 2001, IEEE Phot. Technol. Lett. 13, 565
4. Burrows A., Marley M., Hubbard W. et al. 1997, ApJ 491, 856

5. Choi, H.-J., Soon, W., Donahue, R.A., Baliunas, S.L., Henry, G.W., 1995, PASP 107, 744

6. Chow E., Lin S.Y., Wendt J.R., Johnson S.G., Joannopoulos J.D., 2001, Optics Letters 26, 286

7. de Foresto C.V., J.-M. Mariotti, G. Perrin, 1997, in: Science with the VLT Interferometer, F. Paresce, Springer Verlag 1997, 86

8. Frink, S., Mitchell, D.S., Quirrenbach, A., Fischer, D.A., Marcy, G.W., Butler, R.P. 2002 ApJ 576, 478

9. Hatzes, A.P., Cochran, W.D. 1993 ApJ 413, 339

10. Hempel, M., Schmitt, J.H.M.M. 2003, A&A, 408, 971

11. Hempel, M., Robrade, J, Ness, J.-U., Schmitt, J.H.M.M. 2005, A&A, 440, 727

12. Lalanne P., 2002, IEEE J. Quantum Electron. QE-38, 800

13. Malbet F., P. Kern, I. Schanen-Duport, J.-P. Berger, K. Rousselet-Perraut, P. Benech, 1999, Astron. Astrophys. Supp. Ser. 138, 135

14. Murphy E.J., 1999, Integrated optical circuits and components, Marcel Dekker

15. Neuhäuser R., Guenther E.W., Wuchterl G., Mugrauer M., Bedalov A., Hauschildt P., 2005, A&A 435, L13

16. Olivier S., Rattier M., Benisty H., Weisbuch C., Smith C.J.M., DeLaRue R.M., Krauss T.F., Oesterle U., Houdre R., 2001, Phys. Rev. B 63, 113311

17. Perrin G., V. C. de Foresto, S.T. Ridgway, J.-M. Mariotti, N.P. Carleton, W. Traub, 1997, in: Science with the VLT Interferometer, F. Paresce, Springer Verlag 1997, p. 318

18. Ruske J.-P., M. Rottschalk und B. Unterschütz, 1995b, Proceedings of the 7th European Conference on Integrated Optics ECIO 95, Delft, Niederlande, 03.-06. April 1995, 383

19. Ruske J.-P., B. Zeitner, W. Biehlig, E. Werner, A. Tünnermann, 1999, LaserOpto 31 (2), 40

20. Ruske J.-P., M. Rottschalk, B. Zeitner, V. Gröber, A. Rasch, 1998, Electronics Letters 34 (4), S. 363

21. Ruske J.-P., Zeitner B., Tünnermann A., Rasch A.S., 2003a, Electron. Lett. 39, 1048

22. Ruske J.-P., Werner E.A., Zeitner B., Tünnermann A., 2003b, Electron. Lett. 39, 1442

23. Rottschalk M., J.-P. Ruske, St. Steinberg, K. Hornig, G. Hagner und A. Rasch, 1995, Journal of Lightwave Technology 13 (10), 2041

24. Rottschalk M., J.-P. Ruske, B. Unterschütz, A. Rasch, V. Gröber, 1997, J. Appl. Phys. 81 (6), 2504

25. Setiawan, J., Hatzes, A.P., von der Lühe, O., Pasquini, L., Naef, D., da Silva, L., Udry, S., Queloz, D., Girardi, L. 2003 A&A 398, L19

26. Setiawan, J., Pasquini, L., da Silva, L., Hatzes, A.P., von der Lühe, O., Girardi, L., de Medeiros, J.R., Guenther, E. 2004 A&A 421, 241

27. Shaklan S., F. Roddier, 1988, Appl. Opt. 27 (11), 2334

28. Tünnermann A., Schreiber T., Augustin M., Limbert J., Will M., Nolte S., Zellmer H., Iliew R., Peschel U., Lederer F., 2004, Photonic crystals in ultrafast optics, In: Advances in Solid States Physics, B. Kramer (Ed.) Vol. 44, 117 Springer Berlin

29. Werner E.A., Ruske J.-P., Zeitner B., Biehlig W., Tünnermann A., 2003, Opt. Commun. 221, 9

30. Wuchterl, G., Tscharnuter, W.M., 2003 A&A 398, 1081

31. Zeitner B., Ruske J.-P., Werner E., Rottschalk M., Rasch A., Tünnermann A., 1999, Abschlussbericht BMBF-Verbundprojekt FABIAN (FKZ 16 SV 404/9, Laufzeit: 01.01.1996 - 31.07.1999): *Integriert-optischer Farbmischer für ein Laserfarbbilderzeugungssystem*

GRAVITY: The AO-Assisted, Two-Object Beam-Combiner Instrument for the VLTI

F. Eisenhauer[1], G. Perrin[2], S. Rabien[1], A. Eckart[3], P. Léna[2], R. Genzel[1,4], R. Abuter[1], T. Paumard[1], and W. Brandner[5]

[1] Max-Planck-Institut für extraterrestrische Physik (MPE), Garching, Germany
[2] Observatoire de Paris – site de Meudon, France
[3] 1. Physikalisches Institut der Universität Köln, Germany
[4] Department of Physics, University of California, Berkeley, USA
[5] Max-Planck-Institut für Astronomie (MPIA), Heidelberg, Germany

Summary. We present the proposal for the infrared adaptive optics (AO) assisted, two-object, high-throughput, multiple-beam-combiner GRAVITY for the VLTI. This instrument will be optimized for phase-referenced interferometric imaging and narrow-angle astrometry of faint, red objects. Following the scientific drivers, we analyze the VLTI infrastructure, and subsequently derive the requirements and concept for the optimum instrument. The analysis can be summarized with the need for highest sensitivity, phase referenced imaging and astrometry of two objects in the VLTI beam, and infrared wavefront-sensing. Consequently our proposed instrument allows the observations of faint, red objects with its internal infrared wavefront sensor, pushes the optical throughput by restricting observations to K-band at low and medium spectral resolution, and is fully enclosed in a cryostat for optimum background suppression and stability. Our instrument will thus increase the sensitivity of the VLTI significantly beyond the present capabilities. With its two fibers per telescope beam, GRAVITY will not only allow the simultaneous observations of two objects, but will also push the astrometric accuracy for UTs to 10 μas, and provide simultaneous astrometry for up to six baselines.

1 Introduction

GRAVITY is a general purpose instrument for the VLTI (Glindemann et al. 2000) with two main capabilities: phase-referenced imaging, and narrow-angle astrometry. A non-exhaustive list of possible applications is listed in section 2. The design of the instrument is driven by a set of specific science cases, that also set its name (*General Relativity Analysis via* V$_{LT}$ *InTerferometrY*, or GRAVITY), namely to explore the so far untested regime of strong gravity encountered at a few hundred down to a few Schwarzschild radii (R_S) from supermassive objects.

2 Scientific Justification

2.1 The Center of the Milky Way

The Galactic Center is the closest galactic nucleus, and the best candidate for a supermassive black hole (BH). However, the final proof that Sgr A* is smaller than its event horizon is still pending. Its Schwarzschild radius is $R_S \simeq 9$ μas. With GRAVITY we aim at probing the space-time around this object down to a few R_S. This goal is detailed in the contribution to this workshop by Paumard et al. (2007) and summarized hereby. Unfortunately, the adaptive optics (AO) correction with a visible wavefront-sensor is only modest (Strehl $\approx$ 10% in K-band), because no bright visible guide star is nearby. In contrast, wavefront-sensing at infrared (IR) wavelengths allows observations close to the diffraction limit at 2 μm (Strehl $\approx$ 50%). Another complication is that the emission from the central 100 mas around Sgr A* is quite low ($m_K \gtrsim 17.5$), and that Sgr A* lies in a crowded region (a dozen of stars with $m_K \simeq 15$ within a radius of 0.5″, a few stars with $m_K \simeq 10$–11 at $\simeq 1''$). To overcome all these difficulties, it is necessary to have a photon-efficient instrument and IR wavefront-sensing, i.e. GRAVITY. On the other hand, the high density of stars can be turned into an advantage: the 2″ field of view (FoV) of a single VLTI beam will contain a sufficiently bright star for phase referencing, which will allow us to achieve an astrometric accuracy of approx. 10 μas in only 1–2 min integration time.

Probing Space-Time around the Supermassive Black Hole: Sgr A* is known for exhibiting so-called flares a few times a day. These events last for about 1 h, and their light-curves show significant variations on a typical timescale of 17 min in the IR. From the typical raise time of the substructures in the light-curve, they have to come from regions smaller than about 10 light-minutes, or $\simeq 17$ $R_S \simeq 150$ μas. The emitting region cannot remain static in the potential well of the BH, but its velocity will be comparable to the Keplerian circular velocity $v \simeq (r/R_S)^{-1/2} \times 10\,\mu\mathrm{as\,min}^{-1} \gtrsim 1\mu\mathrm{as\,min}^{-1}$. Therefore measuring the 2D astrometry of flares with 10 μas accuracy and a time resolution of a few minutes will allow not only determining the location of the flares with respect to the BH, but also their proper motion. This will allow us to understand their nature, and provide a probe in the potential well of Sgr A* at a few to 100 R_S. If the flares come from very close to the BH, the 17 min periodicity may be caused by the beaming of the radiations emitted by particles on the last stable orbit of a Kerr BH of spin parameter 0.52 (Genzel et al. 2003). If this is the case, then 10 μas-accuracy, minute-sampling astrometry will allow us to trace the lensed image of the last stable orbit, and therefore to probe the space-time down to the photon-sphere of the BH (Broderick & Loeb 2005, Paumard et al. 2006). This experiment absolutely requires the simultaneous, multi-baseline, narrow-angle astrometry capability of GRAVITY.

Probing Stellar Dynamics in the Regime of General Relativity: The current best estimates of the mass of the central BH and distance to the Galactic Center are obtained through orbit-fitting of stars in the central arcsecond of the Galaxy, the so-called S-stars. But stellar counts predict that a few faint stars ($17.5 \lesssim m_K \lesssim 19.5$) should reside even within the central 100 mas of the Galaxy. These stars have orbital periods of order one year, periapses of order 1 mas $\simeq 100~R_S$, and travel at relativistic velocities during their periapse passages. The repeated interferometric imaging of these stars allows to test relativistic effects, in particular the prograde periastron shift. In addition, it will give the mass enclosed within $\simeq 100~R_S$, while the S-stars measure the mass enclosed within $\simeq 1000~R_S$: comparing the two numbers will give a measurement of the mass of the stellar cusp. Note that the observations for this project also provide the monitoring for flares to perform the above "flare" experiment. The main advantage of GRAVITY over competing instruments is its IR wavefront sensor, and its optimized K-band throughput.

Astrometry of S-Stars: The astrometric capability of GRAVITY will also allow deriving very accurate orbits for the more distant S-stars. Specifically when they pass periapse, we will get a significant improvement on the determination of the position, mass, and distance of Sgr A*. These improvements are indeed required to further constrain the modelling in the two aforementioned experiments. On a longer time-scale (decade), the astrometry of the S-stars will also allow finding the General Relativistic effects, and probing for the extended mass component at the scale of one arcsecond. Compared to the general PRIMA astrometric facility, where the light from the phase reference star and the S-stars is unecessarily seperated at the UT Coude focus, GRAVITY will profit from its narrow-angle astometric mode with phase referencing within the 2" single interferometric beam.

2.2 Intermediate Mass Black Holes

There is compelling evidence for a very compact cluster (GCIRS 13E) that may contain an intermediate mass black hole (IMBH, Maillard et al. 2004). Interferometric imaging of the cluster will provide reliable proper motions for the core stars, and hence dynamically test the IMBH hypothesis. Another place to look for IMBHs is globular clusters. The current main limitation on the ability to measure the mass of these putative dark objects is the low number of bright stars in the core of the Globular Clusters suitable for radial velocity measurements (Baumgardt et al. 2005). GRAVITY will improve the situation by allowing high accuracy proper motion measurements. At 8 kpc, 1 km/s $\simeq 26~\mu$as yr^{-1}, meaning that this accuracy will be reached in 1 yr with narrow-angle astrometry or a few years through imaging. In addition, the acceleration of a star orbiting a 1000 $M_\odot$ BH at 4 mpc (100 mas at 8 kpc) is $\simeq 7~\mu$as yr^{-2}, so that for such stars the acceleration would be detected after a few years in astrometric mode. The same methods can also be applied to search IMBHs in young Galactic starburst clusters (e.g. Arches). Like the experiment on the GC S-star orbits, this science case profits tremendously from the unique narrow-angle astrometry capability of GRAVITY.

2.3 Stellar Orbits around Extragalactic Supermassive Black Holes

The nucleus of M31 is made of a blue disk of 200 Myr old stars orbiting a $\simeq 1.4 \times 10^8 \, M_\odot$ BH (Bender et al. 2005). The half-power radius of this disk is $\simeq 60$ mas $= 0.2$ pc, and it should contain of order 10 bright red evolved stars, that would be resolvable by GRAVITY. Their proper motion, of order $\simeq 1.7 \times 10^3 \, \text{km/s} \simeq 0.5 \, \text{mas yr}^{-1}$ at 0.76 Mpc, will be measurable within only a few years by VLTI imaging. The same proper motion measurements will also be possible in other nearby galaxies. Compared to other instruments with strong spectroscopic capabilities, GRAVITY will be superior for this project because of its optimisation for single band, low-spectral resolution imaging.

2.4 Active Galactic Nuclei

Many active galactic nuclei (AGN) are deeply dust-embedded, and faint in the optical. For these object the IR wavefront-sensing capability of GRAVITY is mandatory. High spectral resolution is not necessary for many projects, because the spectral features of AGNs are comparably broad. More important is the high sensitivity of GRAVITY, a consequence of the optimization for single-band operation at low spectral-resolution. In the standard model of AGNs, the centers of galaxies host supermassive BHs with masses in the range $10^6 - 10^9 M_\odot$. The large luminosity of the core is of gravitational origin, with large amounts of material orbiting in an accretion disk feeding the central BH. The disk and BH are enshrouded in a dust torus beyond the condensation radius for dust. The central part of the AGN contains the Broad Line Region (BLR), a compact region in which the gas velocity can be as large as a few 1000 km/s. Typical scales in Seyfert galaxies for these basic components of the AGN theory are as follows: BH Schwarzschild radius $R_S = 10^{-5} \left(\frac{M_{BH}}{10^8 M_\odot} \right)$ pc, jets up to several kpc, outer edge of accretion disk typically 1000 R_S, size of BLR about 0.003–0.3 pc, dust torus inner radius ~ 1 pc. The accretion disk is of the order of 0.2 mas (0.01 pc) and is beyond reach for VLTI. Up to now the BLR sizes are indirectly derived from reverberation mapping measurements (e.g. Kaspi et al. 2000) in the blue part of the visible spectrum. The sizes of dust tori are guessed from the distance for which graphite and silicate grains can condense (e.g. Barvainis 1987) or by modeling the IR part of the SED, in both cases indirect methods. The size of the torus of NGC 1068 has only recently been directly measured with MIDI at VLTI (Jaffe et al. 2004). Near-infrared (NIR) observations are sensitive to both the dust torus, whose emission in that wavelength range at least for Seyfert 2 remains large, and the more compact central source. Spectroscopically resolved observations can allow to disentangle between regions with spectral features such as the BLR and continuum sources such as the dusty torus or the central engine. In the BLR, lines are as broad as a few 1000 km/s, and a medium spectral resolution of 750 makes the BLR identification easy. The dust torus will a priori be completely resolved for sources closer than 100 Mpc. The largest BLRs are within reach at NIR wavelengths with angular scales of 2 mas for sources closer than 100 Mpc. The Brγ and Paα emission lines can be used to detect the BLR in the K-band. Brγ might also trace any shock in the circum-nuclear environment. This may provide an efficient tool to detect the base of the jets where they are launched into orbiting material. The coupling of imaging and spectroscopy as proposed by Woillez et al. 2007 will allow to combine imaging and reverberation mapping to obtain, through tomography, the 3D structure of the BLR. A by-product is the direct measurement of the distance of AGNs by comparing the angular and linear sizes of the BLR, a measurement of cosmological significance (Elvis & Karovska 2002). A preliminary source selection based on distance, brightness and nearby reference source availability points towards four main sources (NGC 1068, Circinus, NGC 3783, NGC 3758) whose core magnitudes are brighter than K=13. All four sources have bright reference stars (K brighter than 9.4) less than 2 arcmin away.

2.5 Stars and Starformation

Masses of the Most Massive Stars: There still exists a discrepancy by up to a factor of 2 in the mass estimates for the most massive main-sequence stars. Comparison of spectra with atmospheric models yields upper mass limits in the range of $60\,M_\odot$, whereas evolutionary tracks and observed luminosity suggest a mass of up to $120\,M_\odot$ for stars of spectral type O2V and O3V. Clearly, dynamical mass estimates for early O-type main-sequence stars are required. Luckily, quite a number of spectroscopic binary O-stars are known in the cores of Galactic starburst clusters like Arches, Quintuplet of NGC 3603, or extragalactic starbursts like 30 Dor. With a nominal resolution of 4 mas at a wavelength of $2\,\mu m$, Gravity could resolve some of the longer period spectroscopic binaries, and monitor the astrometric motion of the photocenter for the shorter period, closer binaries. Astrometric orbits for these deeply embedded binary stars will hence directly yield dynamical mass estimates. The unique narrow-angle astrometry mode of GRAVITY is ideal for these dynamical studies in crowded regions. Several objects (e.g. Arches, Quintuplet) can not be observed without the IR wavefront sensing provided by GRAVITY.

Circumstellar Disks and Jets around Young Stars: Circumstellar disks and outflows are closely linked to the star formation process. The presence of a circumstellar disk is also the pre-requisite for the formation of planetary systems. The gravitational interaction between a planet and a disk should manifest itself in the occurrence of spiral structures, wakes and gaps. Such structures have indeed already been observed in a number of cases (e.g. GG Tau, Formalhaut, etc.). The relative faintness of a young giant planet compared to the high-surface brightness of a typical circumstellar disk thus far prevented its direct detection in diffraction limited observations with 8m class telescopes in the NIR. GRAVITY's 4 mas resolution (compared to 60 mas resolution for one UT) drastically improves the contrast between a disk and its embedded planet by a factor of $(60/4)^2 = 225$. Hence it should be possible to probe for young giant planets which are almost 6 mag fainter than what is currently achievable. While the ubiquity of jets in star formating regions has been well established, the physics behind the formation of jets, and in particular the launching mechanism, is still poorly understood. Models (e.g. Turner et al. 1999) suggest that first angular momentum gets transferred along horizontal magnetic field lines from the disk to the central material, which then gets accelerated along a vertical pressure gradient, ultimately forming a collimated jet. The important processes seem to take place within less than 0.5 AU from the star, which at typical distances to the nearest star forming regions of 150 pc translates into an angular size of less than 30 mas. At 4 mas resolution, GRAVITY will be able to resolve the central jet formation engine around young, nearby stars. Furthermore, at a distance of 150 pc, an astrometric precision of 10 μas corresponds to a transversal velocity of ≈ 60 km/s. Hence high-velocity outflows, and the formation and evolution of jets from T Tauri stars with typical velocities of 150 km/s can be resolved and traced in real-time. These observations will put tight constraints on jet formation models and the role of magnetic fields.

2.6 Planets and Multiple Systems

Substellar Objects in Multiple System - Dynamical Masses and Calibration of Theoretical Models: Recent claims on the direct detection of planetary mass companions to young brown dwarfs (2MASSW J1207334-393254, Chauvin et al. 2005) and low-mass stars (GQ Lup, Neuhäuser et al. 2005) are based on evolutionary models and model atmospheres, which have not yet been accurately calibrated to observations. Dynamical mass estimates for ultra-cool dwarfs and brown dwarfs have thus only been derived for a handful of objects (LHS 1070, GJ 569Bab, 2MASS J0746+2000, AB Dor C). In general, the observed masses for substellar objects with ages larger than a few 100 Myr seem to be in good agreement with theoretical models. AB Dor C, however, the close companion to the K7 Zero-Age-Main Sequence star AB Dor, is among the youngest very-low-mass objects for which a precise mass estimate has been obtained (Close et al. 2005). Quite surprisingly, AB Dor C turned out to be about twice as massive as would have been expected from theoretical models for an age of $\simeq 40$ Myr. If this result should hold true, it has profound implications on the substellar mass functions of young stellar systems, and would also put the existence of the so-called "cluster planets" into doubt (71 sig Ori, e.g., would then well be in the brown dwarf and not the planetary mass regime). Clearly, more dynamical (astrometric) mass estimates for young, very-low mass and substellar objects are required in order to calibrate the theoretical tracks and model atmospheres. GRAVITY could probe many more multiple systems like AB Dor, which have at least one very-low mass or substellar component. GRAVITY will yield astrometric orbits, and derive the individual component masses. Furthermore, GRAVITY allows one to probe the substellar companions themselves for binarity. In the case of AB Dor C, the only way to reconcile present day evolutionary tracks with the dynamical mass would be if AB Dor C would be a close binary, consisting of two 40 M_{Jup} brown dwarfs. Finally, GRAVITY could even resolve the motion of details (such as atmospheric clouds) on the planet surface, provided this detail dominates the photo-center of the planet: 1 Jupiter diameter would be 10 μas at 100 pc.

Planets in binary systems: The Sun's wobbling due to Jupiter is of order 1 $R_\odot$, or 1 mas at 10 pc, or 10 μas at 1 kpc. If the Sun were a double star (true binary or in projection), so that the companion could be used as a phase reference, GRAVITY would be able to detect Jupiter from 1 kpc away in one Jupiter period (12 yr). The wobble of a Sun due to a hot Jupiter could be determined at distances up to $\simeq$ 10 pc in a few days. GRAVITY offers the possibility to search for exoplanets at kpc distances even with the ATs thanks to its optimised thoughput. Finally, if the planet does not have the same color as the star, the photo-center should not be the same at both ends of the K band. This difference would give a handle on the linear scale of the system.

2.7 Microlensing

Gravitational microlensing events occur when a point-like source and a massive object (lens) are almost aligned on the line of sight. This causes three effects: the luminosity of the source appears enhanced, its photo-center is slightly displaced, and a secondary image appears on the other side of the lens. These events are interesting to probe the mass spectrum of compact objects in the Galaxy, but the light-curves currently provided by the dedicated surveys are not enough to determine the mass: 2D spatial information (either 2D track of the photo-center or separation between the two images) is necessary to disentangle it from the other event parameters. GRAVITY would be very efficient in following events reported by photometric surveys, because its high sensitivity would put a lot of events in reach of the ATs, and most of them in reach of the UTs. The usual events have a simple geometry and are often long (> 10 days). However, when the lens is not simple but has a companion (similar dark object or planet), caustic crossings can happen, characterized by multiple images and complex wobbling of the photo-center. Both astrometry and imaging can give access to the mass of the companion. The second of two crossings (which usually go by pair) can be predicted a few days in advance, but it lasts only a few hours. Therefore GRAVITY's ability to measure high precision phases and visibilities in minutes is highly valuable.

3 Top Level Requirements in the Context of the VLTI Facility

In this section we detail the top-level requirements from the scientific justification (section 2), and outline that the present VLTI with its first generation of instruments and the upcoming PRIMA facility (Delplancke et al. 2000) does not fulfill these functional and performance needs (see Wilhelm et al. (2002) for the functional description of the VLTI, and Delplancke (2004b) for a PRIMA executive summary).

3.1 Functional Requirements

- IR wavefront sensing for observations of red objects: The four UTs are equipped with the MACAO adaptive optics (Arsenault et al. 2003), which sense the wavefront at visible wavelengths. No IR wavefront-sensor is available at the UTs. The ATs have no high order AO at all.
- Simultaneous astrometry for multiple baselines: The present PRIMA facility offers two star-separators (STSs) for UTs (plus two for ATs), two fringe sensor units (FSUs), and two differential delay lines (DDLs). In the near future (2006), at least three UTs will be equipped with STSs, but the astrometric mode of PRIMA will still be limited to a single baseline per observation. The future upgrade of PRIMA for the second generation instruments with four FSUs and four DDLs potentially allows simultaneous multiple baseline astrometry, but details are not settled.
- Narrow-angle astrometry for distances less than two arcseconds: The PRIMA star-separators are optimized for angular separations between the primary star and the secondary object larger than $2''$. The star-separators are not optimum for narrow-angle astrometry. The $0–2''$ range can be reached as well, but only with the risk of cross-talk between the two channels (Delplancke 2004b).
- Phase-referenced imaging interferometry using multiple telescopes: The present PRIMA facility will allow phase-referenced imaging for three telescopes. The 2nd generation PRIMA will then co-phase four telescopes, necessary for phase-referenced imaging on six baselines.

3.2 Performance Requirements

- AO corrections achieving near-diffraction limited performance (Strehl $\simeq$ 50%) for stars with a K-band magnitude $m_K \geq 10$.

– Narrow-angle astrometry with $\simeq 10$ μas accuracy in five minutes observing time: While the astrometric accuracy with the ATs will be $\simeq 10$ μas, the more complex and longer non-common optical light path for UTs reduces the expected accuracy to $\simeq 100$ μas (Delplancke 2004b). The astrometric accuracy of PRIMA using the UTs is thus expected to fail our top level requirements by approximately a factor 10. The random atmospheric differential OPD residuals do not limit significantly the accuracy for narrow-angle astrometry. Even for observations as short as 3 minutes, the typical average atmospheric OPD residual is less than 5 nm or 10 μas.

– Phase-referenced imaging interferometry with a point source limiting K-band magnitude of $m_K \geq 19$ in one hour observing time: The limiting magnitude of the first generation AMBER instrument (Petrov et al. 2000) in K-band without external fringe tracking is presently $m_K \simeq 7$ in 25 ms (Rantakyrö 2005). Even with external fringe-fracking with PRIMA, the present VLTI and its 1st generation instruments will miss our sensitivity requirement for phase-referenced imaging ($m_K \simeq 19$ in one hour) by approximately 2 magnitudes. The main reason for the limited performance is the small optical throughput of $\simeq 2\%$ (Rantakyrö 2005) of AMBER, resulting from the trade-off necessary to provide a multi-purpose, multi-wavelength instrument for reasonable cost. In contrast, a simple throughput-optimized broad-band beam-combiner like the PRIMA FSU is expected to go down to 19th mag (Delplancke 2004b).

We conclude that only a dedicated instrument will allow the astronomical and physical key-experiments proposed in section 2. In short, we need an IR-AO assisted, two-object beam-combiner instrument, optimized for highest optical throughput, most accurate multi-baseline narrow-angle astrometry, and phase-referenced imaging.

4 GRAVITY: Concept and Observing Modes

4.1 An IR AO Assisted, Two-Object Beam-Combiner Instrument for the VLTI

The concept of GRAVITY is directly derived from the top-level requirements. The instrument will be installed in the VLTI laboratory. It will use the 18 mm diameter input-beams of the four telescopes as provided by the beam compressor unit (Koehler & Gitton 2002), and physical access to the exit pupils of the VLTI. The whole instrument will be enclosed in a cryostat and evacuated. This will not only dramatically improve the stability and cleanliness of the instrument, but will also allow for optimum baffling and suppression of thermal background. In our concept we foresee the option to separate the AO module for use with other VLTI instruments. In its baseline design, GRAVITY will only have IR wavefront-sensors, and command the deformable mirrors

(DMs) of the MACAOs. Alternatively we consider a full AO system including DMs in the VLTI laboratory. The second major sub-system of GRAVITY is the field-selector unit, and is best described as a highly compact PRIMA facility working on a single interferometric beam. This unit picks two objects from the 2″ FoV and couples the light to single-mode fibers for beam cleaning. The relative optical path difference (OPD) of the two objects will be compensated by stretching the fibers. An internal metrology system will measure the OPD of the two beams. In contrast to the PRIMA facility, however, the OPD control of the GRAVITY field-selector is significantly easier, because in GRAVITY the accuracy and stability of 5 nm (corresponding to an astrometric accuracy of 10 μas) is only required within a single interferometric beam, and not between the two beams of the PRIMA facility. The feasibility of 10 μas narrow angle astrometry with fringe-tracking on a single beam has already been demonstrated at the Palomar Testbed Interferometer (Lane & Muterspaugh 2004). The light from the eight fibers (2 objects from 4 telescopes) is then fed to the GRAVITY beam-combiner, the third major sub-system of the instrument. Three options are presently investigated for optimum throughput, stability, cost, and complexity: Integrated optics, fiber X coupler, and bulk-optics beam-combiners. The last major sub-system in GRAVITY is the camera unit. Optimized for highest possible throughput, we will only implement transmissive prisms and grisms for low (R $\simeq$ 20) and medium (R $\gtrsim$ 500) spectral resolution. Various detector options are presently investigated, with focus on large pixels, high quantum-efficiency, and low noise.

4.2 Basic Observing Modes

GRAVITY will provide two basic observing modes: **Simultaneously phase-referenced imaging of up to two objects**, and **narrow-angle astrometry within the 2″ FoV of the VLTI**. GRAVITY will provide **internal fringe-tracking** for reference stars within the 2" FoV, but can also be operated using external fringe-tracking with PRIMA for more distant reference stars. GRAVITYs advantage over the PRIMA astrometric mode is the ability of simultaneously measuring multiple baselines, and the largely increased accuracy for UT operation because of the restriction to small distances and operation in a single interferometric beam. Because the instrument will measure the visibility, phases and differential OPD of the two objects for **all six baselines**, GRAVITY will also track on significant fainter sources than PRIMA, which operates only on the minimum subset of baselines. GRAVITYs advantage over competitive instruments for phase-referenced imaging is the sensitivity-optimized concept. We do not compromise for multi-mode, multi-wavelength and high spectral resolution. GRAVITY also has the multiplex advantage from two objects per FoV. This implies the possibility to use one of the two objects for simultaneous visibility calibration. And GRAVITY will have the IR wavefront-sensors for dust-obscured objects.

5 Key Components of GRAVITY

In this section we briefly outline the possible options for the various key-components of GRAVITY. The technology for all sub-systems is well advanced, and (semi-) commercial devices are available. There is no technological show-stopper.

5.1 Infrared Wavefront-Sensor and Deformable Mirrors

IR wavefront sensors are already installed in various astronomical facilities (e.g. NAOS, Gendron et al. 2003), typically as Shack-Hartmann sensors using lenslet-arrays. Also the concept of pyramid wavefront-sensors is promising for GRAVITY. A first IR device has recently been installed at the Calar Alto observatory (Costa et al. 2004). Curvature wavefront sensors could be advantageous, because they could be matched optimally to the MACAO DM geometry (GRAVITY option without additional DMs). The default detector for the wavefront sensor would be the Rockwell HAWAII detector with a read-noise of typically 10 electrons (Finger et al. 2002). The recent developments for large-pixel, low-noise detectors (e.g. Rockwell CALICO prototype, Finger et al. 2004) for IR wavefront-sensors may allow to increase the sensitivity by a factor of two (0.75 mag). Optionally GRAVITY will be equipped with its own DMs (not commanding the MACAO DMs). In this case micro-machined DMs will be used directly in the 18 mm VLTI beam. Potential technologies include magnetic DMs (e.g. from LAOG/LETI, Cougat et al. 2001), or piezo-electrically driven mirrors (e.g. from OKO Technologies, Dayton et al. 2000).

5.2 Field-Selector, Pathlength-Compensation, and Metrology

In order to not move the fibers during observation, we foresee a K-mirror for de-rotation, a tip-tilt for laterally moving the field, and a device to adjust the projected separation of the two fibers. A fiber streching unit will introduce a differential OPD to one of the objects. All motions may be implemented for cryogenic operation, for example using stepper motors (e.g. Berger Lahr) and inductive resolvers (e.g. LTN Servotechnik) for the K-mirror, piezo stacks with capacity sensors for the tip-tilt mirror and OPD control (e.g. PI Ceramic), and piezo-electric translation stages (e.g. Attocube Systems) for fiber coupling. The standard solution for measuring the differential OPD is a dual wavelength laser metrology (e.g. PRIMET, Leveque et al. 2003). The metrology beams will be launched directly at the beam-combiner. Another option for the metrology is to launch the laser at the center of the pupil at wavelengths just above and below the operational wavelength range, and track its phase with the science beam combiner. For that we would make use of the flexible read-out technology of IR detectors, which allow kHz sampling for the metrology pixel, while integrating on the science object.

5.3 Optimization for Faint Objects

The strategy for minimizing light loss is two-fold: reduce the number of optical elements and optimize their losses: first we use reflective optics instead of transmissive optics, and combine several functions in one unit (e.g. field-selector is also the fiber-coupler optics). Second, we restrict our wavelength range to K-band (with possible extension to H-band). We therefore avoid any unnecessary splitting of light with dichroics. The narrow wavelength range also improves the performance of multi-layer coatings and dichroics typically by a factor of two. GRAVITY will use fluoride glass fibers optimized for K-band operation. The attenuation of these fibers (3 dB/km) is negligible for our fiber-length of a few meters, and the fibers have already been successfully used in other interferometric projects (e.g. 'OHANA, Perrin et al. 2004). GRAVITY will only provide low and medium spectral resolution, because we can use high-throughput transmissive dispersion elements for that. It is important to note that GRAVITY does not intend to extend its beam combination to more than 4 telescopes (4+4 strategy for VLTI), because any additional combination is accompanied by according light losses. In order to avoid extra background, a cold stop will be installed in the fiber-coupler unit. In addition, the beam-combiner and the subsequent camera / spectrometer will be operated at cryogenic temperatures. Several options are already available for the science detector (e.g. HAWAII from Rockwell Inc., Finger et al. 2002). However, new developments for IR AO wavefront-sensors (e.g. Rockwell CALICO, Finger et al. 2004) will provide larger pixels at even lower read-noise. With such detectors only one detector pixel is necessary per beam-combiner output and spectral channel, reducing the effective read-noise per information element accordingly. In addition, our restriction to K-band will allow to optimize and select detectors for best quantum efficiency at 2 μm. GRAVITY will enclose all major components in an evacuated cryostat for optimum cleanliness and stability. Also GRAVITY will be free of temperature- and pressure variations, which minimizes the need for calibration, and thus significantly increases the observing efficiency.

5.4 Beam Combiner Options

Three beam-combiner options are considered for GRAVITY: classical bulk-optic combination, integrated-optics and fiber X-couplers. The baseline is integrated optics. Widely used in the telecommunication NIR bands (e.g. for beam switching), the technique has already been applied in astronomical interferometry (IONIC, Berger et al. 2003). Also the extension to the K-band has been demonstrated (Laurent et al. 2002). A comparison of various concepts for the combination of four beams can be found in Le Bouquin et al. (2004). Beam combination in a bulk combiner can be designed to be very efficient, and has successfully been used in astronomical interferometry (e.g. COAST, Haniff et al. 2004). Dust contamination and thermal drifts that might degrade the performance would be avoided in the cryostat. The third option would be fiber X-couplers. The usage of fluoride-glass fiber combiners for the K-band has been demonstrated (e.g. FLUOR, Coudé du Foresto et al. 1998; VINCI, Kervella et. al. 2000).

6 Conclusions

GRAVITY will open a new window for unique science with the VLTI. In addition to a wide variety of astronomical key-observations, the instrument will — for the first time — directly probe the regime of strong gravity close to black holes. The key-components of GRAVITY are an IR wavefront-sensor, a two-object beam-combiner, and high-throughput optics. As such GRAVITY will complement the multi-mode, multi-wavelength, high spectral-resolution instruments with simultaneous multi-baseline astrometry, optimum sensitivity for faint objects, and a 10 times better astrometric accuracy for the UTs. Thanks to its internal phase-referencing, GRAVITY will be able to start operation already with the present PRIMA facility.

References

1. R. Arsenault et al: Proc. SPIE, **4839**, 174 (2003)
2. J.-P. Berger et al.: Proc. SPIE, **4838**, 1099 (2003)
3. R. Barvainis: ApJ, **320**, 537 (1987)
4. H. Baumgardt et al.: ApJ, **620**, 238 (2005)
5. R. Bender et al.: ApJ, 631, 280 (2005)
6. A.E. Broderick & A. Loeb: MNRAS, 363, 353 (2005)
7. G. Chauvin et al.: A&A, **438**, 25 (2005)
8. L.M. Close et al.: Nature, **433**, 286 (2005)
9. V. Coudé du Foresto et al.: Proc. SPIE, **3350**, 856 (1998)
10. O. Cugat et al.: Sensors and Actuators, **A 89**, 1 (2001)
11. S. Dayton et al.: Optics Communications, 339, 345 (2000)
12. S. Dayton et al.: OSA Optics Express, **8 No. 1**, 17 (2001)
13. F. Delplancke et al.: Proc. SPIE, **4006**, 41 (2000)
14. F. Delplancke et al.: Reference missions for PRIMA, Report prepared by the VLTI Implementation Committee (2004)
15. F. Delplancke: PRIMA Executive Summary, private communication (2004)
16. F. Eisenhauer et al.: ApJ, 628, 246 (2005)
17. M. Elvis & M. Karovska: ApJ, **581**, L67 (2002)
18. G. Finger et al.: Proc. SPIE, **4841**, 89 (2002)
19. G. Finger et al.: Proc. SPIE, **5499**, 97 (2004)
20. E. Gendron et al.: Proc. SPIE, **4839**, 195 (2003)
21. R. Genzel et al.: Nature, **425**, 934 (2003)
22. A. Glindemann et al.: Proc. SPIE, **5491**, 447 (2004)
23. C.A. Haniff et al.: Proc. SPIE, **5491**, 59 (2004)
24. W. Jaffe et al.: Nature, **429**, 47 (2004)

25. S. Kaspi et al.: ApJ, **533**, 631 (2000)
26. P. Kervella et al.: Proc SPIE, **4006**, 31 (2000)
27. B. Koehler & P. Gitton: Interface Control Document between VLTI and its Instruments, VLT-ICD-ESO-15000-1826, Issue 3.0 (2002)
28. B.F. Lane & M.W. Muterspaugh: ApJ, **601**, 1129 (2004)
29. E. Laurent et al.: A&A **390**, 1171 (2002)
30. S. Leveque et al.: Proc. SPIE, **4838**, 983 (2003)
31. J.B. Le Bouquin et al.: Proc. SPIE, **5491**, 1362 (2004)
32. J.P. Maillard et al.: A&A, **423**, 155 (2004)
33. F. Malbet A&AS , **138**, 135 (1999)
34. R. Neuhäuser et al.: A&A, **435**, 13 (2005)
35. T. Ott et al.: ApJ, **523**, 248 (1999)
36. T. Paumard et al.: this proceedings (2007)
37. T. Paumard et al.: ApJ, 643, 1011 (2006)
38. G. Perrin et al.: Proc. SPIE, **5491**, 391 (2004)
39. R. Petrov et al.: Proc. SPIE, **4006**, 68 (2000)
40. F. Rantakyrö: AMBER user manual, VLT-MAN-ESO-15830-3522, Issue 1.2 (2005)
41. N.J. Turner, et al.: ApJ, **524**, 129 (1999)
42. R. Wilhelm et al.: Functional Description of the VLTI, VLT-ICD-ESO-15000-1918, Issue 2.0 (2002)
43. J. Woillez et al.: in preparation (2007)

GENIE: a Ground-Based European Nulling Instrument at ESO Very Large Telescope Interferometer

P. Gondoin[1], R. den Hartog[1], M. Fridlund[1], P. Fabry[1], A. Stankov[1],
A. Peacock[1], S. Volonte[1], F. Puech[2], F. Delplancke[2], P. Gitton[2],
A. Glindemann[2], F. Paresce[2], A. Richichi[2], M. Barillot[3], O. Absil[4],
F. Cassaing[5], V. Coudé du Foresto[6], P. Kervella[6], G. Perrin[6], C. Ruilier[3],
R. Flatscher[7], H. Bokhove[8], K. Ergenzinger[7], A. Quirrenbach[9],
O. Wallner[7], J. Alves[2], T. Herbst[10], D. Mourard[11], R. Neuhäuser[12],
D. Ségransan[13], R. Waters[14], and G. J. White[15]

[1] European Space Agency, P.O. Box 299, 2200AG Noordwijk, Netherlands
[2] European Southern Observatory, D-85748 Garching bei Munchen, Germany
[3] Alcatel Space, 06156 Cannes La Bocca, France
[4] University of Liege, B-4000 Sart-Tilman, Belgium
[5] Office National d'Etudes et de Recherches Aerospatiales, 92322 France
[6] Observatoire de Paris-Meudon, 92195 Meudon, France
[7] Astrium Gmbh, D-88039 Friedrichshafen, Germany
[8] TNO institute of Applied Physics, 2600 AD Delft, Netherlands
[9] Leiden University, 2300 RA Leiden, Netherlands
[10] Max-Planck-Institut fur Astronomie, 69117 Heidelberg, Germany
[11] Observatoire de la Cote d'Azur, F-06460 St. Vallier de Thiey, France
[12] Universität Jena, 07795 Jena, Germany
[13] Observatoire de Geneve, 1290 Sauverny, Switzerland
[14] University of Amsterdam, 1098 SJ Amsterdam, Netherlands
[15] University of Kent, Canterbury, Kent CT2 7NR, United Kingdom

Summary. Darwin is one of the most challenging space projects ever considered by the European Space Agency (ESA). Its principal objectives are to detect Earth-like planets around nearby stars, to analyze the composition of their atmospheres and to assess their ability to sustain life as we know it. Darwin is conceived as a space "nulling interferometer" which makes use of on-axis destructive interferences to extinguish the stellar light while keeping the off-axis signal of the orbiting planet. Within the frame of the Darwin program, definition studies of a Ground based European Nulling Interferometry Experiment, called GENIE, were completed in 2005. This instrument built around the Very Large Telescope Interferometer (VLTI) in Paranal will test some of the key technologies required for the Darwin Infrared Space Interferometer. GENIE will operate in the L' band around 3.8 microns as a single Bracewell nulling interferometer using either two Auxiliary Telescopes (ATs) or two 8m Unit Telescopes (UTs). Its science objectives include the detection and characterization of dust disks and low-mass companions around nearby stars.

1 Introduction

Understanding the principles and processes that created the Earth, and allowed the development and evolution of life forms to take place is a key scientific objective which has been brought as a very high priority to the attention not only of the European Science Community but also of the European Space Agency. In particular, the successful detection of Earth-like planets possessing environments benign to life would answer central questions such as "How unique is the Earth as a planet ?" and "How unique is life in the Universe ?". To achieve these objectives, the Darwin mission [1] of the European Space Agency will survey a large sample of nearby stars and search for Earth-size planets within their "habitable zone". Darwin will measure their spectra in order to infer the presence of an atmosphere and search for biomarkers. Detection of Earth-size bodies orbiting nearby stars is extremely difficult because of the weak planetary signal emitted within a fraction of an arc-second from an overwhelmingly bright star. In the most favorable infrared spectral range, the contrast between the star and an Earth-like planet is still higher than 10^6. Only the planetary signal, a millionth of the stellar light, should remain in the input feed of a spectrograph in order to register a planet spectrum in a reasonable time. To accomplish such an extinguishing of light at the relevant spatial scales, the technique of "nulling interferometry" has been selected for Darwin.

2 GENIE objectives

The primary objective of the GENIE nulling experiment is to gain experience on the design, manufacture and science operation of a nulling interferometer using Darwin representative concept and technology. The GENIE experiment will combine all optical functions foreseen into the future Darwin Infrared Space Interferometer. The VLTI [2] is the most appropriate infrastructure on-ground to operate such an instrument in an automated mode representative of space operation. GENIE will benefit from the existing VLTI infrastructure, including the telescope adaptive optics, delay lines, fringe sensors and the beam combiner laboratory. GENIE optical bench within the VLTI laboratory will provide the functions specific to the nulling interferometry technique, namely intensity control, phase shifting, beam combination, internal modulation, spatial filtering, spectrometry, detection, electronics and cryogenics.

A second objective of GENIE is to prepare the Darwin science program through a systematic survey of Darwin candidate targets [3]. The solar zodiacal cloud, a sparse disk of $10-100~\mu$ diameter silicate grains, is the most luminous component of the solar system after the Sun. Its optical depth is only $\approx 10^{-7}$, but a patch of the solar zodiacal cloud 0.3 AU across has roughly the same emitting area as an Earth sized planet. Similar and even

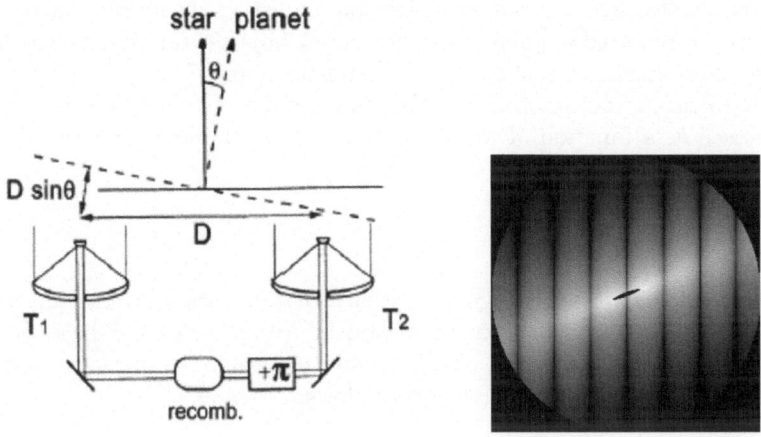

Fig. 1. Left: principle of the Bracewell interferometer concept [4] used in GENIE design. Right: IR emission of an exo-zodiacal dust disk folded through the transmission map of GENIE. The overwhelming flux emitted by the central star is nulled by destructive interference. The transmitted signal is mainly coming from the inner region of the dust disk surrounding the star.

brighter clouds may be common in other planetary systems and present a severe obstacle for the direct detection of extra-solar terrestrial planets. A systematic survey of Darwin candidate targets will screen-out those stars for which circum-stellar dust prevents the detection of Earth-like planets. The location of the VLTI at low latitudes is optimum to survey most (90 %) of Darwin targets which are located in directions close to the ecliptic plane. Bright exo-zodiacal clouds are easier to detect than extra-solar terrestrial planets, but finding an exo-zodiacal cloud is still difficult. Single-dish space telescopes in the mid-infrared such as Spitzer or JWST have the required sensitivity. However, only an interferometer can provide the spatial resolution needed to constrain the dust density in the inner regions of exo-zodiacal dust disks, within the habitable zone of nearby stars. GENIE operating with two 8m UT telescopes has both the required sensitivity and the required spatial resolution.

3 GENIE design

GENIE will operate as a two-telescopes Bracewell interferometer(see Fig. 1). By applying a π phase shift between the two arms of the interferometer, destructive interference is achieved on-axis while interference is constructive for small off-axis angles. The equivalent transmission map of the nulling interferometer is a set of interference fringes with a null in the centre. By placing the central star under this dark fringe, and adjusting the interferometer baseline to the required angular resolution, faint objects can be detected in the close environment of a star. The transmission map $T(\theta,\phi)$ of a diffraction limited Bracewell interferometer with two circular entrance apertures can be expressed as a function of wavelength, telescope diameter D, and projected baseline B as follows:

$$T(\theta,\phi) = 2 \times \frac{J_1(\pi\theta D/\lambda)}{\pi\theta D/\lambda^2} \times sin^2(\pi\theta cos(\phi)B/\lambda) \tag{1}$$

θ,ϕ are respectively the angular distance to the boresight and the azimut angle in the plan of the sky with respect to the projected baseline. The output signal of a Bracewell interferometer pointing to a star surrounded by a dust disk can then be expressed as follows:

$$S(\theta,\phi) = A_{\text{eff}} \int_\theta \int_\phi T(\theta,\phi) \times (O_s(\theta,\phi) + O_z(\theta,\phi))d\theta d\phi + Bgd \tag{2}$$

$O_s(\theta,\phi)$ and $O_z(\theta,\phi)$ are the brightness distributions in the stellar disk (including e.g. limb darkening) and in the exo-zodiacal dust disk. Bgd is the incoherent background signal which includes the thermal emission from the sky and from the telescope and instrument optics. In practice, the central null of the transmission map can be degraded by unequal amplitudes and by a variable optical path difference (OPD) between the arms of the interferometer. Hence, GENIE design includes an intensity control loop, a high accuracy OPD control loop and a dispersion correction subsystem. Furthermore, since wavefront distortions induced by atmospheric turbulence are not fully compensated by the VLTI adaptive optics, GENIE uses monomode optical fibers to filter residual wavefront errors at high spatial frequencies [5].

In 2004-5, two candidate designs for the GENIE instrument were established by two Consortia of European Industries and Scientific Institutes lead by Alcatel Space and EADS Astrium, respectively. The main components of these designs are:

- an achromatic phase shifter consisting of ZnSe or CaF2 dispersive wedges (see Fig. 3 left) that provides the achromatic π phase shift in nulling mode and participates to the compensation for longitudinal atmospheric dispersion,
- two knife edges actuators mounted on voice coils that equalize the intensity of the beams before recombination,
- two high accuracy optical delay lines possibly consisting of cat-eye systems (see Fig. 4) with piezo-actuated mirrors for fast response. The delay lines could be mounted on magnetic bearing and operate at low frequencies in combination with the dispersive wedges in order to compensate for longitudinal atmospheric dispersion.
- a modified Mach-Zender beam combiner (see Fig. 3 right),

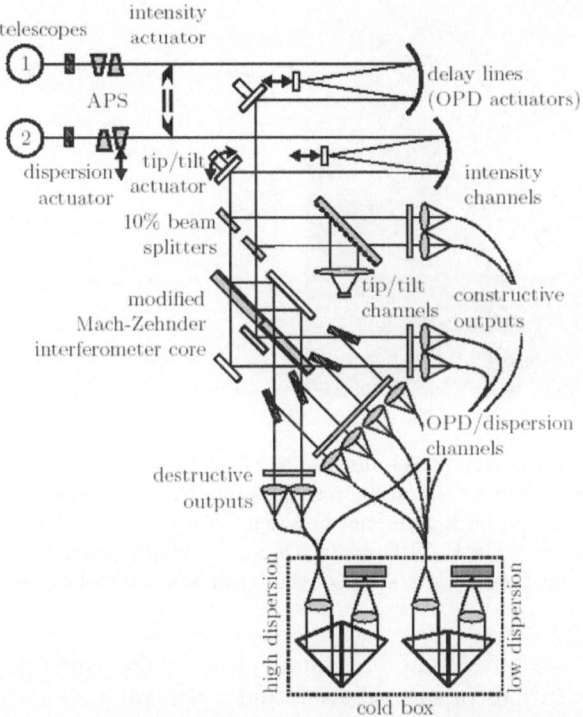

Fig. 2. A candidate design [6] for GENIE.

- fluoride or chalcogenide glass momomode fibers used for spatial filtering of the wavefront after beam recombination and for transportation of the beam to the entrance feed of the spectrometer,

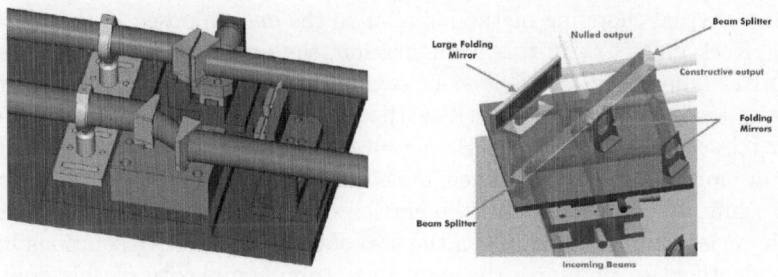

Fig. 3. Left: candidate design of a GENIE achromatic phase shifter using movable ZnSe prismatic plates. Right: a modified Mach-Zender beam combiner for Genie.

 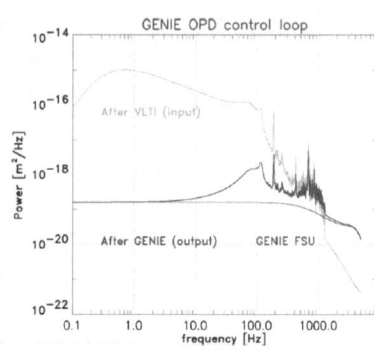

Fig. 4. Left: cateyes delay-lines built by TNO for an ESA nulling laboratory bread-board, a GENIE precursor operating in the H band. Right: simulated power spectral densities of the optical path difference between GENIE beams before and after correction by the fast GENIE OPD control loop. The high frequency peaks around 800 Hz result from the bimorph effect of the VLTI MACAO adaptive optic system.

- a low ($\lambda/\delta\lambda \approx 6$ to 50) and medium resolution spectrometer including a collimator, various prisms or grisms and a camera,
- a detection unit consisting of an HgCdTe array operated with the ESO IRACE control electronics.

GENIE performance depend on the characteristics of the instrument control loops (see Table 1) and on the background calibration accuracy. For background subtraction, GENIE uses the virtual chopping technique based on the ability of the VLTI to transmit a small field of view. The null and the background signals are measured simultaneously on either side of the target without opening the control loops, thus maximizing the observation efficiency of the instrument. For background subtraction, GENIE could also use the internal chopping method similar to the one proposed in the context of the Keck nuller [7]. In this configuration, the image pupils of the UTs in the VLTI laboratory are divided in two parts. Each half-pupil of one telescope is recombined destructively with the corresponding half-pupil of the other telescope, such that two Bracewell interferometers with parallel baselines are formed (see Fig.5). Their nulled output is then added with a $\pm\pi/2$ phase shift. Chopping can thus be carried out at high frequencies by alternately registering the signal from the two outputs. Innovative solutions have been identified to minimize the implementation complexity of this double-Bracewell configuration.

Table 1. Performance range of Genie dispersion, intensity and OPD control loops operating with bright (L ≈ 4 -5) Darwin target stars.

Control loop	Actuators	Bandpass	RMS residuals
Dispersion compensation	phase shifter	200 – 300 Hz	$\sigma_\phi = 7 - 29$ mrad
Intensity equalization	knife edges	1000 Hz	$\sigma_I \approx 4\%$
OPD control	delay lines	13 – 20 kHz	$\sigma_{OPD} = 6 - 17$ nm

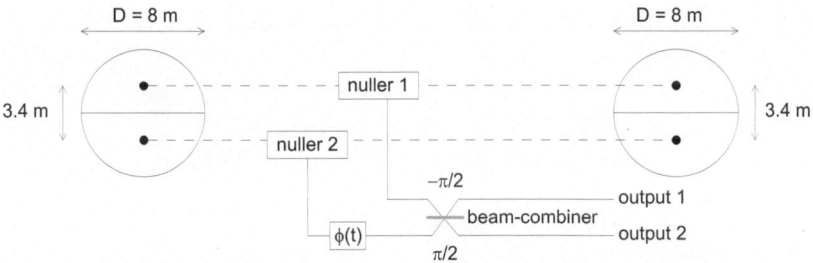

Fig. 5. Principle of a double Bracewell configuration using two telescopes with split pupils. The transmission of the exo-zodiacal light around the target star is modulated by the variable phase shift between the two Bracewell interferometers while the incoherent background contribution is independent of the phase shift.

4 GENIE performance

When observing nearby stars with spatially resolved diameters, GENIE performance are limited by the starlight leakage through the transmission map. For a diffraction limited Bracewell interferometer and a uniform stellar disk O_s, this stellar leakage can be expressed as a function of the operating wavelength λ, the interferometer baseline B and the angular radius θ_s of the star:

$$S_{leak} \approx A_{eff} \times O_s \times (\pi^2/8) \times (\theta_s B/\lambda)^2 \qquad (3)$$

When averaged over the detector frame read-out time, the stellar leakage adds an important offset to the zodiacal light signal. The detection capability of the instrument thus depends not only on the background subtraction accuracy but also on the calibration accuracy of the stellar leakage which depends itself on the angular diameter of the star (see Eq. 3). The uniform disk diameters of Darwin stars could be known with an accuracy better than 0.5 % in most cases. For the larger ($\theta_{target} > 1$ mas) stars, both direct interferometric J band measurements with VLTI/VISA and indirect photometric estimates could be used. For smaller stars, angular diameters could be derived from surface brightness-color relationship [8]. Correction for limb darkening could be achieved by extending stellar atmosphere models to the L' band and by verifying these models on the Sun and through accurate visibility measurements in the L' band of stars with large angular diameters.

Table 2. Minimum 5 σ detectable fluxes (with UTs) of exo-zodiacal dust disks around a solar-type star as a function of its distance assuming a 1% calibration accuracy of the target and calibrator star diameters.

d (pc)	$m_{L,star}$	θ_{star} (mas)	F_{star} (Jy)	$N_{zodi}(5\,\sigma)$	$F_{zodi}(5\sigma)$ (mJy)	F_{zodi}/F_{star}	$m_{L,zodi}$- $m_{L,star}$
10	3.3	0.46	12.5	142	8.5	6.8×10^{-4}	7.9
15	4.2	0.31	5.58	119	3.2	5.7×10^{-4}	8.1
20	4.8	0.23	3.13	86	1.4	4.4×10^{-4}	8.4
25	5.3	0.19	2.01	102	1.0	5.0×10^{-4}	8.2

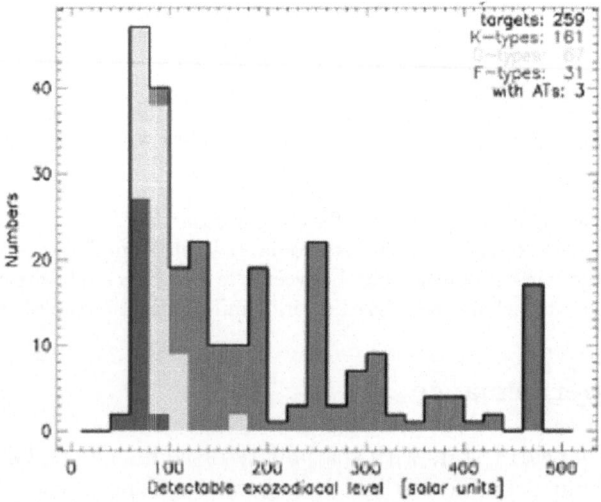

Fig. 6. Histogram of the minimum level of detectable exo-zodiacal dust (in solar zodiacal dust density unit) around Darwin targets assuming a 1% calibration accuracy of stellar diameters.

GENIE performance were calculated using a software model of the instrument [9]. The code emulates the brightness distribution in the source including the star and the circumstellar disk. The noise contribution from the atmosphere and the VLTI is added to the signal. The OPD induced by the atmospheric turbulences is described by a Kolmogorov power spectrum. The VLTI-GENIE instrument is described by the sensors, controllers and actuators transfer functions of the different control loops including MACAO adaptive optics and PRIMA fringe sensor unit. The output signal of the simulator consists in time series of fluxes calculated in different spectral channels. These representative data series are then processed by a prototype pipeline software that uses calibration stars measurements to correct for instrument imperfections. The simulation results of dust disks observations are presented

Table 3. GENIE magnitude limits in constructive mode (see text) for different spectral resolution.

$\lambda/\delta\lambda$	UT's	AT's
6	14.5	11.6
100	13.0	10.2
1000	11.9	8.2

in Table 2 and Fig.6. Assuming a conservative 1% calibration accuracy of stellar diameters, GENIE could detect exo-zodiacal dust disks 40 to 120 times denser than the solar zodiacal dust around F and G stars located within 25 parsecs from the Sun.

The use of dispersive wedges for achromatic phase shifting enable GENIE to be easily configured for visibility measurements in a constructive mode. As a performance indication, Table 3 gives the magnitude limits for visibility measurements at 5 σ of an unresolved K star in 10 mn of integration assuming 50 fringes per OPD scan of 16 seconds duration. Table 3 indicates that GENIE constructive mode will be of special interest not only for many stellar astrophysics studies (e.g. MIRAs, Scutis, AGB giants, Cepheids ...) but also for some extragalactic works (e.g. nearby AGNs).

5 GENIE: a general user instrument

The nulling interferometry technique in the L band is well suited to study any faint cool object located in the immediate environment of a bright astrophysical source. A wide range of astrophysical problems can benefit from this technique, including in particular studies of circumstellar material (shells, ejecta, dust disks) and cool stellar companions.

5.1 The L band: a spectral band of astrophysical interest

The L-band is particularly rich in diagnostics of gas and dust in circumstellar and interstellar environments. At the same time, its diagnostic value is complementary to the information that can be obtained at shorter (J,H,K) and longer (M,N) spectral bands accessible from the VLTI. In particular, while at H and K scattering by dust is still important, at L it is mostly dominated by thermal emission from hot dust. The L band allows deeper penetration into highly obscured regions in e.g. galactic nuclei and star forming regions, because dust opacity is low. The L band contains many transitions of ions, atoms and molecules. Most important transitions are recombination lines of neutral hydrogen and helium, including the high series members of the HI humphreys series. This provides an important diagnostic of ionized gas covering a wide range of densities, in particular when combined with HI Pfund γ, Pfund δ

and Brackett α, all in the L-band. Several important molecular ro-vibratonal bands of abundant molecules are located in the L band, such as H_2O, OH, C_2H_2 and HCN. These bands probe gas at temperatures of 500-2000 K, usually found in the environment of evolved stars, and in the inner regions of the envelopes of young stars. Apart from gas-phase species, the L band also covers the O-H stretch band of H_2O water ice, both amorphous and crystalline. Water ice is expected to condense in the outflows of evolved oxygen-rich stars, and has been detected in several high mass loss rate asymptotic giant branch stars. Furthermore, water ice is present in proto-planetary disks, and probes the icy outer regions of these disks. Perhaps the most important dust band in the L band is the aromatic C-H stretch resonance of Polycyclic Aromatic Hydrocarbons (PAH molecules), ubiquitous in interstellar space, molecular clouds, proto-planetary disks and C-rich (proto-) planetary nebulae.

5.2 Exo-zodiacal dust, proto-planetary-systems and debris disks

As previously mentioned, one main objective of GENIE is to survey nearby stars for the presence of exo-zodiacal dust disks. In our own Solar system, it is known that the dust in different regions is ephemeral. The Poynting-Robertson effect, and solar photon pressure effectively remove the dust from the inner part of the solar system in a short time, unless it is continuously renewed by some mechanisms as e.g. comets evaporating dust or collision of asteroids. The detection of exo-zodiacal dust would thus be a strong indicator for the present of bodies other than the star itself in an extra-solar system. High levels of exo-zodiacal dust would imply intermediary stages between our own system and e.g. β Pictoris or Vega type systems. The survey of the Darwin target stars will therefore result in a valuable database. A comprehensive set of metallicities, ages and other stellar parameters could be modeled against the level of dust and other dust parameters that could be inferred from GENIE spectra. Modeling of the evolution of such systems would then become possible.

Star forming disks probably account for many of the unusual characteristics of young stars and play a role in early stellar evolution, the formation of binary or multiple star systems, and the formation of planets. Accretion onto the star is driven by the transfer of angular momentum from the disk, outwards in bipolar molecular outflows, and from the inner regions through atomic or ionized jets. However, processes that occur close to the stars are not clear and lack direct observations to be confronted with theory. GENIE, in both constructive and nulling mode, operating in the L'-band, will provide the spatial and spectral resolution needed to reveal the bulk flows of material and the physical conditions (density, temperature, magnetic field strength, and chemical abundances) in dense molecular cores, proto-stars, proto-planetary systems, and debris disks [10].

Table 4. Properties of the six shortest period EGPs accessible from Paranal.

target	star sp. type	dist. [pc]	planet per. [day]	a [AU]	temp. [K]	B_{proj} [m]	SNR
HD 73256	G8V	36.5	2.55	0.037	1200	129.5	3.1
HD 83443	K0V	43.5	2.99	0.035	1117	128.6	2.5
HD 179949	F8V	27.0	3.09	0.045	1383	102.4	3.7
Tau Boo	F7V	15.0	3.31	0.047	1388	60.9	3.2
HD 75289	G0V	28.9	3.51	0.047	1265	126.7	3.5
HD 162020	K2V	31.3	8.42	0.059	784	129.0	3.1

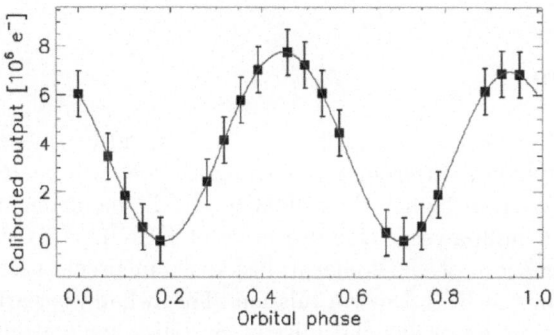

Fig. 7. Modulation of GENIE null signal along observations of HD 73256 during 20 consecutive nights.

5.3 Extra-solar giant planets and brown dwarfs

Since 1995, more than 140 Extra-solar Giant Planets (EGPs) were discovered indirectly, mainly in high-precision Doppler surveys. By nulling the light from the central star, GENIE can detect directly the IR thermal emission from a faint companion in a way similar to coronography but with higher rejection on smaller angular scales. Table 4 list six stars with short periods EGPs that are visible from Paranal. These so-called hot Jupiters are expected to be bright in the L band and could be detected by GENIE in about 4 hours. Observations of these objects during several nights could reveal a modulated null signal induced by the EGP orbital motion across the transmission map of the instrument (see Fig. 7). This could be used to retrieve the orbital parameters in the plane of the sky, and thus - together with the radial velocity data - to determine the masses of these faint companions.

Radial velocity searches for planets find massive planets in close orbits around 5 to 10% of G-type stars. By extrapolating, one would expect massive planets at any orbital separation around 22% of stars [11] including young stars. Young planets like young stars are self-luminous due to ongoing contraction and accretion, mostly in the thermal IR [12]. Hence, young planets could also be detected by GENIE, e.g. in newly found young nearby associations like TW Hya, Tucana-Horologium, the β Pic moving group, and the Her-Lyr association. Isolated young nearby stars, (e.g. Gl 182; d= 27 pc; age= 20 Myr) are also prime targets for detection of young planets with GENIE. A fortiori, GENIE could detect brown dwarf companions at close separations not only around G-type stars (as with radial velocity), but also around all stars inaccessible for different reasons to radial velocities survey, such as early-type stars, very late-type stars, brown dwarfs, young stars, close multiples, white dwarfs and giants. GENIE could thus probe the brown dwarf desert in other stellar sub-samples and at any given age.

6 Conclusion

GENIE at VLTI can provide European Scientists and Engineers with a first experience in nulling interferometry. This instrument will be able to achieve nulling rejection ratio higher than 1000 in the L' band on angular scales smaller than 10 milli-arcsec. One objective of GENIE is to survey nearby stars for the presence of exo-zodiacal dust disks in preparation for the science program of the ESA Darwin mission. The L-band is particularly rich in spectral diagnostics of gas and dust in circumstellar and interstellar environments. Hence, GENIE nulling mode will enable scientists to conduct research programs on a variety of sources including proto-planetary systems, debris disks and low mass companions around nearby stars. Moreover, GENIE ability to operate in a constructive mode will complement the suite of ESO VLTI instruments that operate at shorter and longer spectral bands for a wide range of galactic (e.g. MIRAs, Cepheids, Scuti, AGB giants ...) and extragalactic (e.g. nearby AGNs) studies.

References

1. M. Fridlund & P. Gondoin 2003, SPIE Vol. 4852, p. 394
2. M. Schöller & A. Glindemann 2003, ESA SP-539, p.109.
3. C. Eiroa, M. Fridlund & L. Kaltenegger 2003, ESA SP-539, p.403.
4. R. N. Bracewell 1978, Nature 274, 780
5. M. Ollivier & J. M. Mariotti J.M., 1997, App. Opt. 36 5340
6. O. Wallner, R. Flatscher & K. Ergenzinger 2005, A&A submitted.
7. E. Serabyn 2003, ESA SP-539, p.91.
8. P. Kervella, F. Thevenin, E. Di Folco et al. 2005, A&A 426, 297
9. O. Absil, R. den Hartog, C. Erd et al. 2003, ESA SP-539, p.317.
10. O. Absil, J.C. Augereau, R. den hartog et al. 2005, In: these proceedings.
11. C. H. Lineweaver & D. Grether 2003, ApJ 598, 1350L
12. G. Wuchterl & W. M. Tscharnuter 2003, A&A 398, 1081

Part VII

Instrumentation: Concepts for future
interferometric instrumentation

Multiple Anamorphic Beam Combination

E. N. Ribak[1], M. Gai[2], D. Gardiol[2], D. Loreggia[2], and S. G. Lipson[1]

[1] Physics Dept., Technion - Israel Institute of Technology, Haifa 32000, Israel
   eribak@physics.technion.ac.il
[2] Osservatorio Astronomico di Torino,Via Osservatorio 20, 10025 Pino Torinese,
   Italy

Summary. We suggest a new approach to the problem of simultaneous combination of many beams in an optical or infra-red stellar pupil-plane interferometer. All the beams are combined with all other beams. First all beams are stretched anamorphically to create a comb of light, then this comb is interfered with itself rotated at right angle. The diagonal, with self-interfering beams, provides also the individual intensity calibration.

1 Various combination schemes

In multiple-beam stellar interferometry, one has to combine and interfere many beams. Several configurations may arise. The first is when the beams arrive from different telescopes with the same size (e.g. Keck I and II, COAST), or or different sizes (VLTI, OHANA). In another case the beams arrive from openings inside the telescope aperture, in what is called aperture masking.

The approaches taken to create interference are classified as the Fizeau and the Michelson realisations. In the Fizeau approach, real images of the stellar object with identical magnifications are superimposed and interfere on an imaging detector. Because of the large number of pixels over which the image falls, this scheme is not so efficient in star light. The modified Fizeau approach, or the densified pupil, maintains only the positions of the original telescopes on the secondary plane. At the same time, the sizes of the beams are made relatively larger by a constant factor. This allows plugging more photons into the central lobe of the image and improves its signal to noise ratio.

The Michelson combination approach requires far fewer detector pixels, and is thus more light efficient. Here each beam is split many ways by a cascade of beam splitters, and these beams are combined by a similar arrangement of beam combiners. In addition, intensity fluctuations in the beams need to be measured simultaneously with the interference of the same beams for calibration purposes. For n beams, the number of splitters and combiners is of order n^2, and the number of detector pixels is also of the same order. When the number of beams is large, the arrangement of all beams splitters and mirrors becomes very cumbersome. Because of the large number of reflections and divisions, which may be lossy and also change polarizations, the resulting efficiency is poor. It should be stressed that it is necessary to combine several beams at the same time, rather than pair-wise, since this allows the use of phase closure between every three or more beams.

In many cases it is possible to narrow the band width of the interfering light. In these cases fibre optics can be employed to lead the beams from the telescope foci to the combination station and to the detectors. Using fibres also allows using fibre splitters and combiners, as well as planar optics which simplify the scheme considerably. The main limitation here is the requirement that the fibres are single mode and polarisation preserving, limiting again the available photons.

Thus one has to choose between the Fizeau approach, which tends to be wasteful in pixels and photons, and the Michelson approach, which is very difficult to realise for large numbers of apertures. Is it possible to avoid the cumbersome and wasteful n^2 beams splitters?

2 Matrix multiplication of beams.

Another option for beam division is by spatial beam splitting, as opposed to amplitude splitting [1]. This is similar to Young's double slit experiment, where the wave front is divided by a screen and openings into different beams. In the present method each of the telescope beams is interfered piecewise with each of the other beams. A part of the wave front of each telescope is interfered with a part of a wave front from another telescope The geometrical way this is achieved is by making a matrix of beams. Each column in the matrix is one of the beams, and each row is also one of the same beams.

In order to obtain a matrix of beams, the inputs are first lined up with minimum spacing between them. Then, using anamorphic optics, all the beams are stretched normal to the line by a factor equal to their number. Hence a square comb is prepared, where each comb line is one stretched beam. Now this comb is split into two equal combs by one beam splitter. Each comb is rotated at 45^0 in opposite directions. The two combs are then combined by a second beam splitter and made to interfere.

3 Optical design and realisation

In a fully engineered design, two processes must be applied to obtain the matrix of rays. The first is the anamorphic stretch, which has been designed by software and not tested yet. The second is the comb rotation and combination, which still has to be fully designed for the stellar light but was already realised in the laboratory using a rotation shear interferometer.

Two anamorphic designs were considered. The first used two mirrors in Gregorian configuration in one axis, where the other axis was flat. The two toroidal (conic) mirrors had 200 and 2000 mm in one direction (Fig. 1). For a

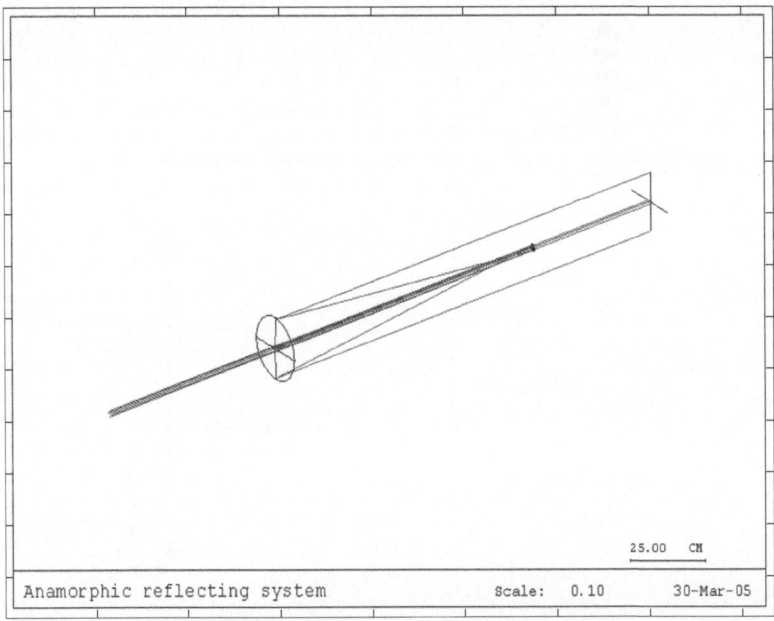

Anamorphic reflecting system   Scale: 0.10   30-Mar-05

Fig. 1. Cylindrical Gregorian beam stretcher.

modest stretching of ten we obtained a maximum phase error of 250 nm, and all of this error was in the margins of the ellipse. When these are excluded, the error drops to 3 nm. In another design two anamorphic achromatic lenses served as telescope in one axis, using cylindrical doublets. The performance was slightly worse, and the band width was limited by the lenses, as opposed to the mirror solution.

The idea was tested in a simple experiment, using a square Sagnac interferometer. A dove prism, rotated at 22.5° turned the beams about their axes by 45° in opposite directions, and the two images interfered with 90° rotation between them. Two independent lasers, combined by a beam splitter, served as a source, the angle between them being such that they were barely resolvable by a lens a few metres away. Light was collimated again by a second lens, and fed into the interferometer. We punched a number of holes in an opaque card to simulate the beams just behind the collecting lens, and placed a cylindrical lens at its focus to stretch them anamorphically by a factor of about twenty (Fig. 2).

The results are shown in Fig. 3 for one laser and two. Since the first is unresolved, all fringes have high contrast. This changes when the coherence function drops to zero at some base lines by the addition of a second laser, making the system resolved. Notice the high contrast along the diagonal, where each beam interferes with itself. This also provides intensity calibration for each beam, especially important for unequal telescopes and for fibre beam transport.

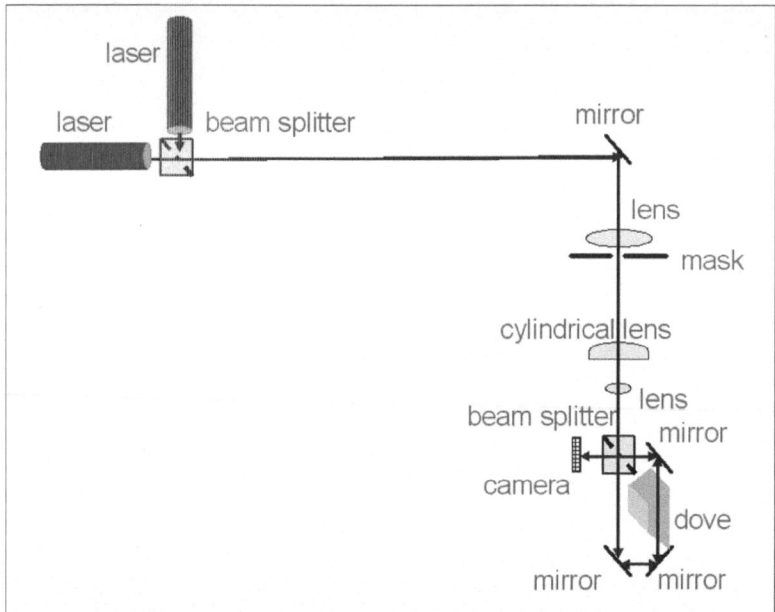

Fig. 2. Optical setup, including two independent sources, a telescope, mask to create separate beams, cylindrical lens for their stretching, and a Sagnac rotational shear interferometer.

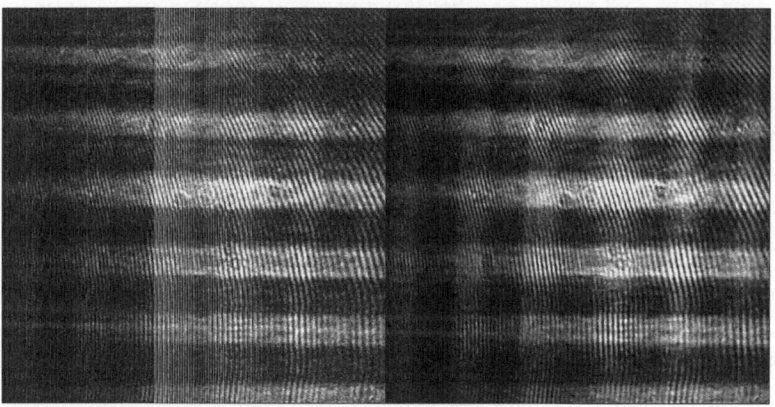

Fig. 3. A single unresolved laser (left) producing a matrix of interfering beams. When another laser is added, some of the off-diagonal beams lose their contrast (right). Artificial fringes are created by misalignment of the telescope and interferometer.

4 A few comparisons and perspectives

Naturally we start with the advantages of the tested scheme. In the first place, the simplicity of measurement: the two output patterns need only use a single camera each. The number of pixels may also be limited, to essentially twice to four times the number of fringes. In addition, every two beams give four fringe patterns for calculation of both the contrast and the phase between them. If more than a single fringe is measured in such a matrix element, it also allows fringe phase tracking. In cases where the intensity in each beam is not constant, there is no need for a calibration bleeder, namely a separate light pick up channel. This is easily achieved by the matrix diagonal which allows the intensity calibration of each beam with itself.

Other optical advantages will be the use of mirrors in a final design (excluding the beam splitter), reducing significantly the dispersion and permitting measurement at most wave lengths, and reducing polarisation losses. The losses are minimal, since there is only one beam-splitter (used twice), independent of the number of beams, with a further five mirrors (three being in a reflective equivalent of a dove prism). The volume is also much more compact.

On the negative side, we mention that some light is lost between the stripes. If we choose to have many fringes, we need many pixels, considerably more than the number of single pixel detectors in the Michelson approach. Also, the readout noise and time are worse then the single detectors. In some shearing interferometers the symmetric output is difficult to access. The anamorphic stretch is limited to about twenty beams, and it limits the spectral capability.

Currently we have the optical design performed in Turin, and the initial results obtained in Haifa. We are testing a new shear interferometer which is both stable, has fewer reflections, and has two accessible outputs. We need to finalise the optical design, and perform a light budget study and a full comparison as mentioned above.

Reference

1. E Ribak, E B Hochberg, N A Page, S P Synnott, and J B Breckinridge 1988: Beam combination in a multi-telescope, monolithic interferometer. *Proc ESO* **29**, 1105-15, Munich.

The Potential of IR-Heterodyne Spectroscopy

R. Schieder[1], D. Wirtz[2], G. Sonnabend[3], A. Eckart[1]

[1] I. Physikalisches Institut der Universität zu Köln, Cologne, Germany
   schieder@ph1.uni-koeln.de, eckart@ph1.uni-koeln.de
[2] Phillips Research Laboratories,Hamburg, Germany
   daniel.wirtz@phillips.com
[3] Goddard Space Flight Center/NASA, Greenbelt/MA, USA
   samstag@ph1.uni-koeln.de

Summary. Infrared Heterodyne spectroscopy is a very attractive method for combining high frequency with high spatial resolution. At frequency resolutions $R > 10^5$ the sensitivity of heterodyne observations becomes comparable or even superior to direct detection methods. This is particularly of interest for detailed investigations of the cold interstellar gas for example. Important species like CH_4, C_2H_2, or H_2 can be observed which are not detectable in the radio-wave regime. With tuneable Quantum Cascade lasers we have demonstrated that mid-IR heterodyne spectroscopy can be a very powerful method for astronomy. For antenna arrays heterodyne detection has particular advantages, since the losses for the distribution of the received signal into the many baselines in a multi-telescope arrangement are avoided. In addition, other losses like those in optical delay-lines do also not occur.

1 Introduction

Heterodyne spectroscopy is a well established remote sensing method for the detection of atoms and molecules in the interstellar medium or in planetary atmospheres. The technique is proven to be very powerful in the radio-, mm-submm- or far infrared wavelength region, but the mid-infrared is dominated by direct detection methods. This is explainable because of a couple of arguments:

i) Contrary to the radio regime, infrared heterodyne instrumentation is not so easily available due to the lack of suitable local oscillators and efficient broad band mixers.
ii) The sensitivity of heterodyne detection is limited by the so called quantum limit ($T_Q = h\nu/k_B = 1440K$ at $10\mu m$ wavelength), when expressed in terms of the noise brightness temperature of the instrument.
iii) For many scientific topics the required frequency resolution is often lower than heterodyne detection is capable to provide. Or, when turning the argument around, the potential merits of high frequency resolution are mostly neglected.

It should be emphasized that in principle heterodyne detection has big advantages when studying specific scientific problems. The important information about the velocity structure of the interstellar medium is only available by means of high frequency resolution. If the interstellar gas is very cold, total line widths of atoms and molecules near one km/sec are common, which calls for a frequency resolution $R = \nu/\delta\nu \geq 10^6$, if such lines are resolved by a few resolution elements per line width. With low resolution instruments, narrow lines tend to disappear due to the inevitable frequency dilution effects. But, even in regions with higher velocity dispersion, like planet forming disks for example, it is an interesting question, how the dynamics or the chemical composition is looking like as a function of radial distance to the star. Other questions are related to the line shapes of molecular lines in planetary atmospheres, which can provide detailed information about the height or temperature distribution of the species. This might become an important task in future when investigating atmospheres of exoplanets for example.

A lot of such information can be gathered in the radio-, mm-, and submm-frequency range, but, due to the longer wavelength, the spatial resolution is generally rather poor. In addition, many important species like CH_4, C_2H_2, C_2H_4, C_2H_6 etc. are not observable at radio frequencies due to their missing permanent dipole moment. The situation is different in the mid-infrared, where practically all molecular species are detectable by their vibrational transitions. And, most important, spatial resolution in the sub arcsec range is already obtained with a medium size telescope of 3m diameter (at $\lambda = 10\mu m$). This resolution would correspond to a 900m diameter telescope at a wavelength of 3mm. From this it is obvious that mid infrared heterodyne detection is an ideal tool to combine high spatial with high frequency resolution.

Clearly, the brightness temperatures of molecules in the cold interstellar gas are rather small according to Planck so that the detection of emission lines becomes difficult. But there are many sufficiently bright background sources available which can be used to detect absorption signatures from the cold gas. For example, the rotational quadrupole lines of molecular Hydrogen at 17 and 28μm are very good candidates for such experiments, and important information about the abundance of molecular Hydrogen would become available. At high spatial resolution, i.e. large collecting area, there are many sources available with a flux of 10Jy or above, which is sufficient for these observations. In any case, such program is an excellent example for the unique capabilities of heterodyne spectroscopy.

One question remains: How do the sensitivities of heterodyne and direct detection compare? Although there is a quantum limit for heterodyne detectors, the sensitivity becomes very competitive at very high frequency resolution since the spectral analysis does not reduce the instrumental efficiency. Whereas high resolution direct detection instruments suffer from low throughput, typically in the range of a few percent only, heterodyne systems are only limited by the detector quantum efficiency itself, which is usually in the range of 50% or higher. At resolutions above $R = 10^5$ gratings become unpractical, since the required size of a grating would have to be larger than

1m. When considering a Fabry-Perot interferometer the situation becomes even worse, since the throughput approaches zero at very high frequency resolution. The conclusion is, at high efficiency requirements the heterodyne method is practically the only option when looking for resolutions better than 0.01cm^{-1} (300 MHz), as is valid almost independent on wavelength.

2 Present experimental situation

Since many years attempts have been made in several laboratories to introduce tunable infrared Lead salt lasers (TDL) as local oscillators in mid-infrared heterodyne receivers. Unfortunately, such lasers provide only little single mode power ($\approx$ 0.1mWatt) and are difficult to control in amplitude and frequency. Nevertheless, some successful realizations of tunable heterodyne systems have been reported, which are mainly used for atmospheric studies [1, 2, 3]. Due to the low power of the TDLs the sensitivity of these systems is rather poor, in our case we reached double sideband noise temperatures near $T_{sys}(DSB) = 15,000\text{K}$ at $10\mu m$ wavelength, which is about a factor of 10 above the quantum limit. This corresponds to an NEP of about $3.7 \cdot 10^{-15}$ at a resolution of $R = 10^6$, so that the usability of such receiver is rather limited. It is well established that CO_2-laser pumped infrared receivers can do much better, typical values are near twice the quantum limit ($T_{sys}(DSB) = 3000\text{K}$), which enables one to observe signals below 1Jy with an 8m diameter telescope within 1 hour. Impressive results have been obtained during observations of planetary atmospheres [4], but also extra-solar signals have been seen [5, 6, 7].

Recently, new lasers have been introduced, the quantum cascade lasers (QCL), which provide power in the 10mWatt range or above. They are tunable and can be operated in single mode by means of distributed feedback structures or by coupling to an external cavity. They are very competitive with CO_2-lasers with respect to linewidth, amplitude and frequency stability. We also found noise temperatures near 3000K, and the performance corresponds closely to what we could achieve with a CO_2-laser [8]. Therefore, these new lasers provide an excellent opportunity for exciting science in the mid infrared.

First measurements at the McMath-Pierce solar observatory on Kitt Peak/AZ were done in Nov/Dec 2002 and 2003 to demonstrate the usability of a QCL pumped heterodyne receiver for the first time. Details of the system are described in [9]. The frequency stability of the QCL frequency in our system is found to be better than 1MHz rms within several hours. This corresponds to a relative frequency stability of $R = 3 \cdot 10^7$. Similar, the linewidth of the QCL has been determined much below 1MHz as well. Switching between signal- and reference-position on sky as well as two temperature loads is done on the timescale of 1 second. The amplitude stability of the system is characterized by the Allan variance test (see Fig.1), which determines the noise resulting from the difference of two contiguous measurements of equal time length as a function of the integration time (see e.g. [10]). The minimum of the plot determines the point, where the radiometric noise of the receiver becomes dominated by drift noise in the system, which is inevitably present due to thermal or other effects. The time at minimum therefore characterizes the maximum integration time the system allows before the measurement becomes heavily deteriorated by the drift noise in the system. With a minimum time of the order of 50 seconds the excellent stability of the instrument is well demonstrated so that astronomical observations are clearly feasible.

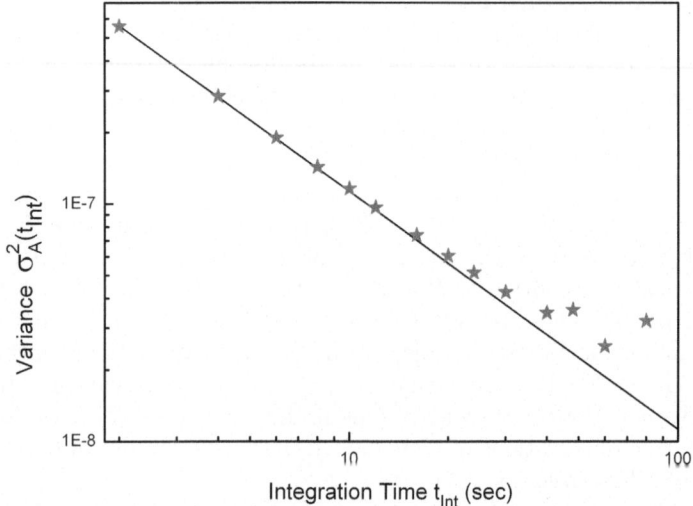

Fig. 1. Allan variance plot of THIS. The data were taken with an ambient temperature load. The minimum time near 50 seconds is and excellent proof of the stability of the system. The straight line indicates the radiometric expectation at a frequency resolution of $\delta\nu = 1.4 MHz$. Generally, observations are done using the beam switch method with switch times of the order of one second. Therefore there is zero contribution of instrumental drift noise.

In Fig.2 an absorption line of highly excited H_2O in the solar atmosphere above a sunspot is depicted. The high frequency resolution allows a fit of a Doppler profile to the line, which leads to a gas temperature of about $5440\pm240K$. This is to be compared with the sunspot brightness temperature of about 3900K (brightness temperature outside the line profile), which is equivalent to 4600K physical temperature. The peak absorption at the center of the line indicates that the gas temperature must be significantly lower than 4400K (3680K brightness temperature), since the line is not saturated. Lower temperatures are required anyway because otherwise the molecules would dissociate completely. This is a clear indication that the line width can not be determined by the thermal motion of the gas, but is influenced in addition

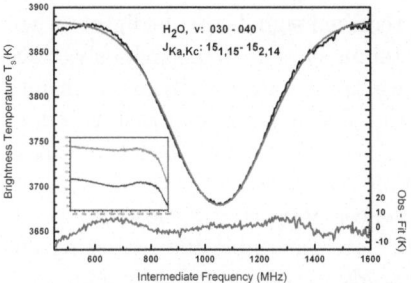

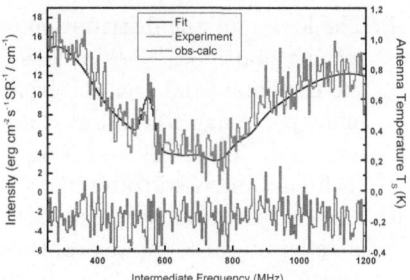

Fig. 2. Line of highly excited water above a sun spot at $1088.92511cm^{-1}$. The two curves in the insert (bottom left) are the signals seen on the sun (top) and on the sun spot (bottom). The steep drop of signal at higher frequency is due to atmospheric Ozone absorption. The calibrated difference spectrum represents a Gaussian profile, which can be fitted to a Doppler profile for an effective gas temperature of 5440K.

Fig. 3. Broad feature of the 10347.4341 P(30) line of CO_2 seen in absorption against the warmer Mars surface. The narrow dip is due to non-LTE emission in the stratosphere. The dip is shifted in frequency indicating a wind speed of about 70m/sec. The lower trace is a plot of the signal minus a fit.

turbulence due to convective flows and solar oscillations in the neutral gas [11]. Zeeman splitting does not contribute since the effect is negligible for a neutral species like H_2O. Fig.3 depicts the result of an observation of non-LTE emission of CO_2 in the Mars atmosphere at high rotation number (J=30) of a line in the $9.6\mu m$ CO_2 laser band. The frequency resolution is 4MHz. The antenna temperature of the narrow feature is 300mK, seen after roughly 40 minutes of total integration, and it indicates that the instrument has the assumed sensitivity. With a point source, the same signal to noise ratio would be achieved with a 1.6Jy signal when observed with an 8m diameter telescope under similar conditions otherwise. In summary we can conclude that the sensitivity of our system is as predicted. This was also verified by the detection of continuum signals from the Moon, Mercury and Betelgeuse for example.

3 Applications in interferometry

One of the most tempting applications of heterodyne spectroscopy is in interferometry, when combining extreme spatial resolution with very high frequency resolution. The feasibility of mid-infrared interferometry is well demonstrated by [12, 13, 14] even for a relatively coarse resolution in the GHz range. In fact, the heterodyne method has some particular advantages.

i) The losses for the distribution of the received signals are negligible when serving many baselines of a multi-antenna system simultaneously, since the IF-signal from each receiver can be amplified arbitrarily and split into many portions without any change in signal to noise. In contrast, this does not apply for direct detection systems, as long as the interferograms of each baseline need to be investigated independently.

ii) The compensation of pathlength differences between the telescopes is not affected by throughput problems, which are inevitable and rather significant for optical delay lines. In the heterodyne case the delay is done electronically, which is standard technique at various radio-interferometers around the world. In fact, the required accuracy of the delay for an accurate measurement of the visibility is inverse proportional to the resolution bandwidth. Thus, at a resolution of, say, 30MHz at $10\mu m$ wavelength, the control of the optical delay becomes very simple, since the coherence length at such resolution is of the order of $10\mu m$. This reduces the necessary effort drastically.

If we consider the situation of a VLT interferometer with four UTs, we have 6 baselines plus 4 photometric channels to consider. Each antenna output must be split into 4 portions so that the signal strength per telescope becomes seriously reduced. With equal power in all channels the resulting signal of each baseline corresponds to that of one pair of telescopes with a diameter of only 4.1m each. This is certainly rather disappointing, because it reduces the sensitivity of the interferometer drastically. (It might be even less demanding to use 6 pairs of telescopes with 4.1m diameter each in order to obtain the identical result.) In addition the sensitivity becomes further reduced by the reduced throughput of the optical delay lines, which might by in the range of 30% only. In total, one looses more than 86% of the fringe signal when comparing with the ideal configuration of a single pair of UTs at zero delay line loss. Consequently, the necessary observing time increases by a factor of approximately 50 for a given signal to noise, when comparing with a heterodyne setup of equal sensitivity and resolution.

On the other hand, heterodyne receivers suffer from the quantum limit, as was mentioned before. This means that the sensitivity of a single heterodyne receiver is reduced by about one order in magnitude, when comparing with the presently best reported sensitivities of a high resolution direct detection system at a resolution of $R = 10^5$ (see e.g. [15]). Nevertheless, already at this frequency resolution heterodyne reception becomes superior in sensitivity when comparing with direct detection methods in our case. The situation changes even more to the advantage of heterodyne instrumentation when considering the full VLTI arrangement with 8 telescopes and 36 signal channels. At the same time, the best direct detection instruments still have about one order in magnitude less frequency resolution than is desirable for particular observations, and any attempt to increase the frequency resolution would drastically reduce the sensitivity of such instruments. It is therefore evident that heterodyne spectroscopy in combination with interferometry is a very attractive choice when considering both, very high frequency as well as spatial resolution.

Acknowledgements We are grateful for the support during our observations at the McMath-Pierce solar observatory on Kitt Peak. This work was funded through the "Deutsche Forschungsgemeinschaft" through special grant SFB494.

References

1. B. Parvitte, C. Thiébeaux, D. Courtois; Spectrochimica Acta, Part A: Molecular and Biomolecular Spectroscopy 55 (10), 2027 (1999)
2. H. Fukunishi, S. Okano, M. Taguchi, T. Ohnuma; Applied Optics 20, (18), 2722 (1990)
3. F. Schmülling, B. Klumb, M. Harter, R. Schieder, B. Vowinkel, G. Winnewisser; Applied Optics 37 (4), 5771 (1998)
4. T. Kostiuk; Infrared Physics and Technology 35 (2-3), 243 (1994)
5. A.L. Betz; ApJ 244, L103 (1981)
6. T. Kostiuk, M.J. Mumma, J.J. Hillman, D. Buhl, L.W. Brown, J.L. Faros, D.L. Spears, Infrared Phys. 17 (6), 431 (1977)
7. D.M. Goldhaber, AL..Betz; ApJ 279, L55 (1984)
8. G. Sonnabend, D. Wirtz, V. Vetterle, R. Schieder; A&A Vol. 435 No. 3, 1181 (2005)
9. G. Sonnabend, D. Wirtz, R. Schieder, A. Eckart; Proc. of the ESO Workshop "Scientific Drivers for ESO Future VLT/VLTI Instrumentation" held in Garching, Germany, 11-15 June 2001, J. Bergeron, G. Monnet eds., 225, (2001)
10. see e.g. in R. Schieder, C. Kramer; A&A 373, 746 (2001)
11. M. Asplund, A. Nordlund, R. Trmpedach, C. Allende Prieto, R.F. Stein; A&A 3359, 729, (2000)
12. D.S. Hale, M. Bester, W.C. Danchi, W. Fitelson, S. Hoss, E.A. Lipman, J.D. Monnier, P.G. Tuthill, C.H. Townes; ApJ 537, 998 (2000).
13. J.D. Monnier, W.C. Danchi, D.S. Hale, E.A. Lipman, P.G. Tuthill, C.H. Townes; ApJ 543, 861 (2000)
14. J.D. Monnier, W.C. Danchi, D.S. Hale, P.G. Tuthill, C.H. Townes; ApJ 543, 868 (2000)
15. J.H. Lacey, M.J. Richter, T.K. Greathouse, D.T. Jaffe, Q. Zhu; Publications of the Astronomical Society of the Pacific 174, 153 (2002)

Part VIII

Posters — Science

Study of the Projection Factor to Break the Frontier of Accuracy in Cepheid Distance Determination

N. Nardetto[1], D. Mourard[1], Ph. Mathias[1], and A. Fokin[1,2]

[1] Observatoire de la Côte d'Azur, Dpt. Gemini, UMR 6203, F-06130 Grasse, France, Nicolas.Nardetto@obs-azur.fr
[2] Institute of Astronomy of the Russian Academy of Sciences, 48 Pjatnitskaya Str., Moscow 109017 Russia

The Interferometric Baade-Wesselink method is affected by a significant bias: the projection factor. Hydrodynamical calculations show that an error of 6% can be done on the distance if the *wrong* projection factor is used.

1 Scientific rationale

The distance of galactic Cepheids can be derived through the interferometric Baade-Wesselink method (IBW). Recent observations with VINCI/VLTI have provided the distance of seven Cepheids (Kervella et al. 2004a [1]). From these results it is possible to calibrate in a new geometric way the Period-Luminosity (Kervella et al. 2004b [2]) and the Surface-Brightness relations (Kervella et al. 2004c [3]), which are both very important for extragalactic distances determination.

The IBW method combines spectrometric and interferometric observations. The interferometric measurements lead to limb-darkened angular diameter estimations over the whole pulsation period, while the stellar radius variations can be deduced from the integration of the pulsation velocity (Vpuls). The latter is linked to the observationnal velocity (Vrad) deduced from line profiles by the so-called *projection factor p*. In this method angular and linear estimations of the Cepheid dimension have to correspond to the same layer in the star to provide a correct estimate of the distance.

2 The impact of the projection factor

The spectral line profile, in particular its asymmetry, contains the whole physics present in the dynamical atmosphere of the Cepheid : photospheric pulsation velocity, velocity gradients, limb-darkening, rotation, turbulence... All these effects, except the rotation velocity, are supposed to vary with the pulsation phase. When measuring radial velocities from line profiles, we include the integration in two directions over the surface, throught limb-darkening, and over the radius, throught velocity gradients. All these physical effects are currently gathered in one specific quantity, generally considered as constant with time: the projection factor p. It is defined as $V_{puls} = p * V_{rad}$. The knowledge of p is currently an important limiting factor for this method of distance determination.

A self-consistent and time-dependent model of the star δ Cep is computed in order to study the dynamical structure of its atmosphere together with the induced line profile. Different kinds of pulsation velocities are then derived :

1. the *gas* velocity corresponding to $\tau = 2/3$ in the line
2. the velocity of the layer corresponding to $\tau = 2/3$ in the spectral center of the line
3. the velocity of the photospheric layer ($\tau = 2/3$ in the continuum)

We compile a suitable average value for the projection factor related to different observational techniques, such as spectrometry [case 1, $p = 1.35$], and spectral-line [case 2, $p = 1.32$] or wide-band [case 3, $p = 1.27$] interferometry. These results are given when using the gaussian fit method to derive the radial velocity. We show that the impact on the average projection factor and consequently on the final distance deduced from this method is of the order of 6%. This systematic error is comparable to the statistical error corresponding to interferometric observations : with VINCI/VLTI we obtain a relative precision on l Car of 5%. We also study the impact of a constant or variable p-factor on the Cepheid distance determination. We conclude on this last point that if the average value of the projection factor is correct, then the influence of the time dependence is not significant as the error in the final distance is of the order of 0.2% (see Nardetto et al. 2004 [4]).

3 Conclusion

The projection factor is currently the most important bias of the IBW method for cepheid distance determination. We find a difference of 6% depending on the definition adopted for the projection factor. This theoretical work has now to be confirmed by observations. A study of the limb-darkening and the projection factor through HARPS high resolution spectrometric observations of ten galactic Cepheids is currently in progress (see Nardetto et al. 2005 [5]).

References

1. P. Kervella, et al., 2004a, A&A, 416, 941
2. P. Kervella, et al., 2004b, A&A, 423, 327
3. P. Kervella, et al., 2004c, A&A, 428, 587
4. N. Nardetto, et al., 2004, A&A, 428, 131
5. N. Nardetto, et al., 2006, A&A, 453, 309

LMC Cepheids with the VLTI

D. Mourard[1], N. Nardetto[1], S. Lagarde[1], R. Petrov[2], D. Bonneau[1], and
F. Millour[2]

[1] Observatoire de la Côte d'Azur, Dépt. GEMINI, Avenue Copernic, 06130
Grasse, France
[2] Université de Nice, LUAN, Parc Valrose, 06000 Nice, France

Summary. We investigate the feasibility of applying the Interferometric Baade
Wesselink method on LMC cepheids with the VLTI.

1 Scientific rationale

The Period-Luminosity (P-L) relationship of Cepheids is a fundamental link
between the short (Milky Way) and large (LMC) to very large (the visible
Universe) distances. Galactic Cepheids have already been measured by different interferometers such as GI2T, PTI, NPOI, VLTI and CHARA. A recent
work [1] shows first attempts to the absolute calibration of the 0 point of the
P-L relationship. Direct access to Cepheids in the LMC will allow a direct
calibration of secondary distance indicators, which are the most useful for
the distance scales in the Universe.

2 Main characteristics of LMC cepheids

We have used a sample of LMC Cepheids extracted from a recent work done
by Persson and collaborators [2]. This sample gives access to J, H and K
magnitudes of about 90 cepheids with periods ranging from 3 to 48 days.
In order to estimate angular diameter for each star of this sample, we have
used empirical relations based on the (J,J-K) parameters. The results are
presented in Fig. 1. Half of the sample have a K magnitude between 10 and
12 and an angular diameter between 20 and 40μas.

3 Observing strategy and estimation of performances

These very small angular diameters are very far of being resolved by a 200
meters baseline in the near infrared. However, such small diameters could
be measured by the Differential Interferometry technique [3]. Indeed, these
stars present interesting spectral lines with the mixing of pulsational and
rotational velocity fields. Photocenter displacements are then expected. Such
effects are detectable by the VLTI through small phase shifts in spectral lines.

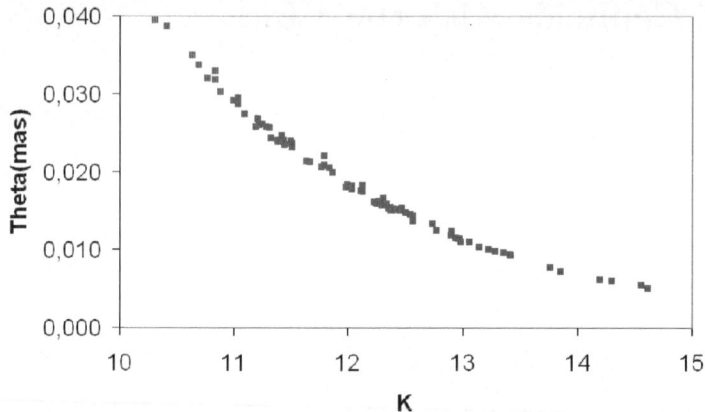

Fig. 1. Angular diameters (in mas) of our LMC cepheids sample as a function of the K magnitude.

Preliminary estimations show that the photocenter displacement through a spectral line is typically of 15% of the angular diameter, so around 4.5μas. The variation of the angular diameter during the pulsation cycle beeing estimated to 15% of the diameter, the pulsation will then lead to an amplitude of variation of the photocenter displacement of 0.7μas.

For a 200m baseline, a wavelength of 1μm and a precision of 10^{-3} on the phase measurements, one can reach a sensitivity of 1μas. One can estimate that this precision could be reached within a few hours of observation.

4 Conclusion

Direct determination of angular diameters of LMC cepheids and thus direct distance determination seems within the reach of an enhanced version of the VLTI, allowing measurements of photocenter displacements at the level of 1μas, preferably in the visible domain in order to increase the angular resolution and with spectral resolutions of the order of 10^4. Such performances are compatible with the current design of the $VEGA$ proposition [4].

References

[1] Kervella P., Bersier D. et al., 2004, A&A, 423, 327
[2] Persson S.E., Madore B.F. et al., 2004, Astron.J., 128
[3] Petrov R., Roddier F. and Aime C., 1986, J. Opt. Soc. Am. A, Vol. 3, No. 5
[4] Mourard D., Antonelli P. et al. this conference

Evolutionary Modeling of Nearby Stars Using Asteroseismic and Interferometric Constraints

P. Kervella[1] and F. Thévenin[2]

[1] LESIA, UMR 8109, Observatoire de Paris-Meudon, 5, place Jules Janssen,
   F-92195 Meudon Cedex, France, pierre.kervella@obspm.fr
[2] Département Cassiopée, UMR 6202, Observatoire de la Côte d'Azur, BP 4229,
   06304 Nice Cedex 4, France

Summary. We present an innovative approach to model nearby stars for which asteroseismic frequencies and interferometric diameter measurements are both available. Using the VLTI-VINCI instrument, we measured the angular diameters of several nearby stars with asteroseismic frequencies: α Cen A (G2V), α Cen B (K1V), Procyon (F5IV-V), δ Eri (K0IV), η Boo (G0IV), ξ Hya (G7III). The linear diameters are deduced based the Hipparcos parallaxes, that are very precise for these bright and nearby stars. Using classical spectro-photometric constraints, as well as the radius and seismic frequencies, we then build CESAM evolution models that enable us to derive in particular the mass, age and internal structure of these stars.

For all our interferometric observations, we used the VLT Interferometer with its commissioning instrument, VINCI ([1]), a two telescopes beam combiner operating in the K band (2.0-2.2 μm). This instrument measures the squared visibility (V^2) of the interferometric fringes. It is related to the angular diameter of the star through the Zernike-Van Cittert theorem. On α Cen A and B, we obtained the following limb darkened angular sizes: $\theta_{LD} = 8.511 \pm 0.020$ and 6.001 ± 0.034 mas ([2]). Coupled with the Hipparcos parallax of 747.1 ± 1.2 mas ([5]), this translates into radii of R= $1.224 \pm 0.003\,R_{\odot}$ and $0.863 \pm 0.005\,R_{\odot}$. Similarly, we obtain an angular diameter of $\theta_{LD} = 5.448 \pm 0.053$ mas, and a linear radius of $R = 2.048 \pm 0.025\,R_{\odot}$ for Procyon ([3]). For the three remaining stars ([6]), we get $\theta_{LD}(\delta Eri) = 2.39 \pm 0.03$ mas ($R = 2.33 \pm 0.03\,R_{\odot}$), $\theta_{LD}(\eta Boo) = 2.20 \pm 0.03$ mas ($R = 2.68 \pm 0.05\,R_{\odot}$), $\theta_{LD}(\xi Hya) = 2.39 \pm 0.02$ mas ($R = 10.3 \pm 0.3\,R_{\odot}$).

We have computed a series of models using the CESAM evolutionary code ([4]). The evolution of each star was modeled starting from the homogeneous ZAMS corresponding to their pre-estimated masses. For both α Cen and Procyon, which are visual binary stars with well known orbits, it was possible to use a precise value of their masses for the input of our models. Several models were computed, leading to different evolutionary tracks in the HR diagram. They were considered acceptable when the evolutionary track reached the center of the uncertainty domain defined by the photometric, spectroscopic and interferometric constraints. The essential added value of interferometry is that the constraint imposed by the linear radius R and the effective temperature $T_{\rm eff}$ is much tighter than the classical L-T uncertainty box. This results in particular in a better estimation of the age (Fig.1).

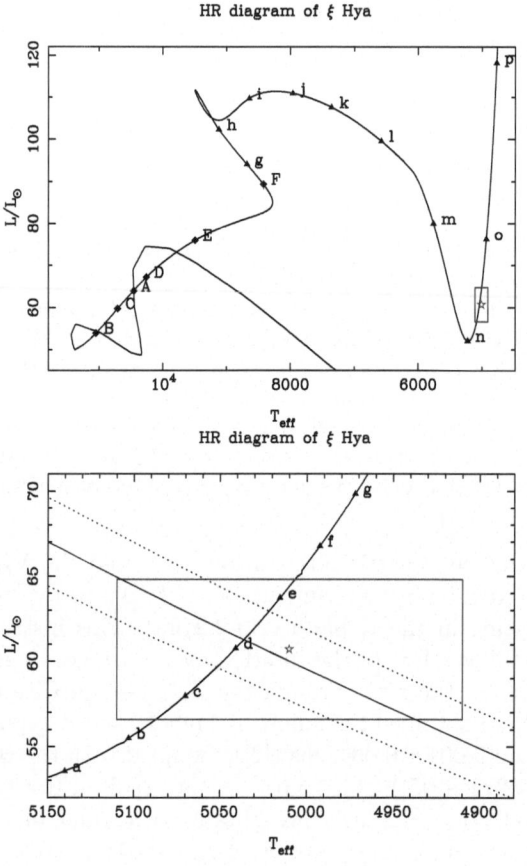

Fig. 1. CESAM evolution track of ξ Hya (top, zoom on bottom), with the classical uncertainty box (rectangle) and diameter constraint (diagonal zone). Each mark is separated by 0.1 Myr: point d gives a best fit age of 509.5 ± 0.1 Myr.

References

1. Kervella, P., Ségransan, D. & Coudé du Foresto, V.: A&A **425**, 1161 (2004)
2. Kervella, P., Thévenin, F., Ségransan, D., et al.: A&A **404**, 1087 (2003)
3. Kervella, P., Thévenin F., Morel, P., et al.: A&A **413**, 251 (2004)
4. Morel, P.: A&AS **124**, 597 (1997)
5. Söderhjelm, S.: A&A **341**, 121 (1999)
6. Thévenin, F., Kervella, P., Pichon, B., et al.: A&A **436**, 253 (2005)

Diameter Determinations from VINCI Using Global Calibration Solutions

Jeffrey Meisner

Leiden Observatory, P.O. Box 9513, NL-2300 RA Leiden, The Netherlands
meisner@strw.leidenuniv.nl

Summary. Although primarily built for testing and alignment purposes, the K band instrument VINCI at the VLTI has produced numerous scientific results on individual targets. The present work consists of a global analysis of the entire VINCI data archive. A visibility estimator using coherent integration has successfully obtained some 15500 raw visibilities on 293 objects. Those results are interpreted using a global solution algorithm which simultaneously solves for calibration (transfer function) fluctuations and for stellar diameters. Stars which do not follow the standard uniform disk visibility curve are automatically or manually flagged and are avoided in the solutions for nightly calibrations.

The standard approach to calibration based on observations of "calibrator" stars having assumed diameters, is rejected in the present work. No *a priori* diameters are input to the algorithm, which relies on baseline diversity to simultaneously solve both for diameters of stars having well-behaved characteristics, and a quasi-static transfer function subject to various hardware fluctuations. The success of this approach depends on a diversified schedule of observing the same target on different nights in conjunction with a mixture of other targets, so that all visibilities can be "cross-calibrated."

This is practical in the case of the VINCI data set where half of the 666 observing nights produced at least 22 successful visibilities on a median of 7 different targets. Half of the 293 objects were successfully observed at least 18 times and on at least 6 separate nights, with observations over a median baseline range of 4.8:1. Uniform disk diameters are a by-product of the cross-calibration procedure which uses this database to solve for an evolving transfer function on a nightly basis.

1 Solving for calibrations and fitting of visibilities

In Fig. 1 we see the solution for the transfer function over a period of 9 months. Also shown is a set of 94 visibility points for observations of β Gru which is fit as a uniform disk of diameter 26.25 mas. The fitting procedure employs a cost function appropriate for gaussian distributed errors among 95% of the data points, and up to 5% "bad" datapoints whose possible large deviations from the fit will thus not degrade that fit; in this example 3 aberrant visibility points can be seen which have effectively been ignored in the diameter determination. The raw results of these diameter solutions are posted on

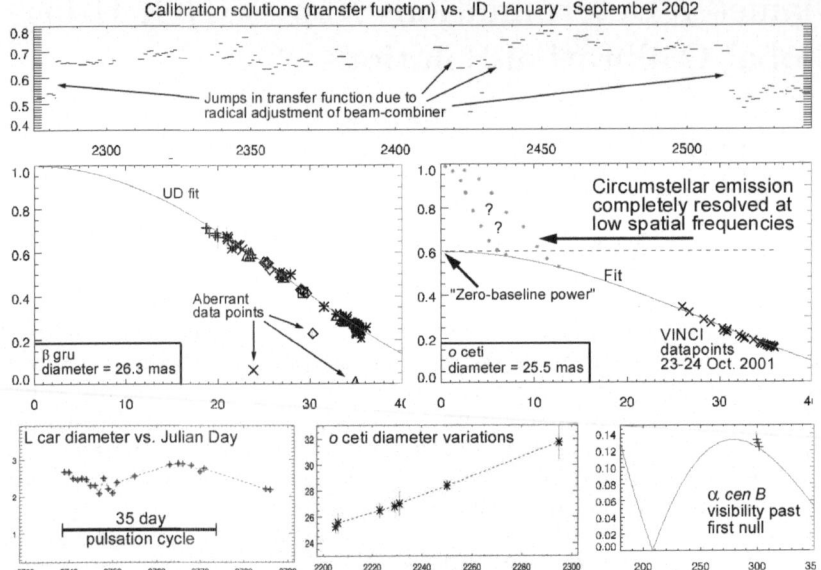

Fig. 1. Top: Solutions for VINCI transfer function over 9 months. **Center left:** Calibrated visibility vs. spatial frequency (cycles/ arcsec) for a typical star. **Center right:** Visibility model for a star with circumstellar emission, fit using a "zero-baseline power" of .6. **Bottom:** Diameter solution for each night for L Car and o Ceti (taking its reduced zero-baseline power into account). **Bottom right:** Three visibility points for α Cen B past first null with UD fit for D=5.87 mas.

the web at:
http://www.strw.leidenuniv.nl/~nevec/VINCI/meisner/

Improvements in the diameter determinations over, typically, 50 iterations of the algorithm, result in increasingly precise determinations of the evolving transfer function, and thus more precise calibrated visibilities. We have examined the statistics of the residuals of these calibrated visibilities in fitting uniform disk functions as in Fig. 1. If we take, for each object, the median of the magnitude of residuals divided by their visibilities, then taking the median over all objects observed at least 10 times, we can obtain a relative error as low as 1.1%, reflecting both the inherent precision of VINCI, and the success of this approach to fitting these data.

In several cases (apparent) diameter variations over time are detected. Solving for a nightly diameter, we have plotted this variation in Fig. 1 for L Car, a cepheid, and the long period variable o Ceti. Also, in at least two cases, a convincing variation of apparent diameter with position angle, contradicting circular symmetry, is detected.

2 Reduced zero-baseline power

In a number of interesting cases a portion of the "stellar" emission comes not from the surface of the star, but from elsewhere in the telescopes' field-of-view. If that emission, whose luminosity is L_2, is overresolved at the baselines at which the star is observed, then the measured visibility will be reduced by the factor $L_1/(L_1 + L_2)$ where L_1 is the luminosity due to the photospheric emission itself. This (over-)simplified model may well approximate Mira's and other stars with extended molecular atmospheres.

By extrapolating the observed visibility curve to zero spatial frequency, we can often detect a "zero-baseline power" significantly reduced from unity. This is illustrated in Fig. 1 for observations of o Ceti at one epoch; the visibility must approach unity at zero spatial frequency, but at medium spatial frequencies we consider the circumstellar emission to be completely resolved, which reduces the star's visibility by the above cited factor. The algorithm attempts to measure this quantity routinely, from which we have obtained a number of apparent detections. Thus we obtain an estimate of the relative circumstellar emission, L_2/L_1 for the following stars: α Ori .065, o Ceti .53, R Aqr .04 (with up to .18 resolved using the longer baseline), R Leo 1.1, R Lep .19, R Scl .04, RR Aql .16, and α PsA, .015. These results were all reported with rms error bars of 15% to 25%, and in several cases anomalies in the visibility fits and/or pulsation of the star over the observational period call the numerical accuracy of the results further into question. We believe that these are actual detections of circumstellar emission, whereas a number of other formally significant results obtained remain questionable at best.

3 Observations past the first visibility null

13 stars were observed at baselines beyond the first visibility null. Among these, 5 had visibility points consistent with a uniform disk visibility curve, and whose visibility offset from that curve is within .01 (α Cma) or .005 (α Cen B, 2 Cen, α Hya, λ Vel). These data for α Cen B are shown in Fig. 1. Three stars, ψ Phe, W Hya and α Ceti, had visibilities clearly lower than the uniform disk curve indicating limb-darkening; additionally RX Lep and σ Lib had formally significant visibility reductions of less than .01 which we treat cautiously. The interpretation of visibilities of R Aqr obtained beyond the first null were complicated by the lack of a simple diameter solution for this star, while the visibilities of α Sco and α Ori past the first null were anomalous.

Towards the Interferometric Imaging of Red Supergiants

Hans-Günter Ludwig[1] and Jacques Beckers[2]

[1] Lund Observatory, Lund University, Box 43, 22100 Lund, Sweden
hgl@astro.lu.se
[2] Astronomy and Astrophysics Department, University of Chicago, U.S.A

To promote efforts to characterise stars beyond simple limb-darkened circular disks we derived interferometric observables expected for red supergiants with emphasis on *small scale surface structure*. Based on detailed radiation-hydrodynamcis models of Freytag et al. (AN 323, p.213, 2002) we calculated visibilities and closure phases. Here, we present the results for a typical instance in time encountered during the temporal evolution of the convective flow (see Fig. 1). Absolute model dimensions have been scaled to approximately match the angular diameter of the red supergiant α Sco (Antares). Antares has a blue, unresolved bright companion (separation 2.9 as) which in the future should allow phase tracking. Visibilities and closure phases carry clear signatures of deviations from circular symmetry. Observations of Antares and other supergiants will enable the exploration and comparison of various imaging techniques. Moreover, such observations are relevant for validating present models, and for pushing further model developments.

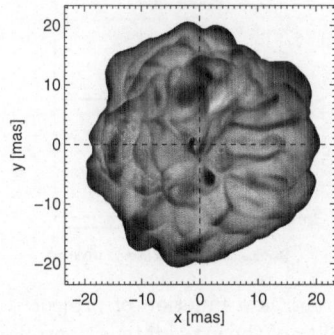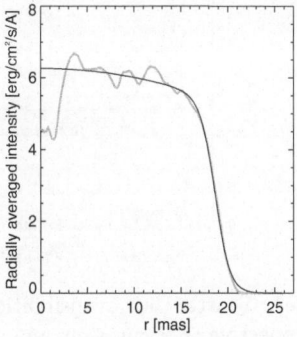

Fig. 1. Left: Snapshot of the approximate monochromatic intensity at 2.1 μm taken during the evolution of the model. The model exhibits substantial deviations from spherical symmetry and shows small scale, convection-related features which evolve on a time scale of ≈ 1 month. Right: Radially averaged intensity profile (grey, thick line) fitted by an expression of the form $a_0 \left[1 - (a_1 r)^2 - (a_2 r)^4 \right] \tanh \left[a_3 (r - a_4) \right]$ (black, thin line). The analytical expression provides a close match to the profile at the stellar limb. The fitted profile serves as reference in later figures.

Main findings: i) convection-related surface structures are indeed observable (Fig. 2) ii) visibilities are influenced by deviations from axisymmetry already close to the first null (Fig. 3, left panel) iii) closure phases are commonly not zero or $\pm\pi$ (Fig. 3, right panel).

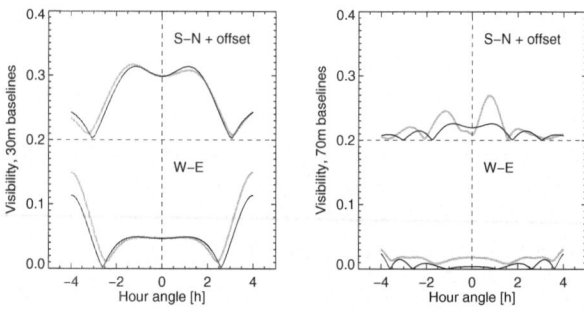

Fig. 2. Left: Visibilities for two 30 m baselines oriented S→N and W→E as a function of hour angle. The star is assumed to be located at declination $-26.4°$, the observing site at latitude $-24.6°$. The visibility of the snapshots is depicted in grey, of the average profile in black. Right: same as left panel but for 70 m baselines. The visibility of the S→N baseline shows an asymmetric behaviour with respect to hour angle zero.

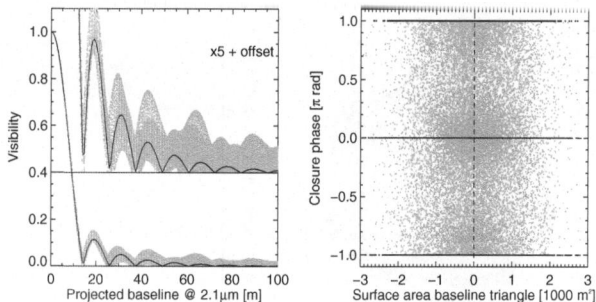

Fig. 3. Left: Scatter-plot of visibilities (grey) as a function of projected baseline length irrespective of orientation, and visibility of the fitted intensity profile (black). The small scale features lead to a significant scatter around the average visibility beyond the first null, and dominate the signal beyond the third null. Right: Scatter-plot of closure phases of 20 000 random baseline triangles with a maximum linear extension of 100 m as a function of their (signed) surface area. The statistical distributions related to the model snapshots (grey) shows clear deviations from the axisymmetric case (black).

Interferometric Aperture Synthesis of Altair: Gravity Darkening and Inclination Angle

A. Domiciano de Souza[1], P. Kervella[2], S. Jankov[3,4], F. Vakili[3], N. Ohishi[5], T. E. Nordgren[6], and L. Abe[5]

[1] Max-Planck-Institut für Radioastronomie, Auf dem Hügel 69, 53121 Bonn, Germany adomicia@mpifr-bonn.mpg.de
[2] LESIA, UMR 8109, 5 place Jules Janssen, 92195 Meudon Cedex, France
[3] LUAN, UMR 6525, UNSA, Parc Valrose, 06108 Nice Cedex 02, France
[4] Astronomical Observatory Belgrade, MNTRS 1940, Volgina 7, 11050 Beograd, Serbia and Montenegro
[5] National Astronomical Observatory of Japan, 2-21-1 Osawa, Mitaka, Tokyo 181-8588, Japan
[6] Department of Physics, University of Redlands, 1200 East Colton Avenue, Redlands, CA 92373, USA

Summary. We perform a physically consistent analysis of all interferometric data available for Altair in order to estimate two important parameters for this rapid rotator : (1) a gravity-darkening compatible with the expected value for hot stars ($T_{\rm eff} \propto g^{0.25}$; von Zeipel effect) and (2) an intermediate inclination for the rotation axis ($i = 55° \pm 8°$ for our best model).

1 Introduction and observations

Although Altair (α Aql, HR 7557, HD 187642) is known to be a rotationally flattened star, a self-consistent determination of the fundamental and apparent parameters related to its rapid rotation has yet to be performed.

Several interferometric observations of Altair exist. They include squared visibilities V^2 in the H and K bands from VLTI/VINCI (Domiciano de Souza et al. 2005; see also Domiciano de Souza in these proceedings), V^2 in the K band from PTI (van Belle et al. 2001), and V^2, triple amplitudes, and closure phases in the visible from NPOI (Ohishi, Nordgren & Hutter 2004).

2 Results

Figure 1 shows the results of our physically consistent analysis of these rich data set using our interferometry-oriented model for fast rotators. This model includes Roche approximation, limb-darkening from Claret (2000), and a von Zeipel-like gravity-darkening law: $T_{\rm eff} \propto g^{\beta}$, with $\beta = 0.25$ for hot stars with radiative external layers (von Zeipel 1924) and $\beta = 0.08$ for cold stars with convective external layers (Lucy 1967). Further details on the model are given by Domiciano de Souza et al. (2002) and the complete description of this analysis of Altair's data is given by Domiciano de Souza et al. (2005).

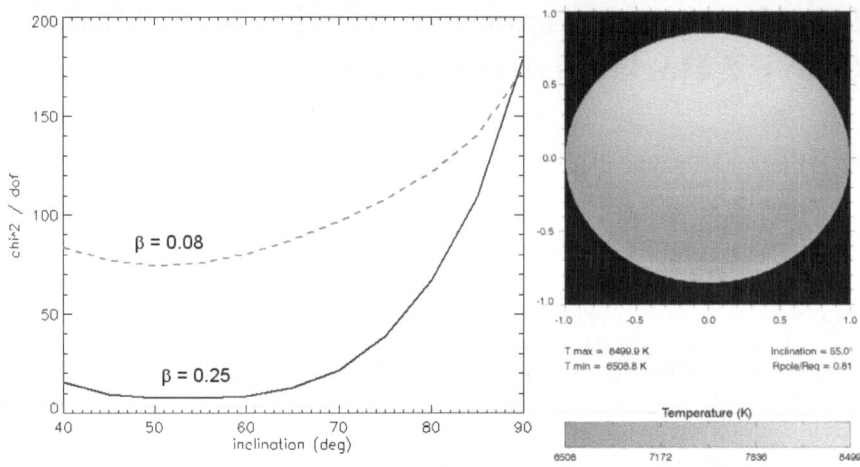

Fig. 1. *Left*: Reduced χ^2 (χ^2/dof) as a function of the stellar inclination i computed from all interferometric observations of Altair for the two test models (radiative and convective limits for the gravity-darkening; $T_{\mathrm{eff}} \propto g^{\beta}$). These values correspond already to the best equatorial angular diameter and major axis orientation for a given i. Clearly, models with $\beta = 0.25$ (solid curve) are preferred compared to models with $\beta = 0.08$ (dashed curve). The minimum χ^2/dof ($\chi^2_{\min}/\mathrm{dof} = 7.3$) is obtained for $\beta = 0.25$ and $i = 55° \pm 8°$. *Right*: Derived effective temperature map corresponding to $\chi^2_{\min}/\mathrm{dof}$. Note that 3 main effects are present in this best model: the stellar flattening (ratio between the equatorial and polar radii $R_{\mathrm{eq}}/R_{\mathrm{p}} = 1.24$), the stellar inclination ($i = 55° \pm 8°$), and the gravity-darkening ($T_{\mathrm{eff}} \propto g^{0.25}$).

References

1. Claret, A. 2000, A&A, 363, 1081
2. Domiciano de Souza, A., Vakili, F., Jankov, S., Janot-Pacheco, E. & Abe, L. 2002, A&A, 393, 345
3. Domiciano de Souza, A., Kervella, P., Jankov, S., Vakili, F., Ohishi, N., Nordgren, T. E. & Abe, L. 2005, A&A, 442, 567
4. Lucy, L.B. 1967, Z.Astrophys., 65, 89
5. Ohishi, N., Nordgren, T. E. & Hutter, D. J. 2004, ApJ, 612, 463
6. van Belle, G. T., Ciardi, D. R., Thompson, R. R., Akeson, R. L. & Lada, E. A. 2001, ApJ, 559, 1155
7. von Zeipel, H. 1924, MNRAS 84, 665

Spectroscopic and Interferometric Tests of Stellar Atmosphere Models: UVES and VINCI Measurements of the M-giant α Cet

V. Roccatagliata[1], M. Wittkowski[1], J. P. Aufdenberg[2], T. Driebe[3], B. Wolff[1], and F. Paresce[1]

[1] ESO, Garching, Germany, vroccata@eso.org
[2] NOAO, Tucson, USA
[3] MPIfR, Bonn, Germany

Summary. We present VLT/UVES spectroscopic and VLTI/VINCI interferometric observations of the cool giant α Cet (M2 III). Spherically symmetric PHOENIX stellar atmosphere models are tested by comparison with our spectroscopic and interferometric observations. The high spectral resolution of UVES allows us to constrain the effective temperature and the surface gravity of the star by comparing observed and model predicted bands that are temperature and gravity indicators. High angular resolution and high precision VLTI/VINCI observations directly measure the strength of the limb darkening effect of α Cet in the K-band. We derive fundamental stellar parameters, namely a Rosseland diameter of 12.08 ± 0.18 mas corresponding to a Rosseland linear radius of $88 \pm 6\,R_\odot$, and an effective temperature of 3805^{+95}_{-109} K, using the Hipparcos parallax and the bolometric flux.

1 Comparison of the PHOENIX models with observations

We initially contrained T_{eff} and θ_{Ross} (3754 ± 134 K, 12.35 ± 0.21 mas) by comparing PHOENIX models [4] with spectrophotometry from the literature [2, 3, 5].

We compared our UVES spectrum with the synthetic spectrum generated by the PHOENIX model with the same parameters that fit the spectrophotometry. Our analysis focused on the singly and doubly ionized metals to constrain the ionization equilibrium and hence T_{eff} and $\log g$. Fig. 1 shows an example of this analysis. Finally, we compared the PHOENIX model predictions to our VLTI/VINCI interferometric observations (Fig. 2), see also [1, 6], and obtained $\theta_{\mathrm{Ross}} = 12.08 \pm 0.18$ mas. Together with the parallax (14.82 ± 0.83 mas) and the bolometric flux ($1.02 \pm 0.14 \cdot 10^{-12}$ Wcm^{-2}), these values correspond to a Rosseland linear radius of $88 \pm 6\,R_\odot$ and an effective temperature of 3805^{+95}_{-109} K.

2 Summary, results and outlook

We derived high precision stellar parameters of α Cet and found a good agreement between the PHOENIX model predictions and the spectrophotometric,

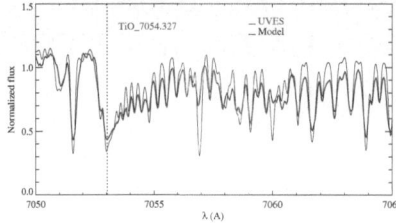 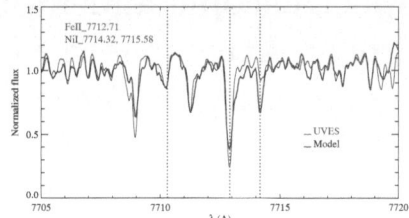

Fig. 1. Examples of surface gravity (FeI & NiI lines) and temperature (TiO band) diagnostics: the UVES spectrum (in red) is in good agreement with the synthetic spectrum (PHOENIX model: T_{eff}=3800 K, $\log g$=1.0, $M/M_\odot$=2.3).

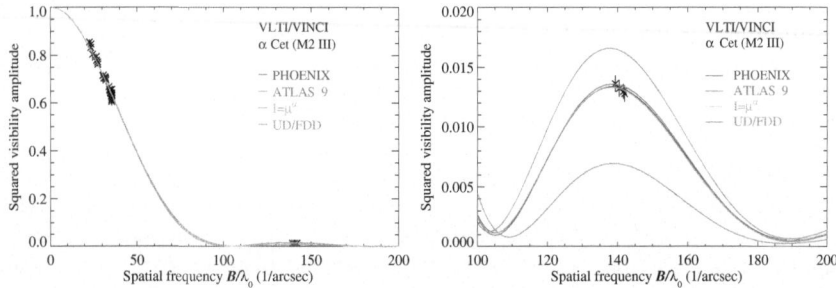

Fig. 2. V^2 of α Cet obtained with VLTI/VINCI, together with best fitting models of a uniform disk (upper grey line), of a fully darkened disk (lower grey line), of I=μ^α with α=0.23 (green line) and of PHOENIX and ATLAS 9 models (red and orange lines). The left panel shows the full range of V^2, while the right panel is an enlargement of the low V^2 in the 2nd lobe. Our measurements are significantly different from UD and FDD models, and consistent with the predictions by PHOENIX and ATLAS 9 models.

spectroscopic and interferometric data. We are going to better constrain the effective temperature and the surface gravity by a more detailed comparison with the UVES spectrum. AMBER and MIDI observations can probe the wavelength dependence of the limb darkening effect.

References

1. J. P. Aufdenberg, P. H. Hauschildt: Proc. SPIE **4838**, 193 (2003)
2. I. N. Glushneva et al.: VizieR Online Data Catalog **3207** (1998)
3. I. N. Glushneva et al., VizieR Online Data Catalog **3208** (1998)
4. P. H. Hauschildt et al.: ApJ **525**, 871 (1999)
5. H. L. Johnson: Communications of the Lunar and Planetary Laboratory **3**, 73 (1965)
6. M. Wittkowski, J. P. Aufdenberg, P. Kervella: A&A **413**, 711 (2004)

The Equatorial Disk at the Center of the Planetary Nebula CPD-568032

O. Chesneau[1], O. de Marco[2], A. Collioud[1], A. Rothkopf[3], A. Zijlstra[4], S. Wolf[3], A. Acker[5], and G. Clayton[6]

[1] Observatoire de la Côte d'Azur-CNRS-UMR 6203, Dept. Gemini, Avenue Copernic, F-06130 Grasse, France Olivier.Chesneau@obs-azur.fr
[2] Department of Astrophysics, American Museum of Natural History, Central Park West at 79th Street, New York, NY 10024, USA
[3] Max-Planck-Institut für Astronomie, Königstuhl 17, 69117 Heidelberg, Germany
[4] Observatoire de Strasbourg, 11 rue de l'Université, 67000 Strasbourg, France
[5] Department of Physics and Astronomy Louisiana State University Baton Rouge, LA 70803, USA

Summary. We present observations of the dusty emission from the planetary nebulae CPD-568032 by VLTI/MIDI. The dusty environment of CPD-568032 exhibits a bright unresolved core and a more diffuse environment. From MIDI acquisition images at 8.7 micron (dominated by PAHs emission), the extension and geometry of the core have been estimated and compared to the geometry of the nebula and the equatorial disk observed by the HST (De Marco et al., 1997 and 2002). The UT2 and UT3 telescopes were used providing projected baselines between 40 and 45 meters. The bright infrared core is almost fully resolved with these baselines although high SNR fringes at low level have been detected. This clear signal reveals a ring structure interpreted as the bright inner rim of the equatorial disk exposed to the flux from the Wolf-Rayet star at the center of the system. These observations bring a new insight of the mechanism at the origin of the dust observed at the center of some asymmetric planetary nebulae.

1 Introduction

Wolf-Rayet ([WC]) central stars of Planetary Nebulae (PN) are Hydrogen deficient stars that exhibit strong ionic emission lines from their dense stellar winds. Among the coolest stars (less than 10^4K) in this group are the [WC10] CPD-568032 (He3-1333) and He2-113 (He3-1044). Their ISO spectra shows the simultaneous presence of C-rich and O-rich dust[1]. De Marco, Barlow & Cohen [2], presented the first direct evidence for an edge-on disk/torus around CPD -568032, as revealed by recent HST/STIS spectroscopy. Their HST spectra show a spatially resolved continuum split into two bright peaks separated by 0".10 and interpreted to be stellar light reflected above and below an obscuring dust disk. From these HST observations, the disk thickness of CPD -568032 is deduced to be 134 AU, which, at 1.35 kpc translates into about 100 mas.

2 MIDI observations

The data have been obtained in open time (Program 073.D-0130) under good atmospheric conditions. With all the baselines the object is very resolved but a low level sinusoidal variation is clearly visible which can immediately be attributed to a binary or ring-like object. The error bars represent a 2s errors dominated by achromatic photometry fluctuations but the chromatic errors are much less than indicated in this graph. The two first visibility spectra have been obtained with almost identical projected baselines and similar PA angle. They differ mostly by a phase change of the sinusoid wave. The third one has been obtained with a slightly less projected baseline with a clear change of PA angle. This last curve is the key one to discard the binary hypothesis (already not probable in view of HST and MIDI acquisition data) and constrain the aspect ratio of the ring-like structure.

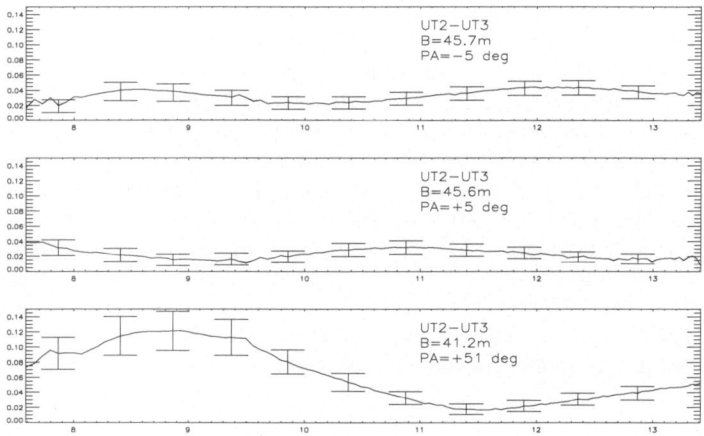

Fig. 1. MIDI dispersed visibilities.

Simple models are under study which provide already satisfactory matches to the data. The proposed object is a dusty (oxygen-rich) disk inclined by about 30°, heavily resolved by the interferometer. The disk geometry consists in a 140 mas diameter circular inner rim and a rapidly decreasing although important flux in the outer regions. The direction of the long axis is consistent with the HST images and spectroscopy.

References

1. Waters, L.B.F.M., Beintema, D.A., Zijlstra, A.A. et al., A&A, **331**, L61 (1998)
2. De Marco, O., Barlow, M.J. & Cohen, M., ApJ, **574**, L83 (2002)

First MIDI Observations of a Be Star: α Ara

A. Meilland[1], O Chesneau[1], T. Rivinius[2], and Ph. Stee[1]

[1] Observatoire de la Côte d'Azur,6203 Avenue Copernic, Grasse, France
anthony.meilland@obs-azur.fr
[2] Landessternwarte Heidelberg, Königstuhl 12, 69117 Heidelberg, Germany

Be stars are hot stars exhibiting hydrogen lines in emission and infrared excess. Optical/infrared and ultraviolet observations of Be stars have been widely interpreted as evidence for two quite distinct regions in their circumstellar environment : a rotating, dense equatorial region, and a diluted polar region which expands with velocities that may reach 2000 km.s^{-1}.

The VLTI/MIDI interferometric observations of α Arae were done during the nights of June, 16th and 17th 2003. We obtained the 8-13μm flux and the visibility modulus for both projected angles. Complementary spectra have also been recorded quasi-simultaneously with the VLTI run in order to know whether α Arae has shown strong emission lines. We observed the star in the J2 band (1.22-1.29μm) at the Observatorio do Pico dos Dias (Brazil). We were able to observe and reduce the Paβ line profile. We also used in our modelling, with some precautions due to the non simultaneity of the other measurements, Hα and Hβ line profile obtained in april 1999 with the HEROS spectrograph at la Silla (Chile). Other HEROS spectra taken over 69 nights in May to July 1999 were used to investigated variability of α Arae Variation of the Hα equivalent width (EW) was also studied using data from 1979 to 1993 taken from [1] and [2]. An ISO spectrum was also used in addition to the 8-13μm VLTI/MIDI Spectral Energy Distribution.

The V/R-ratio of the violet and red peaks of the higher Balmer lines have made one full cycle during the 69 days of the HEROS observations. At the same time, the radial velocity of the emission component of the Balmer lines was changing in a cyclic way as well. Such behaviour, known in a few other Be stars, has been linked to binarity. A search for any spectral contribution of such a hypothetical companion in the phase-binned HEROS spectra of α Arae did not return a positive result. Assuming a mass of 10 $M_\odot$, a 70 day period would give a radius of about 154 $R_\odot$, in the approximation of a circular orbit of a companion with negligible mass. With $R_\odot$= 4.8 $R_\odot$, this corresponds to about 32 stellar radii.

In our modelling of the circumstellar envrionment of α Arae we have used the SIMECA code, which has been already described in [3] and [4]. It computes spectroscopic and photometric observables, but also intensity maps in hydrogen lines and in the continuum in order to obtain theoretical visibility curves which can be directly compared to high angular resolution data.

α Arae is a B3Ve star with a $4.8R_\odot$ stellar radius and a $9.6M_\odot$ mass. Its effective temperature is 18000K and its luminosity is 2200 $L_\odot$. From the fit of the SED we have estimated that the distance of the star should be 105 parsecs. In order to obtain the general envelope parameters, the Hα and Hβ line profiles are fitted simultaneously. With the parameters reproducing well the shape of these 1999 line profiles, the modelled Paβ line profile has a greater intensity than actually observed in 2003. In the following, we assumed that the only parameter that might have changed between 1999 and 2003 is the density at the base of the envelope. In fact, with a base density decreased by about 25%. The parameters obtained by the above modelling of the emission line profiles, and the density required to fit the 2003 Paβ line, can now be used to compute the expected visibility curves for a distance of 105 pc, as estimated from the fit of the SED. The resulting visibility curves clearly shows that the modelled envelope should be well resolved at 79m and 102m (V = 0.63) whereas the VLTI/MIDI, without any doubt have hardly resolved the target. To obtain simultaneously an unresolved envelope and quite strong emission line, and assuming that the distance is well determined, we have computed a model with a truncated disc. The radius where the truncation occurs was set to $22R_*$.

From our model fitting we obtained an inclination angle of about 45° and we derive the envelope position in the sky from the polarization measurement (PA=172°). The truncation of the disc at $22R_*$, needed to fit the visibility curve, is fully compatible with the possible presence of a companion at $32R_*$ with a period of 70 days and a mass less than $2M_\odot$. The envelope of α Arae seems to be confined within the orbit of the companion. The density of the disc must be large enough to produce the strong emission lines. At the base of the photosphere, and assuming the hypothesis of the truncated disc, the density is about 10^{-11} g.cm^{-3}. The terminal velocity of the wind at the pole is 2000 km.s^{-1} , whereas at the equator this value falls to 170 km.s^{-1}. The envelope is nearly spherical and we estimated that the opening angle is about 160°.

α Arae was observed with VLTI/AMBER during two SDT nights with 2 and 3 telescopes. There is no doubts that this AMBER data, which are under reduction, will be very useful to constrain the circumstellar environment of this Be star.

References

1. Dachs, J., Rohe, D. & Loose, A.S., 1990 A&A, 238, 227
2. Hanuschik, R.W., Hummel, W., Dietle, O., Sutorius, E. 1995 A&A, 300, 163
3. Stee, Ph. & de Araùjo, F.X., 1994 A&A, 292, 221
4. Stee, Ph., de Araùjo, F.X., Vakili F. et al. 1995 A&A, 300, 219

Probing the Outer Atmosphere of Mira Variables and the Effects of Chemical Composition on the Mid-Infrared Visibility

K. Ohnaka[1], T. Driebe[1], K.-H. Hofmann[1], Th. Preibisch[1], D. Schertl[1], G. Weigelt[1] and M. Wittkowski[2]

[1] Max-Planck-Institut für Radioastronomie, Bonn, Germany
   kohnaka@mpifr-bonn.mpg.de
[2] European Southern Observatory, Garching, Germany

1 Introduction

Intermediate- and low-mass stars experience slow, but massive mass loss, while moving along the asymptotic giant branch (AGB) during the late stage of their evolution. Mira-type AGB stars are of particular interest, because they exhibit complicated physical and chemical processes such as large-amplitude stellar pulsation, shock waves, grain formation, and momentum transfer from photons to dust grains. Our MIDI observations of the oxygen-rich Mira variable RR Sco have revealed that optically thick emission from dense, warm molecular layers (H_2O and SiO) as well as dust thermal emission can make the N-band angular size more than twice as large as that in the K band (Ohnaka et al. [1]). Here we present the results of MIDI observations of Mira variables RS Lib and RT Sco. While RS Lib is an oxygen-rich Mira variable, RT Sco is classified as an S-type Mira variable. S stars are characterized by C/O ratios close to unity, which means that neither oxygen-bearing molecules nor carbon-bearing molecules are abundant in their photosphere. Such peculiar chemical composition is expected to affect the structure of the outer atmosphere, where molecule/dust formation takes place. Therefore, a comparative study on Mira variables with different chemical compositions is useful for better understanding the physical properties of the outer atmosphere as well as of the innermost region of the dust shell.

2 MIDI observations

RS Lib was observed with MIDI (prism mode, $\lambda/\Delta\lambda \simeq 30$) using UT2 and UT3 at two different epochs: 2004 April 09 and 11 (phase $\sim$0.2–0.3) and 2004 July 29 (phase $\sim$0.7–0.8). The projected baseline lengths range from 45 to 46 m. Three and two data sets were obtained in April 2004 and July 2004, respectively. We fitted observed visibility points with a uniform-disk at each wavelength, and the resulting uniform-disk diameters are plotted in Figs. 1a and 1b as a function of wavelength.

RT Sco (S-type) was observed on 2004 June 05 and 06 (near maximum light) using the grism mode with $\lambda/\Delta\lambda \simeq 230$. These observations were carried out in the framework of the Science Demonstration Time of the MIDI instrument. The projected baseline lengths ranged from 96 to 102 m, using UT1 and UT3. The uniform-disk diameters derived from these MIDI observations are plotted in Fig. 1c.

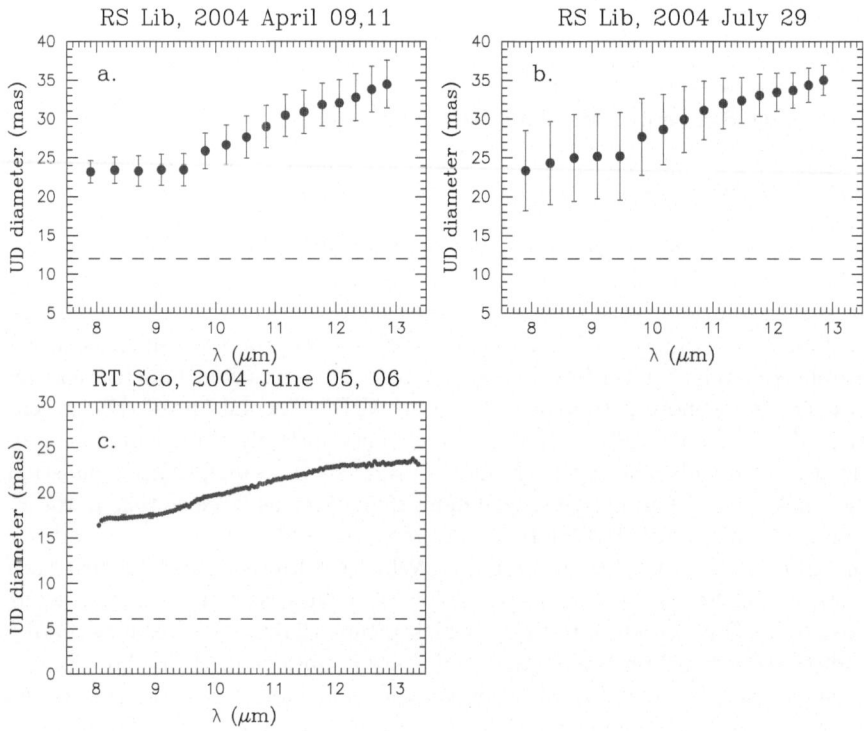

Fig. 1. a: Uniform disk diameters observed for RS Lib in April 2004. **b:** Uniform disk diameters observed for RS Lib in July 2004. **c:** Uniform disk diameters observed for RT Sco (S-type Mira) in June 2004.

Reference

1. Ohnaka, K., Bergeat, J., Driebe, T., et al.: A&A, **429**, 1057 (2005)

Temporal Variation of the Warm Molecular Layers around the Mira Variable RR Sco Detected with the VLTI/MIDI Instrument

K. Ohnaka[1], T. Driebe[1], K.-H. Hofmann[1], D. Schertl[1], G. Weigelt[1] and M. Wittkowski[2]

[1] Max-Planck-Institut für Radioastronomie, Bonn, Germany
   kohnaka@mpifr-bonn.mpg.de
[2] European Southern Observatory, Garching, Germany

1 Introduction

Infrared spectroscopic observations of asymptotic giant branch (AGB) stars with the Infrared Space Observatory (ISO) have revealed the existence of a quasi-static, warm, dense molecular envelope close to the star (e.g., Tsuji et al. [4], Yamamura et al. [5], Cami et al. [1], Matsuura et al. [2]). Although this warm molecular envelope is most likely to play an important role in mass loss, its formation mechanism is not yet understood. The first mid-infrared spectro-interferometric observations of the Mira variable RR Sco (period 281 days, distance 320 pc) using MIDI have revealed that optically thick emission from dense, warm molecular layers (H_2O and SiO) as well as dust thermal emission can make the N-band angular size more than twice as large as that in the K band (Ohnaka et al. [3]). The physical properties of the warm molecular layers derived from the MIDI observations ($T_{mol} \sim 1400$ K, $R_{mol} \sim 2.3\ R_\star$, H_2O column density $\simeq 3 \times 10^{21}$ cm^{-2}, SiO column density $\simeq 10^{20}$ cm^{-2}) have turned out to be consistent with those derived for other Mira variables from ISO observations. Here we present the results of MIDI observations of RR Sco at the second epoch.

2 MIDI observations

RR Sco was observed with MIDI (prism mode, $\lambda/\Delta\lambda \simeq 30$) in 2004 July as part of the Open Time proposal (P.I.: K. Ohnaka, 073.D-0347(A)). We used two different MIDI data reduction packages, MIA developed in the Max-Planck-Institute für Astronomie and EWS developed at the Leiden Observatory, and took the average of the visibilities obtained with both reduction packages.

Figure 1 shows the results of MIDI observations at the second epoch (phase 0.0). The wavelength dependence of the uniform-disk diameter observed at phase 0.0 reveals a marked difference compared to the observation at phase 0.6 presented by Ohnaka et al. ([3]). The uniform-disk diameters at phase 0.0 are ∼40% larger than that observed at phase 0.6. The wavelength dependence of the uniform-disk diameter observed at phase 0.0 shows a monotonic increase from 8 to 13 μm, while that observed at phase 0.6 is characterized by the constant part between 8 and 10 μm and a gradual increase longward of 10 μm. However, it should be noted that the visibilities at the second epoch were obtained with baseline lengths ranging from 42 to 47 m, while the data at the first epoch were obtained with baseline lengths of 74–100 m. This large difference in the baseline lengths makes the intepretation of the data less straightfoward, and it is necessary to carry out model calculations in order to derive physical properties of the warm H_2O +SiO layers and their temporal variation between maximum and minimum light.

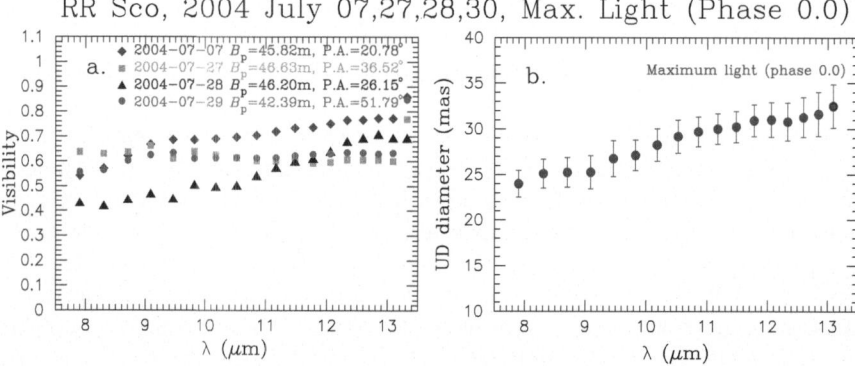

Fig. 1. a: Visibilities measured at four different projected baseline lengths are plotted as a function of wavelength. The errors of the observed visibilities are typically ±10–20%, but the error bars are omitted in this panel for the sake of visual clarity. **b:** Uniform-disk diameter as a function of wavelength. The diameters are derived from uniform-disk fits using all four visibility data points at each wavelength.

References

1. Cami, J., Yamamura, I., de Jong, T., et al.: A&A, **360**, 562 (2000)
2. Matsuura, M., Yamamura, I., Cami, J., Onaka, T., & Murakami, H.: A&A, **383**, 972 (2002)
3. Ohnaka, K., Bergeat, J., Driebe, T., et al.: A&A, **429**, 1057 (2005)
4. Tsuji, T., Ohnaka K., Aoki, W., & Yamamura, I.: A&A, **320**, L1 (1997)
5. Yamamura, I., de Jong, T., & Cami, J.: A&A, **348**, L55 (1999)

N-Band Observation of the Silicate Carbon Star IRAS08002-3803 (Hen 38) with VLTI/MIDI

K. Ohnaka[1], T. Driebe[1], K.-H. Hofmann[1], Th. Preibisch[1], D. Schertl[1], G. Weigelt[1] and M. Wittkowski[2]

[1] Max-Planck-Institut für Radioastronomie, Bonn, Germany
   kohnaka@mpifr-bonn.mpg.de
[2] European Southern Observatory, Garching, Germany

1 Introduction

Silicate carbon stars are a puzzle to date, because they exhibit prominent silicate emission features at ~10 and ~18 μm despite their carbon-rich photospheres. Since their discovery in the IRAS LRS by Little-Marenin [1] and Willems & de Jong [5], several scenarios have been proposed for these peculiar objects, and the scenarios widely accepted at the moment suggest that silicate carbon stars have a companion, possibly a main-sequence star, and that oxygen-rich material is shed by mass loss when the primary star was an M giant and this oxygen-rich material is stored in a circumbinary disk (Morris [3]; Lloyd-Evans [2]) or in a circumstellar disk around the companion (Yamamura et al. [6]) until the primary star becomes a carbon star. Observations with high spatial resolution within the silicate emission feature in the mid-infrared would be the most direct approach for investigating the dust environment around silicate carbon stars. VLTI/MIDI provides us with an excellent opportunity to directly study the circumstellar environment of silicate carbon stars in the 10 μm region, exactly where silicate emission from the oxygen-rich reservoir is located. Here we present the results of MIDI observations of the silicate carbon star Hen 38 (IRAS08002-3803).

2 MIDI observations

Hen 38 was observed with MIDI on three consecutive nights in February 2004 within the framework of the Science Demonstration Time (SDT) program. Four data sets were obtained using the 47 m baseline between the telescopes UT2 and UT3. Figure 1a shows the calibrated visibilities of Hen 38 derived from the four data sets in the wavelength region between 8 and 13 μm. Our observation of Hen 38 has spatially resolved the reservoir of oxygen-rich dust around a silicate carbon star for the first time. We also extracted the *N*-band spectrum of Hen 38 from the MIDI data, and the absolutely calibrated spectrum of Hen 38 is plotted in Fig. 1b. The good agreement with IRAS LRS

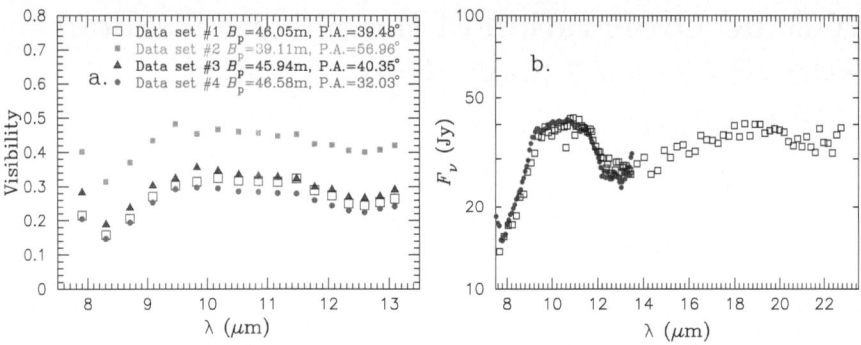

Fig. 1. Left: Visibilities measured at four different projected baseline lengths are plotted as a function of wavelength. The errors of the observed visibilities are typically ±10–15%, but the error bars are omitted in this panel for the sake of visual clarity. **Right:** Observed spectra of Hen 38. The MIDI spectrum is plotted with the filled circles, while the IRAS LRS is plotted with the open squares.

illustrates the stability of the silicate emission feature in the last 21 years. Figure 1 reveals that all visibilities derived from the four data sets show a distinct wavelength dependence: a steady increase from 8 to ∼10 μm and a nearly constant part longward of 10 μm.

These N-band visibilities are totally unexpected for an object showing pronounced silicate emission. The N-band visibility expected for such an object is characterized by a rather steep decrease from 8 to 10 μm and a gradual increase longward of 10 μm, which shows marked contrast to the visibilities observed for Hen 38. Model calculations with our Monte Carlo radiative transfer code show that neither spherical shell models nor axisymmetric disk models consisting of silicate grains alone can simultaneously explain the observed wavelength dependence of the visibility and the observed spectral energy distribution (SED). We propose that the circumstellar environment of Hen 38 consists of two grain species with different spatial distributions: silicate and a second grain species, for which we consider amorphous carbon, corundum, metallic iron, and large silicate grains. The details of our models are described in Ohnaka et al. [4].

References

1. Little-Marenin, I. R.: ApJ, **307**, L15 (1986)
2. Lloyd-Evans, T.: MNRAS, **243**, 336 (1990)
3. Morris, M.: PASP, **99**, 1115 (1987)
4. Ohnaka, K., Driebe, T., Hofmann, K.-H., et al.: A&A, **445**, 1015 (2006)
5. Willems, F., & de Jong, T.: ApJ, **309**, L39 (1986)
6. Yamamura, I., Dominik, C., de Jong, T., Waters, L. B. F. M., & Molster, F. J.: A&A, **363**, 629 (2000)

Preparing Observations of π^1 Gruis Dust Shell with the VLTI/AMBER

S. Sacuto[1] and P. Cruzalèbes[1]

Observatoire de la Côte d'Azur-CNRS-UMR 6203, Dept. Gemini, Avenue Copernic, F-06130 Grasse, France stephane.sacuto@obs-azur.fr, pierre.cruzalebes.@obs-azur.fr

Summary. The aim of this work is to help for choosing the configuration of the VLTI for the observation of dust shell inner boundary of the S star π^1 Gru. We show that this depends on the choice of the used geometrical model. Different models are compared, based on the output intensity distribution produced by the DUSTY spherical transfer radiative numerical code fitting wide-band spectro-photometric data. The JMMC-ASPRO software is used to calculate the visibility points. The optimal configurations will minimize the error on the inner shell radius.

1 Astrophysical context

S stars are evolved bright giants, cooler than 3500 K and from 10^2 to 10^4 $L_\odot$. They are traditionally considered as transition objects between oxygen-rich M and carbon-rich C stars on the AGB. Irregular variable of type Srb, from 5.4 to 6.7 visual magnitude in 150 days, π^1 Gru is one of the brightest intrinsic S stars having a dust shell [4], distant from 153 pc.

2 Geometrical models of π^1 Gru

2.1 Physico-chemical parameters obtained by DUSTY spectro-photometry fitting

The DUSTY code [3] solves the problem of radiation transport in a circumstellar dusty environment [2]. Given the physical parameters of the star and its dust shell, the code produces a wide-band spectrum, a radial profile and using a fast Hankel transform algorithm yields a model visibility curve. A comparison between the reprocessed flux and the observations strongly fixed τ_{10}=0.010 at 10 μm and the effective central star temperature around 3000 K. We obtain a best fit to the dust emission feature for only 10% silicate grains and 90% graphite. We find a best fit model for a dust condensation temperature of 1700 $\pm$ 100 K. We obtain best fits for a simple r^{-2} distribution corresponding to a uniform dust flow velocity. Finaly a single grain size distribution with a=0.05 μm properly fits the 10 μm silicate feature.

2.2 Thin dust shell geometrical models

If we want to prepare interferometric observations, we need to get an analytical description of the object visibility as close as possible to the expected one. Dirac ring, uniform disk, uniform ring and a combination of a uniform disk and a uniform ring are built in order to represent the intensity distribution of the shell model generated with DUSTY.

3 Optimal configurations vs geometrical models

A χ^2 least square analysis is used to minimize deviations between data given by ASPRO [1] and the geometrical model dependent on one of the most significant parameter in the physical mechanisms which cause the mass loss on the AGB : the inner shell radius. After inserting the intensity distribution model, ASPRO generates visibility points thanks to the UV coverage that is imposed by the choice of the interferometric configurations. We find that

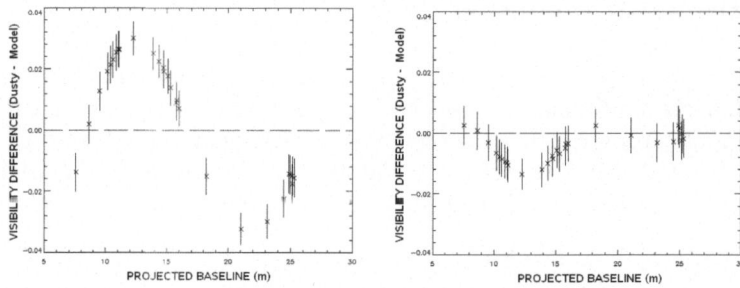

Fig. 1. Example of the visibility difference between measurements and models of Dirac ring and Uniform disk + Uniform ring for the D1-C3-B4 configuration

configurations which minimize error on the inner shell boundary for 2 and 3 ATs are B0-C0 and A1-B2-C2 for the Dirac ring model, B0-C1 and B0-D0-C1 for the uniform disk, B2-C3 and A1-B1-C2 for the uniform ring, A1-C2 and A1-C2-B4 for the combination of uniform disk and uniform ring.

References

1. G. Duvert, P. Berio, F. Malbet: User Manual for ASPRO (2001)
2. Z. Ivezić, M. Elitzur: MNRAS **279**, 1011 (1996)
3. Z. Ivezić, M. Nenkova, M. Elitzur: User Manual for DUSTY (1999)
4. G.R. Knapp, K. Young, M. Crosas: A&A **346**, 175 (1999)

Interferometric Observations of the Mira Star o Ceti with the VLTI/VINCI Instrument in the Near-Infrared

T. Driebe[1], H. C. Woodruff[1], M. Eberhardt[1], K.-H. Hofmann[1],
K. Ohnaka[1], A. Richichi[2], D. Schertl[1], M. Schöller[2], M. Scholz[3,4],
G. Weigelt[1], M. Wittkowski[2], and P. R. Wood[5]

[1] Max Planck Institute for Radioastronomy, Bonn (driebe@mpifr-bonn.mpg.de)
[2] European Southern Observatory, Garching/Santiago
[3] Institute for Theoretical Astrophysics, University of Heidelberg,
[4] Institute of Astronomy, School of Physics, University of Sydney,
[5] Research School of Astron. & Astrophys., Australian Nat. Univ., Weston Creek

Summary. We present K-band commissioning observations of the Mira star prototype o Cet obtained at the ESO Very Large Telescope Interferometer (VLTI) with the VINCI instrument and two siderostats mounted to the VLTI stations EO and GO, forming an unprojected baseline length of 16 m. Rosseland angular radii were derived from the measured visibilities by fitting theoretical visibility functions obtained from center-to-limb intensity variations (CLVs) of different Mira star models, and the phase dependence of the visibility function and the apparent diameter have been investigated. Comparison of the derived Rosseland radii, effective temperatures, and the shape of the observed visibility functions with model predictions suggests that o Cet is a fundamental mode pulsator.

1 Observations

Mira stars are long-period variables which evolve along the asymptotic giant branch (AGB) with well-defined pulsation periods (80-1000 days). Because of their large visual light curve amplitudes, they are easily identified and classified. The change of CLV and spectral type with phase and cycle, as well as the brief time these stars remain in the variable stage, pose interesting problems for observation and modeling.

Although VINCI was designed only for the VLTI commissioning program, it provided enough scientific data to allow investigations of several southern Mira stars, many of them being first-time visibility determinations. o Ceti, the prototype of Mira stars, is an ideal target for infrared interferometry because of its large photospheric size and its relatively small distance from Earth (HIPPARCOS: 107 pc), together with a substantial infrared flux. The observations were carried out between 2001 October and December, in 2002 January and December, and in 2003 January using the test siderostats and the baseline E0-G0 with an unprojected length of 16m (see [5] for details). The measurements obtained at variability phase $\Phi = 0.13$ together with different model fits are shown in Fig. 1.

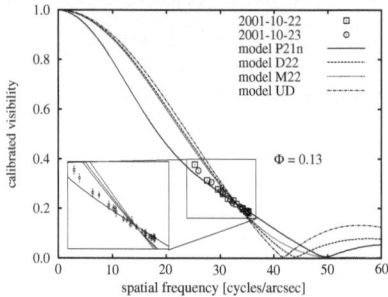

Fig. 1. VINCI measurements of the K-band visibility of o Ceti at phase $\Phi = 0.13$. The inset shows an enlargement of the relevant spatial frequency range. Here, the error bars are included. The various curves show fits with different Mira star models. The model CLVs represented by the solid, dashed, and dotted lines are taken from [4, 1, 2], respectively, while the dash-dotted line shows the CLV corresponding to a simple uniform-disk model. As the figure reveals, the model P21n from [4] shows best agreement with the measurements.

2 Results

As Fig. 1 illustrates, the visibilities measured for o Ceti with VINCI at phase 0.13 (observation date: 22/23 Oct. 2001) are best fitted with the model P21n from [4]. This particular model was calculated for o Ceti at phase 0.2 in the second cycle of computation. When comparing the visibilities measured with VINCI to different atmosphere models of Mira stars, we found that the visibilities and apparent sizes measured for o Ceti are generally best fitted with the P series fundamental mode pulsator models from [4] and [3] (see Fig. 1). Furthermore, we investigated the variation of visibility function, apparent diameter, and effective temperature with phase. We found that the Rosseland angular diameter of o Cet increased from 28.9 ± 0.3 mas ($= 332\pm38$ $R_\odot$ for an adopted distance of $D = 107\pm12$ pc) at $\phi = 0.13$ to 34.9 ± 0.4 mas ($402 \pm 46\,R_\odot$) at $\phi = 0.4$. The error of the Rosseland linear radius almost entirely results from the error of the parallax, since the error of the angular diameter is only approximately 1%. For the effective temperature we found a decrease from $T_{\text{eff}} = 3192 \pm 200$ K at $\Phi = 0.13$ to 2918 ± 183 K at $\Phi = 0.26$. For more details the reader is referred to [5].

References

1. M. S. Bessell, M. Scholz, & P. R. Wood: A&A **307**, 481 (1996)
2. K.-H. Hofmann, M. Scholz, & P. R. Wood: A&A **339**, 846 (1998)
3. M. Ireland, M. Scholz,& P. R. Wood: A&A **352**, 318 (2004)
4. A. Tej, A. Lancon, M. Scholz & P. R. Wood: A&A **412**, 481 (2003)
5. C.H. Woodruff, M. Eberhardt, T. Driebe et al.: A&A **421**, 703 (2004)

High-Resolution Near-Infrared Speckle Interferometry and Radiative Transfer Modeling of the OH/IR Star OH 104.9+2.4

D. Riechers[1], T. Driebe[2], Y. Y. Balega[3], K.-H. Hofmann[2], A. B. Men'shchikov[4], and G. Weigelt[2]

[1] Max Planck Institute for Astronomy, Heidelberg `riechers@mpia-hd.mpg.de`
[2] Max Planck Institute for Radioastronomy, Bonn
[3] Special Astrophys. Observatory, Nizhnij Arkhyz, Karachaevo-Cherkesia, Russia
[4] Institute for Computational Astrophysics, Saint Mary's Univ., Halifax, Canada

Summary. We present near-infrared speckle interferometry of the OH/IR star OH 104.9+2.4 in the K' band obtained with the 6m telescope of the Special Astrophysical Observatory (SAO) in Sep. 2002 and Oct. 2003. At a wavelength of $\lambda = 2.13\,\mu m$ the diffraction-limited resolution of 74 mas was attained. The reconstructed visibility reveals a spherically symmetric, circumstellar dust shell (CDS) surrounding the central star. The visibility function shows that the stellar contribution to the total flux at $\lambda = 2.13\,\mu m$ is less than 30% at all phases, indicating a rather large optical depth of the CDS. To determine the structure and the properties of the CDS of OH 104.9+2.4, radiative transfer calculations using the code DUSTY [1] were performed to simultaneously model its visibility and the spectral energy distribution (SED). Since OH 104.9+2.4 is highly variable, the observational data taken into consideration for the modeling correspond to different phases of the object's variability cycle. This offers the possibility to derive several physical parameters of the central star and its CDS as a function of phase.

1 Observations

The majority of OH/IR stars are long-period variables (LPVs) of variability type Me, extending the sequence of optical Mira variables towards longer periods, larger optical depths, and higher mass-loss rates. As a consequence of their high mass loss, OH/IR stars are surrounded by massive, optically and geometrically thick circumstellar envelopes composed of gas and dust.

For OH 104.9+2.4, a highly dust-enshrouded OH/IR type II-A class star, we obtained visibilities from speckle-interferometric observations with the SAO 6 m telescope on Sep 22, 2002 and Oct 11, 2003 by applying the speckle interferometry method [2]. The measurements were accomplished with a K'-band filter at $\lambda = 2.13\,\mu m$ ($FWHM = 0.11\,\mu m$). Although obtained at different epochs, both visibilities exhibit striking similarity and reveal that the CDS is fully resolved by our measurements (see Fig. 1a). From the 2D-visibilities no major deviation of the CDS from spherical symmetry could be detected.

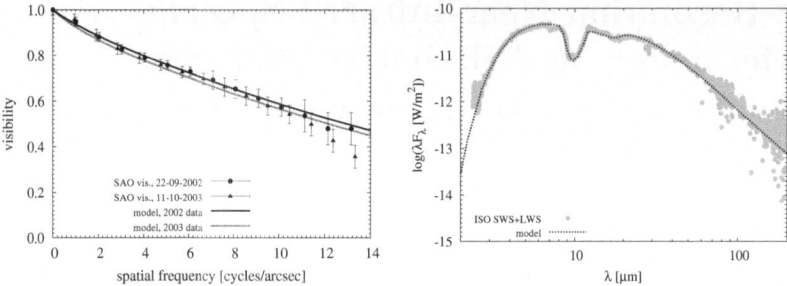

Fig. 1. Comparison of K'-band visibility (**left**) and SED (**right**) from our best-fitting model (solid lines) to the SAO (left) and ISO (right) measurements of OH 104.9+2.4. Different bolometric flux values have been used for the SED and visibility model in order to account for the different epochs/phases of the observations. For more details on the model parameters see [4].

2 Results

Our goal was to simultaneously model the K'-band visibilities and the SED of OH 104.9+2.4 measured at different epochs to determine the temporal change of some physical parameters of the CDS. To accomplish this goal, we used the 1D radiative transfer code DUSTY [1] and calculated several 10^5 models to scan large fractions of the corresponding parameter space.

According to our final model (see also Fig. 1) the effective temperature of the central star increases from $T_{\text{eff}} = 2250$ K at minimum phase ($\Phi = 0.5$) to $T_{\text{eff}} = 3150$ K at maximum phase ($\Phi = 0.0$), while the stellar radius decreases from R $= 730$ $R_\odot$ at $\Phi = 0.5$ to 675 $R_\odot$ at $\Phi = 0.0$. For the CDS, we found that the inner boundary of the dust shell is located at 8.3 $R_\star$ at minimum phase and approximately a factor of two further away at maximum phase ($R_{\text{in}}/R_\star = 17.5$). The optical depth at 2.2 μm decreases from 8.5 to 3.5 between minimum and maximum phase. Our detailed analysis demonstrates the potential of dust shell modeling constrained by both the SED and visibilities obtained from interferometric measurements. For further details on the modeling the reader is kindly refered to [3] and [4].

References

1. Z. Ivezić, M. Elitzur: ApJ **445**, 415 (1995)
2. A. Labeyrie: A&A **6**, 85 (1970)
3. D. Riechers, Y. Y. Balega, T. Driebe et al.: A&A **424**, 165 (2004)
4. D. Riechers, Y. Y. Balega, T. Driebe et al.: A&A **436**, 925 (2005)

Mid-Infrared Long-Baseline Interferometry of the Symbiotic Mira Star RX Pup with the VLTI/MIDI Instrument

T. Driebe[1], K.-H. Hofmann[1], K. Ohnaka[1], D. Schertl[1], G. Weigelt[1], and M. Wittkowski[2]

[1] Max Planck Institute for Radioastronomy, Bonn (driebe@mpifr-bonn.mpg.de)
[2] European Southern Observatory, Garching/Santiago

Abstract We present mid-infrared long-baseline interferometric observations of the symbiotic Mira star RX Pup obtained with the VLTI/MIDI instrument in prism mode within the framework of the Science Demonstration Time (SDT) program in Feb. 2004. Four visibility measurements have been carried out using the unit telescopes UT2 and UT3, with projected baseline lengths ranging from 34.7 to 46.5 m. As we show by means of radiative transfer modeling with the code DUSTY [3], the wavelength dependence of the visibility and the N-band spectrum measured with MIDI can be interpreted as the signature of a circumstellar dust shell which is dominated by silicate dust.

1 Observations

RX Pup belongs to the class of so-called D-type symbiotic Miras [1]. These objects are interacting binary stars with a Mira star as primary component and a hot compact component, presumably a white dwarf, as secondary component. While the optical, UV, and X-ray part of the spectrum are dominated by emission from the hot component, the infrared regime of the SED is governed by the cool Mira star and its circumstellar dust shell. RX Pup exhibited strong brightness changes in the visual and infrared regime during the last decades with a remarkable overall decline in the near-infrared brightness over the last 30 years. For a comprehensive review on RX Pup we refer to [6].

RX Pup was observed with MIDI in prism mode on three consecutive nights in February 2004 within the framework of the SDT program. In total, four observations were carried out using the 89 m baseline between UT2 and UT3. Due to projection effects, the projected baseline lengths range between 34.7 and 46.5 m. As Fig. 1a illustrates, all visibility measurements show a distinct wavelength dependence: A rather steep decrease between 8 and 10 μm, and a shallower monotonic increase longward of 10 μm. For the corresponding uniform disk diameter, this visibility shape translates into a diameter increase by a factor of 2 from 25 to 50 mas between 8 and 10 μm, and an almost wavelength independent diameter between 10 and 13 μm (see Fig. 1b).

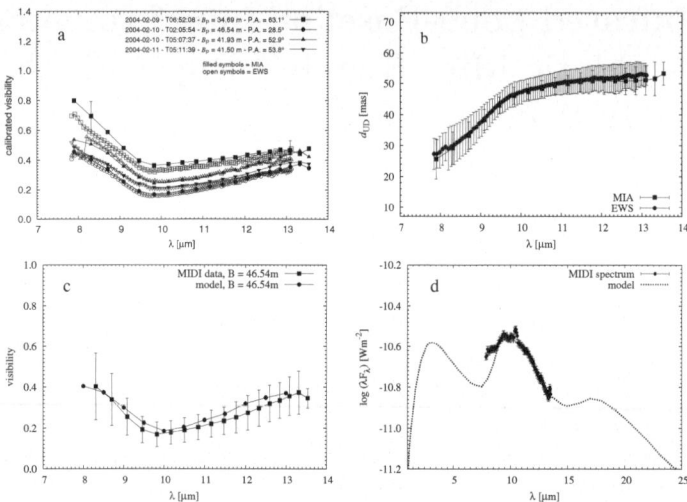

Fig. 1. (a) Calibrated MIDI visibilities of RX Pup. Data represented by filled and open symbols were obtained using the MIA [5] and EWS [4] software package, respectively. For the sake of clarity, only three representative error bars are shown. (b) Uniform disk (UD) diameter of RX Pup as a function of wavelength. The UD diameters were obtained from simultaneous fits of all 4 visibility measurements shown in panel (a) using the MIA and EWS sofare packages. (c,d) Comparison of the calibrated visibility (panel c) and N-band spectrum (panel d) measured with VLTI/MIDI to the results from our best-fitting radiative transfer model. For more details on the modeling and the model parameters see [2].

2 Results

We carried out 1D radiative transfer calculations with the code DUSTY ([3] to model the spectrum and the wavelength dependence of the visibility measured with VLTI/MIDI. We found that this wavelength dependence can be interpreted as the mid-infrared signature of a circumstellar dust shell which is dominated by silicate dust. For both, the visibilities as well as the N-band spectrum we find good agreement between the MIDI measurements and our model (see Fig. 1c,d). For details on the modeling we refer the reader to [2].

References

1. K. Belczyński, J. Mikolajewska, U. Munari et al.: A&AS **146**, 407 (2000)
2. T. Driebe, K.-H. Hofmann, K. Ohnaka et al., AN, **362**, 649 (2005)
3. Z. Ivezić, M. Elitzur: APJ **445**, 415 (1995)
4. W. Jaffe, K. Meisenheimer, H. J. A. Röttgering et al.: Nat. **429**,47 (2004)
5. Ch. Leinert, R. van Boekel, R., L. B. F. M. Waters et al.: **423**,537 (2004)
6. J. Mikolajewska, E. Brandi, W. Hack et al.: MNRAS **305**,190 (1999)

High-Resolution Near-Infrared Speckle Interferometry and Radiative Transfer Modeling of the OH/IR Star OH 26.5+0.6

T. Driebe[1], D. Riechers[2], Y. Y. Balega[3], K.-H. Hofmann[1], A. B. Men'shchikov[4], and G. Weigelt[1]

[1] Max Planck Institute for Radioastronomy, Bonn driebe@mpifr-bonn.mpg.de
[2] Max Planck Institute for Astronomy, Heidelberg
[3] Special Astrophys. Observatory, Nizhnij Arkhyz, Karachaevo-Cherkesia, Russia
[4] Institute for Computational Astrophysics, Saint Mary's Univ., Halifax, Canada

Summary. We present near-infrared speckle interferometry of the OH/IR star OH 26.5+0.6 in the K' band obtained with the 6m telescope of the Special Astrophysical Observatory (SAO) in Oct. 2003. At a wavelength of $\lambda = 2.13\,\mu m$ the diffraction-limited resolution of 74 mas was attained. The reconstructed visibility reveals a spherically symmetric, circumstellar dust shell (CDS) surrounding the central star. In accordance with the deep silicate absorption feature in the spectral energy distribution (SED), the drop of the visibility function to a value of 0.36 at the cutoff frequency indicates a rather large optical depth of the CDS. To determine the structure and the properties of the CDS of OH 26.5+0.6, radiative transfer calculations using the code DUSTY [3] were performed to simultaneously model its visibility and the SED. As in the case of another OH/IR star, OH 104.9+2.4 (see [5] and Riechers et al., this volume), we used these observational constraints at different epochs to derive several physical parameters of the central star and the CDS of OH 26.5+0.6 as a function of phase.

1 Observations

The majority of OH/IR stars are long-period variables (LPVs) of variability type Me, extending the sequence of optical Mira variables towards longer periods, larger optical depths, and higher mass-loss rates. As a consequence of their high mass loss, OH/IR stars are surrounded by massive, optically and geometrically thick circumstellar envelopes composed of gas and dust.

For OH 26.5+0.6, a highly dust-enshrouded OH/IR type II-A class star, we obtained visibilities from speckle-interferometric observations with the SAO 6 m telescope on Oct 15, 2003 by applying the speckle interferometry method [4]. The measurements were carried out with a K'-band filter at $\lambda = 2.13\,\mu m$ ($FWHM = 0.11\,\mu m$). The drop of the visibility down to ~ 0.36 at the cutoff frequency reveals that the CDS is fully resolved by our measurements (see Fig. 1a). Other than in the case of the recent mid-infrared measurements using VLTI/MIDI [1], from the 2D K'-band visibilities no major deviation of the CDS from spherical symmetry could be detected.

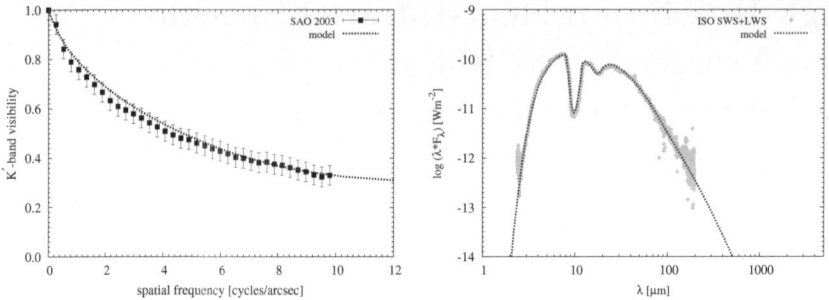

Fig. 1. Comparison of K'-band visibility (**left**) and SED (**right**) from our best-fitting model (dotted lines) to the SAO measurement of OH 26.5+0.6 (left) and the ISO SWS and LWS (right). Different bolometric flux values have been used for the SED and visibility model in order to account for the different epochs/phases of the observations. For more details on the model parameters see text and [2].

2 Results

Our goal was to find a radiative transfer model to simultaneously explain the K'-band visibility measurement from Oct. 2003 and the SED of OH 26.5+0.6 measured at different epochs to determine the temporal change of some physical parameters of the CDS as a function of phase. After deriving the pulsation period ($P = 1560$ d) and the phases of the relevant measurements from a fit of the K-band lightcurve, we used the 1D radiative transfer code DUSTY [3] and calculated several 10^5 models to scan large fractions of the corresponding parameter space.

According to our best-fitting model, OH 26.5+0.6 is surrounded by an optically thick silicate dust shell ($\tau_{9.7\mu m} = 10 \ldots 18$) where the inner shell boundary moves from $R_{in} = 10.1\,R_*$ ($= 29.3$ mas $= 40.1$ AU for an adopted distance of 1.37 kpc) at minimum phase to $R_{in} = 27.0\,R_*$ ($= 69.5$ mas $= 95.2$ AU) at maximum phase. Between minimum and maximum phase, the central star luminosity changes by a factor of 4, and the mass-loss rate increases from $\dot{M} = 2.9 \cdot 10^{-5}\,M_\odot/\text{yr}$ to $\dot{M} = 1.2 \cdot 10^{-4}\,M_\odot/\text{yr}$. For further details the reader is referred to [2].

References

1. O. Chesneau, T. Verhoelst, B. Lopez et al.: A&A **435**, 563 (2005)
2. T. Driebe, D. Riechers, Y. Y. Balega, et al.: AN, **326**, 648 (2005)
3. Z. Ivezić, M. Elitzur: ApJ **445**, 415 (1995)
4. A. Labeyrie: A&A **6**, 85 (1970)
5. D. Riechers, Y. Y. Balega, T. Driebe et al.: A&A **436**, 925 (2005)

Near-Infrared Keck Interferometer and IOTA Closure Phase Observations of Wolf-Rayet stars

J. Rajagopal[1], D. Wallace[1], R. Barry[1], L.J. Richardson[1], W. Traub[2] and W.C. Danchi[1]

[1] NASA Goddard Space Flight Center, Greenbelt, MD 20771 USA
jayadev@iri1.gsfc.nasa.gov
[2] Harvard-Smithsonian Center for Astrophysics, Cambridge, MA 02138 USA
wtraub@cfa.harvard.edu

Summary. We present first results from observations of a small sample of IR-bright Wolf-Rayet stars with the Keck Interferometer in the near-infrared, and with the IONIC beam three-telescope beam combiner at the Infrared and Optical Telescope Array (IOTA) observatory. The former results were obtained as part of shared-risk observations in commissioning the Keck Interferometer and form a subset of a high-resolution study of dust around Wolf-Rayet stars using multiple interferometers in progress in our group. The latter results are the first closure phase observations of these stars in the near-infrared in a separated telescope interferometer. Earlier aperture-masking observations with the Keck-I telescope provide strong evidence that dust-formation in late-type WC stars are a result of wind-wind collision in short-period binaries. Our program with the Keck interferometer seeks to further examine this paradigm at much higher resolution. We have spatially resolved the binary in the prototypical dusty WC type star WR 140. WR 137, another episodic dust-producing star, has been partially resolved for the first time, providing the first direct clue to its possible binary nature. We also include WN stars in our sample to investigate circumstellar dust in this other main sub-type of WRs. We have been unable to resolve any of these, indicating a lack of extended dust. Complementary observations using the MIDI instrument on the VLTI in the mid-infrared are presented in another contribution to this workshop.

1 Introduction

Wolf-Rayet stars are evolved massive stars ($M_i > 20$ $M_\odot$ at solar metallicities and small rotation rates) characterized by distinct emission lines of He, C, and N. These strong lines are the result of powerful winds and heavy mass loss. Some WR stars with strong carbon lines (late WC-type) show evidence of dust formation, while those showing strong nitrogen lines (WN-type) do not. Many of these late WC-type stars are persistent dust producers, while others show episodes of dust formation. Aperture-masking images in the near IR with the Keck I telescope (Tuthill et al. 1999, Monnier et al. 1999) show a spiral pin-wheel structure for the WC stars WR 104 and WR 98a, strongly indicative of dust production at the site of a binary wind-wind interaction.

We initiated an interferometric study of WR stars to better understand their dust production. We present here first results from observations of two WC and three WN stars with the Keck Interferometer as well as preliminary results from a sample of WR stars with the 3 telescope IOTA interferometer in the near IR.

2 WR 140

WR 140 is the prototypical colliding wind binary WC star producing intermittent dust. Careful long term IR photometry has established a period of 7.9 yrs for the dust production and IR maxima (Williams et al. 1990). This work, combined with radial velocity measurements has yielded a spectroscopic orbit for this system. The most recent dust-forming episode was in 2001, during periastron passage (Monnier et al. 2002). The dust production was believed to last for about 4 months. We observed WR 140 at a few epochs in August and October 2003 with the Keck Interferometer. The squared modulus of visibility amplitude (V^2) was measured in the near IR (K band) as part of a shared-risk first observations program (Danchi PI). The Keck Interferometer offers a baseline of $\sim$ 85 meters, with a resolution of $\sim$ 5 mas at this wavelength.

Fig. 1 is the V^2 plot for all the epochs we have on WR 140. The sinusoidal nature of variation in the visibility (diamonds) is clear, showing that we have spatially resolved the binary. The error bars account for both the internal scatter in the data as well as a 5 % systematic error, the nominal value adopted for the Keck Interferometer (Colavita ct al. 2003). The squares are a preliminary fit to a static binary model, where the components of the binary are assumed to be stationary over the period of these observations. Since we have only 6 data points on the curve to solve for three parameters (the separation, position angle and ratio of intensities), we note that the fit is only weakly constrained, with χ^2 showing shallow minima; we intend to explore the parameter space further. The fitted parameters are as below. Our estimate for the position angle is consistent with a recent IOTA (Monnier et al. 2004) measurement of this system, though the separation and intensity ratios differ from their values.

- Separation: 29 ($\pm$1.0) milliarcseconds
- Position angle: 167 degrees
- Intensity Ratio: 3.6 ($\pm$0.6)

3 WR 137

WR 137 is another well known dust-producing WC type star. The dust production sequence is very different from WR 140 with a gradual build up. The binary nature of this star has been an open question with no definite spectroscopic indications (Williams et al. 2001). The differences between these two WC stars are difficult to explain, highlighting the poor understanding of the dust production mechanism. We obtained data at two epochs (in August 2003) on this star with the Keck Interferometer. Fig. 2 shows the large variation in V^2 between the two. We believe that the significant drop in visibility shows the presence of a binary, resolved here for the first time. Unfortunately, given the constraints of the shared-risk program, we were unable to further sample the visibility. The minimum separation is ∼2.6 mas (assuming the first minimum of visibility amplitude to be at 84.7 m baseline).

4 IOTA Data

We are attempting observations of WC type stars with the 3 telescope IOTA interferometer at Mt. Hopkins, Arizona. Most of these objects are at the near-IR sensitivity limit for IOTA (6-7 magnitudes in H), and this is very much a work-in-progress. We have measured closure phases on WR 137, with a telescope configuration with baselines between 18 and 38 m. No significant deviation from zero phase was seen after accounting for instrumental phase.

4.1 WN Stars

We also included 5 WN stars in our sample for comparison. These stars show IR excess, but no evidence for dust formation. Indeed, we were unable to resolve WR 134, WR 135, WR 136 or WR 148. We place an upper limit of ∼ 5 milliarcseconds on their sizes.

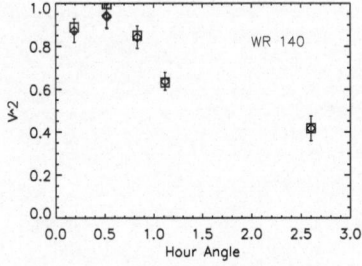

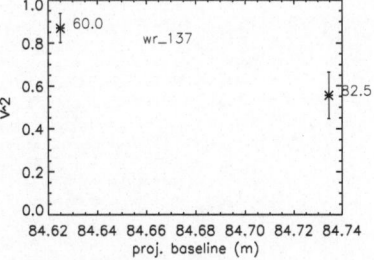

Fig. 1. Observations of WR 140 with the Keck Interferometer in Aug-Nov 2003. Squared visibility is shown by the diamonds, and the squares are the fit to a binary model. The reduced χ^2 is 0.7

Fig. 2. Squared visibility for WR 137. The position angle of the baseline is is given next to each data point

References

1. Colavita, M. et al. : ApJ, **592**, L83 (2003)
2. Monnier, J.D. et al. : ApJ, **602**, L57 (2004)
3. Monnier, J. D., Tuthill, P.G., Danchi, W. C.: ApJ, **567**, L137 (2002)
4. Monnier, J. D., Tuthill, P.G., Danchi, W. C.: ApJ, **525**, L97 (1999)
5. Tuthill, P. G., Monnier, J. D., Danchi, W. C.: Nature, **398**, 487 (1999)
6. Williams, P.M. et al. : MNRAS, **324**, 156 (2001)
7. Williams, P.M. et al. : MNRAS, **243**, 662 (1990)

The Shape of the Inner Rim in Proto-Planetary Disks

Andrea Isella[1,2], Antonella Natta[1], and Leonardo Testi[1]

[1] INAF-Osservatorio Astrofisico di Arcetri, Largo E. Fermi 5, 50125 Firenze, Italy
[2] Dipartimento di Fisica, Universitá di Milano, Via Celoria 16, 20133 Milano, Italy isella@arcetri.astro.it

Summary. We discuss the properties of the inner puffed-up rim that forms in circumstellar disks when dust evaporates. We argue that the rim shape is controlled by a fundamental property of circumstellar disks, namely their very large vertical density gradient, through the dependence of grain evaporation temperature on gas density. Numerical models show that the bright side of the rim is naturally curved, rather than vertical as expected when a constant evaporation temperature is assumed. The rim emits most of its radiation in the near and mid-IR, and provides a simple explanation to the observed values of the near-IR excess (the " $3\mu m$ bump" of Herbig Ae stars). We have computed synthetic images of the curved rim. Face-on rims are seen as bright, centrally symmetric rings on the sky; increasing the inclination, the rim takes an elliptical shape, with one side brighter than the other.

1 Introduction and Model

The interferometric results provide strong support to the idea that the inner disk structure deviates substantially from that of a flared disk: dust evaporation introduces a strong discontinuity in the opacity that results in a "puffed-up" rim at the dust destruction radius. This idea was proposed by Natta et al (2001) and developed further by Dullemond et al. (2001) for Herbig Ae stars, to account for the shape of the near infrared excess of these stars (the "$3\mu m$ bump"). These authors pointed out that the rim had the right properties to explain also the early interferometric results of Millan-Gabet et al. (2001). The observed IR emission depends on the inclination of the rim with respect to the observer and is strongly dependent on the shape of the bright side of rim. In the model of Dullemond et al. (2001) the rim is supposed "vertical", as expected when the dust evaporation temperature is constant. Such a model has the disadvantage that the rim emission vanishes for objects seen face-on, for which the projection on the line of sight of the rim surface is null, and for objects seen edge-on, where the rim obscures its own emission. This is clearly inconsistent with observations of the SED, which show that all the Herbig Ae stars with disks have similar near-IR excess, regardless of their inferred inclination.

Our model for the inner part of passive irradiated flaring disks gives a self consistent solution of the dust opacity structure in the dust evaporation

region. The temperature in the rim atmosphere is determined using the analytical solution of the problem of the radiation transfer as in Calvet et al.
(1992), neglecting the heating term due to the mass accretion. The vertical
structure of the rim is then computed in a way derived from Dullemond et
al. (2001), adding a relation between the dust vaporization temperature and
the gas density as proposed in Pollack et al. (1994). As a result we obtain a
curved model for the bright side of rim (Fig. 1), whose features are described
in detail in Isella and Natta (2005).

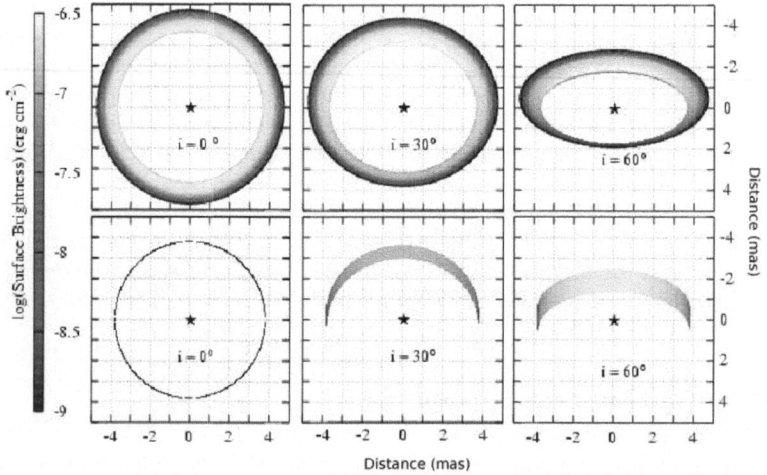

Fig. 1. The upper panels show the images of the curved rim surface, calculated for
silicate grains with radius $a = 1.3\mu m$. The lower panels show the images for the
verical rim calculated for the same grain size. The surface brightness of the rim,
plotted in colors, is computed for a wavelength of $2.2\mu m$. The model is computed for
a stellar temperature of 10000K, luminosity of $50L_{sun}$ and a distance of $d = 144pc$.
The vertical rim for $i = 0°$ has zero surface brightness.

References

1. N. Calvet et al.: Rev. Mexicana Astron. Astrof. 24, 27 (1992)
2. C.P. Dullemond, C. Dominik, A. Natta: ApJ 560, 957 (2001)
3. A. Isella, A. Natta: A&A 438, 899 (2005)
4. R. Millan-Gabet et al.: ApJ 546, 348 (2001)
5. A. Natta, T. Prusti, R. Neri et al: A&A 371, 186 (2001)
6. J.B. Pollack, D. Hollenbach, S. Beckwith, et al.: ApJ 421, 615 (1994)

Dust and Winds from Evolved Stars

P. Mathias[1], B. Lopez[1], J. L. Menut[1], O. Chesneau[1], F. Przygodda[1],
N. Berruyer[1], and S. Wolf[2]

[1] Observatoire de la Côte d'Azur - Dept. GEMINI - F06304 Nice Cedex 4
mathias@obs-nice.fr
[2] Max Planck Institute for Astronomy, Königstuhl 17, D69117 Heidelberg

Summary. Until now, most interpretations have been limited to spherical geometry (or axisymmetric one when a dust torus is seen for instance). However, many measurements (polarisation, high resolution imagery...) have already showed that dust shells may deviate from spherical symmetry. We present here some scientific cases concerning evolved stars. The APreS-MIDI beam combiner (presented elsewhere in these proceedings: Lopez et al.) will allow image reconstruction of these circumstellar environments of different families of evolved stars.

AGB: clumpy environment?

Departures from spherical symmetry are often noted. Pure hydrodynamical collimation provided by dense equatorial disks or torii, and/or magnetohydrodynamical collimation can explain the development of the extreme bipolar geometries observed. In addition to these larges scale structures, clumps of dust are probably present in the environment of AGB stars. For Mira itself, Lopez et al. [4] modelled their observations with clumps of typical size a few stellar radius and located some tens radius away from the central object.

RSG: bipolar outflow?

The wide range of measured inner dust radius may be due to asymmetries in the envelopes caused by supergranulation. The stream velocities in the hotspots could both shock and eject portions of the atmosphere, possibly initiating mass loss [6]. Due to the finite lifetime of the spots, mass loss can be episodic. For instance, VY CMa presents 3 arcs that are kinematically distinct from the surrounding stationary circumstellar material. Their random orientations suggest that they were ejected at different times *and* by localized events from different active regions on the star. In addition, numerous clumps of small knots or condensation are observed relatively close to the star (500 mas) [2].

Post-AGB, RV Tau: geometry of the disk torus

During this short evolutionary phase, stars are pulsationally variable. The presence of a disk seems to be the common ingredient in the best studied individual post-AGB binary stars. Also, most RV Tau stars are thought to have an embedded companion, and the system is inside a circumbinary disk [5]. The measurement of the circumbinary torus with APreS-MIDI will provide decisive information about the inner radius (to be compared with the component separation) and the vertical extent of the torus.

Planetary Nebulae (PN): disk geometry

The ISO spectra of [WC] stars showed the simultaneous presence of C-rich and O-rich dust, which in the context of a single-star scenario, would point to a recent transition between the O-rich and the C-rich chemistries. An alternative scenario envisages these systems as binaries [1], in which the O-rich silicates are trapped in a disk as a result of a past mass transfer event, with the C-rich particles being more widely distributed in the nebula. At the moment, MIDI has observed a few of PN disks but APreS-MIDI is much better suited for the study of these geometrically complex objects. It must be noted, however, that their complex chemistry implies that a minimum spectral resolving power of 3-5 is required.

Symbiotic stars/Novae: role of binarity

Many interaction processes take place in symbiotic binaries. Dust grains can form around the cool component, in the interactive zone between both components, or during a possible thermonuclear runaway of the white dwarf component. Tuthill et al. [7] emphasize that serious inadequacies with both interferometric and IR spectrometric data are revealed, indicating an asymmetric geometry. These unpredictable events, that may occur on very different time scales, implies that the APreS-MIDI observations of Novae have to be considered only in the frame of "Target of Opportunity" observations.

RCrB: localisation of dust clouds formation

Photometric observations of these stars present fadings generally attributed to the formation of a cloud of carbon soot. None of these stars are known to be binaries, and they all seem to be pulsational variables. The localization of the dust forming region is not understood since the pulsation wave should be quite spherical, and the hot photosphere should prevent large scale convection motions. Therefore, the precise localization of dust forming region with APreS-MIDI will provide a deep insight into this problem.

References

1. Cohen, M., Barlow, M.J., Sylvester, R.J., Liu, X.-W., Cox, P., et al. 1999, ApJ, 513, L135
2. Humphreys, R.M., Davidson, K., Ruch, G. 2005, AJ, 129, 492
3. Kato, M., Hachisu, I. 2003, ApJ, 587, L42
4. Lopez, B., Danchi, W.C., Bester, M., Hale, D.D.S., Lipman, E.A., et al. 1997, ApJ, 488, 807
5. Maas, T., Van Winckel, H., Lloyd, Evans T. 2005, A&A, 429, 297
6. Tuthill, P.G., Haniff, C.A., Baldwin, J.E. 1997, MNRAS, 285, 529
7. Tuthill, P.G., Danchi, W.C., Hale, D.S., Monnier, J.D., Townes, C.H. 2000, ApJ, 534, 907

VV CrA - The Dusty Environment of an Infrared Companion

Th. Ratzka, Ch. Leinert, F. Przygodda, and S. Wolf

Max-Planck-Institute for Astronomy, Heidelberg, Germany ratzka@mpia.de

Summary. VV CrA is one of the rare binary systems harbouring a so-called infrared companion (see contribution by Th. Ratzka & Ch. Leinert). While in 1987 this companion dominated at all infrared wavelengths, it is now fainter than the primary even in the N-band. This 'fading' may provide the chance to reveal the underlying properties causing the existence of infrared companions. In the following we will present a preliminary interpretation of the interferometric observations obtained with MIDI that provide insights into the geometry of the dusty environment. They seem to favour a geometrical explanation for the different apppearances of the two components of VV CrA.

1 Properties of a Southern Binary

VV CrA has been classified in the HBC catalogue as a T Tauri star. The spectral type is K7. The binarity of the system was found by J. Frogel (unpublished). Both components are surrounded by circumstellar disks. The north-eastern infrared companion VV CrA NE at a projected separation of 270 AU is deeper embedded in the circumstellar material. This material can be directly seen in the MIDI spectra as silicate absorption (Fig. 1). Both components show strong emission lines [1] that are direct indicators of shocked gas and thus of active accretion. Even jets have been observed.

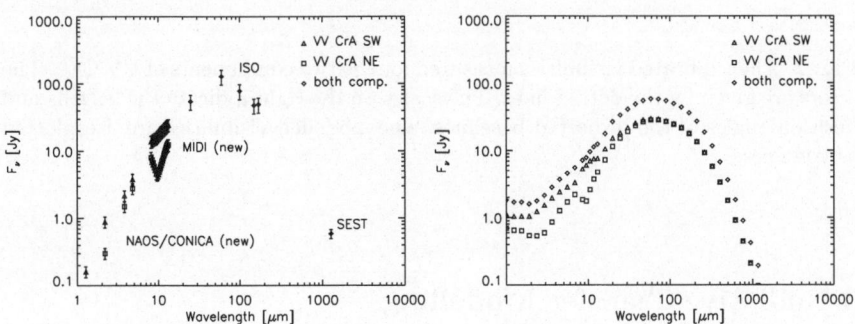

Fig. 1. The measured (left) and the modelled (right) fluxes.

2 Interferometric Observations

We observed separately the components of VV CrA with MIDI at the VLTI within our GTO programme. All observations have been performed in the so-called HIGH SENS mode, i.e. the photometry was recorded after finishing the interferometric measurements. A prism with $\lambda/\Delta\lambda \approx 30$ was used to obtain dispersed visibilities. The calibrated visibilities presented in Fig. 2 are derived by using the software package MIA that analyses the power spectrum (see contribution by R. Köhler). A detailed description of this method can be found in [2].

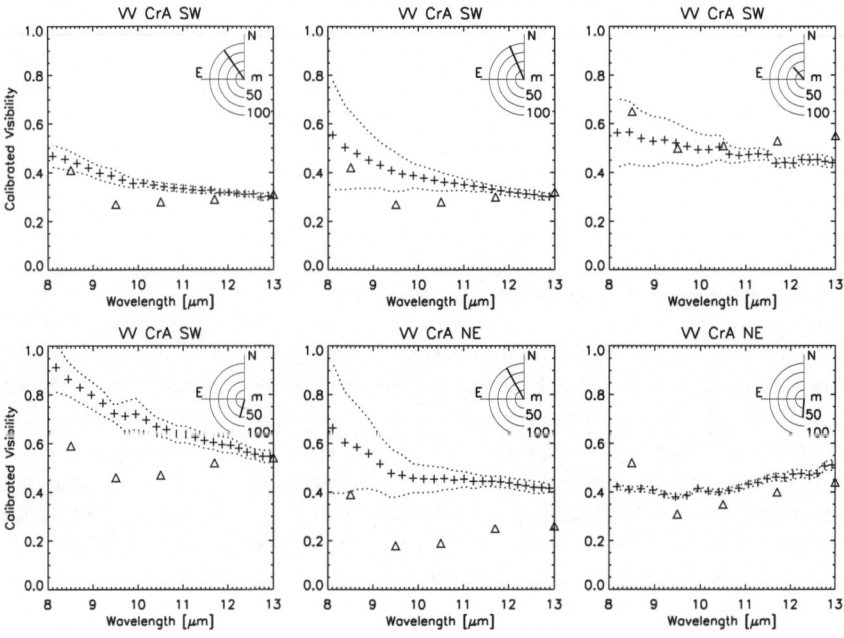

Fig. 2. The calibrated visibilities measured for the two components of VV CrA. The errors are given by the dotted lines. The scales on the right indicate the lengths and position angles of the projected baselines. The modelled visibilities are overplotted as triangles.

3 Radiative Transfer Modelling

The fluxes derived from a MC3D [3] model are shown in the right panel of Fig. 1. This model has been selected, because it does not only fit the spectra of the two components reasonably well when assuming that the fluxes longwards of the N-band are contributed equally by the two stars, but also the visibilities. In this model each of the central sources is represented by a star of 5500 K with a radius $R_* = 4.3\,R_\odot$. These sources are surrounded by disks with a size of 150 AU containing $0.1\,M_\odot$ of dust and gas each. The density ϱ of the disks as function of the cylindrical coordinates (r, z) is given by

$$\varrho(r, z) = \varrho_0 \left(\frac{R_*}{r}\right)^{\frac{15}{8}} \exp\left[-\frac{1}{2}\left(\frac{z}{h}\right)^2\right] \quad \text{with} \quad h(r) = h_0 \left(\frac{r}{R_*}\right)^{\frac{9}{8}}.$$

The scaling factor h_0 was chosen in such a way that $h(100\,\text{AU}) = 30\,\text{AU}$. The only parameter that is different for the two components is the disk inclination (SW: 43°, NE: 45°). Simulated wavelength-dependent images have been used to determine theoretical visibilities. They are overplotted in Fig. 2 after fitting the position angles of the semi-major-axes of the disks (SW: 90°, NE: 40°).

4 The Future

Additional observations and more detailed studies will show whether and how good a geometrical interpretation focusing on the orientation of the disks can explain the differences between the two binary components. New photometric measurements in the visual regime will clarify whether the discrepancy between observations and theory at these wavelengths can be interpreted as extinction (Fig. 1). Furthermore, the planned mm-measurements will allow to determine the amount and spatial distribution of the dust in the system.

References

1. L. Prato, T.P. Greene, M. Simon: ApJ, **584**, 853 (2003)
2. Th. Ratzka: High Spatial Resolution Observations of Young Stellar Binaries. PhD Thesis, Ruperto-Carola University, Heidelberg, Germany (2005)
3. S. Wolf, D.L. Padgett, K.R. Stapelfeldt: ApJ, **588**, 373 (2003)

GENIE: High-Resolution Study of Debris Disks

Olivier Absil[1], Jean-Charles Augereau[2,3], Roland den Hartog[4], Emilie Herwats[1,2], Philippe Gondoin[4], and Malcolm Fridlund[4]

[1] Institut d'Astrophysique et de Géophysique, Université de Liège, Belgium
[2] Laboratoire d'Astrophysique de Grenoble, France
[3] Sterrewacht Leiden, University of Leiden, The Netherlands
[4] ESA/ESTEC, Noordwijk, The Netherlands

1 Simulated observations of the dust disk around ζ Lep

In order to evaluate the scientific potential of GENIE [4] in the context of debris disks, we have simulated future observations of ζ Lep (A2V, 21.5 pc), a prototypical Vega-type star expected to harbour warm dust in the inner part of its debris disk [3].

We have used the 3D debris disk model developed by Augereau et al. [2] to compute synthetic L'-band images of the optically thin dust disk around ζ Lep, using a variety of physical parameters. Both the thermal emission and the scattered light are included in the model. Two types of grains have been considered: smoothed astronomical silicates [7] and cold-coagulation type grains [6]with a porosity up to 90% when the H_2O-dominated ices are sublimated (< 6 AU). The grain size distribution is assumed to follow an $a^{-3.5}$ power law between a minimum and maximum grain size, and the surface density to fade with the distance from the star as $r^{-0.6}$. Sublimation happens at 1700 K, at a distance that depends on the grain composition and size. For each particular grain distribution, we compute the SED of the dust disk and adjust the surface density so that our SED gives the best possible fit to the observed mid-IR excesses. Synthetic images of the disk are then generated for different inclination angles.

The simulated observations are obtained with the GENIEsim software [1], using various AT-AT baselines ranging between 8 and 80 m on the main East-West VLTI track. Our simulations take into account all the expected noise sources, including residual atmospheric turbulence after the control loops. Each individual observation consists in a 1000 s exposure, followed by the observation of a calibrator star with similar H-K-L magnitudes (η Lep, F1V at 15 pc) to evaluate and subtract the contribution of instrumental leakage. The contribution of geometric leakage, due to the finite extent of the stellar disk, is removed analytically, using a surface brightness model developed by [5] to evaluate the stellar diameter (estimated LD diameter: 0.756±0.015 mas). The final output of the nuller is the nulling ratio, measured as the ratio between the constructive and destructive outputs.

2 Influence of the disk parameters on the nulling ratio

The evolution of the sublimation radius as a function of grain size is given in Fig. 1. Large grains are less susceptible to sublimation and are thus more concentrated close to the star. The global SED of the disk is therefore shifted towards shorter wavelengths, inducing a larger flux in the L' band for the same mid-IR excess. This leads to a larger amount of transmitted light in the nulled output. Moreover, the nulling ratio reaches a local maximum when the chosen baseline makes the first bright fringe coincide with the bright inner rim of the disk ($\lambda/2B = r_{in}$), which happens at 80 m for the largest grains and at 24 m for the smallest grains.

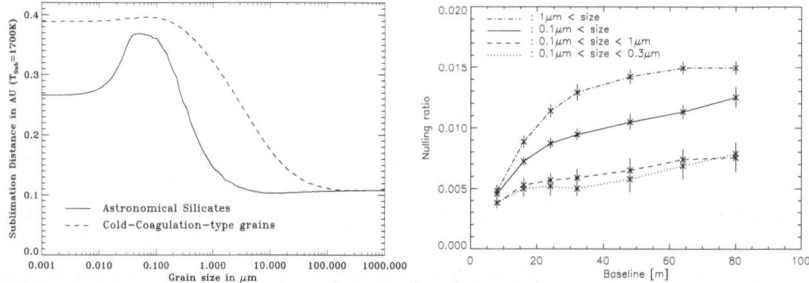

Fig. 1. Left: Sublimation distance of dust grains as a function of their size, for two types of grains around ζ Lep. The heating process is more efficient for small grains, which reach the sublimation temperature further from the star. Right: Influence of the size distribution of dust grains on the observed nulling ratio with GENIE, using different AT-AT baselines. Error bars take into account all noise sources including calibration processes. The disk is supposed to be face-on.

The influence of grain type is investigated in Fig. 2, using either silicate or porous grains. The latter are less effective than silicate grains to re-emit a thermal radiation. A larger density of porous grains is thus required to produce the same mid-IR excess, which leads to an increased L'-band flux with respect to the silicate grains. The nulling curve is thus translated upwards. A second effect comes from the better resistance of silicate grains to sublimation due to their compact nature. The thermal emission is therefore more concentrated close to the star in the case of silicate grains. This can be clearly seen in Fig. 2: the first maximum in the nulling ratio is obtained for a smaller baseline in the case of porous grains. This means that the first bright fringe reaches more rapidly the inner rim of the disk as the baseline is increased, revealing a larger inner rim in the case of porous grains.

Finally, another important effect on the nulling ratio is the orientation under which the dust disk is seen. We have simulated observations of inclined disks, using angles between 10° (highly inclined) and 90° (face-on).

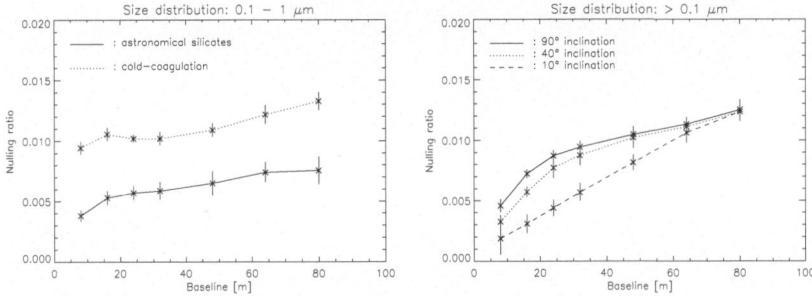

Fig. 2. Left: Influence of the physical composition of the dust grains on the observed nulling ratio. Cold-coagulation grains are highly porous and thus more efficiently heated. The inner rim is therefore shifted outwards in this case. Right: Influence of the disk inclination on the nulling ratio. We have assumed the major axis of the projected dust disk to be aligned with the main AT-AT track in this simulation. Highly inclined disks are more attenuated by the central dark fringe.

The baselines are perpendicular to the major axis of the projected image of the disk (fringe pattern parallel with the major axis). Highly inclined disks are less transmitted by the nuller because most of the disk emission falls onto the dark fringe, especially for short baselines. This effect is clearly evidenced in Fig. 2, showing a great sensitivity of the nuller to inclination. Note that the contribution of scattered light (only a few percent of the total disk emission in L' band) is more extended and thus less affected by inclination.

In conclusion, an L'-band nuller such as GENIE is well adapted to study the thermal emission from the inner parts of debris disks. A few baselines are sufficient to constrain the grain composition and major disk parameters.

References

1. O. Absil, R. den Hartog, C. Erd et al: GENIEsim, the GENIE simulation software. In: *Towards Other Earths*, ed B. Battrick (ESA-SP 539, 2003), 317
2. J.-C. Augereau, A.-M. Lagrange, D. Mouillet et al.: A&A **348**, 557 (1999)
3. S. Fajardo-Acosta, C. Telesco and R. Knacke: AJ **115**, 2101 (1998)
4. P. Gondoin, O. Absil, R. den Hartog et al: Proc. SPIE **5491**, 775 (2004)
5. P. Kervella, F. Thévenin, E. Di Folco, D. Ségransan: A&A **426**, 297 (2004)
6. A. Li, J.I. Lunine: ApJ **590**, 368
7. J.C. Weingartner, B.T. Draine: ApJ **548**, 296

Circumstellar Structures around Herbig AeBe Stars: A Direct Interferometric Insight

S. Antoniucci[1,2], G. Li Causi[1], D. Lorenzetti[1], B. Nisini[1], T. Giannini[1], F. Strafella[3], D. Elia[3], and F. Paresce[4]

[1] INAF - Osservatorio Astronomico di Roma, via di Frascati 33, I-00040 Monteporzio Catone, Italy antoniucci@mporzio.astro.it
[2] Università degli Studi di Roma "Tor Vergata", via della Ricerca Scientifica 1, I-00133 Roma, Italy
[3] Università di Lecce - Dipartimento di Fisica, Via Arnesano, I-73100 Lecce, Italy
[4] ESO-European Southern Observatory, Karl-Schwarzschild-Straße 2,D-85748 Garching bei München, Germany

The geometric distribution of circumstellar matter around young stars more massive than the Sun, i.e. the Herbig AeBe (HAeBe) stars, has great relevance in the evolutionary process of these sources toward the main sequence. According to a widely accepted scenario, the circumstellar matter should be organised in disks in the less massive and more evolved HAe stars, while the earlier HBe should still retain a dusty envelope [4, 3]. Observations of MIDI at VLTI can provide a direct test for this scenario. Indeed, first MIDI results obtained on a sample of HAe stars [2] indicate physical sizes which can be reproduced by disk models, although the match of such models with the MIDI visibility spectrum still needs to be refined.

In this context, we took into account a model consisting of a spherical dusty envelope with cavities [5, 1]. This model is able to reproduce the observed Spectral Energy Distribution (SED) for a large (more than 40 objects) sample of HAeBe stars over a wide range of wavelengths (from UV to radio). We calculated the expected visibility values in the MIDI spectral range for different sources of both Ae and Be types and then compared such visibilities with VLTI-MIDI interferometric data of a few HAe stars, available from the ESO archive. In Fig. 1 and 2 we show the comparison between the observed SED and visibility and the expected ones for the star HD144432. The visibilities predicted by the model are too low with respect to the observations, thus indicating that the spherical envelope model, in spite of the good fit of the SED it provides, fails to reproduce the compact (10-20 mas) structure observed around the source at 10μm.

This conclusion gives an additional support to the current view that Herbig Ae stars are surrounded by circumstellar disks. MIDI observations of Herbig stars of earlier type (HBe), planned for the near future, will be used to check if flattened structures play a major role also for more massive pre-main sequence stars.

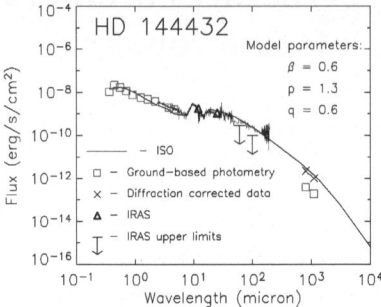

Fig. 1. The observed SED of HD144432 and the relative best fit model. The observations (from UV to radio) include ground-based photometry, IRAS, ISO-SWS and LWS spectrophotometric data. Model parameters are indicated in the figure: b is the dust opacity index, p and q are the spectral indexes of the density and temperature distribution, respectively.

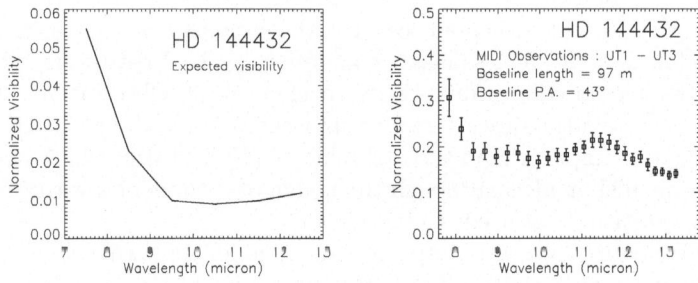

Fig. 2. Comparison between the expected (left) and observed (right) MIDI N-band visibility of HD144432. The predicted visibility was derived from the model which best matches the SED, considering a slit of 0.5x2.0 arcsec and a baseline of 97 m, in order to reproduce the VLTI-MIDI observations. These measurements (ESO Archive) were carried out using the VLT UT1 and UT3; the visibility spectrum is the average between the curves obtained using three calibrators (two observations of υ Lib and one of HD139997), the error bars beeing the standard deviation. The reduction and analysis were performed using the Leiden EWS software (http://www.strw.leidenuniv.nl/~nevec/workshop_2004/packages).

References

1. Elia D., Strafella F., Campeggio L. et al.: ApJ **601**, 1000 (2004)
2. Leinert Ch., van Boekel R., Waters L.B.F.M. et al.: A&A **423**, 537 (2004)
3. Meeus G., Waters L.B.F.M., Bouwman J. et al.: A&A **365**, 476 (2001)
4. Natta A., Grinin V.P., Mannings V. In: *Protostars and Planets IV* ed. by V.Mannings, A.P.Boss and S.S.Russel (Univ. of Arizona Press 2000), p. 559
5. Pezzuto S., Strafella F., Lorenzetti D.: ApJ **485**, 290 (1997)

Observations of Circumstellar Disks

M. Hempel

Astrophysikalisches Institut und Universitäts-Sternwarte, Uni Jena, Germany
marc@astro.uni-jena.de

1 Introduction

IRAS observations first showed that a sizable fraction of the nearby stars are (still) surrounded by dust disks (Cheng et al. [3]), the most prominent prototypes of this class of dusty stars being Vega, β Pic, and α PsA. Such dust or debris disks have generated particular interest in the framework of young earth-like planetary systems. The analysis of both gas and dust in circumstellar disks is of fundamental importance for any theoretical CS disk model, for an investigation of the CS disks dynamics, and for a determination of the disk masses.

2 Detection of circumstellar gas

Infrared (IR) observations are sensitive only to dust but not to gas, which constitutes only a minor fraction of the total mass if the interstellar gas-to-dust ratio of $\approx$ 100:1 applies. What about the gas content of circumstellar disks? How can the disk gas be diagnosed? Do CS disks (still) contain gas? And if they do, how much? What is the chemical composition of the gas? What are the timescales on which the gas is expelled? A common method for the detection of gas is to search for abundant molecules like H_2 and CO Thi et al. [9]. Direct detection of molecular hydrogen is difficult because of the zero dipole moment of H_2 Bary et al. [1]. CO observations around A-type stars pose problems as well. As shown by Kamp & Bertoldi [7], the radiation field of A stars with an effective temperature comparable to Vega ($T_{\mathrm{eff,Vega}} = 9500\,\mathrm{K}$) can destroy CO molecules in the CS environment, and thus the absence of CO need not imply the absence of any gas in the disk. Alternatively, the presence of gas can be diagnosed by both CS absorption (Holweger& Rentzsch-Holm [5], Holweger, Hempel, Kamp [6], Hempel & Schmitt [4]) or disk emission lines in the optical (Brandeker et al. [2]).

3 Results

We initiated a comprehensive study of all suitable nearby stars with disks to systematically search for the signatures of gas. Our studies of IR sources (cf., Holweger& Rentzsch-Holm [5], Holweger, Hempel, Kamp [6], Hempel & Schmitt [4]) show that about 30% of the investigated A stars do show narrow absorption features in optical lines like Ca K and Na D indicating the presence of CS gas. It is evident that VLTI instruments such as MIDI are especially suited to study and resolve suchlike disks to analyse the dust counterparts in detail (see, e.g., Leinert et al [8]).

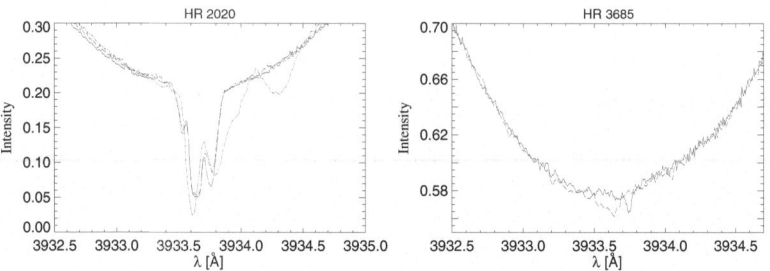

Fig. 1. *Left panel:* HR 2020, obtained on 2002 Mar 25, UT 23:37 (solid line), on 2002 Mar 26, UT 23:25 (dashed line), and on 1996 Feb 29, UT 00:11 (dashed-dotted line). Note the nightly variation at ∼3933.8 Å in the 2002 spectra which is attributed to falling evaporating bodies. *Right panel:* The Ca II K line of HR 3685 recorded on 2002 Mar 26, UT 00:13 (solid line) and 1996 Feb 29, UT 03:30 (dashed line). A change in shape and radial velocity of the circumstellar absorption is clearly discernible.

References

1. Bary, J.S., Weintraub, D.A., Kastner, J.H. 2002, ApJ, 576, L73
2. Brandeker, A., Lieseau, R., Olofsson, G., Fridlund, M., 2004, A&A, 413, 681
3. Cheng, K.P., Bruhweiler, F.C., Kondo, Y., Grady, C.A. 1992, ApJ, 396, L83
4. Hempel, M., Schmitt, J.H.M.M., 2003, A&A, 408, 971
5. Holweger, H., Rentzsch-Holm, I. 1995, A&A, 303, 819
6. Holweger, H., Hempel, M., Kamp, I. 1999, A&A, 350, 603
7. Kamp, I., Bertoldi, F. 2000, A&A, 353, 276
8. Leinert, Ch., van Boekel, R., Waters, L.B.F.M., Chesneau, O., et al. 2004, A&A, 423, 537
9. Thi, W.F., van Dishoeck, E.F., Blake, G.A., van Zadelhoff, G.J., Horn, J., Becklin, E.E., Mannings, V., Sargent, A.I., van den Ancker, M.E., Natta, A., Kessler, J., 2001, ApJ, 561, 1074

Fundamental Parameters of *Delta Velorum*, a Quintuple Stellar System.

A. Kellerer[1], M. Petr-Gotzens[1], P. Kervella[2] and V. Coudé du Foresto[2]

[1] E.S.O., Karl Schwarzschild Str. 2, D-85748 Garching akellere@eso.org
[2] Observatoire de Paris, 5 place J. Janssen, F-92195 Meudon Cedex

Delta Velorum (δVel) is a nearby ($\approx 24pc$) quintuple stellar system. Its most luminous component (Aa+Ab) was discovered only in 2000 to be the brightest eclipsing binary in the sky, even brighter than Algol. The system offers the rare opportunity to derive fundamental parameters for at least five stars that are of same age and span a spectral range from A1 to K0. Our observations of Aa+Ab, using the VLT interferometer with VINCI, spatially resolved the components and suggest stellar diameters substantially larger than expected for main sequence stars. This could be due either to the fact that one of the stars is just leaving the main sequence or to the existence of a circumstellar dust disc.

1 A nearby system ($\approx 24pc$) containing the brightest eclipsing binary in the sky:

δVel has a hierarchical structure, consisting of two common proper motion pairs, AB (sep$\approx$ 0.7") and CD (sep$\approx$ 6"), separated approximately by 70" [5]. The component δVel A itself was found to be an eclipsing binary with a period of 45.15 days [2], hence making AB actually a triple (Aa+Ab+B). Amazingly, among known eclipsing binaries, δVel A is now the brightest and nearest in the whole sky!

2 VINCI and AMBER observations of the eclipsing binary Aa+Ab:

In May 2003 we had the opportunity to perform VLTI/VINCI 2.20μm observations of the eclipsing binary Ab-Ab (angular semi-major axis of $\approx 10 - 15$mas) at four orbital phases. The combination of these interferometric observations (17 squared visibilities (V^2) measurements at 4 epochs, spread over 3 weeks) together with the existing photometric data from the literature provided preliminary values of the physical properties of Aa and Ab.

Surprisingly, our interferometric results suggest that Aa and Ab have angular diameters of $1.6-2.0$ mas, and $1.0-1.3$ mas respectively, which is larger, or significantly larger (in case of Aa) than expected for main-sequence dwarfs of spectral types A1 V and A5 V: expected diameters for an A1 V and A5 V star at 24pc distance are 0.9mas and 0.7mas respectively.

We propose the following interpretations:

- The interferometric observations might have been affected by the presence of circumstellar material. This is supported by the fact that an infrared (IR) excess associated with δVel AB has been reported by the *Infrared Astronomy Satellite* [4] and more recently confirmed with 10μm observations carried out by Jayawardhana et al [1]. The observations by Jaywardhana et al. were made with an angular resolution of ≈ 1.2", excluding the possibility that the IRAS excess is due to an unrelated background source. The beam was, however, too large to resolve the δVel AB system and they reported a total, *unresolved* 10μm flux of 6.46±0.60 Jy. *If* the IR-excess is associated with δVel A, it would indicate the presence of a, probably Vega-type, dust disc around Aa (or Ab). This seems to be a a clear possibility, since $\approx$20% of the nearby A stars are surrounded by dust discs [3]. Yet, Aa and Ab are close, their semi-major axis being ≤ 1 AU and in such a configuration a circumbinary rather than a circumstellar disc seems more likely. Clearly, the circumbinary disc would have to posess a hot inner rim in order to affect the VINCI observations at 2.2μm. K-band observations of the system are unfortunately not available and we cannot rule out the possibility of a near-IR excess.

- Another reason for the more massive component Aa ($\approx$2.7 M$_\odot$) being significantly larger than expected, might be that Aa is leaving the main-sequence to become a sub-giant, while the less massive Ab ($\approx$2.0 M$_\odot$), is still on the main-sequence. As the evolution of the stellar radius is extremely fast during this transition phase, this would also permit to precisely trace the age of the system and hence would very tightly constrain evolutionary models. In this scenario the IR-excess flux, detected from the stellar system as a whole, is more likely to be associated with component B. IR-excess emission from this star B, which is apparently a single star, may then indicate a circumstellar dust disc.

References

1. Cheng et al., ApJ **396**, 396 (1992)
2. IRAS NASAR **1190** (1988)
3. Jayawardhana et al., AJ **122**, 2047 (2001)
4. Jeffers et al., Pub.Lick Obs. **21** (1963)
5. Otero et al., IBVS **4999**, 1 (2000)

VLTI/MIDI Measurements of Extended Mid-Infrared Emission in the Galactic Center

J.-U. Pott[1,2], A. Eckart[2], A. Glindemann[1], T. Viehmann[2], and Ch. Leinert[3]

[1] ESO, Karl-Schwarzschildstr. 2, 75478 Garching b.M., Germany jpott@eso.org
[2] I. Physik. Institut, Univ. zu Köln, Zülpicher Str. 77, 50937 Cologne, Germany
[3] Max-Planck-Institut for Astronomy, Königstuhl 17, 69117 Heidelberg, Germany

We investigated with MIDI the extension of dusty mid-infrared excess sources (IRS 1W, IRS 10W, IRS 2, IRS 8) in immediate vicinity to the black hole (BH) at the GC. We derive 3σ upper limits of the correlated fluxes of our target sources which give direct constraints on the size of the emitting regions. Most probably the emission originates from bow shocks generated by windy stars ploughing through the dense matter of the Northern MiniSpiral.

Why is the dust morphology of interest?

At the Center of our Galaxy (GC) star formation (SF) close to a supermassive BH can be studied at unique linear scale.

The existence of massive stars contributing more than 50% of the ionizing luminosity within the GC confines the latest SF activity to happen at maximum a few Myr ago [2]. Simulations show, that a parental molecular cloud can spiral into the very center during the lifetime of a massive star, providing in particular good conditions for the more massive stars. For lower mass stars matter is less bound and will be easily removed in the vicinity of the massive BH [1].

On the other hand the enigmatic featureless infrared excess sources within the MiniSpiral (e.g. 1W and 10W) only a few arsec away of SgrA* (1"~39 mpc) could also indicate embedded young stellar objects. Recent near infrared AO imaging suggested a different explanation for these sources. Tanner et al. (2005) [4] found bow-like morphologies. A more thorough analysis revealed that these bows can be explained by heated dust which is shocked through the interaction of a strong stellar wind ($v_\infty \leq 1000$ km s^{-1}) from a massive star (most favourable Wolf-Rayet type) ploughing through the MiniSpiral.

Observations & Results

We conducted a VLTI observation [3] in the mid infrared (8-12 μm) which is totally dominated by thermal re-radiation of the stellar UV luminosity by dust. Low resolution dispersed visibility moduli at 20 mas angular scale were obtained in July 2004 using MIDI at the UT2-UT3(47m) baseline. The image motion was corrected with a tip-tilt unit guiding on a 25" distant optical foreground guide star.

MIDI provides internal fringe tracking on the basis of group delay fringe tracking. The group delay is estimated by the Fourier transform of the dispersed interferometric spectrum. An optical path difference (OPD) between the two interfering light beams results in a cos-pattern in the interferometric spectrum as long as the OPD is shorter than the coherence length.

No fringes could be detected on the embedded sources in the MiniSpiral (1W, 2 and 10W). Because the total flux densities of all these sources are above the detection limit, the negative fringe detections showed, that either the VLTI resolved out the entire flux density of the source or any compact unresolved source component is weaker than the upper limit of the correlated flux density listed in Table 1.

We quantified that statement by giving an upper limit of the correlated flux densities of the different measurements, depending on the actual observing conditions in Table 1. Furthermore we assumed for two sources a dust morphology similar to the near infrared findings [4]. Then the lower limit of the width of the bowshock feature can be fitted to the upper limit of the correlated flux. The estimated lower limits of the bowshock widths are well within the near infrared findings. A further constraint on the extension of 1W was achieved by deconvolution of the acquisition image. We could derive an overall extension scale of ~ 350mas which contains the entire bowshock and indicates a significant increase in size of the radiating structures at MIR wavelengths with respect to the near infrared.

Outlook

The new higher order adaptive optics system MACAO has improved the stability of the beams and concentrates more light into the interferometric field-of-view. The widths found in the NIR are of the order of the lower limits presented here. Therefore positive fringe detection may be within reach already without the external fringe tracker FINITO.

Table 1. Upper correlated flux limits give lower limits of the source size.

Sources with existing bowshock models:		IRS 1W	IRS 10W
Total flux at 8.6micron (VISIR), (extinct. Av=25)	[Jy]	4.6	2.7
Upper limit of the correlated flux density	[Jy]	< 0.3	< 0.25
Visibility limits	[1]	< 0.06	< 0.09
Width of bowshock models	[mas/AU]	> 30/240	> 20/160
Sources without existing bowshock models:		IRS 2	IRS 8
Upper limit of the correlated flux density	[Jy]	< 0.4	< 0.35

References

1. O. Gerhard: ApJ **546**, 39 (2001)
2. F. Najarro, A. Krabbe, R. Genzel, et al. A&A **325**, 700 (1997)
3. J.-U. Pott, A. Eckart, A. Glindemann, et al. ESO Msngr **119**, 43 (2005)
4. A. Tanner, A. Ghez, M. Morris, et al. ApJ **624**, 742 (2005)

Impact of High Spectral Resolution on Stellar Interferometry

S. Jankov[1,4], F. Vakili[1], A. Domiciano de Souza[3], R. G. Petrov[1],
F.-X. Schmider[1], S. Robbe-Dubois[1], and P. Mathias[2]

[1] Laboratoire Universitaire d'Astrophysique de Nice, UMR 6525 Parc Valrose,
F-06108 Nice Cedex 02, France Slobodan.Jankov@unice.fr
[2] Observatoire de la Côte d'Azur, Dpt. Gemini, UMR 6203, F-06304, Nice
Cedex 04, France
[3] Max Planck Institut für Radioastronomie, Auf dem Hügel 69, 53121 Bonn,
Germany
[4] Astronomical Observatory Beograd, Volgina 7, Belgrade, Serbia

Summary. In this contribution we review a selection of outstanding problems in
stellar physics showing the potential of new methods which combine the classical
spectroscopy and Long Baseline Interferometry, providing informations that cannot
be obtained otherwise with each of these techniques taken separately.

1 Introduction

Traditionally, optical interferometry has been considered as a tool to determine the fundamental properties of stars, namely their effective temperatures,
radii, luminosities and masses, by the combination of angular diameters, with
complementary photometric, and spectroscopic measurements, made with
conventional telescopes. But, the impact of interferometry on stellar physics
extends beyond classical applications. For instance, the differential interferometry (Beckers 1982, Petrov 1988) makes it possible to measure the shift
of the stellar photometric barycenter (photocenter) of an unresolved star as
a function of wavelength, providing the first order moment of the spatial
brightness distribution, in addition to the zero order moment spectroscopic
information and allowing better spatial resolution of stellar atmospheres when
compared to the classical Doppler Imaging (Jankov & Foing 1987, Jankov et
al. 2001).

The new generation of ground-based instruments for high angular resolution from infrared and optical interferometry should provide a qualitatively
new information for improving our understanding of stellar physics, through
the comparison of observational results with the predictions of theoretical
models of stellar interiors and atmospheres. However, with the classical array
and/or earth rotation synthesis it will be possible to spatially resolve only a
limited number of stars. On the other hand, the High Resolution Spectroscopy
allows (through Doppler Mapping) indirect observational information on atmospheric structures of an unresolved star by modeling the observed flux
distribution across the spectral lines.

2 Some applications

In the application to *stellar non-radial pulsations*, Jankov et al. (2001) considered the artificial star tilted by i=85°, with surface brightness perturbation due to the $\ell = 5$, $m = 4$ mode. The input image could not be reconstructed by classical Doppler Imaging since the solution is intrinsically non-unique and practically the artifacts due to mirroring effect prevented the mapping from spectral line profiles as well as from photocenter shift orthogonal to rotation axis. However, in the image reconstruction from photocenter shift parallel to rotation the north-south ambiguity was removed and the input image was correctly reproduced.

It was shown recently that the spectrally resolved interferometric signal introduces crucial improvements for the study of many outstanding problems in stellar physics. Examples can be found for *stellar surface structure* (Jankov et al. 2003), *differential rotation* (Domiciano de Souza et al. 2004), *magnetic fields* (Perraut-Rousselet et al. 2004) as well as for *gravity darkening* (Domiciano de Souza et al. 2002).

3 Conclusion

Spectrally resolved interferometry provides qualitatively new informations that cannot be obtained from classical spectroscopy alone, and is a mandatory requirement for next generation VLTI instrumentation.

References

1. Beckers J.M.: 1982, Optica Acta Vol. **29**, 361 (1982)
2. Domiciano, A., Vakili, F., Jankov, S., Janot-Pacheco, E., Abe, L.: A&A, **393**, 345 (2002)
3. Domiciano de Souza, A., Zorec, J., Jankov, S., Vakili, F., Abe., L., Janot Pacheco, E.: A&A, **418**, 781 (2004)
4. Jankov, S., Foing, B.: in *Cool Stars, Stellar Systems, and the Sun*, eds. J.L. Linsky and R.E. Stencel (Springer-Verlag, Berlin Heidelberg 1987) p. 528
5. Jankov S., Vakili F. Dominiciano de Souza Jr., Janot-Pacheco E.: A&A, **377**, 721 (2001)
6. Jankov, S., Domiciano de Souza, A., Stehle, C., Vakili, F., Perraut-Rousselet, K., Chesneau, O.: Interferometry for Optical Astronomy II, SPIE, **4838**, p. 587 (2003)
7. Petrov, R. G.: in *Diffraction-Limited Imaging with Very Large Telescopes*, eds. D.M. Alloin, J.M. Mariotti (Kluwer, 1988) p. 249
8. Rousselet-Perraut, K., Stehl C., Lanz, T., Le Bouquin, J. B., Boudoyen, T., Kilbinger, M., Kochukhov, O., Jankov, S.: A&A, **422**, 193 (2004)

Deploying an Antarctic Interferometer

M. R. Swain[1], D. Roche[2], M. Guillon[2], E. Lanford[2], K. Knepper[2], V. Olson
and P. Little[2]

[1] Laboratory of Astrophysics, Observatory of Grenoble
   mark.swain@obs.ujf-grenoble.fr
[2] Harvey Mudd College, Claremont, California plittle@hmc.edu

1 Introduction

Because Dome C permits major improvements in interferometer performance
[1], several interferometer concepts [2, 3] are being developed for possible deployment at Antarctic Dome C. The deployment model explored here is based
on a four step process. **(1)** Build a modular interferometer packaged in standard shipping containers. **(2)** Assemble and validate the interferometer in the
Northern hemisphere. **(3)** Ship the interferometer, with no disassembly beyond the container level, to Dome C Antarctica, using the established Concordia station supply traverse system. **(4)** Connect the containerized modules at
Dome C and begin observations. The deployment model assumes the instrument will not be shipped to Antarctica until it can operate in the Northern
hemisphere exactly the way it will operate in Antarctica (within the limitations imposed by the test site). Recent atmospheric models [4] show that
the excellent atmosphere at Dome C lies above a ~ 30 m thick, turbulent
boundary layer containing strong seeing. In the absence of laser guide stars,
interferometer and adaptive optics system sensitivity is limited by the coherence volume, and thus it is necessary to elevate the interferometer primary
telescopes above the strong seeing layer.

2 Measurements and Design Results

The regular supply traverse to the Concordia station routinely transports
shipping containers to Dome C; however, the shock loads associated with
this transport process were largely unknown. To measure the shock loads,
an automonous, battery powered accelerometer data logging system was designed and built. This instrument was then attached to a supply container for
the final 2004-5 season traverse. The instrument captured acceleration data
for the major (and essentially uncharacterized) segments of the Dome C supply route. With one exception, the shock loads experienced by the container
were 5 Gs or less. The exception occurred when the container was off-loaded
from the ship to the ice. Integration of the acceleration data indicates that
the container was dropped from approximately 1.5 m onto the ice surface,
resulting in a 10 G load.

An isolated truss, inserted into a container, was used to support the model interferometer delaylines and optical benches. Finite Element Analysis (FEA) methods found it was feasible to provide isolation for a 5 G load using commercially available isolation components. Assuming the containers can be offloaded from the ship with sufficient care, modular packaging of a working interferometer in shipping containers appears eminently feasible.

A concept for a deployable tower, packaged in a container envelope, was also developed. The tower design inputs included a 30 m height, a 10,000 Kg telescope, and model wind gust data. Figure 1 shows a concept for the deployable tower. A dynamical analysis of this concept found the bending modes for the tower were around 1 Hz and had typical deflections of around 10 cm. Although the deflections are large, the low frequency suggests that active servo compensation would be highly effective at correcting the piston and tilt components of this motion.

This work was carried out by a team of undergraduate students participating in the Harvey-Mudd College (HMC) Engineering Clinic program. This work shows there is broader scope for the meaningful participation of undergraduate engineering teams in substantial astronomical engineering projects.

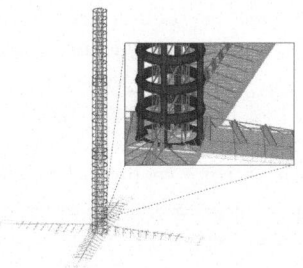

Fig. 1. Deployable 30 m tower concept designed to support 10,000 Kg. The external envelope is that of a standard container during transport. For deployment, the external portions of the structure fold outward forming the tower foundation. The first bending modes of the tower are of order 10 cm at 1 Hz for typical surface winds at Dome C.

References

1. Swain, M.R.: EAS **14**, 147S (2005).
2. Swain, M.R., et al.: SPIE **5491**, 176 (2004).
3. Vakili, F., et al.: SPIE **5491**, 1580 (2004).
4. Swain, M.R., and Galleé, H.: PASP **118**, 1190 (2006).

Direct Detection of Exo-Planets: GQ Lupi

Ralph Neuhäuser[1], Eike Guenther[2], and Peter Hauschildt[3]

[1] Astrophysikalisches Institut und Universitäts-Sternwarte, Schillergäßchen 2,
   D-07745 Jena, Germany rne@astro.uni-jena.de
[2] Thüringer Landessternwarte Tautenburg, D-07778 Tautenburg, Germany
[3] Hamburger Sternwarte, Gojenbergsweg 112, D-21029 Hamburg, Germany

1 Introduction: GQ Lupi and its companion

Since several years, we have been searching for sub-stellar companions around young (up to 100 Myrs) nearby (up to 150 pc) stars, both brown dwarfs and giant planets. Young sub-stellar objects are self-luminous due to ongoing contraction and accretion, so that young stars are good targets. We have found several brown dwarf companion candidates and confirmed three of them by both proper motion and spectroscopy: TWA-5 B, HR 7329 B, and GSC 8047 B, all being few Myr young late M-type brown dwarfs.

With K-band imaging using VLT/NaCo, Subaru/CIAO, and HST/PC, we detected a 6 mag fainter object $0.7''$ west of the classical T Tauri star GQ Lup, which is a clearly co-moving companion ($\geq 10\ \sigma$), but orbital motion was not yet detectable. The NaCo K-band spectrum yielded $\sim$ L1-2 (M9-L4) as spectral type. At 140 ± 50 pc distance (Lupus I cloud), it can be placed into the H-R diagram. For more details, see Neuhäuser et al. (2005). According to our own calculations following Wuchterl & Tscharnuter (2003), it has 1 to 3 M_{jup}, but according to Baraffe et al. (2002) and Burrows et al. (1997) models, it is anywhere between 3 and 42 M_{jup}. The latter models are not applicable for the young GQ Lup and its companion, while the former does take into account the formation and collapse; the Wuchterl & Tscharnuter (2003) convection model is calibrated on the Sun.

2 Comparison with model atmospheres

We use the new so-called GAIA-dusty grid (Brott & Hauschildt 2005), which is an updated version of the models from Allard et al. (2001). The updates include improved molecular dissociation constants as well as more dust species and their opacities. In addition, the models were computed for a convective mixing length parameter of 1.5 times the pressure scale height H_p. It uses spherical symmetry, which is most important for young and sub-stellar objects as GQ Lup A and b with low gravities. The Allard et al. (2001) AMES-dusty sometimes had problems at very low gravity.

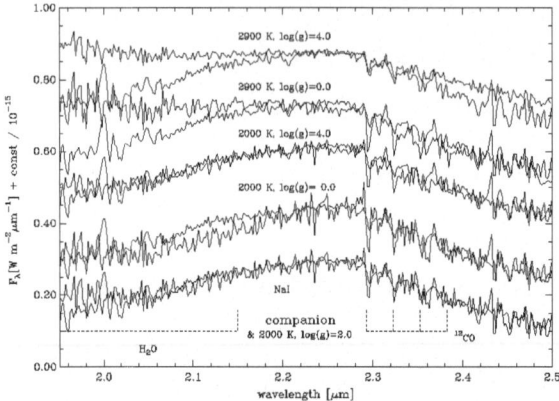

Fig. 1. Comparison of our GAIA-dusty model atmosphere grid with the observed spectrum overplotted. The high T=2900 K in the upper two spectra do not reproduce the water steam absorption in the blue. The model with T=2000 K and log g =2 fits best (bottom), see text. H_2O, Na, and 12CO are indicated.

We compare our K-band spectrum (see Neuhäuser et al. 2005) with the GAIA grid for temperatures T_{eff} = 2000 and 2900 K and for gravities of log g = 0, 2, and 4 (g in cgs units). See figure 1 for the comparison.

The model spectrum with T_{eff} = 2000 K is much better than the hotter temperature (where there is no water vapour absorption in the blue part), again indicating an early L spectral type. For the gravities, log g = 2 fits best (log g = 4 is better than log g = 0). This gravity is fully consistent with the gravity-sensitive CO index measured in our spectrum to be 0.862 ± 0.035, yielding log g = 2.5 ± 0.8 according to Gorlova et al. (2003). We conclude log $g \simeq 2.0$ to 3.3. The good fit indicates that the new spherically symmetric GAIA-dusty model is better for low gravities than the former AMES-dusty.

It is important to note that the GAIA-dusty models are applicable: They are independant of age and stand-alone, even without interior models like the Baraffe et al. (2002) models. They are used, however, as outer boundary conditions by, e.g., Baraffe et al. (2002) to provide their evolution models with a more realistic description of how an interior model loses energy through the atmosphere. Models of the solar atmosphere usually used are also stand-alone and independant of the interior and age.

3 Discussion: Mass estimate for the GQ Lup companion

At 140 ± 50 pc distance, with T_{eff} = 2050 ± 450 K and the flux of the companion (K = 13.10 ± 0.15 mag), we can estimate its radius to be $1.2 \pm$

0.5 R_{jup}. With this radius and log $g \simeq 2.0$ to 3.3, we can estimate its mass to be ≤ 1 M_{jup} (for log $g \simeq 4$ and 2 R_{jup}, its ~ 6 M_{jup}).

The Wuchterl & Tscharnuter (2003) calculations (Fig. 4 in Neuhäuser et al. 2005) indicate ~ 1 to 3 M_{jup}, co-eval with the star at ~ 1 Myr.

Mohanty et al. (2004a) measured gravities for isolated young brown dwarfs and free-floating planetary mass objects. Their coolest objects have spectral type M7.5 and gravities as low as $\log g = 3.125$ (GG Tau Bb). This lead Mohanty et al. (2004b) to mass estimations as low as ~ 10 M_{jup}. GQ Lup A is younger than the Mohanty et al. Upper Sco objects. Its companion is at least as late in spectral type, probably even cooler. An object younger and cooler must be lower in mass. The GQ Lup companion is fainter than the faintest Mohanty et al. object (USco 128, ~ 9 M_{jup}), so that the mass estimate for the GQ Lup companion is ≤ 8 M_{jup}.

According to Burrows et al. (1997) and Baraffe et al. (2002), the mass of the companion could be anywhere between 3 and 42 M_{jup}, as given in Neuhäuser et al. (2005). However, these models are not valid at the young age of GQ Lup, so that they are not applicable. According to both Mohanty et al. (2004b) and Close et al. (2005), the Baraffe et al. (2002) models overestimate the masses of young sub-stellar objects below ~ 30 M_{jup}, but underestimate them above ~ 40 M_{jup}. This is consistent with our results.

Hence, according to all valid estimations, the mass of the GQ Lup companion is ~ 1 to 8 M_{jup}, i.e. significantly below ~ 13 M_{jup}, hence almost certainly a planet imaged directly, to be called GQ Lup b.

Acknowledgements. We would like to thank Subu Mohanty and Gibor Basri for very fruitfull discussion about GQ Lupi b. Especially, we would like to acknowledge Gibor Basri for pointing us to the Mohanty et al. work.

References

1. Allard F., Hauschildt P.H., Alexander D.R., Tamanai A., Schweitzer A., 2001, ApJ, 556, 357
2. Baraffe I., Chabrier G., Allard F., Hauschildt P.H. 2002, A&A 382, 563
3. Brott, I, Hauschildt, P. H. 2005, in *The Three-Dimensional Universe with Gaia*, C. Turon, K.S. O'Flaherty, M.A.C. Perryman, eds., Nordwijk: ESA, 565
4. Burrows A., Marley M., Hubbard W. et al. 1997, ApJ 491, 856
5. Close L.M., Lenzen R., Guirado J.C., et al., 2005, Nature 433, 286
6. Gorlova N.I., Meyer M.R., Rieke G.H., Liebert J., 2003, ApJ 593, 1074
7. Mohanty S., Basri G., Jayawardhana R., Allard F., Hauschildt P., Ardila D., 2004a, ApJ 609, 854
8. Mohanty S., Jayawardhana R., Basri G., 2004b, ApJ 609, 885
9. Neuhäuser R., Guenther E.W., Wuchterl G., Mugrauer M., Bedalov A., Hauschildt P., 2005, A&A 435, L13
10. Wuchterl G., Tscharnuter W.M. 2003, A&A 398, 1081

Antarctic Interferometry: Science Demonstration from the Concordia Station

V. Coudé du Foresto[1], J. Monnier[2], M. Swain[3], and F. Vakili[4]

[1] LESIA, UMR 8109, Observatoire de Paris-Meudon, F-92195 Meudon Cedex, France, vincent.foresto@obspm.fr
[2] LAOG, UMR 5571, U. Joseph Fourier, F-38041 Grenoble
[3] U. Michigan, Ann Arbor, MI 48109 USA
[4] LUAN, UMR 6525, U. Nice, Parc Valrose, F-06108 Nice Cedex 2

Summary. The high antarctic plateau has recently been identified as the best accessible ground-based location for high angular infrared observations: extremely dry atmosphere, low sky emission, low and slow turbulence, large isoplanetic angle, are properties that make it possible to consider observations that could otherwise only be carried out from space (such as the detection of exoplanets in the habitable zone of nearby stars). This would require a major infrastucture which needs to be prepared by a lower scale pathfinder, for which we present here one option.

1 Introduction

Recent campaigns at the Dome C - Concordia station have confirmed that the high antarctic plateau combines characteristics which are optimal for infrared astronomy at high angular resolution, most notably:

- A dry ($250\,\mu$m PWV) and cold (200K) atmosphere which results in improved transparency windows and limited background emission for the mid-IR;
- An exceptional seeing above the ground layer ([1],[2]), with both benign and slow turbulence, resulting in a large isoplanetic patch even at ground level.

Taken together, these features are unique for a ground-based site and can dramatically enhance the performance of an interferometer. In interferometry as well as for adaptive optics, the final performance is determined by the total number of photons available in the coherence volume of the atmosphere. The gain provided by the antarctic environment is twofold:

- The very large r_0 (median night time seeing measured at 0.27 arcsec) and τ_0 increase the volume of coherence by a factor 12 compared to the best temperate sites (Paranal);
- A 25× photometric gain (in the most favorable bands, ie K and L) due to the improved transparency and reduced background of the atmosphere densifies the number of photons within that volume of coherence.

The combination of those two factors provides a gain of up to ×300 with respect to the best temperate sites for AO and interferometry. On bright sources, the increased number of photons in the coherence volume and slower time evolution of the control loops results in better phase correction.

2 An antarctic interferometer demonstrator

While it is recognized that one of the major long term goals for an antarctic facility is the detection of exoplanets in the habitable zone of nearby stars, this will require either a large scale interferometer such as KEOPS [4] or API [5] or a large monolithic coronographic telescope equipped with extreme AO. Such a major facility will probably necessitate a dedicated infrastructure on the (yet to be determined) optimal antarctic site, as its logistic needs will be well beyond what the Concordia station can offer at the moment. However, in a first stage a more modest facility can be operated from Dome C. Such a pathfinder would provide invaluable insight into antarctic interferometry operation and could demonstrate the scientific potential of the site by providing unique science results in the field of extrasolar planet research, such a as the infrared spectrophotometry of a few hot extrasolar giant planets ("Pegasides") or the characterization of exozodys around main sequence nearby stars – a required precursor science for Darwin.

An example of a design that could be used for fast track deployment of a 1.65 m collector on Dome C can be found in the telescopes of the Infrared Space Interferometer developed by Ch. Townes and collaborators of UC Berkeley [3]. Each of the three collectors of ISI is installed in a truck trailer for easy transport and relocation and consists in a plane siderostat feeding an horizontal telescope whose primary is a 1.65m, f/3.14 paraboloid. A single such telescope at Dome C would have the depth of a 8m telescope on a temperate site, while being diffraction limited in the K and L bands.

References

1. Aristidi, E., Agabi, A., Vernin, J., Azouit, M., Martin, F., Ziad, A., Fossat, E.: A&A **406**, L19 (2003)
2. Lawrence, J., Ashley, M., Tokovinin, A., Travouillon, T.: Nature **431**, 278 (2004)
3. Hale, D., Bester, M., Danchi, W. C., Fitelson, W., Hoss, S., Lipman, E., Monnier, J., Tuthill, P., Townes, C.: ApJ **537**, 998 (2000)
4. Vakili, F., et al.: SPIE **5491**, 1580
5. Swain, M., et al.: SPIE **5491**, 176

Color-Differential Interferometry with AMBER/VLTI. Data Processing and Precision from Early Observations.

M. Vannier[1], F. Millour[2,3], R. G. Petrov[2], and F. Rantakyrö[1]

[1] European Southern Observatory, Casilla 19001, Alonso de Cordova 3107, Santiago, Chile. mvannier@eso.org
[2] LUAN, Université de Nice-Sophia-Antipolis, France
[3] LAOG, Université Joseph Fourier, Saint Martin d'Hères, France

1 Data processing for color-differential observables

AMBER, the near-infrared dispersive monomode recombiner at the VLTI, has recently had succesfully its first GTO and SDT runs. The general scheme of data reduction (internal calibrations, extraction of the complex visibility from the raw data,...) was described by Milour et al. (SPIE conf. 5491, 2004). We present hereafter the principles the color-differential data processing and its early performances from sky observations. The reference goal is to get a precision on the differential phase close to limits of the fundamental noises (namely the photon noise) in order to make possible such demanding applications as the spectroscopy of hot giant exoplanet (Vannier et al., MNRAS, 367, 825). The main steps of the high-precision processing can be summarized as follow:

• **Getting un-pistoned differential observables on each frame.** For each of the individual exposure frames (whose integration time is, typically, a few tens of ms), the complex visibilities are extracted from the interferometric channels and calibrated using the photometric channels and the internal calibration files. The piston p between the beams is first estimated from the visibilities, and a correction phaser $e^{-2i\pi p/\lambda}$ is applied on each frame. From this, the color-differential phase and visibility are calculated, for a given spectral channel λ, with respect to a set of reference channel(s). We chose the reference channel to consist in the whole spectral bandwidth except the considered channel λ.

  • **Selection and weighting of the frames.** A fringe SNR estimate is computed to select the frames above a chosen threshold and to weight them for calculating $\langle \Delta\Phi(\lambda) \rangle$ over the frame serie. The standard deviation $\sigma_{\Delta\Phi(\lambda)}$ of the residual phase over the frame serie uses the same weighting.

2 Results

In Medium Resolution (R=1500), the standard deviation of $\Delta\Phi(t, \lambda)$ over the time is a factor 3 above the photon noise for good-quality observations, thus

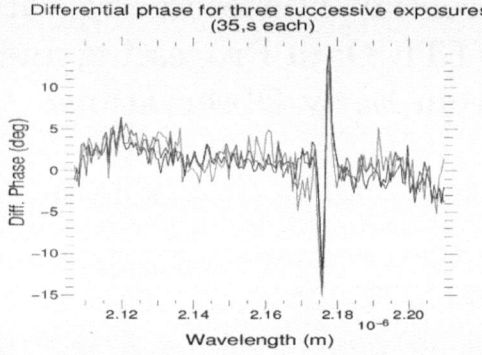

Fig. 1. Color-differential phases of Alfa Ara from preliminary SDT data, which is being evaluated prior to public release. $\Delta\Phi(\lambda)$ is measured here in medium-resolution from 3 series of 500 successive frames, separated by about 100 s. The high peak of phase varation is due to a rotating emission-line structure. The smaller wriggles are artifacts, most of which may be calibrated using a fast internal spatial modulation at a timescale of about 60 s and/or using a calibrator star.

yielding a precision of ≈ 0.5 deg over 1 minute of integration (Fig. 1). In low resolution, we could reach a precision of ≈ 0.3 deg over 1 mn, i.e. a factor 15 over the photon noise in K band. Similar results were obtained on each baseline (for our 3-baselines observations), whereas the differential closure phase $\sum \Delta\Phi$ shows a standard deviation equivalent to their quadratic sum.

3 Discussion

We can currently achieve an equivalent accuracy better than 10^{-3} rad (0.06 deg) in one hour of exposure, which allows most of the scientific applications addressed by the differential phase mode. It is still necessary to gain a factor 10 to achieve the extrasolar planets goal. The error is not yet dominated by the atmospheric and chromatic effects, which would be canceled in the closure phase. Instead, both differential and closure phases seem dominated by detection errors, which may include a number of causes: data processing errors, crosstalk between correction of instantaneous piston, time delay between spectral channels during detector read-out, variations of the detector gain or offset and/or unperfect monomode guiding of the fibers.

Acknowledgements: The authors deeply acknowledge the AMBER consortium[4] and the ESO/VLTI team for their efforts, which permitted to obtain these results.

---

[4] See list of consortium members at:
http://amber.obs.ujf-grenoble.fr/article.php3?id_article=45

The Fourier-Kelvin Stellar Interferometer: A Progress Report and Preliminary Results from our Laboratory Testbed

R. K. Barry[1], W. C. Danchi[1], J. Rajagopal[1], L. J. Richardson[1],
M. Kuchner[1], D. Wallace[1], V. J. Chambers[1], A. Martino[1], W. Traub[2],
H. Ford[2], R. J. Allen[3], and S. Seager[4]

[1] NASA Goddard Space Flight Center, Greenbelt, MD 20771 USA
    richard.k.barry@nasa.gov
[2] Harvard-Smithsonian Center for Astrophysics, Cambridge, MA 02138 USA
[3] The Johns Hopkins University, Baltimore, MD 21218 USA
[4] Carnegie Institution of Washington, Washington, D.C. 20005 USA

Summary. The Fourier-Kelvin Stellar Interferometer (FKSI) is a passively cooled, space-borne interferometer that has been developed for high angular resolution astrophysics in the near-to-mid IR wavelength range (3-8 microns). The scientific emphasis of the mission is on the direct detection of extrasolar giant planets, characterization of their atmospheres, and observation of secular changes in their atmospheric spectra. FKSI will also facilitate the search for brown dwarfs and Jupiter and sub-Jupiter mass planets and the study of the evolution of protostellar systems from the collapse of the precursor molecular cloud core through the formation of the disk surrounding the protostar, the formation of planets in the disk, and subsequent dispersal of the disk material. FKSI could also play a powerful role in the investigation of the structure of active galactic nuclei and extra-galactic star formation. While FKSI is a high angular resolution system complementary to JWST, it is also an important scientific and technological pathfinder for the TPF Interferometer and Darwin missions as well as NASA Vision missions SPIRIT, SPECS, and SAFIR.

1 Scientific Emphasis of the Mission

The spatial and spectroscopic resolving capability of the FKSI instrument together with its calculated sensitivity will position it as an important facility for the study of a range of astronomical phenomena. Its science objectives are, very broadly, to directly detect extrasolar giant planets (EGP), study the evolution of protostellar and evolved stellar systems, and to facilitate the study of extra-galactic star formation regions and the extended neighborhoods of AGN.

The detection and study of EGPs will be greatly facilitated by a dedicated interferometer on a satellite platform operating at infrared wavelengths [1]. To answer key questions about these planets FKSI be able to detect 25 of the known EGPs and make precise determination of their orbits. FKSI could also be used to search for exoplanets around nearby stars not planned for observations with TPF such as evolved stars. It has been shown [2]that such post-main sequence stars have expanding habitable zones, which could encompass outer planetary companions if extant. These would be scientifically important targets in any survey for terrestrial-type planets.

Stellar physics will also benefit from the capabilities of FKSI. In particular, the instrument will allow the conduct of a infrared-excess survey of nearby M dwarfs (greater than 30 such stars) at distances within 10 pc together with an infrared-excess and debris disk survey of nearby F, G, K giants and subgiants within 30 pc (greater than 50 such stars). Circumstellar material may be studied with this instrument as well as measurements of the resonances in exozodiacal dust and debris disks each of which are associated with planets.

2 Design Realization

Our team, consisting of scientists from a broad array of institutions together with engineering support from GSFC, has expended significant effort to develop a range of design options for FKSI. We have studied various beam combination techniques and array architectures in preparation for submission of FKSI as a Discovery-class mission. These design studies were conducted initially at GSFCs Instrument Synthesis and Analysis Laboratory and the Integrated Mission Design Center - important functions within GSFCs infrastructure used to facilitate rapid vetting of various mission and instrument design concepts. These studies were then augmented by the work of a larger, focused team of experience scientists and engineers dedicated to the FKSI mission. The resulting design is a nulling interferometer configuration with an optical system consisting of two 0.5 m telescopes on a 12.5 m boom feeding a Mach-Zehnder beam combiner. A null tracker and fiber wavefront error reducer further augment the system and allow it to produce the required 10^{-4} null of the central starlight [3].

The FKSI testbed is now fully funded and is being built in the Horizontal Flow Facility at NASAs Goddard Space Flight Center in Greenbelt, Maryland. This testbed will allow the instruments designers to evaluate technically challenging aspects of the design. In particular, the instruments nulling architecture will be examined together with a novel Ditherless Quadrature fringe tracking approach. Significant progress has already been made on the testbed, which is near transition from visible to monochromatic IR testing. A mathematical model of the experiment is also in work, which will allow the experimenters to predict behaviors of the instrument to various adjustments and perturbations when we move to the more challenging IR wavelengths. A high-fidelity instrument imaging simulator has also been produced [4] and is being used to test various operational conditions.

References

1. Danchi, W.C., Deming, D., Kuchner, M.J., Seager, S., ApJ **579**, L57 (2003)
2. Lopez, B., Schneider, J., Danchi, W.C., ApJ **627**, 974 (2005)
3. Danchi, W.C., et. al., Proc. SPIE **5491**, 236 (2004)
4. Rajagopal, J., et. al., The Second TPF/Darwin International Conference, manuscript in preparation

The PRIMA Astrometric Planet Search Project

R. Launhardt[1], E. J. Bakker[2], P. Ballester[7], H. Baumeister[1],
P. Bizenberger[1], H. Bleuler[4], R. Dändliker[5], F. Delplancke[7], F. Derie[7],
M. Fleury[3], A. Glindemann[7], D. Gillet[4], H. Hanenburg[6], Th. Henning[1],
W. Jaffe[2], J. A. de Jong[2], R. Köhler[2], C. Maire[3], R. J. Mathar[2],
D. Mégevand[3], Y. Michellod[4], P. Müllhaupt[4], K. Murakawa[6], F. Pepe[3],
R. S. Le Poole[2], J. Pragt[6], D. Queloz[3], A. Quirrenbach[2], S. Reffert[2],
L. Sache[4], Y. Salvadé[5], O. Scherler[5], D. Ségransan[3], J. Setiawan[1],
D. Sosnowska[3], R. N. Tubbs[2], L. Venema[6], K. Wagner[1], L. Weber[3], and
R. Wüthrich[4]

[1] Max Planck Institut für Astronomie (MPIA), Heidelberg, Germany rl@mpia.de
[2] Sterrewacht Leiden, P.O. Box 9513, NL-2300 RA Leiden, The Netherlands
[3] Observatoire de Genève, 51 Ch. des Maillettes, CH-1290 Sauverny, Switzerland
[4] Ecole Polytechnique Fédérale de Lausanne (EPFL), Switzerland
[5] Ecole d'Ingénieurs ARC, St-Imier, Switzerland
[6] ASTRON, P.O. Box 2, NL-7990 AA Dwingeloo, The Netherlands
[7] European Southern Observatory (ESO), Garching, Germany

1 Narrow-angle astrometry with the VLTI

PRIMA, the instrument for Phase Referenced Imaging and Micro-arcsecond
Astrometry, is currently being developed at ESO. PRIMA will implement the
dual-feed capability at the VLTI for both UTs and ATs to enable simultaneous interferometric observations of two objects that are separated by up
to 2 arcmin, without requiring a large continuous field of view. PRIMA will
be composed of four major sub-systems: Star Separators, Differential Delay
Lines (DDLs), a laser metrology system, and Fringe Sensor Units (FSU). The
system is designed to perform high-accuracy (10 μas) narrow-angle differential astrometry in K-band with two FSUs and, with one FSU in combination
with AMBER or MIDI, phase-referenced aperture synthesis imaging. The
purpose of the DDLs in differential astrometry is to increase the astrometric
accuracy by separating the large OPD correction terms which are common for
the two stars from the small differential terms and to increase the sensitivity
by stabilizing the fringe pattern (in a closed loop with the laser metrology)
and thus allow for longer integrations.

2 Differential Delay Lines and Astrometric Software for PRIMA

In order to speed up the full implementation of the 10 μas astrometric capability of the VLTI and to carry out a large astrometric planet search program, a consortium lead by the Observatoire de Genève (Switzerland), the Max Planck Institute for Astronomy in Heidelberg (Germany), and the University of Leiden/NOVA (The Netherlands) agreed with ESO to build and deliver the Differential Delay Lines for PRIMA and to provide all necessary operation and software tools to perform narrow-angle astrometry at the 10 μas level. This includes developing and building the DDLs, the construction and analysis of an astrometric error budget, the establishment of an operations and calibration strategy, and the development of observation preparation and data reduction software:

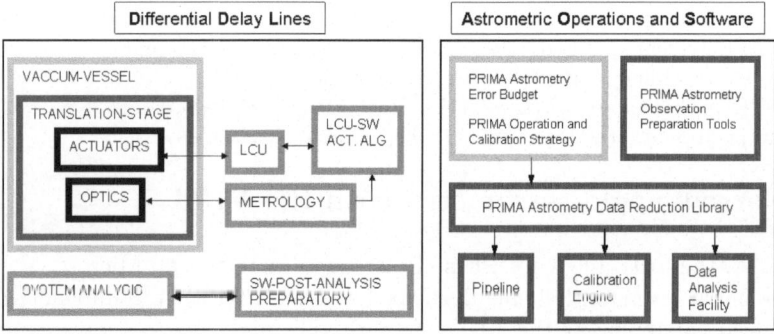

Fig. 1. The PRIMA-DDL project: breakdown of hard and software components

3 The astrometric planet search program

When completed in 2007, we will use the upgraded PRIMA facility to detect and characterize extra-solar planets through the reflex motions of their host stars in the plane of the sky. Two core programs are planned to be carried out over a duration of at least three years:

1) Observe stars with known radial-velocity planets. We will resolve the sin i uncertainty of the planet masses and thus constrain the uncertain upper end of the planetary mass function. For stars with multiple planetary systems we will derive the relative inclination of the orbits. We will follow up long-term radial velocity trends and search for new planets in longer-period orbits for which astrometry is more sensitive than the radial velocity method.

2) Search for planets around nearby stars of different mass and age without known planets. The search for planets by the radial-velocity technique is restricted to stars with narrow and stable spectral lines, thus excluding pre-main sequence stars as well as A and most F stars. Our astrometric planet search program will explicitly include such stars. For nearby (<20pc) late-type (F-M) main sequence stars, the primary new discovery space opened by such an astrometric facility would be Saturn down to Uranus-mass planets with orbital periods of a few years (a=1-5AU).

Scientific Requirements for the Darwin Mission

Malcolm Fridlund

Research and Scientific Support Department, European Space Agency
malcolm.fridlund@esa.int

Summary. In the European Space Agency's (ESA's) planning for the new Cosmic Vision 2015-2025 scientific program, the search for and study of exo-planets – particularly terrestrial ones – have a dominating position. The merging of the astronomical objectives, with a number of other sciences such as biology, chemistry, geophysics and planetology will undoubtedly begin a new science - Comparative Planetology. The study of how planets - and especially Earth-like ones - are formed, how they evolve in interaction with the star that they orbit and under different circumstances, how often they give rise to conditions that could in principle be benevolent (if benevolence is required?) enough for the origin of life to occur, and even if life as we know it has arisen on any world near the Earth. Since 1997, ESA has carried out detailed studies and preparatory work for a nulling space interferometer – designated Darwin. An important factor in the planning is a precise definition of the scientific objectives. In this contribution, these factors are summarised.

1 High level scientific requirements of Darwin

The high level objectives of the Darwin mission are:

- Perform a survey of nearby stars for terrestrial planets (i.e. planets similar to Venus, Earth and Mars in our Solar System), orbiting within the habitable zone of each star.
- Carry out a detailed study of any planets found, especially as what concerns composition, orbital mechanics, geophysical conditions (atmospheric composition, evolutionary status), and to determine its evolutionary status
- Carry out a spectroscopic analysis of any planetary atmosphere found for signs of biological activity as we would define it (commonly called bio-markers), or pre-cursors of the same.

2 Detailed scientific objectives and definitions

- Terrestrial planet definition: A minimum of 0.5 Earth radii and a maximum of 2 Earth radii. Earth albedo, Earth emissivity, Effective temperature between 260K and 373K. A density > 3g cm^{-3} and < 7g cm^{-3}.

- The Habitable Zone (HZ) definition: The area where liquid water could exist on a planetary surface of terrestrial type orbiting a solar type star. Considered to be a uniform zone with radii 0.7AU to 1.5AU scaled by the square root of the stellar luminosity (Sun = 1).
- Continuously Habitable Zone (CHZ) definition: The region where conditions have been benevolent to life as we know it for all of the Sun's evolutionary history. For the solar system, currently believed to be 0.9AU to 1.1AU scaled by the square root of the stellar luminosity (Sun = 1).

The number of terrestrial planets existing in the galaxy is completely unknown at the present. The excellent results obtained through radial velocity measurements, have determined the prevalence of giant planets in orbits relatively close to solar type stars, in our part of the galaxy, to be between 5% (all stars) and 15% (high metallicity stars). Determining this number for terrestrial planets is one of the main objectives of Darwin, and it can be defined accordingly: Either it is the prevalence of terrestrial type planets ("rocky planets") per solar type star. In the solar system this number is > 3. Alternatively it is also understood in terms of what percentage of solar-type stars have Earth-like planets (η_{Earth}).

A solar type star (in the context of the Darwin mission) is defined as being of main sequence F5 - K9 class. This means the same order of evolutionary time scales and no bound rotation in HZ. Essentially all identified exoplanets are found orbiting stars of high metallicity, but so far NO solar system analogue has been found dispite claims to the contrary.

3 Number of planets required to constitute success

The task of Darwin, as defined and required by the community is to: *"Carry out searches for and studies of terrestrial type exo-planets, particularly as what concerns their ability to host life as we know it"*. A success parameter to detect and study 15 systems (derived from evolutionary timescales and statistical consideration) with rocky planets. The detection of at least one such system must be considered a partial success.

Another important success parameter is what do constitute a meaningful *negative* result? Clearly, Darwin need to survey a large sample with a high degree of confidence (at least 90%) that no "Earth" has gone undetected. Darwin can survey all 500 solar type single stars out to 25pc if there is no zodiacal dust in the systems, and at least 150 systems if a significant (3-10 times the solar system value) dust level is present. *It should be pointed out that all models today require at least significant cometary or asteroidal presence for dust to be present around mature main sequence solar type stars.*

In summary, it is one main requirement for the mission to be able to detect an Earth size planet with a representative temperature, orbiting a solar type star at a distance of 1 AU scaled with luminosity, with a s/n of at least 5, at a distance of 25pc. Similarly, the mission must have the capability of determining if planets does NOT exist around a particular star, and within the HZ with a high degree of confidence. This requires repeated observations since the location of the planet could in any instance be where it is unobservable (e.g. behind the star), and depending on the inclination of the system.

4 Spectral range

The spectral range is determined by two key factors, namely, where the contrast between the star and any exo-planet is the least. For an effective temperature of 300 ± 50 K of a planet, this is at 12μm wavelength. Within the chosen range, signatures of the atmospheric characteristic and so-called bio-markers (i.e. indications of biological activity must be found at relevant conditions (temperature, pressure, etc). Such signatures are found in the Earth's atmosphere at wavelengths near 12μm, specifically CO_2 at 15.8μm O_3 at 9.3 μm and flanked by H_2O at 6 μm and beyond 18 μm. Further we find CH_4 (important bio-marker in the early atmosphere of the Earth) at a wavelength of 7.5μm.

5 Baseline mission

Darwin will search for terrestrial planets around at least 150 stars, if the stars are surrounded by 10 exo-zodies of dust, and significantly more (300 - > 500) if the stars are surrounded by less or no dust. This mission will detect $> 15 - 50$ systems under the assumption of $\eta_{Earth} = 0.1$. The baseline mission uses the Earth orbiting at 1 AU around a G2V star at a distance of 25pc with 10 times more exo-zodiacal dust than in our solar system as its measuring quantity for the detection of a planet. It is assumed that no dust (or very little) will be present if there are no planets (if the star is older $> 5 \times 10^8$ years). A signal to noise ratio of > 5 is required to qualify for detection. Colours have to be measured in at least 3 bands with a precision of 0.1 magnitude. An identified planet should be observed at least 3 times over the mission in order to characterize the orbit. A non-detection will have to have a confidence of 90%. A spectrum will be recorded (of each planet in a specfic detected system) with sufficient signal-to-noise to be able to measure the equivalent width of CO_2, H_2O and O_3 with a precision of 20%.

Searching for Faint Companions with MIDI Colour Differential Phase Measurements

Robert N. Tubbs[1]

Sterrewacht, Leiden University, 2300 RA Leiden, The Netherlands
tubbs@strw.leidenuniv.nl

Summary. Long observations of calibrator stars with the VLTI-MIDI instrument indicate that companions 200 times fainter than the parent star could be detected using colour differential phase measurements. I discuss the accuracy achieved with MIDI, the level of improvement required in order to detect large planets, and suggest improved observing strategies.

1 Introduction

There is currently a lot of interest in the detection of faint companions around nearby stars using mid-infrared interferometry. At mid-infrared wavelengths there is a lower contrast between sub-stellar companions and their parent stars, and the resolving power of long-baseline interferometers allows the inner few AU around nearby stars to be probed directly.

VLTI-MIDI is an excellent testbed for the technologies required for the detection of faint companions, including the colour differential phase technique. This method detects faint companions from the effect they have on the fringe phase when plotted as a function of spectral channel.

2 Analysis and Results

Two 5.5-minute calibrator observations taken with MIDI in Oct 2004 were reduced using the differential phase approach described in [1].

The residual phases from these observations are highly correlated, showing that phase calibration should be possible. The required calibration frequency is set by the temporal properties of the phase deviations. To investigate these, the data from both runs were split into 22-s segments, and phases were calculated for each segment. The mean square difference in phase between two segments was plotted for every pair of data segments, allowing a fit for the time-dependence. The fit parameters showed that each 22-s segment has a random variance of 0.50 deg^2 per channel, and that an optimum calibrated phase accuracy of 0.11 degrees should be obtained if 7 min observations are taken of source and calibrator, assuming 25-minutes are required to acquire each target. This would allow the detection of companions over 200 times fainter than the parent star with only a few observations (Fig. 1 b) and c)).

Fig. 2 shows observational results from the low-contrast binary Z CMa. The binary signature is visible in both the amplitude and phase, although the amplitude data suggest a slightly more complex source structure.

3 Conclusions

The present sensitivity of MIDI to faint companions should allow it to set upper mass limits on known faint sub-stellar companions around nearby stars, and to directly detect brown dwarfs.

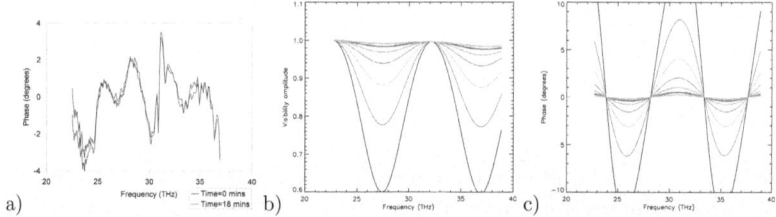

Fig. 1. Panel a) shows the residual phases in two calibrator observations. Panel b) shows the visibility amplitude from simulations of binaries with flux ratios of $1 : 2^n$ for $n = 2 \ldots 8$. Panel c) shows the corresponding differential phases.

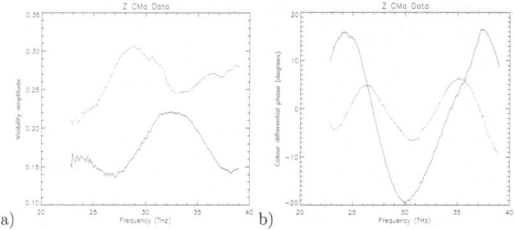

Fig. 2. Panel a) shows visibility amplitudes measured in two MIDI observations of Z CMa, and panel b) shows the corresponding differential phases.

Reference

1. R. N. Tubbs, J. A. Meisner et al: Proc SPIE **5491**, 588 (2004)

Infrared Light Curves and the Detectability of Close-In Extrasolar Giant Planets

L. J. Richardson[1], S. Seager[2], D. Deming[3], J. Harrington[4], R. K. Barry[1], J. Rajagopal[1], W. C. Danchi[1]

[1] Exoplanets and Stellar Astrophysics Laboratory, NASA/GSFC, Greenbelt, MD 20771 Lee.J.Richardson.1@gsfc.nasa.gov
[2] Carnegie Institute of Washington, Department of Terrestrial Magnetism, 5241 Broad Branch Rd. NW, Washington, DC 20015 seager@dtm.ciw.edu
[3] Planetary Systems Laboratory, NASA/GSFC, Greenbelt, MD 20771 ddeming@pop600.gsfc.nasa.gov
[4] Cornell University, 326 Space Sciences Bldg., Ithaca, NY 14853 jh@oobleck.astro.cornell.edu

1 Light Curves

We have developed a set of routines to calculate the infrared light curve for a given extrasolar planet. An outline of the process is as follows:

1. Read the physical parameters for the star and planet and determine the orbital elements. We are currently studying the well-known transiting extrasolar planet HD 209458 b, with parameters from [2].
2. Treat the planet as a blackbody and assume a non-zero temperature asymmetry between the day and night sides of the planet. This could occur if the planet is tidally locked and the atmosphere does not efficiently redistribute the absorbed stellar irradiation.
3. Integrate over the visible portion of the planetary disk as a function of its position in its orbit around the star. This gives the emitted radiation of the planet, relative to the star, as a function of time.

2 Theoretical Planetary Spectra

We ultimately plan to replace the blackbody assumption with a theoretical spectrum of the planetary atmosphere, calculated as outlined below:

1. Input temperature-pressure profile for the model atmosphere, calculated by solving the radiative transfer equation. We adopt profiles for HD 209458 b from [6], and we compare the results to updated models to be described in a future paper (Seager, in prep, 2005).
2. Tabulate line opacities for H2O [5], CH4 [1], and CO [4].
3. Finally, we integrate over the optical depth of the layers of the model atmosphere to obtain the emergent flux density.

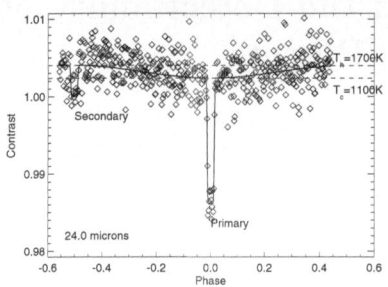
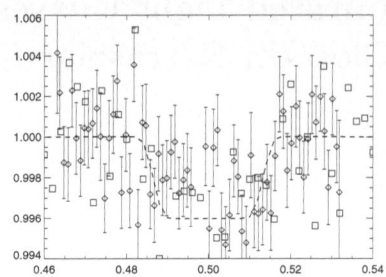

Fig. 1. Theoretical light curve at 24 microns, with stellar photon noise included. The ideal light curve is overplotted as a solid line. We assume a day-side blackbody $T = 1700$ K and a night-side $T = 1100$ K, as indicated by the dashed lines.

Fig. 2. Actual data (diamonds with error bars) from the [3] detection of the secondary eclipse from Spitzer. Noisy theoretical calculation (squares) from left, as well as ideal curve (solid line) are overplotted.

3 Future Work

The next step in this program is to combine the spectral synthesis and light curve routines. The spectral synthesis code must be expanded to include more extensive line opacity data in order to extend the calculation to 24 microns. We can then integrate the spectrum over a given bandpass for a particular instrument, and use this as input to the light curve routine. Finally, we will estimate noise from all sources, add it to the calculated light curves, and quantify to what extent the light curve is recoverable from the noisy synthetic data, allowing a prediction of what features (e.g., secondary eclipse, temperature asymmetry) are detectable for the known extrasolar planets.

References

1. Brown, L. R., Dulick, M., & Devi, V. M. 2001a, AAS/Division for Planetary Sciences Meeting, 33, 5611
2. Brown, T. M., Charbonneau, D., Gilliland, R. L., Noyes, R. W., & Burrows, A. 2001b, ApJ, 552, 699
3. Deming, D., Seager, S., Richardson, L. J., & Harrington, J. 2005, Nature, 434, 740
4. Goorvitch, D. 1994, ApJS, 95, 535
5. Partridge, H. & Schwenke, D. W. 1997, JCP, 106, 4618
6. Seager, S. & Sasselov, D. D. 2000, ApJ, 537, 916

Searching for VLTI Calibrators with the JMMC's Search Calibrators Tool

D. Bonneau[1], X. Delfosse[2], S. Cetre[2], J.-M. Clausse[1], G. Zins[2], O. Chesneau[1], G. Duvert[2], D. Mourard[1], and P. Cruzalèbes[1]

[1] Observatoire de la Côte d'Azur, GEMINI Daniel.Bonneau@obs-azur.fr
[2] Laboratoire d'Astrophysique de l'Observatoire de Grenoble

Summary. A new version of ASPRO "Virtual observatory" companion tool looking for possible calibrators in the vicinity of a science object has been released. This user dedicated software creates an evolutive catalog of stars giving the useful information for the selection of calibrators with respect to the requirements of the astrophysical program and of the instrumental configuration. SearchCal (for Search Calibrators) includes an K-Band (for AMBER) and an N-Band (for MIDI) calibrator search. It can be found on the JMMC Web site: $http://mariotti.fr/aspro\_page.htm$

1 The Problem

In stellar interferometry, the raw fringe contrast must be calibrated to obtain the true object visibility which will give the object parameters that can be interpreted in term of astrophysical parameters. Ideally, a calibrator must be a point source giving a fringe visibility equal to 1.0., but in practice, the smaller the calibrators the lesser the sensibility of the angular diameter determination to their intrinsic visibility or sources of instabilities. With the installation of AMBER and MIDI instruments at the VLTI, the selection of suitable calibration stars is crucial to reach the ultimate precision of the interferometric observations.

2 The Method

To create a dynamical catalog of calibrators surrounding the science object, a method of the virtual observatory type is adopted to develop the SearchCal software. The calibrator field is defined as a rectangular box centered on the scientific object in the ASPRO software environment [1] using astrophysical requirements. For the near IR case, an on-line request to CDS is created to extract relevant astronomical parameters from stellar catalogs available in the data base VisieR [2]. For the mid IR, we use the list of potential calibrators for the MIDI instrument provided by the MIDI Consortium [3]. The result is a list of stars with known parameters (astrometry, spectral type, photometry, indication of variability and multiplicity, measured angular diameter).

For each star, corrections for interstellar absorption are applied and missing photometry is completed from the compilation of published color-luminosity class-spectral type relation. The angular diameter are computed using a surface brightness method based on (B-V), (V-R) and (V-K) color index [4]. The choice of the calibrators can be refined by changing, a posteriori, the selection criteria (size of the field, magnitude range, spectral type and luminosity class, accuracy on the calibrator visibility, variability and multiplicity flags).

3 Technical aspects

SearchCal tool is a distributed application that sifts through CDS-based stellar catalogs to find a list of stars useable as calibrators for optical interferometry observations. It is built around a server that handles user requests and queries CDS stellar catalogs according to predefined scenarios. The resulting VO-Tables (standardized XML-based format defined for the exchange of tabular data in the context of the Virtual Observatory) are parsed to select the possible calibration stars. Its Graphical User Interface (GUI), developed using XML to Java Toolkit to be fully integrated in the JMMC's ASPRO Web software, allows user to display, sort, filter and save the catalog of calibration stars. The application is written in C++ using a flexible and scalable object-oriented methodology, promoting the reuse, extensibility and maintainability of software. This design is able to follow the improvements of scientific knowledge used in the scenarios as well as changes in the web queries and data format evolutions.

4 Toward fainter calibrators

To anticipate the gain in sensitivity expected with the instrument AMBER and PRIMA on the VLTI we have undertaken to develop an extended version of SearcCal allowing to reach $K > 5$. For faint calibrators search, it is planned to use the catalogues 2MASS and DENIS as primary catalogs and the size of calibrator field will be automatically adapted on the base of stellar count computed using a predictive model of the stellar population in the Milky Way [5] to limit the number of returns from CDS request.

References

[1] G. Duchene,J-P. Berger, G. Duvert, G. Zins, G. Mella, SPIE **5491**, 611, 2004
[2] F. Ochsenbein, P. Bauer and J. Marcout, A&AS **143**, 23, 2000
[3] http://www.eso.org/observing/dfo/quality/MIDI/qc/calibrators_obs.html
[4] X. Delfosse and D. Bonneau, In: SF2A Scientific Highlights 2004, ed by F. Combes et al., (EDP Sciences), 181, 2004
[5] A.C., Robin, C., Reylé, S., Derriére and S., Picaud, A&A **409**, 523, 2003

Observation of Asteroids with the VLTI

D. Loreggia[1], M. Delbo[1], M. Gai[1], M. G. Lattanzi[1], S. Ligori[1], L. Saba[1], M. Wittkowski[2], and A. Cellino[1]

[1] INAF-Osservatorio Astronomico di Torino, Strada Osservatorio 20, 10025 Pino Torinese (TO), Italy delbo@to.astro.it
[2] European Southern Observatory, Karl-Schwarzschild-Str. 2, D-85748 Garching bei Muenchen

Summary. We briefly discuss the capabilities of VLTI to obtain direct measurement of the sizes of asteroids. VLTI can play a crucial role in the study of interesting cases and to assess the reliability of thermal infrared, polarimetric and radar models.

1 Introduction

Information on the sizes, albedos, and surface characteristics of asteroids is essential for the knowledge of their nature and to constrain the models of formation and evolution. Direct techniques for measuring asteroid diameters like the measurement of their seeing disks and the observation of stellar occultation, are severely limited by the apparent angular sizes of the targets and by the difficulties in predicting and observing occultation events. So, indirect methods are mainly used in practice, based on radiometry and polarimetry. However, several cases of asteroids with anomalous albedos are known to exist: e.g. *a*) albedos derived from radiometry in disagreement with the expected taxonomic classification; *b*) disagreement between albedos derived from polarimetry and radiometry. VLTI has the capability of measuring directly the sizes of asteroids (see Fig.1). Moreover, the use of the mid-infrared instrument, MIDI, in disperse mode allows us to obtain visibility information at different wavelengths. Since the visibility is directly related to the compactness of the thermal infrared source, this makes it possible to obtain information about the surface temperature distribution of the object and to compare it with the predictions from thermal models.

2 Direct measurement of asteroid sizes with the VLTI

On the basis of the van Cittert-Zernike theorem, if we assume that the brightness distribution of an asteroid can be represented by a uniform disk of angular diameter θ, the fringe visibility function is $V = 2J_1(\pi B\theta/\lambda)/(\pi B\theta/\lambda)$, where J_1 is the first order Bessel function, B is the length of the projected baseline and λ the wavelength. In this hypothesis, the angular diameter, and thus the size, of an asteroid can be derived by a single visibility measurement.

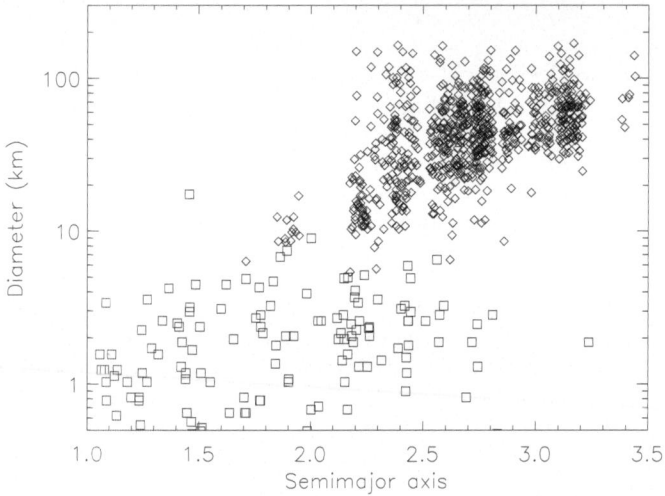

Fig. 1. Asteroids for which MIDI can provide direct size determination, according to the present instrument requirements: i.e *i*) a correlated thermal infrared flux (flux × visibility) at 11.8 *μm* greater or equal to 1 Jy and *ii*) a visibility greater or equal to 0.1. Diamonds: main belt asteroids (MBAs). Squares: near-Earth asteroids (NEAs). Orbital element are from the Minor Planet Center (MPC). Diameters have been calculated from the MPC H values assuming an average albedo of 0.2. Ephemerides have been calculated every 5 days for NEAs and every 15 days for MBAs for a time span of 15 years starting from June 1, 2005, to check for those bodies meeting conditions *i* and *ii*. For the calculation of thermal infrared flux, we have used the Standard Thermal Model of Lebofsky et al., 1986, Icarus 68, 235-251.

Fig.1 shows the distribution of those asteroids for which MIDI can provide direct size determination (see Fig.1 caption for more details). In October 2004, our team carried on the first VLTI observation run on asteroids (Science Demonstration Time). A second run has been proposed for Open Time, in December 2005. Interesting results obtained in Oct.'04 for the asteroid 1459 Magnya, the only V-type asteroid in the outer main belt, are presented by Delbo et al. (in preparation, expected to be published in 2005).

3 Conclusions

VLTI has the unique capability of measuring directly the sizes of small asteroids (down to 1 km in diameter in the NEA population and to 10 km in the main belt). This kind of observation, never performed with interferometric techniques before, is crucial to verify the reliability of thermal infrared, radar and polarimetric models.

Extraterrestrial Search Through Temporal Coherence

Erez N. Ribak

Physics Department, Technion - Israel Institute of Technology, Haifa 32000, Israel
eribak@physics.technion.ac.il

Summary. It is proposed to probe the temporal coherence of celestial objects to locate yield natural coherent sources, such as masers. But we might also see gravitationally lensed objects, ionospheres of planets, and occasionally intelligent life forms that use lasers. A simple realisation is by an imaging interferometer operating at large optical path difference, and is rather easy on a single telescope with a zero-shear interferometer. In a survey mode, a very wide field is measured, yielding all the coherent sources in it. Past experience shows that coherent signals can be detected against their incoherent bright host objects. The contrast ratio demonstrated is one thousand, and at known wave length, more than a million.

There might be some clues and suggestions as to where it might be possible to find temporal coherence in astrophysical processes. Among others, synchrotron radiation can be also coherent over the scale of electron groups or clumps inside the groups. Extremely strong scintillation was discovered in quasars which can be explained by small spatial dimensions but also by temporal coherence. In masers, emission occurs because of population inversion and hence photon amplification. The effect can also occur in the visible, such as in planetary nebulae or regions near intense sources. Coherence in quasars might be due to stimulated emission from rapidly cooled plasma. Stimulated emission might explain unusual iron lines intensities in the ultra violet. Another interesting possibility is in gravitational lensing by a small object, where the paths to the observer do not differ significantly (an Einstein ring). Lensing surveys did not look for coherence of the beams. No longitudinal coherence was observed in celestial sources until today. Masers are usually discovered by their narrow lines and polarization, but the effect of temporal coherence is generally not considered, perhaps because of the assumption of maser line broadening.

In radio interferometry, measurement of temporal coherence is by correlation of long lag, for every dish and between dishes. As the wave length becomes shorter, the number of photons per mode decreases, and near 10μm it drops to a few. This makes correlations inefficient as compared to amplitude interferometry. Still, it was proposed recently to utilize heterodyne intensity interferometry at 1μm to detect coherent radiation. A similar idea is to employ intensity interferometry on extremely large telescopes towards this end, as causality (such as in stimulated emission) can only be discovered in testing for photon arrival times. Perhaps simple amplitude interferometry will be able to detect temporal coherence very efficiently, even at smaller telescope sizes. The field of view to be searched is thus imaged twice, through two unequal arms of the interferometer, and the images are made to overlap with each other (zero-shear interferometer). Objects with a shorter coherence length will have a minimum mutual coherence, and their contrast will be depressed, while those with coherence longer than the delay will still interfere, and their contrast will be measurable.

Zero-shear interferometers can perform temporal coherence measurements, provided they have a very wide field of view, and the delay can be changed easily (for a temporal modulation) or be introduced into the stellar images (for spatial coding of the fringes). Fabry-Perot etalons with extended cavities are also good candidates. As was realized very early on, detection of laser radiation in the battlefield is an essential weapon, and thus there is a wealth of military devices for coherence detection. Many of these use only polarized light and must be modified for astronomy.

A lateral-shear interferometer illustrates the principle. Between the beam splitter and combiner we insert two different delays for the two beams, a fixed delay and an oscillating one. Two complementary images of the stellar field are formed on the detector with mirror symmetry. Inequality of intensity in corresponding pixels will signify coherence longer than the delay. Spatial fringes can also be formed on purposefully extended stellar images. A Fourier transform of the whole image will detect the modulation in space or time.

Experiments performed with military devices already give us some very good clues as to the ratio of the coherent signal to the incoherent clutter (here, the celestial field). A 100% coherent beam, at a known wave length, was discernible against an incoherent one 10^5 brighter and under favorable conditions even 10^7 brighter. Extrapolating from these, it is expected to be able to detect coherent signals on a 4 m telescope within hours on large patches of the sky.

Great effort is being invested in the search for extra-terrestrial intelligence (SETI), which might be using pulsed beacons. Historically initiated in the radio, the visible regime is a serious contender. A recent search for pulsed optical signals found no candidates. It might make sense to look for unintended such signals, modulated or not. The fingerprint of such signals is their temporal coherence and perhaps, in addition, their time sequence signatures. If such beings have indeed produced a laser, they might also have tried to use it at shorter wave lengths, down to x-rays. These should be detected on a much quieter background, such as that introduced by a nearby sun. Thus the survey for natural temporal coherence should supplement SETI, with the great advantage that it does not have to be directed at a specially chosen source. Further details, magnitude estimates, and references will be provided in an upcoming paper on this subject.

MIA+EWS, the Software for MIDI Data-Reduction

Rainer Köhler[1] and Walter Jaffe[1]

Sterrewacht Leiden, Niels Bohrweg 2, NL-2333 CA Leiden, The Netherlands,
koehler@strw.leidenuniv.nl, jaffe@strw.leidenuniv.nl

Summary. We present the data reduction software for MIDI written at Sterrewacht Leiden and MPIA Heidelberg. It uses power-spectrum analysis or the information contained in the spectrally-dispersed fringe measurements, in order to estimate the correlated flux and the visibility as function of wavelength in the N-band.

1 Introduction

For the first commissioning and scientific observations with the Mid-Infrared Interferometric Instrument MIDI [1], two data-reduction software packages were developed, one called *"MIDI Interactive Analysis" (MIA)* at MPIA Heidelberg, and another called *"Expert Work-Station" (EWS)* at the Sterrewacht Leiden. In 2005, we combined both packages into one with the somewhat uninspired name *"MIA+EWS"*.

2 Supported MIDI Modes

MIDI is a complicated instrument with a lot of different observing modes. Of course, our aim is to support all of them, but so far MIA+EWS supports only a subset.

- **Dispersive Elements**
  PRISM: fully supported
  GRISM: supported, but not fully tested
- **Number of Channels**
  HIGH_SENS: fully supported
  SCI_PHOT: work in progress
- **Tracking mode**
  undispersed: fully supported
  dispersed: supported, but not fully tested

3 Download and Installation

MIA+EWS is written in C and IDL, it runs on Linux, Solaris, and HP-UX. Porting it to other Unix-variants should be quite easy. We published it under the GNU Public License and make the full source available, so you can check if the code actually works the way we claim it does. We also encourage improvements and additions to the program by others.

You can download the complete package from

```
http://www.strw.leidenuniv.nl/~nevec/MIDI
```

4 Algorithms

MIDI is a Michelson or pupil interferometer. To measure the fringe amplitude, the optical path difference (*OPD*) is varied with the help of MIDI's internal delay lines. This produces a quasi-sinusoidal variation of the signal on the detector. Due to atmospheric fluctuations, the OPD between the two beams varies on timescales of milliseconds. Therefore, the fringes of different scans are not aligned and can not simply be added to improve the signal to noise ratio. In MIA+EWS, two different algorithms are implemented to solve this.

4.1 Incoherent Analysis

Here we use the power spectra of the scans. Because of the OPD variations, the fringe gives a signal within a known frequency range. The integral over these frequencies gives the square of the correlated flux. The contribution of noise sources is measured in scans far away from OPD = 0 and subtracted.

4.2 Coherent Analysis

This method uses the fact that fringes are not only visible in the OPD scans, but also in each single spectrum, which allows to measure the real OPD in each frame. Then, we can "rotate" the fringes back to OPD = 0 and add them. Here we sum the signal linear, not quadratic, so noise is zero on average. A full description of this method can be found in [2].

References

1. Ch. Leinert et al.: Ten-micron instrument MIDI: getting ready for observations on the VLTI, SPIE **4838**, 893 (2003)
2. W. Jaffe: Coherent fringe tracking and visibility estimation for MIDI, SPIE **5491**, 715 (2004)

Understanding Cross Talk on the NPOI Multibeam Combiner

H. R. Schmitt[1,2], J. T. Armstrong[1], R. B. Hindsley[1] and T. A. Pauls[1]

[1] Remote Sensing Division, Code 7215, Naval Research Laboratory, 4555 Overlook Avenue SW, Washington, DC20375, USA
hschmitt@ccs.nrl.navy.mil
[2] Interferometrics Inc., 13454 Sunrise Valley Drive, Suite 240, Herndon, VA20171, USA

Summary. We present the results of tests done with the Navy Prototype Optical Interferometer (NPOI) multibeam combiner, designed to study the effects of fringe cross talk resulting from multiple baselines being recorded by the same spectrograph. We find that in most cases cross talk is not a significant issue, except when the fringes are separated by only one wavenumber.

1 Introduction

The NPOI consists of an imaging sub-array composed of 6 movable telescopes, and an astrometric sub-array composed of 4 fixed telescopes (Armstrong et al. 1998). The light from up to 6 telescopes can be combined pairwise by the beam combiner. The output of the beam combiner is recorded in 3 spectrograph, where the light from up to 4 input beams, representing 6 baselines, can be recorded in each spectrograph at the same time. Each input beam is modulated by a 500 Hz triangular delay, with amplitudes that are integers in the range -4μm to 4μm. The demodulation of the signal is done by measuring the photon count rates in 64 bins per modulation period. A Fourier transform of these bins is generated and the power is measured in frequencies corresponding to 1, 2, ... fringes per modulation. When 4 stations are recorded by the same spectrograph, we have 6 fringes and 8 possible fringe scanning frequencies. For the fringes separated by one wavenumber, possible non-linearities in the delay stroke can create cross talk.

The observations used in these tests took advantage of the beam combiner design, where certain baselines are recorded by more than one spectrograph. We started with 2 stations that were simultaneously recorded by 2 spectrograph. On subsequent nights we observed with these 2 stations plus a 3rd and a 4th station, which could only be recorded by one of the spectrographs. In this way the spectrograph with a single baseline is used as a reference, to analyze the effect of adding multiple stations to the other spectrograph.

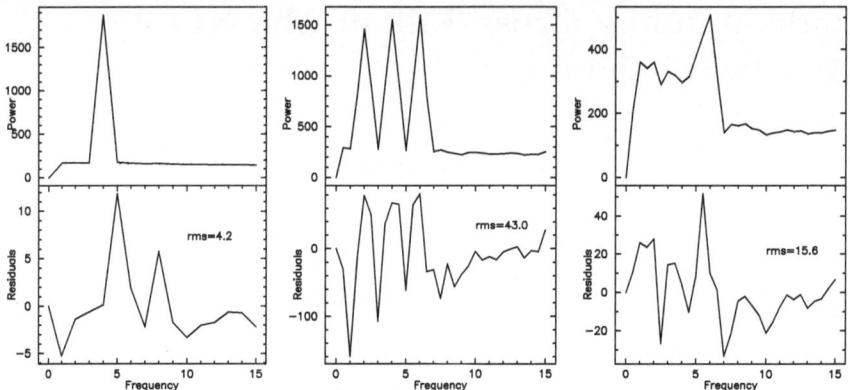

Fig. 1. Fringe power spectrum (top) and residuals from the subtraction of model (bottom). The left, middle and right panels show the cases of 2, 3 and 4 stations (1, 3 and 6 baselines) in the same spectrograph, respectively.

2 Results

Starting with the baseline that was recorded by both spectrographs, we compared the ratio of the squared visibilities measured by the different spectrographs with values predicted based on photometric scans. Overall we find a good agreement between observed and predicted values, with the exception of small deviations for a few channels. These deviations can be attributed to different detector characteristics, which we are currently determining. We also found that the noise increases when multiple baselines are added to the same spectrograph, as expected, especially in the blue part of the spectrum where the sensitivity of the detectors is lower.

A preliminary analysis of the fringe power spectrum and residuals obtained after fitting and subtracting the expected fringe frequencies is presented in Figure 1. The left panel shows the case when only 1 baseline is recorded (k=4), the middle one shows 3 baselines (k=2, 4 and 6) and the right one shows 6 baselines (k=1 through 6). When only one baseline is recorded we do not see significant amounts of power at frequencies other than the expected one, as evidenced by the small residuals. In the case of 3 baselines separated by 2 wavenumbers each, we can recover the power in each one of the peaks without any problem; however, the residuals indicate a small mismatch between the observed and expected frequencies. In the case of 6 baselines we find much larger residuals, suggesting that cross talk is a significant problem in this case, thus showing the limitations of this configuration.

Reference

1. Armstrong, J. T., et al. 1998, ApJ, 496, 550

Part IX

Posters — Instrumentation

A Model Experiment for APreS-MIDI

J.-L. Menut[1], Y. Bresson[1], Y. Hugues[1], S. Flament[1], Pa. Antonelli[1],
A. Roussel[1], N. Schweitzer[1], Pi. Antonelli[1], S. Lagarde[1], M. Dugué[1],
B. Lopez[1], S. Wolf[2], U. Graser[2], S. Jankov[1], T. Ratzka[2], L. Mosoni[2,3],
A. Niedzielski[4], and E. Thiébaut[5]

(1) Observatoire de la Côte d'Azur
(2) Max Planck Institut für Astronomie
(3) Konkoly Observatory
(4) Torun Observatory
(5) Observatoire de Lyon

1 Intoduction

APreS-MIDI is a concept proposed for a $10\mu m$ imaging instrument dedicated to the VLTI. It has the objective to extend the MIDI instrument to four recombined beams (see Lopez et al. [1], Dugué et al. [2]).

One important step in the present phase of the instrumental study is to simulate the principle of recombination of APreS-MIDI with a model experiment. The goal is to produce laboratory data close to the ones that will be generated by APreS-MIDI. This data will be used to prepare the backbone of the future data analysis software.

2 Physical principle and implementation of the model experiment

Based on the same physical principle of APreS-MIDI, the model experiment provides a recombination of three beams by a pupil plane multiaxial method. Figure 1 shows the three basic steps of this method.

On the first step, the wavefront $\hat{O}(u)$ coming from an object $o(\alpha)$ is sampled by three telescopes located in u_1, u_2, u_3. On the second step, the three beams are superimposed (technically, the beam are sent through the same set of mirrors). On the final step, a slight tilt angle is introduced on each pupils before their superimposition onto the detector, where the fringe pattern is obtained (Fig. 2).

The technical implementation of the model is the following (Fig. 3): the flux from a source is collimated by a twenty-centimeter mirror. Three little mirrors sample the wavefront. Three beams are reflected on a multi-faceted mirror which superimposes them, with a tilt angle, on the detector (through some intermediate optics).

To set up the model, we use a laser and a pinhole. A white source and a more complex geometry will be used, in order to prepare the data reduction software and the image reconstruction process.

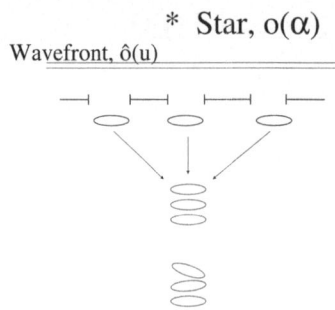

Fig. 1. Physical principle of the recombination

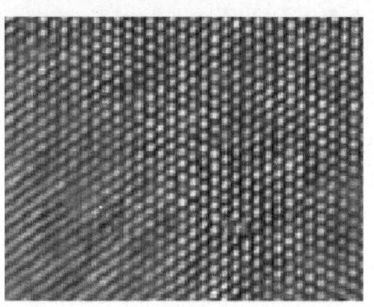

Fig. 2. Fringes pattern oberved onto the detector

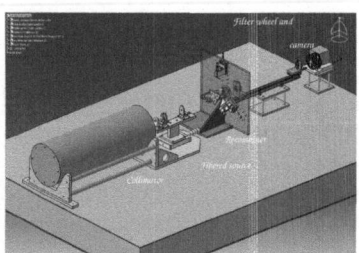

Fig. 3. Catia drawing of the model experiment

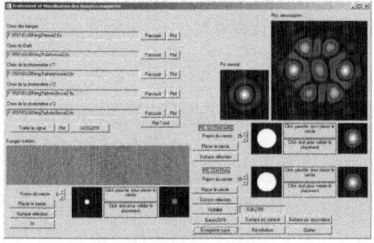

Fig. 4. First version of the data reduction software

3 Data reduction

The data reduction process is still under research but a first version is already running (Fig. 4). For evaluating the complex visibility, the algorithm computes the ratio between a fringe peak in the Fourier conjugated plane and the related photometric signal. Other methods will be based on the fitting of the fringes peaks to extract their parameters, or on direct fitting of the fringes.

References

1. B. Lopez, S. Wolf and the APreS-MIDI team : APreS-MIDI, APerture Synthesis in the Mid Infrared. In this volume.
2. M. Dugué and the APreS-MIDI team : APreS-MIDI, a four beam recombiner. In this volume.

From the VLBI to the VLTI: An APreS-MIDI Image Reconstruction Study

László Mosoni[1,2], Sebastian Wolf[1], Bruno Lopez[3], Frank Przygodda[1], Thorsten Ratzka[1], Jean-Luc Menut[3], and the APreS-MIDI Science Team

[1] Max Planck Institute for Astronomy, Heidelberg, Germany
[2] MTA Konkoly Observatory, Budapest, Hungary `mosoni@konkoly.hu`
[3] Observatoire de la Cote d'Azur, Departement Gemini, Nice, France

Summary. We present results of an image reconstruction study in the framework of the APreS-MIDI Science Case Study [1]. We investigate image reconstruction methods of the Very Long Baseline Interferometry and the possible gains for APreS-MIDI. Different image reconstruction methods are available and can be considered depending on the future VLTI infrastructure and instrumentation.

1 Introduction

APreS-MIDI [1], designed as a 3 or 4-beam combiner, will enable image reconstruction in the mid-infrared wavelength regime at the Very Large Telescope Interferometer (VLTI). It will overcome the ambiguities often existing in the interpretation of simple visibility measurements.

Self-calibration image reconstruction methods are widely used in the Very Long Baseline Interferometry (VLBI) [2]. These methods were developed to make phase unstable VLBI arrays imaging instruments. We applied phase self-calibration algorithms, mainly difference mapping [3], and compared the results to the reconstructed images obtained with a bispectrum algorithm (i.e. Building Block Method [1, 4]).

2 Image Reconstruction Results

For the APreS-MIDI image reconstruction studies, model images were created with a radiative transfer code MC3D [5]. First, we consider the closure phase and the visibility amplitude information only. Phase self-calibration methods need a starting model. In some cases, this model can be a point source and so the procedure is model-independent. However in general, the starting model consists of more complex model components (such as a Gaussian distribution) in order to obtain a reconstructed image which fits well to the visibility data. This fit can be considered as a measure of the reliability of the reconstructed image. Although in many cases in our study, the image reconstruction was successful with difference mapping, the model-independent Building Block Method is more favourable [1].

But in some cases, neither the bispectrum nor the closure phase information is sufficient for a successful image reconstruction. In most of these cases, the image reconstruction is possible with VLBI techniques because the Fourier-phase information can be considered (Fig. 1). In a 3 or 4-telescope configuration, the closure phases and the bispectrum contain much less information on the source structure than the Fourier-phases. Self-calibration algorithms can handle some tens of degrees of Fourier-phase errors.

Different methods have been developed to handle the image reconstruction problem, e.g. deficiencies of the Fourier-plane sampling and the limited phase information. Bispectrum and phase self-calibration methods are available, the image reconstruction will depend on the VLTI infrastructure and instrumentation – what observables will be measured.

One of the basic factors of the interferometric image reconstruction is the uv-coverage, i.e., how well the Fourier-plane is sampled. Our image reconstruction studies show that 3-7 nights of observations with 3-4 Auxiliary Telescopes (ATs, at varying locations) will result in an uv-coverage which is sufficient in order to reconstruct images which allow to address profound science questions [1].

Acknowledgement. L.M. acknowledges the help from Christoph Leinert, Ralf Launhardt, Ulrich Hiller (MPIA) and Zsolt Paragi (Joint Institute for VLBI in Europe). S.W. was supported by the German Research Foundation (DFG) through the Emmy Noether grant WO 857/2-1.

Fig. 1. Example: The test image (left) is a circumstellar disk around a T Tauri star with an inner hole up to 4 AU at the distance of 140 pc. The final reconstructed images considering Fourier-phase information: with a 3 nights 4 ATs (middle) and 5 nights 4 ATs (right) baseline configurations. In the first case (middle), the reconstructed image by a bispectrum method showed unresolved structure.

References

1. S. Wolf, B. Lopez, et al.: *The APreS-MIDI Science Case Study* (2005)
2. T.J. Pearson & A.C.S. Readhead: ARA&A **22**, 97 (1984)
3. M.C. Shepherd: ASP Conf. Ser. Vol. **125**, 77 (1997)
4. K.-H. Hofmann & G. Weigelt: A&A **278**, 328 (1993)
5. S. Wolf, Th. Henning & B. Stecklum: A&A **349**, 839 (1999)

A Numerical Simulator for VITRUV

J.-B. LeBouquin[1], E. Herwats[1], M.-I. Carvalho[2], P. Garcia[2], J.-P. Berger[1], and O. Absil[3]

[1] LAOG-UJF, BP 53, 38041 Grenoble, France
jean-baptiste.lebouquin@obs.ujf-grenoble.fr
[2] DF-FEUP et CAUP, Porto University, Portugal
[3] Institut d'Astrophysique, Allée du 6 Août, 17 B-4000 Liège, Belgique

Summary. VITRUVsim is a numerical tool with *as much as possible* physics included. Inputs are the source parameters (flux, morphology, position...) and outputs are sequences of observed fringes and/or reduced visibilities. VITRUVsim is written in a portable and free language Yorick[4].

1 What and How is the physics currently included ?

The principle is to follow as much as possible the real instrumental architecture, because it provides a major flexibility. VITRUVsim is dedicated to single mode instrument. Each optical element can be summarized by :

Electric transmission and phase: It can depend from beam, wavelength and polarization state. The formalism is a generalization of the coherency matrix used in radio interferometry [1]. Especially, it allows to deal with (de)polarization effects.

Strehl ratio and tip/tilt: Strehl codes the ratio between the whole pupil flux and the fundamental mode energy. Tip and Tilt are angles between the fundamental mode and the optical axis. These three quantities can depend only from the considered beam. This is one of the main limitation of VITRUVsim : the beam profile cannot be different between polarizations and/or wavelengths.

The elements taken into account are the atmosphere, the VLTI optical train and facilities (FINITO, MACAO, STRAP), the fibers, the integrated optics combiners [2] and the detection unit. Many informations about VLTI modelisation have been extracted from the GENIEsim project [3].

2 What can we do with this tool ?

First, it allows to check the instrumental requirements. Then, it provides virtual observations useful to test the expected performances. A simple version of the Data Reduction Algorithm is also under development. At least, it will

---

[4] ftp://ftp-icf.llnl.gov/pub/Yorick/doc/index.html

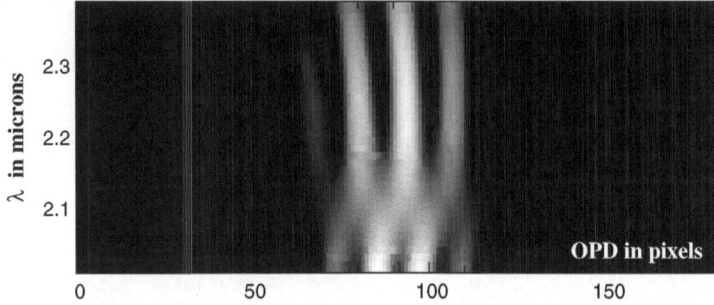

Fig. 1. Instrumental investigations: A difference of few millimeters between the fiber length introduce a strong chromatic variation of the instrumental contrast. VITRUVsim can predict how additional instrumental and/or astrophysical polarization will bias the calibration of this effect.

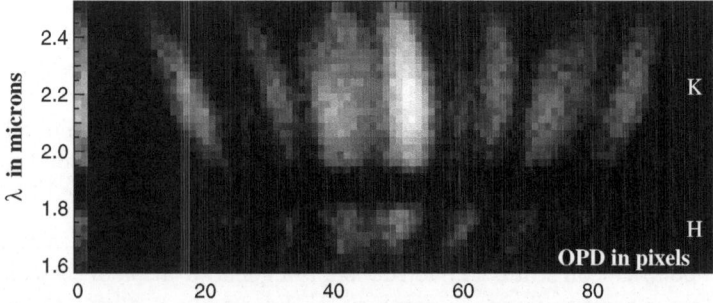

Fig. 2. Observation simulations: This dispersed fringe pattern have been simulated with a 4-beam multi-axial combiner. The contrasts differ because the target is resolved by the baselines. The fringes are curved by the turbulent piston and the dispersion in the delay lines. Since the injection is optimized for the K band, the atmospheric refraction reduce the coupled flux in the H band. Both photon and detector noises degrade the image quality.

help to interpret the laboratory experiments with different kinds of integrated optic combiners.

References

1. Hamaker, J. P., Bregman, J. D., and Sault, R. J.: 1996, *A&AS* **117**, 137
2. LeBouquin, J., Berger, J., Labeye, P., Tatulli, E., Malbet, F., Rousselet-Perraut, K., and Kern, P.: 2004, in *Proceedings of SPIE Volume 5491. Edited by Wesley A. Traub.*, pp 1362−+
3. Absil, O., den Hartog, R., Erd, C., Gondoin, P., Kaltenegger, L., Fridlund, M., Rando, N., and Wilhelm, R.: 2003, in *ESA SP-539: DARWIN/TPF and the Search for Extrasolar Terrestrial Planets*, pp 317–321

Multiway Beam Combiners for VITRUV

M. Benisty[1], J.-P. Berger[1], L. Jocou[1], and P. Labeye[2]

[1] LAOG 414 Rue de la Piscine, 38400 St Martin d'Heres
[2] LETI, CEA, 17 rue des Martyrs, 38054 Grenoble Cedex 9
   Myriam.Benisty@obs.ujf-grenoble.fr

1 Integrated optics and interferometry

It has been proved that integrated optics (IO) technology improves the quality of the interferometric measurements and highly simplifies a multi beam combination instrument (see [4, 6, 7]).

IO combiners are compact, provide high stability and spatial filtering, which combined with photometric calibration improves the visibility accuracy. They have a low sensitivity to external constrains and are particularly well suited for interferometric combination of numerous beams to achieve aperture synthesis imaging.

In this paper, we report the first laboratory measurements obtained with a 4 way IO combiner, similar to the one used at the IOTA 3-telescope (see [2]). We systematically characterized the properties of our beam combiner in the H band: throughput, birefringence and dispersion, through the analysis of the polarizations and of the visibilities.

2 Experimental setup

The interference fringes were acquired using a 3-telescope Mach Zender interferometer with a 4-beam combiner in the H band. The fibers we used have equal lengths and maintain polarization. A Wollaston prism at the IO combiner outputs splits the two polarizations and the interferometric signals are detected on a cooled HgCdTe camera, with a H band filter. A eight-telescope laboratory interferometer, specially developed to simulate the VLTI and the VITRUV instrument, will be used to characterize the next generation combiners (see [3]).

3 The IO beam combiner

The IO combiners are made with the silica on silicon etching technology and are similar to the one used at the IOTA 3-telescope interferometer. This prototype (of size 5mm x 27mm) allows pairwise combination between 4 beams and has 12 outputs (see fig 1 for a sketch and an image of the IO combiner). The output signal is coaxial and therefore temporally encoded.

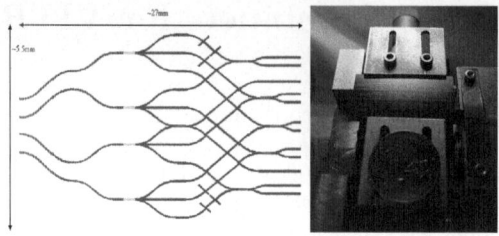

Fig. 1. *Left:* Sketch of the 4-way IO combiner. *Right:* Image of the combiner.

4 Results

With this combiner, we measured a good transmission of about 78%. The visibility measurements of a point source give satisfactory contrasts for a first prototype, from 50% to 60% and a similar behavior for the two polarization states. The contrasts are very stable ($\pm$ 1%) over a 10-day period and are repeatable over a long term period. We measured cross talk flux, and it has been corrected for the second generation prototype.

This optical circuit is one of the building block of the 4-way combiner for the VITRUV instrument and this extensive characterization has led us to design a second generation beam combiner with improved performance. We expect the final instrumental constrast to reach a level similar to the IOTA/IONIC 3T combiner (contrast above 90% in laboratory), by decreasing the cross talk flux and by improving the imaging system.

5 Conclusion

These results extend the capacity of IO combiners to combine 4 beams (e.g. four VLTI UT or AT beams) and confirm that the integrated optics technology is particularly well suited for interferometric combination of multiple beams to achieve aperture synthesis imaging with the VLTI-VITRUV (see [1, 2]).

References

1. F. Malbet et al, these proceedings.
2. J.-P. Berger et al., these proceedings.
3. L. Jocou et al., New Frontiers in Stellar Interferometry, Proceedings of SPIE, Vol. 5491, 2004, p. 1351
4. J.-B. LeBouquin et al., A&A **424**, 719, 2004
5. F. Malbet et al., A&AS **138**, 135, 1999
6. J.-P. Berger et al. A&A **376**, L31, 2001
7. J. Monnier et al., ApJ **602**, L57, 2004

A Laboratory Interferometer for VITRUV

L. Jocou, M. Benisty, P. Gratier, J.P. Berger, E. Le Coarer, A. Delboulbé,
Y. Magnard, P. Kern, K. Perraut, and F. Malbet

LAOG 414, Rue de la piscine - 38400 St Martin d'Heres
laurent.jocou@obs.ujf-grenoble.fr

1 Introduction

This laboratory interferometer is specifically designed to characterize the 4T
to 8T Integrated Optics (IO) combiners developed for VITRUV. The system
(Fig. 1), designed for the H band, is composed of an object simulator, a VLTI
simulator presenting up to 8 telescopes and a spectro-polarimeter as proposed
in the VITRUV design. We report here the design of the bench and the first
results obtained in laboratory.

2 Simulator description

Object simulator : The optical mount (based on a Michelson concept)
allows to obtain two non-coherent luminous spots simulating a binary star.
The separation and the contrast between each spot can be adjusted to study
the reconstruction techniques of complex objects. The image of this object
is placed at the focal plane of a F/5 collimator which produces a 110mm
collimated beam.

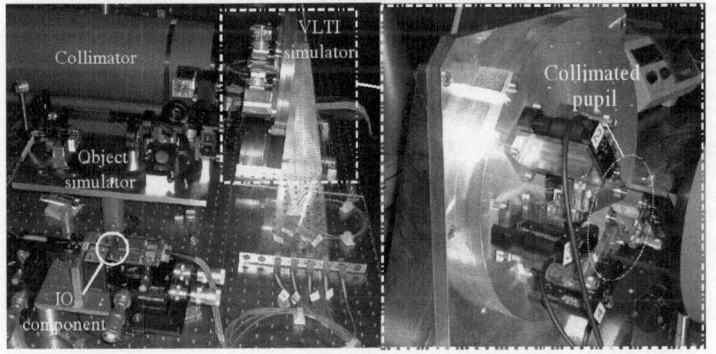

Fig. 1. View of the bench - zoom on VLTI simulator

VLTI simulator :   The aim of this system is to place 8 telescopes in the collimated beam (Fig. 1). The 8mm diameter telescopes can be set on a baseline varying from 20 to 100mm distributed on the field to have a good cover of the UV plan. The telescopes are composed of an F/5 gradium lens feeding a maintaining polarization fiber. Each telescope is mounted on an individual module with the required adjustments (tip-tilt) and a motorized translation stage. This one ensures both OPD adjustment and optical path modulations.

IO combiner environment :   The fibers coming from the telescopes are gathered in a V-groove which feeds the IO combiner input. To guarantee the coupling stability, we have chosen to cement each IO component to its fiber array. This connecting process is performed with a dedicated bench available at LAOG.

Spectro-polarimeter :   The spectrograph is designed to image the waveguide output on the detector with a medium spectral resolution (R=200 - 40 spectral channels on the H band). The imaging function is fulfilled by a couple of achromatic doublets. A Wollaston prism has been implemented in order to have access to a polarization analysis of the fringes. The detector used will be a picnic (Rockwell) 256x256 px.

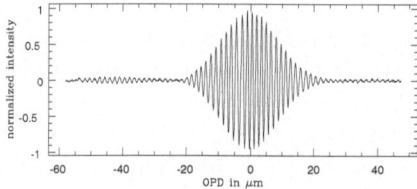

Fig. 2. Fringes obtained with a 4T IO combiner

3 Status and first results

The object and the VLTI simulators are both under characterization while, the spectro-polarimeter and the detector should be operational during summer. The detector used for the current characterization is a single pixel coupled with a single fiber collecting the flux at the waveguide outputs. The first fringes (Fig. 2), have been obtained with a broadband source ($\lambda = 1.6\mu$m - $\delta\lambda$=110nm), a non resolved single object and a 4T IO combiner. The visibility of 95% measured here reveals a sane behavior of the system.

Image Reconstruction with VITRUV

E. Tatulli[1], E. Thiébaut[2], F. Malbet[1], P. Garcia[3], and G. Duvert[1]

[1] LAOG-UJF, BP 53, 38041 Grenoble, France
eric.tatulli@obs.ujf-grenoble.fr
[2] CRAL, Lyon University, France
[3] CAUP, Porto University, Portugal

Summary. Imaging in optical interferometry is definitively the path to follow with the advent of interferometers with large aperture. The VLTI has an important way to play, since it is a unique interferometric site with 4 relocable 2m-class telescopes in addition to the 4 Very Large Telescopes. However moving from model-fitting of visibilities obtained with only 2 or 3 telescopes to image reconstruction with a full array of telescopes is not trivial. Therefore we have made simulations of what we can expect from the VLTI both with AMBER and the 8-way beam combining instrument project called VITRUV. We have taken the environment of young stellar objects (YSOs) as an example in order to quantify the image reconstruction quality.

1 VITRUV: a spectro-imaging instrument for the VLTI

Instrumental Parameters: VITRUV is a spectro-imaging interferometric instrument [1], designed to work on the VLTI in the infrared range, with a possible extension in the visible. It is assumed to be coupled with high-speed fringe tracking, allowing long integration times. VITRUV will recombine simultaneously 4 and 8 telescopes, and will therefore give access to visibility and closure phase measurements. Table 1 gives the instrumental parameters of VITRUV in its typical configuration.

Table 1. Instrumental configuration of VITRUV

Exposure time τ_{dit} (s)	Exposures N_{dit}	Spectral Resolution $\lambda/\Delta\lambda$	Detector Noise $\sigma(e^-/\text{pixel})$	Telescopes N_{tel}
5	100	100/1500/10000	15(IR)/1(Vis)	4/8

Expected Performances: Limiting magnitudes have been computed assuming a visibility SNR of 10 per spectral channel. Average atmospheric parameters of Paranal have been considered, with a Seeing of $\mathcal{S} = 0.8''$ (see Fig. 1). Great pupil diameters combined with spatial-filtering integrated optics recombiner drives to very promising performances [2], that will enable the observation and the imaging of many faint sources

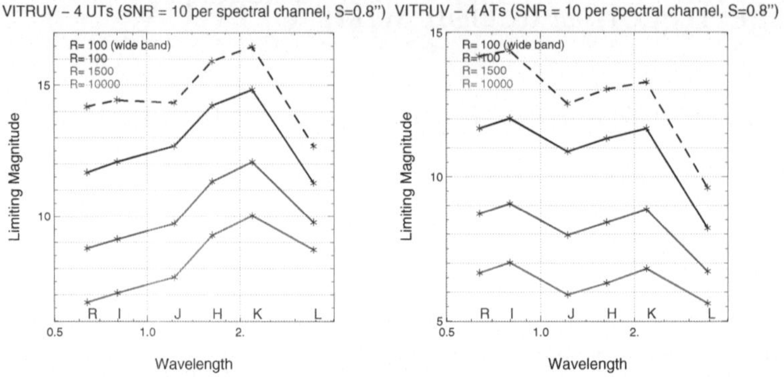

Fig. 1. Limiting magnitudes for the *VITRUV* instrument, using 4 UTs and 4 ATs.

Spatial frequency coverage and imaging potential: VITRUV instrument together with the unique coverage of the VLTI will: (i) bring significant gain in the number of possible complex visibilities. VITRUV is 2 to 6 times faster than AMBER for 4 or 8 telescopes configuration, respectively; (ii) measure closure phase together with visibility measurements, and then will enable to unveil source's asymetries; (iii) conjugate very high spatial and spectral resolution: low spectral resolution for continuum imaging, intermediate for line emission imaging and high for stellar surface imaging and line emission kinematics.

Reconstruction algorithm: The MIRA (Multi-aperture Image Reconstruction Algorithm) algorithm [3] is a imaging process specifically designed for IR interferometric data: (i) it is based on V^2 + closure phases constraint (non-convex criterion), (ii) it uses multiwavelength informations, (iii) it uses constraints such as positivity, MEM regularization and (u, v) plane apodization.

2 Imaging Young Stellar Objects

Structure of inner disks: Image reconstruction of young stellar objects should answer the question about their dusty close environment. Particularily, the shape of inner surface/rim and inner cavity of these disks should be unveiled (see Fig. 2). However if the structure/size/intensity of the dusty torus is expected to be retrieved, the image of the surrounding accretion disk itself should be very difficult to obtain because of the too high dynamical range involved [4].

Launching of jets and winds: The morphology of the emission lines in close environment of young stars will be studied as well by interferometric imaging [3]. It will give precious elements of responses concerning main

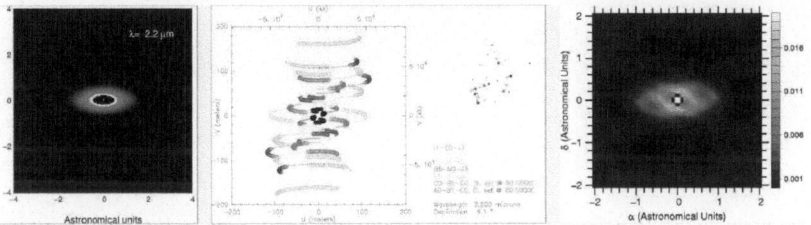

Fig. 2. From left to right: simulated image of a Herbig Ae/Be star with its dusty ring set at the dust sublimation radius, (u, v) coverage and reconstructed image.

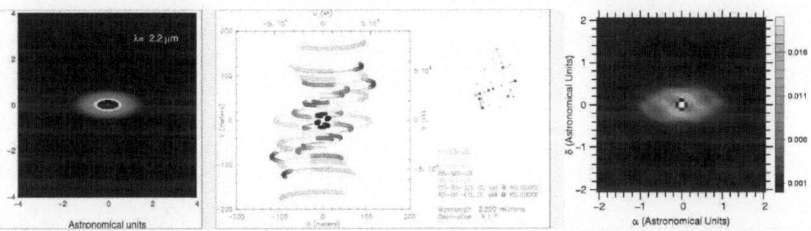

Fig. 3. From left to right: simulation of Paβ emission from the inner region of a MHD disk wind, (u, v) coverage and reconstructed image.

astrophysical aspects such as the origin of the jet (disk/star/disk-star interacting region), the accretion/ejection phenomenon and the physical mechanisms involved, or the relation between the observed dust morphology and the jets/winds (see Fig. 3).

3 Conclusions

The VITRUV instrument, by recombining simultaneously 4 or 8 telescopes with great sensitivity, accuracy and high spectral resolution power, will exploit the full imaging capability of the VLTI infrastructure. It is expected to have a large scientific productivity, not only in the YSO field, but in many astrophysical areas not addressed here, among them stellar surfaces, evolved stars and active galactic nuclei.

References

1. Kern, P. Y., et al.: 2003, *Proceedings of SPIE Volume 4838.*, 312
2. LeBouquin, J., et al.: 2004, in *Proceedings of SPIE Volume 5491.*, 1362
3. Thiébaut E., et al.: 2003, in *Ap&SS, v. 286*, 171
4. Tatulli, E., et al.: 2004, in *Proceedings of SPIE Volume 5491.*, 117

VITRUV Precursors: IONIC2T/IONIC3T

J.-P. Berger[1], P. Haguenauer[2], P. Kern[1], J.-B. Lebouquin[1], L. Jocou[1],
K. Perraut[1], F. Malbet[1], A. Delboulbé[1], M. Benisty[1], P. Labeye[3],
I. Schanen[4], J. Monnier[5], R. Millan-Gabet[6], E. Pedretti[5], W. Traub[7],
M. Schöller[8], and A. Glindemann[8]

[1] LAOG 414 Rue de la Piscine, 38400 St Martin d'Heres
    berger@obs.ujf-grenoble.fr
[2] JPL 4800 Oak Grove Drive, Pasadena, Ca, 9110
[3] LETI, CEA, 17 rue des Martyrs, 38054 Grenoble Cedex 9
[4] IMEP, 38016 Grenoble Cedex 1, France
[5] U. Michigan, Ann-Arbor, MI 48109-1090
[6] Caltech. 770 S. Wilson Av, Pasadena, CA 91125
[7] Harvard-Smithsonian CfA, 60 Garden Street, Cambridge MA 02138
[8] ESO, K. Schwarzschild-Str, D-85748 Garching bei München

1 Instrumental context

LAOG and its partners (LETI and IMEP) have been testing since 1996 the
potential of IO (Integrated Optics) technologies to replace the heart of an
interferometric beam combiner. IO technology allows to integrate in or on
a substrate single/multi mode optical circuits. The motivation behind this
work is our deep conviction that the use of compact IO optical circuits will
result in considerable simplification of the instrument operation and increase
in the quality of measurement.

Our working strategy can be divided into several steps:

- designing an IO circuit well suited for astronomical interferometry appli-
  cations(broadband operation, throughput ...);
- intensive laboratory characterization phase (dispersion, birefringence,
  throughput ...);
- *on-sky experiments.*

The latter of these points is the purpose of this contribution. We have
collaborated with ESO-VLTI and Harvard-Smithsonian CfA-IOTA in order
to test our best 2-way and 3-way H and K beam combiners on operational
instruments. These experiments can be seen as Vitruv precursors.

2 Integrated optics-based instruments

2.1 VLTI 2TH and 2TK combiners

The motivation behind the collaborative work with VLTI was, for LAOG, to
gain experience on the VLTI environment and to measure the performances

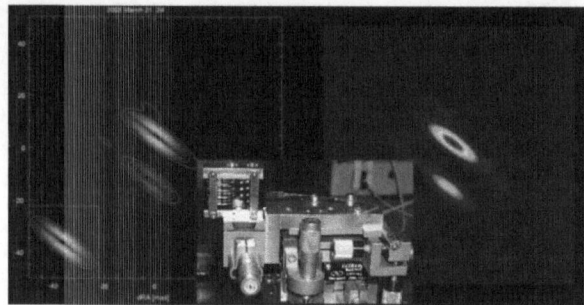

Fig. 1. First reconstructed image of λ Vir with IONIC3/IOTA. From S. Kraus in [7]. A view of the IONIC3/IOTA beam combiner

of an IO combiner associated with VLTI. For VLTI the interest was to have access to alternatives to fiber combiners and extend VINCI capability to the H band.

In a first step (July 2002) we have installed a two telescope combiner operating in the H band (2TH). The combiner was designed by GeeO[9] using ion-exchange IO technologies[3]. The combiner is made of two fibers connected to an IO combiner. The combination is made thanks to a reverse Y junction resulting in a loss of 50% of the interferometric signal. This combiner was designed to allow the switching with the MONA combiner and was not optimized in terms of throughput performances. Analysis of the results showed a stable and high on-sky instrumental contrast[1] (87% on average over 10 days).

In a second project (august 2004) we provided VLTI with an IO combiner operating in the K band (2TK, designed by LETI) and that could replace the ageing MONA combiner. The combiner was designed to have a high throughput (70%) and the two interferometric outputs were recordable thanks to the use of a coupler. The instrumental contrast over a few days was stable and higher than 90%. IONIC2T K was used to obtain the first fringes between the two auxiliary telescopes[2, 4].

2.2 IONIC3-IOTA

A collaborative effort between LAOG and the IOTA Smithsonian CfA team allowed us to install at IOTA an IO 2-way (2000) and 3-way (2002) beam combiner[5, 6]. The latter combiner was designed by LETI. It allows the temporally encoded pairwise combination of the three beams in the H band.

This instrument allowed to measure IOTA first closure phases and reconstruct the first images[7] (see figure 1). The instrumental behaviour is now well characterized after more than three years of operation. Instrumental and

---

[9] 16 ch. du Vieux Chêne, 38240 Meylan, France

visibility and closure phases are remarkably stable and allow to reach accuracies of 1% and 1° respectively. Improvements on these performances are expected by including a polarisation capability to the instrument. The fact that this instrument is in use by observers coming from more than ten institutions worldwide is a tremendous opportunity to gather user feedback and learn how to improve the instrument. Among the science topics tackled at IOTA, young stellar objects, Mira stars, stellar surfaces etc...

3 Perspectives

These different instruments have brought us important information about the behaviour of IO based instrument. Instrument performances have been extensively tested and convinced us of the fundamental interest of IO technologies for interferometric instrumentation.

In preparation to Vitruv we are considering testing our next generation beam combiners - the ones that will be able to combine four to eight beams - at CHARA. The MIRC project, leaded by P. J. Monnier will allow to combine CHARA's 6 beams. It has been designed since its beginning to be compatible with an IO combiner. First prototypes will be tested at LAOG 2005.

References

1. J.-B. LeBouquin et al., A&A **424**, 719, 2004
2. J.-B. LeBouquin et al., A&A **450**, 1259, 2006
3. F. Malbet et al., A&AS **138**, 135, 1999
4. ESO-PR, 06/05
5. J.-P. Berger et al. A&A **376**, L31, 2001
6. J.-P. Berger et al. Interferometry for Optical Astronomy, SPIE **4838**, 1099, 2003
7. J. Monnier et al., ApJ **602**, L57, 2004

First Results on Integrated Optics Developments for Mid-Infrared Interferometry

Lucas Labadie[1], Pierre Labeye[2], Pierre Kern[1], Brahim Arezki[1], Isabelle Schanen[3], Jean-Emmanuel Broquin[3], and Etienne Le Coarer[1]

[1] Laboratoire d'Astrophysique de Grenoble, BP53, F-38041 Grenoble Cédex 9
   lucas.labadie@obs.ujf-grenoble.fr
[2] Laboratoire d'Electronique et des Technologies de l'Information (CEA-Leti), 17 rue des Martyrs, 38054 Grenoble Cédex 9
[3] Institut de Microélectronique, Electromagnétisme et Photonique, 23 rue des Martyrs, 38016 Grenoble Cédex 1

1 Introduction

In terms of high angular resolution and high dynamic range, nulling interferometry [1] is among the most promising instrumental concept for the detection and characterization of telluric exo-planets in the infrared band [4-20 μm]. Achieving a deep "null" (i.e. a very high star cancellation) relies therefore on our ability to recombine perfectly plane wavefronts coming from more than two apertures [2]. Perfectly planar wavefronts requires modal filtering [3] in the mid-infrared range and multi-apertures interferometry can be greatly relaxed by using integrated optics (IO) components [4]. Since we believe single-mode IO can represent a major solution for beam combination for mid-infrared interferometry, we have planed to extend through the ESA-funded program *Integrated Optics for Darwin* the integrated optics solution to longer wavelengths.

2 Development of Dielectric Planar Waveguides

A first direction we have investigate is how to manufacture single-mode waveguides using dielectric materials suitable for the mid-infrared like **Zinc Selenide glass** and **Chalcogenide glasses**. Zinc Selenide is a crystalline material with high transmittance from 0.6μm to 20μm. Our first goal has been to produce a single-mode planar waveguide from which to extract later a single-mode channel waveguide through an appropriate chemical etching process. The modal behavior was assessed by mid-infrared m-lines analysis [5]. The first manufactured slab waveguide has been a ZnSe/ZnS single-mode planar waveguide at λ=10 μm with a thickness of 5.5μm and a refractive index of n_c=2.39. Chalcogenide glasses are amorphous infrared materials transparent in almost the whole range [4 - 20 μm]. Our research has focused on As$_2$Se$_3$-type glasses and TeX (or telluride) glasses. We have been able to synthesize complete and repeatable chalcogenide slab waveguides that could

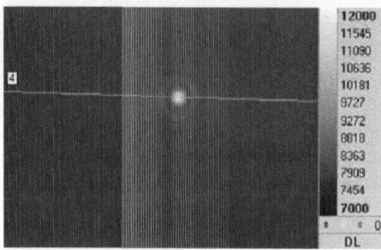 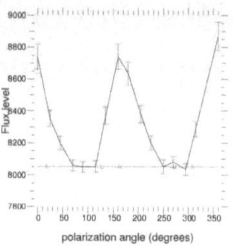

Fig. 1. left: Observed output of a HMW at λ=10μm. right: Polarization curve for a single-mode HMW.

confirm that a reliable technological process has been developed. As a result, we have assessed the guiding properties of As_2Se_3/As_2S_3 samples and $Te_2As_3Se_5/As_2S_3$ samples at λ=10μm. Although ZnSe/ZnS seemed at first a sufficiently mature technology to pursue with the fabrication of mid-infrared channel waveguides, only chalcogenide solutions have been maintained for the second phase due to a better applicable etching process.

3 Development of Hollow Waveguides

An alternative to dielectric waveguides solution are Hollow Metallic Waveguidies (HMW) whose guiding properties are insured by successive reflections on the metallic walls of a rectangular waveguide. This work has led to the fabrication and the characterization of first single-mode HMW at λ=10μm. The polarization properties of rectangular metallic waveguides [6] have permitted to assess the single-mode behavior of our first samples.

Fig. 1-left shows the observed output of a hollow metallic waveguide at λ=10μm, which demonstrates that waveguiding is obtained with this structure. Fig. 1-right is the polarization curve obtained with an expected single-mode waveguide at the same wavelength: because of polarization properties of the rectangular hollow waveguides, the extinction that is reached experimentally at 90° and 270° proves that single-mode hollow metallic waveguides have been synthetized.

References

1. R. N. Bracewell: Nature **274**, 780 (1978)
2. A. Léger et al: Icarus **123**, 249 (1996)
3. B. Mennesson et al: OSA Journal A **19**, 596 (2002)
4. F. Malbet et al: A&AS **138**, 135 (1999)
5. L. Labadie et al: *Proc. of SPIE* **5491**, 1333 (2004) (2003)
6. E. C. Jordan & K. G. Balmain: *Electromagnetic Waves and Radiating Systems*, 2nd edn (Prentice-Hall Inc., New Jersey 1985) pp 636-646

APreS-MIDI, a 4 Beam Recombiner

M. Dugué and the APreS-MIDI team

Observatoire de la Côte d'Azur, BP 4229, F-06304 Nice Cedex 4, France
dugue@obs-nice.fr

1 Introduction

The potential of the VLTI for producing images by aperture synthesis in the mid-infrared region is unique among the world due to the existence of 4 telescopes of large aperture (the Unit Telescopes) that could be coupled by interferometry and by 3-4 auxiliary telescopes (the ATs) relocatable on 30 different stations. For sensitivity reasons, affected by the thermal background in the 10 microns band for instance, the use of large aperture is required for various astrophysical topics like the observations of Active Galactic Nuclei and the study of the circumstellar discs of Young Stellar Objects. In this article, we are presenting a new generation instrument for the VLTI, aiming to produce aperture synthesis images in the 10 microns band.

2 Instrument concept

APreS-MIDI recombines 4 telescope beams in the pupil planes. The use of 4 telescopes allows 6 baseline exploration in one snapshot. The 4 pupils coming from the telescopes are superimposed with small tilt angles between them (see Figure 1). A set of fringes is thus produced. The major interest in this type of recombination is that it can be applied to the MIDI instrument which is used as a "performant camera". APreS-MIDI is a module inserted in front of MIDI allowing to upgrade MIDI capability recombination to 4 instead of 2 beams. The main optical element is a segmented mirror with 4 reflecting faces. Images of the star produced by the 4 different telescopes are focussed on the different faces of the segmented mirror. The role of this mirror is to send the 4 beams along the same axis into the MIDI entrance window. The small tilt angle between the pupils aiming at producing the fringe pattern is directly linked to the angle which remains between the beams after the reflection onto the segmented mirror. APreS-MIDI can be fed by any of the 7-8 telescopes of the VLTI (4 UTs and 3-4 ATs).

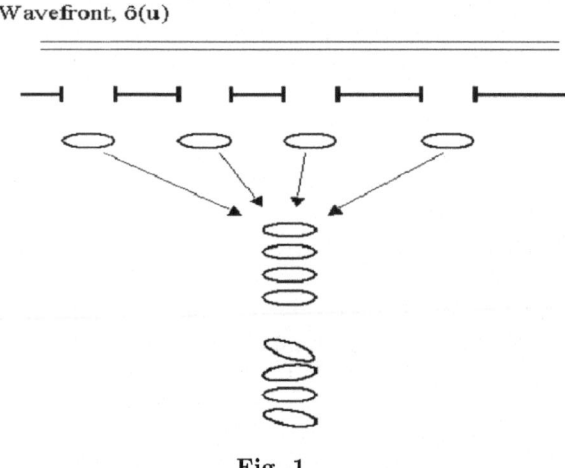

Fig. 1.

3 Expected performances

- Recombination mode: densified images and pupil plane with tilt angle, 3 or 4 telescope beams recombination allowed (4 ATs or 4 UTs).
- Angular resolution: 10 mas at 10 microns
- Spectral mode: using 2-3 filters, R=30. Complementarity of MIDI for spectrometry : R = 230.
- Field view : 'corresponding to Airy disc sizes'. At 10 microns : 0.25 arcsec with UTs; 1 arcsec with ATs.
- An external fringe tracking system is required for their stability. Theoretical following values for the sensitivity are assumed for 3 T mode (3 visibility and one closure phase), Visibility = 1, and for a signal to noise of 50 :
  - Basic mode (10 seconds) : 2.0 Jy with UTs, 40. Jy with ATs
  - Fringe track sequence of 15 minutes : 0.2 Jy with UTs, 4. Jy with ATs
- Expected imaging performance: one aperture synthesis image from 10 to 70 data sets.

Reference

1. B. Lopez, S. Wolf and the APreS-MIDI team: APreS-MIDI, APerture Synthesis in the Mid Infrared at the VLTI. The Power of Optical/IR Interferometry: Recent Scientific Results and 2nd Generation VLTI Instrumentation, this volume.

First Results from SIRIUS, the Interferometric Imaging Demonstrator for VIDA

F. Patru[1], D. Mourard[1], O. Lardière[2], M. M. Clausse[1], P. Antonelli[1], Y. Bresson[1], and S. Lagarde[1]

[1] Observatoire de la Côte d'Azur, Dépt. GEMINI, Avenue Copernic, 06130 Grasse, France
[2] Observatoire de Haute Provence, 04170 Saint Michel l'Observatoire, France

1 The project

The Observatoire de la Côte d'Azur is involved in the VIDA proposal (VLTI Imaging with a Densified Array), a second generation instrument for the VLTI. VIDA is a direct imaging system based on a densified pupil beam combiner [1]. In order to prepare the feasibility studies, a test bench called SIRIUS is developed. A multi-aperture mask reproduces the configuration of the VLTI array, with the four unit telescopes and four auxiliary telescopes on some of the 32 stations. A wavefront sensor (WFS) installed before the mask allows a direct control of the piston on each of the sub-apertures in the imaging beam. It will be useful to specify the accuracy required for the correction of the differential pistons between sub-apertures.

The purpose is to compare the imaging performances of different interferometric beam combination schemes: aperture synthesis imaging, Fizeau and densified pupils direct imaging. The analysis of the Fizeau images allows to quantify the expected stability of the system and to validate the calibration procedures.

2 Image Analysis

The analysis software studies the direct image properties in term of contrast or photometric distribution. This analysis software is the same as the output module of the simulation software. The comparison between simulation and experiment is the key of this work. It is foreseen to study simple objects with the bench, as single stars which are unresolved or partially resolved, and binaries, the purpose being to estimate the accuracy with which we can recover the parameters of the object in different conditions of measurement. This software is based on the classical spectral density analysis which allows us to measure squared visibilities on the different baselines. A second method of analysis has been developed: it intends to analyze the image in terms of maximum of intensities as does the CLEAN algorithm [2].

3 First results and calibration

Monochromatic PSF have been recorded to evaluate the stability of the system. The precision on the parameters of the input object have been estimated: 0.1mas, corresponding to the VLTI scale, on the position of a peak, 1% on the amplitude of a peak and 0.5% on the visibility. However, the accuracy is biased by the residual light and the differential photometry. We are currently implementing dark and photometric calibrations. The WFS allows a precision better than $\lambda/50$ with a stability better than $\lambda/10$ during few hours. Differential piston defaults have been created by adding a phase screen across the entrance beam. Piston variations from $\lambda/20$ up to λ have been measured properly.

A preliminary calibration locates the position of the sub-pupils of the Fizeau mask with respect to the micro-lenses of the wavefront sensor. To take into account the defaults due to the misalignement and the peculiar defaults of the optical elements, an other calibration consists in replacing the imaging camera by the WFS to measure the exit wavefront. It gives the differential map between the entrance and the exit wavefront. Thus, this differential map is substracted to the entrance measurements to correct the defaults of the optical system. This calibration is required to correct the differential piston errors between the beams. Indeed, the WFS will be the estimator of the cophasing devices used with the densification assembly.

4 Densification assembly

A fiber assembly using monomode optical fibers has just been designed. The diameters of the sub-apertures are zoomed without changing the positions of the sub-apertures centers. The densification factor, corresponding to the output/input diameters ratio, can be modified. The spatial filtering properties of the fibers are used to remove the tip-tilt of each sub-aperture. The effects of the fibers in the direct image are under studies. The densification assembly is also dedicated to study other array configurations up to eight beams. Afterwards, it is foreseen to study the installation of a coronograph at the densified focus.

References

1. O. Lardière et al., these proceedings
2. J.A. Högbom: Astron. Astrophys. Suppl. **15**, 417 (1974)

The Dispersed Speckles Cophasing System for Direct Imaging with VIDA

V. Borkowski, A. Labeyrie, F. Martinache, O. Lardière

LISE-Observatoire de Haute-Provence - CNRS, F-04870 Saint-Michel l'Observatoire, France, borkowski@obs-hp.fr

1 Introduction

We studied a cophasing method for the VIDA concept (Lardière O., this proceeding). We recall briefly the principle of this method called *dispersed speckles method* in the first section. In the second one we present the results obtained with simulations.

2 The dispersed speckles principle

The dispersed speckles method is used to measure piston between pairs of telescopes. We record images from an interferometer at several wavelengths. We stack all the recorded images (after correction of the dispersion effects) to build a cube called *dispersed image cube*. A 3 dimensional Fourier transform of this cube gives an output cube in which there are *active* columns. Heights of dots in active columns give piston measures for the corresponding baseline. We studied the possibility to applied this method to cophase the telescopes of VIDA.

3 The results applied to the VIDA configuration

The simulations give the number of photons needed for the precision on the piston reachable. The limiting magnitudes are calculated with the expression:

$$F = N/(1.51 \times 10^7 \times S \times T \times \tau \times q_{eff} \times (\Delta\lambda_0/\lambda_0)) \tag{1}$$

with N the number of photons, S the collecting surface, T the exposure time, τ the optical efficiency and q_{eff} the quantum efficiency. $S = \pi \times R^2 \times n$, with R the radius of the telescope (4.1 m for the Unit Telescopes and 0.9 for the Auxiliary Telescopes of the VLTI) and n the number of telescopes. The values of the parameters are: $T = 5ms$, $\tau = 0.022$, $q_{eff} = 0.6$. And finally:

$$mag = 2.5log(E_0/F) \tag{2}$$

We have studied three cases: 1/ **visible** for which $(\Delta\lambda_0/\lambda_0) = 0.1618\mu m$ and $E_0 = 3950Jy$ (Jansky); 2/ **J band** :$(\Delta\lambda_0/\lambda_0) = 0.24\mu m$ and $E_0 = 1770Jy$; 3/ **K band** : $(\Delta\lambda_0/\lambda_0) = 0.1818\mu m$ and $E_0 = 629Jy$. The results are presented on *figure* 1 for precisions of $\lambda/1000$ and $\lambda/100$. You can noticed that in visible for a coronagraphic precision ($\lambda/100$) you can reach a magnitude of 4. So you can expected to reach 8 magnitude stars for a precision of $\lambda/10$ on the piston.

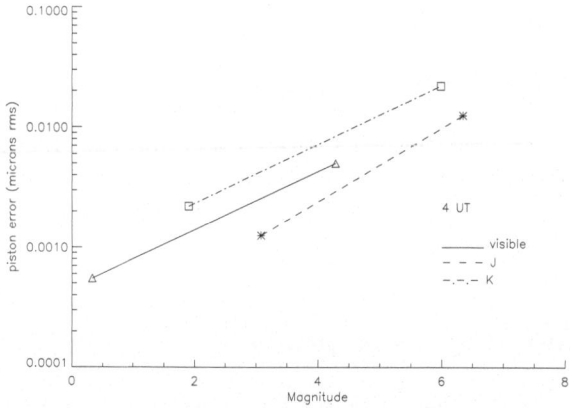

Fig. 1. Piston errors versus the magnitude for a configuration with 4 UT and for spectral bandwidth in visible, in J and in K band.

4 Conclusion

The simulations show that the dispersed speckles method could allow to cophase the telescopes of VIDA without the need of many photons because with $2.6\ 10^6$ photons, a coronagraphic precision is reachable.

References

1. Borkowski V. et al. 2005, A&A **429**, 747
2. Labeyrie A. 2002 Proc. ESLAB 36 Symposium.
3. Lardière O. et al., these proceedings
4. Lardière O. et al. 2003, Proceedings of the SPIE, **4838**, 1018
5. Martinache F. 2004, J. Opt. A: Pure Appl. Optics **6**, 216

A New Generation of Micropositioning Devices for Optical Fiber with Submicrometric Metrology

O. Preis, A. Chetail, B. Neichel, C. Aubert, and B. Rabaud

Laboratoire d'Astrophysique de Grenoble, 414, rue de la Piscine - U J F - 38400
St Martin d'Hères - FRANCE, Tel. : +33 4 76 51 47 89, Fax. : +33 4 76 44 88 21
Olivier.Preis@obs.ujf-grenoble.fr

1 Introduction

A project of interferometric recombination for the eight ESO telescopes of the VLTI, called "VITRUV", is under study at the LAOG. The light, coming from the eight telescopes, will be transmitted by optical fibers linked to an integrated optic beam combiner to obtain multiple interference fringes.

To control the position of every fiber we have developed a new generation of micropositioning devices [1] including mini-sensors for submicrometric position measurements.

2 The principle

This device contains a piezoelectric tube on which there are five metallic electrodes. The electrical voltage applied to the 4 external electrodes will create a transverse motion of the extremity of the tube [2] where the optical fiber is fixed. This motion is about $0.03\,\mu m/V$, but there is a hysteresis effect typical in piezoelectric material [3] and over time creep gradually occurs.

If we want to obtain a high precision on the position of the fiber, it is needed to control the hysteresis and creep as much as possible. A solution is to measure the deformation of the tube on real time with small sensors (fig. 1). The actual speed of the closed-loop, composed of an electronic control system, the tube and the measurement of the deformation, is 10 ms and the accuracy of the fiber position measurement is about $0.15\,\mu m$ rms.

3 The software

There are three specific modes in the software to control the micropositioning device:

Position mode : The user enters the position(s) he wants to reach in a file. Then, the extremity of tube moves to the expected position and the position is maintained even if there is any disturbance.

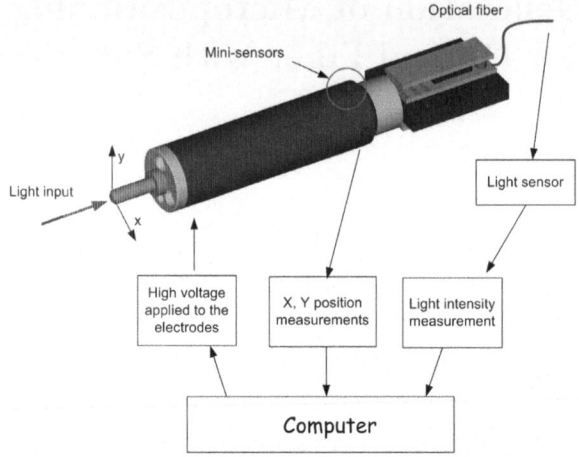

Fig. 1. The micropositioning device with the mini-sensors.

Image of the light source : Because we can move the fiber input on an addressing plane (x, y), with a $0.15\,\mu m$ step, we can establish a bi-dimensional image from the intensity of the light source.

Automatic alignment of fiber input with a light source : This micropositioning device can search automaticaly the position where the light intensity is maximum and tracks this maximum even if it moves.

4 Conclusion

This new device can move the input fiber, which is fixed to the device, in a plane of $20\,\mu m$ range, with an accuracy of $0.15\,\mu m$ rms, to control very precisely the position of every fiber in front of the eight telescopes focus.

References

1. Preis, O., Pichon, L., Delboulbe, A., Kern, P. Y., Magnard, Y., Ventura, N., *SPIE : New Fontiers in Stellar Interferometry*, edited by Wesley A. Traub, **5491**, 1379 (2004)
2. C. Julian Chen: App. Phys. Lett., **60**, 132–134 (1992)
3. N. Setter : *Piezoelectric Materials in Devices*, N. Setter Ed. (2002)

ESO ASTROPHYSICS SYMPOSIA
European Southern Observatory

Series Editor: Bruno Leibundgut